P9-ANZ-975

SPACE SCIENCE AND APPLICATIONS
Progress and Potential

OTHER IEEE PRESS BOOKS

The Calculus Tutoring Book, *By C. Ash and R. Ash*
Next-Generation Computers, *Edited by E. A. Torrero*
Insights Into Personal Computers, *Edited by A. Gupta and H. D. Toong*
The Space Station: An Idea Whose Time Has Come, *By T. R. Simpson*
The Making of a Profession: A Century of Electrical Engineering in America, *By A. M. McMahon*
General and Industrial Management, *By H. Fayol; revised by I. Gray*
Engineers and Electrons, *By J. D. Ryder and D. G. Fink*
Compendium of Communication and Broadcast Satellites: 1958 to 1980, *Edited by M. P. Brown, Jr.*
The Engineer in Transition to Management, *By I. Gray*

SPACE SCIENCE AND APPLICATIONS
Progress and Potential

Senior Editor
John H. McElroy
Space and Communications Group
Hughes Aircraft Company

Specialty Editors
Space Science:
Franklin D. Martin
Director of Space and Earth Science
NASA/Goddard Space Flight Center

Remote Sensing:
Ralph Bernstein
Senior Technical Staff Member
IBM Palo Alto Scientific Center

Materials Processing in Space:
Louis R. Testardi
Chief, Metallurgy Division
National Bureau of Standards

Communications and Navigation:
Robert R. Lovell
Director, Communications Division
NASA Headquarters

Published under the sponsorship of the
IEEE Aerospace and Electronic Systems Society.

The Institute of Electrical and Electronics Engineers, Inc., New York

This book is set in Times Roman

PRINTED IN THE UNITED STATES OF AMERICA

IEEE Order Number: PC01909

Library of Congress Cataloging-in-Publication Data
Main entry under title:

Space science and applications.

"Published under the sponsorship of the IEEE Aerospace and Electronic Systems Society."
Includes index.
1. Space sciences. 2. Space industrialization.
3. Remote sensing. I. McElroy, John H. II. IEEE Aerospace and Electronic Systems Society.
QB500.S6155 1986 500.5 85-23817

ISBN 0-87942-195-9

Contributors

Ralph Bernstein
Senior Technical Staff Member
IBM Palo Alto Scientific Center

Robert L. Bernstein
Associate Research Oceanographer
Scripps Institution of Oceanography

Geoffrey A. Briggs
Director of Solar System Exploration Division
NASA Headquarters

John R. Carruthers
Manager of Components Research
Intel Corporation

Robert N. Colwell
Associate Director of Space Sciences Laboratory
Professor of Forestry, Emeritus
University of California, Berkeley

C. Louis Cuccia
Manager of Advanced Planning
NASA Headquarters

Samuel W. Fordyce
President
Riparian Research Corporation

David Gilman
Senior Staff Scientist of Astrophysics Division
NASA Headquarters

Louis J. Lanzerotti
Distinguished Member of Technical Staff
AT&T Bell Laboratories

Robert R. Lovell
Director of Communications Division
NASA Headquarters

Franklin D. Martin
Director of Space and Earth Sciences
NASA/Goddard Space Flight Center

John H. McElroy
Space and Communications Group
Hughes Aircraft Company

Robert L. McPherron
Professor of Geophysics and Space Physics
University of California, Los Angeles

Robert J. Naumann
Chief of Low-gravity Science Division
NASA/Marshall Space Flight Center

William A. Oran
Chief of Market Development
NASA Headquarters

Charles J. Pellerin, Jr.
Director of Astrophysics Division
NASA Headquarters

Joseph N. Pelton
Director of Strategic Policy
Intelsat

William L. Quaide
Chief of Planetary Science Branch
NASA/Solar System Exploration Division

S. Ichtiaque Rasool
Distinguished Visiting Scientist
Jet Propulsion Laboratory

Fondation de France
Chair in Atmospheric Sciences
Ecole Normale Supérieure

Thomas F. Rogers
President
The Sophron Foundation

Joseph N. Sivo
Executive Vice President
Gateway Technology Associates

(*Contributors continued on next page.*)

Gerald A. Soffen
Associate Director for Space and Earth Sciences
NASA/Goddard Space Flight Center

Payson R. Stevens
President/Creative Director
InterNetwork, Inc.

Louis R. Testardi
Chief of Metallurgy Division
National Bureau of Standards

George C. Weiffenbach
Senior Fellow
Applied Physics Laboratory
Johns Hopkins University

Charles F. Yost
Manager of Commercial Applications
NASA Headquarters

CONTENTS

Contributors **v**

Preface, *John H. McElroy* **ix**

Introduction, *John H. McElroy* **xi**

Part I: Space Science **1**

1 Space Science: An Overview ... 3
Franklin D. Martin

2 Sun-Earth Relations ... 7
L. J. Lanzerotti

3 Solar System Exploration ... 19
Geoffrey A. Briggs and William L. Quaide

4 Astrophysics ... 31
David A. Gilman and Charles J. Pellerin, Jr.

5 NASA's Life Sciences Program ... 55
Gerald A. Soffen

Part II: Remote Sensing from Space **69**

6 Remote Sensing from Space: An Overview ... 71
Ralph Bernstein

7 Land Applications for Remote Sensing from Space ... 77
Robert N. Colwell

8 Ocean Remote Sensing ... 123
Robert L. Bernstein and Payson R. Stevens

9 Geophysical Remote Sensing ... 133
Robert L. McPherron

10 Weather and Atmosphere Remote Sensing ... 143
S. Ichtiaque Rasool

Part III: Materials Processing in Space 151

11 Materials Processing in Space: An Overview 153
Louis R. Testardi

12 Materials Science and Engineering in Space 155
John R. Carruthers

13 Materials Processing in Space: Review of the Early Experiments 159
Robert J. Naumann

14 Current Program to Investigate Phenomena in a Microgravity Environment 173
William A. Oran

15 Commercialization of Materials Processing in Space 177
Charles F. Yost

Part IV: Communications and Navigation 183

16 Communications and Navigation: An Overview 185
Robert R. Lovell

17 Communication Satellite Applications 187
Joseph N. Pelton

18 Communications Spacecraft 201
Samuel W. Fordyce

19 Navigation Satellites 215
George C. Weiffenbach

20 Communications Technology 227
C. Louis Cuccia and Joseph Sivo

21 Future Developments 251
Thomas F. Rogers

Index 255

Editors' Biographies 259

PREFACE

In 1983 NASA celebrated its silver anniversary and in 1984 the Institute of Electrical and Electronics Engineers celebrated its 100th anniversary. The two anniversaries, while one year apart, are inextricably tied through shared achievements and the interdependence of electrical engineering and space activities. On the one hand, the century-long record of achievements associated with the IEEE made possible the space advances. On the other hand, the space program fostered through its manifold demands the advance of the science and technology of electrical engineering.

The concept behind *Space Science and Applications* was originated by Dr. Sajjad H. Durrani, then President of the IEEE Aerospace and Electronic Systems Society. He recognized the strong ties and synergism between the IEEE and the space program and encouraged me to commemorate them with this book.

Special mention must be made of the four part editors—Franklin Martin, Ralph Bernstein, Louis Testardi, and Robert Lovell, who was aided by Louis Cuccia. They carried the heaviest workload in organizing their parts, selecting authors, and editing the diverse contributions. The difficult work of arriving at a final draft rested largely upon these individuals.

Special mention must also be made of Emily Gross and W. Reed Crone of the IEEE PRESS, who had the unenviable job of converting the final draft into publishable form. Their efforts deserve the highest praise.

Finally, the core of *Space Science and Applications* is the series of original articles that were prepared by authors who are genuine experts in their fields. They provide a new synthesis of the progress in space activities made to date and guidance for the progress to come. The authors have invested more than two years in bringing this project to a successful completion. Ultimately, of course, the success of this book rests with them.

It has been a great pleasure to be associated with the outstanding authors, part editors, and members of the staff of the IEEE PRESS. They collectively made the task of being the editor for the project an easy one.

JOHN H. MCELROY
Editor

INTRODUCTION

John H. McElroy

The achievements in space science and applications are a vivid demonstration of the power of the human intellect. No achievements of the space age rank higher in their effect upon mankind. They can be related to three categories: the vantage point provided by a space system, the capabilities produced by space systems, and the use that can be made of the space environment itself.

The new views—both earth-directed and space-directed—obtained from space platforms have permanently changed not only our understanding of the universe, but even the manner in which we think about it. The new capabilities that space systems have produced for meteorology, navigation, and communications are now indispensable elements of day-to-day life nearly everywhere in the world. The very attributes of the space environment—notably microgravity—are leading to a better understanding of materials science, and may someday lead to a spaceborne manufacturing complex. These are the achievements commemorated in this book.

The four parts of this book capture the intellectual challenge and stimulation of the space and earth sciences, and review the broad sweep of applications of space systems. The first part discusses space science, opening with a review of the field by Part I Editor Franklin Martin. Next, there is a review of the complex interrelationships that exist between the sun and the earth. The interrelationships provide intellectual challenge to both the scientists seeking a deeper understanding and the forecasters who must provide alerts of potentially harmful solar events. The third chapter reviews the great voyages of exploration that have been sent into the solar system—and even beyond. It is followed by a review of the great new astronomical observatories that have attained a vantage point above the disruptive influence of the atmosphere, and that are extending mankind's vision toward the edge of the universe. The final chapter in this part discusses the challenging problems in the life sciences program associated with ensuring man's survival and ability to work in the space environment. As we move toward space stations, space colonies, or long-duration manned missions to explore the solar system, major questions concerning human biology must be answered.

The second part of the book turns the spaceborne observing systems back toward the earth itself. From the first pictures taken with hand-held cameras, to the first full-disk earth image, and to the sophisticated multispectral and radar sensors of the 1980's, the views of the earth hold a never-ending fascination. After an introduction by Part II Editor Ralph Bernstein, the second chapter in this part traces the development and progress made in land observations from space. It focuses on sensing the earth's land resources—the timber, agricultural crops, soils, water, etc. The next chapter shows what has been accomplished in ocean observations. It shows how spaceborne sensors can monitor ocean wave heights, sea surface temperature, surface wind and wave velocities, sea ice, phytoplankton concentrations, and other parameters. Chapter 9 addresses geophysical remote sensing and examines the sensing of geologic structure and crustal composition through imaging sensors and then how satellites have been used to map the earth's magnetic and gravitational fields. The final chapter in this part reviews meterological and atmospheric measurements from space, the most operational of all remote sensing activities. Taken collectively, these chapters show how the parts of the great dynamic machine called the earth can be examined from space. They also show the manifold interactions among those parts that will be a fruitful area of research for decades to come.

The third part turns our attention from the science that can be done *from space* to the science that can be done *in space*. There is an introductory chapter on materials processing in space, by Part III Editor Louis Testardi, followed by a review of the role that gravity plays in a variety of fluid-phase processing operations. The third chapter in this part reviews the extensive series of investigations that have been carried out on Skylab, Apollo-Soyuz, and sounding rockets. The expansion of these investigations into the era of the space shuttle is the subject of the fourth chapter, while the budding commercial effort is described in the final chapter of the part. Taken as a whole, the chapters provide a case study of the as yet incomplete evolution of a new discipline from its earliest conception to the threshold of commercial application. It is a fascinating story, even if the final conclusion is yet to be written.

The final part of the book is devoted to the most widely recognized success story in space applications, the creation of communication and navigation satellite systems. It begins where many accounts end, with the applications that communication satellite systems have found. The almost unimaginable international growth of such systems that has occurred in less than 25 years is unparalleled in the history of technology. From concept to private investigations, government investigations, commercial applications, and multiple domestic and international systems, the evolution proceeded with a speed that could not have been predicted even by the most enthusiastic of supporters. An overview is provided by Part IV Editor Robert Lovell. Next is a chapter on applications, followed by one which describes the communications satellites that have made this evolution possible. Paralleling the development of communication satellites was that of navigation satellites, which is the subject of Chapter 19. Underlying both communication and navigation satellites is a common base of technology. It was this base of technology that made the advances not simply possible, but also reliable enough to support operational systems. The final chapter of the book looks toward future developments in communication and navigation and envisions a future evolution almost as exciting as what we have already seen.

The achievements commemorated in this book can be viewed from many perspectives. One perspective is the recognition of the contribution of visionaries who were able to give birth to conceptions far beyond the thinking of their contemporaries. In each of the parts of this book, a fascinating companion narrative could be written describing the opposition to these ideas, and how it was overcome or has yet to be overcome. From another perspective, we can appreciate the incisive scientific thinking that has led the way to new concepts to be tested and to the penetrating interpretation of what was often difficult and ambiguous data.

Finally, and most appropriately for a book published by the IEEE, there is the perspective gained by stepping back and looking in awe at the automated engineering marvels that engineers have created. Engineers, often overlooked in the excitement of scientific research, have sent emissaries from our planet out into space. Our emissaries have extended our vision to hazardous regions we can never visit and to spectral regions where our eyes do not function. They reveal the secrets that our poor limited sensory capabilities can never detect, and expose the secrets that allow us to advance to a deeper and deeper understanding of the universe and its phenomena. In June of 1983 one of our emissaries passed out of our solar system on the start of an unending journey through the universe. We have broken the physical bounds of our birthplace; our explorations broke the mental bounds very early in the space age.

This book celebrates the visions, the science, and the engineering that we associate with *Space Science and Applications*.

PART I: SPACE SCIENCE

1

SPACE SCIENCE: AN OVERVIEW

Franklin D. Martin

Space science is generally defined as investigations that are conducted by scientists using earth-orbiting satellites, deep-space probes, and suborbital systems. Primarily because of organizational approaches to the management of space research by NASA headquarters and because of other considerations, space science has come to include four basic areas of research: sun–earth relations, life sciences, solar system exploration, and astrophysics. Many other areas of earth sciences are considered space applications and are included in Part II of this book, on remote sensing of the earth from space. The ties that hold these diverse elements together involve many individuals, organizations, and interests. These entities include individual research scientists, universities, professional research organizations and societies, the National Academy of Sciences, the NASA headquarters and centers, and industry. All have developed working relationships (as described by Homer Newell in his book *Beyond the Atmosphere*) that have allowed controversies to be aired, debates to be conducted, priorities to be established, and decisions to be made that have led to one of the greatest scientific adventures of all time. In the last quarter of a century, we have explored the near-earth space environment, most of the solar system, opened the wider universe to detailed study, and prepared man to take a major evolutionary step—living and working permanently in space. How did we get to this point?

Space science grew out of the rocket research which followed the end of World War II. The early investigations which began in 1946 concentrated on studies of the earth's atmosphere, solar radiation, and cosmic rays from space. While scientific progress was made during the following decade, not until the launch of Explorer 1 in January of 1958, with the subsequent discovery of the Van Allen belts, did scientific research in the U.S. move into space to stay. Early successes were followed by NASA's first "deep" space probe, Pioneer 1, which traveled over 70 000 mi from the earth. These early flights concentrated on the areas of plasma physics (charged particles and magnetic fields) and laid the foundation for what has become a highly developed multidisciplinary area of research referred to as solar terrestrial relations (sun–earth relations).

In 1960, with the launch of Pioneer 5, we moved away from the earth into interplanetary space—a first tentative step toward the planets. That same year the first orbiting observatory was launched, a Navy satellite which detected solar X-rays. The Solrad mission was followed in 1962 with the launch of NASA's Orbiting Solar Observatory (OSO-1), an earth-orbiting satellite which began a series of nearly continuous studies of the sun from space.

While unmanned probes were changing our view of the near-earth environment, NASA was busy implementing the Mercury program, selecting the first seven astronauts in 1959. In order to prepare for manned flight, a new series of investigations was initiated to understand the impact of the space environment on living creatures and ultimately on man. Primates were flown into space and returned safely. These activities and the actual study of the astronauts, as well as selected biological systems, led to an increased scientific interest in life sciences. The vigor of this area of research ebbed and flowed with the access to manned flight opportunities. However, the mystery of human spaceflight was replaced by confidence that man could conquer space as he had the skies.

The U.S. began the *in situ* study of the planets in 1962 with Mariner 2 which flew near the planet Venus. During the mid-1960's, the Ranger program took the first close-up pictures of the lunar surface; this was followed by the successful landing of the Surveyor 1 in the Oceans of Storms in 1966. In the interim, Mariner 4 provided us our first close encounter with Mars.

The decision by President Kennedy in 1961 to place a man on the moon and return him safely within the decade made the moon not only a prime political target, but also a prime scientific target. The scientific community went "along for the ride" even though it was strongly felt by many influential scientists that the exploration of the moon could be done more cost effectively with robots. This conflict has continued to resurface from time to time with the space shuttle and the recently announced Reagan initiative, the space station, becoming the focal points for passion and debate. In spite of the arguments over how best to spend the nation's resources (there were valid arguments on both sides), the studies of the moon did proceed and yielded many scientific results. Because of the numerous lunar samples, the detailed images, other scientific data, and the basic nature of the lunar environment, we now know more about the history and evolution of the moon than we do about the earth, at a corresponding point in exploration. While this major effort was unfolding, the unmanned exploration of the planets continued with the Mariner 5 flyby of Venus in 1967, and with Mariner 6 providing pictures from Mars in 1969, just nine days after Neil Armstrong set foot on the lunar surface.

The 1970's brought with it the end of the Apollo lunar

program. However, investigation of the near-earth environment continued with a series of atmospheric explorers and three international sun-earth explorers to study the sun's effects on the boundaries of the magnetosphere. The picture that emerged was even far more complex than Van Allen's original discoveries indicated. In short, the earth is surrounded by a dynamic environment of particles and magnetic fields that is strongly affected by the fields and radiation from the sun.

The study of life sciences received an unanticipated boost after the Apollo program, when the Skylab program, which involved three extended manned flights in the early 1970's, allowed the first long-term studies of how U.S. astronauts can live and work in space. These studies, along with results provided by cooperative studies with the Soviets, have helped to lay the foundation for the long-term habitation of the space station, which will be launched in the 1990's.

These efforts have continued into the shuttle era, in which flights are providing a large database of human flight physiology that has allowed us to relax the biomedical requirements in the selection of the crew members. The medical measurements have paid dividends in giving us insight into the processes behind the physiological changes, and we are able to deal with the problems in short-term missions by mission planning and the use of simple countermeasures. Through these efforts, we will be able to extend the productivity of the shuttle, come to grips with the problem of long-term habitability of the space station, and investigate some of the interesting biological problems posed by gravitational physiology.

There is also a biological science connection to the planetary program in terms of understanding the origin of life. While most scientists believe that the earth is the only planet in the solar system that is inhabited, the formation of life on earth is inextricably tied to the chemistry and history of planetary formation.

During the 1970's, not only were the lunar samples for Apollo under intensive study, but the outer reaches of the solar system were brought within our grasp. Pioneer 10 was launched in March 1972, followed by Pioneer 11 in 1973. These spacecraft provided our first close-up look at Jupiter, and Pioneer 11 used a gravity assist from Jupiter to make its way to Saturn and out of the solar system. In the meantime, the inner solar system was being explored by Mariner 10, which conducted flybys of Venus and Mercury, revealing the almost eerie resemblance of the surface of the sun's nearest planet to that of the moon. The comparative study of the planets had become a reality. Then on July 20, 1976, NASA helped the nation celebrate its 200th year with the successful landing of the robot spacecraft, Viking 1, on the surface of Mars. This was followed by the landing of Viking 2 on September 3. Voyagers 1 and 2 were launched in 1977 and were intended initially for flyby's of Jupiter and Saturn. They were equipped with far more sophisticated instruments than were aboard the Pioneer spacecraft. In addition, Pioneer Venus 1 and 2 were launched in 1978, providing the first U.S. atmospheric probes to that planet. By the end of 1980, the atmosphere of Venus had been successfully probed. Voyager 1 and 2 had produced dramatic results at both Saturn and Jupiter, with Voyager 2 headed on to further encounters with Uranus and Neptune later this decade.

In addition to life sciences, sun-earth relations, and solar system exploration, a new space science discipline emerged to compete for support—astrophysics. Aside from the studies of solar physics with the early satellites, a number of astronomers were convinced that if they could get above the atmosphere for extended periods of time they could see the universe unobstructed by the effects of the atmosphere, which blurs images and absorbs most of the radiation emitted by objects and phenomena in the universe. Early efforts by several X-ray groups with sounding rockets, ostensibly flown to look for solar induced X-ray florescences from the moon, accidently discovered the most intense X-ray source in the sky. This was followed by a series of X-ray, gamma ray, and cosmic ray studies on the Small Astronomy Satellite and in the High-Energy Astronomy Observatory (HEAO) series. The latter three missions were launched beginning in 1977. The results were spectacular. The universe was ablaze in X-rays, confirming, what some astronomers had suspected, that the universe is indeed a violent place. On another front, scientific interest in ultraviolet astronomy placed two extremely productive satellites in space, the Orbiting Astronomical Observatory (OAO-3) in 1972 and International Ultraviolet Explorer (IUE) in 1978. These two allowed astronomers to study for the first time the ultraviolet spectra of stars and galaxies as well as the intervening material. In addition, these missions were responsible along with HEAO-2 for attracting ground-based astronomers to the space program. The astrophysics activities were further enhanced with the initiation of the development of the Hubble Space Telescope (HST) in 1978. The HST, a Mount Palomar in the sky, is unique in that not only will it be the most powerful optical facility for observing the universe, but it reflects NASA's commitment to support an on-going astronomical system. The HST is designed to make use of the Space Transportation System for on-orbit maintenance, repair, and installation of next-generation instrumentation. This makes astrophysics somewhat unique among the science disciplines in that the love–hate relationship between science and the manned program is lessened somewhat. The HST as it is currently defined could not be implemented without the space shuttle. Virtually all of the future astrophysics facilities make use of manned presence for extended operations (maintenance and repair) and assembly.

During the past 25 years, the Soviet Union has conducted an aggressive manned space program and intensive studies of the moon, Venus, and Mars. In spite of this competition and impacts of budget decisions on specific programs, the past efforts in space science have been spectacular and the future is bright. In the area of sun–earth relations, a comprehensive international program involving the U.S., Japan, and European Space Agency (ESA) is under definition. This program, when implemented, will provide for the first time a complete look at the solar–terrestrial system, from ejection and transport of radiation from the sun through space connecting to the earth's magnetosphere and on through to the upper atmosphere. The Solar System Exploration Program is moving toward the launch of Galileo in 1986, a mission to conduct an extensive study of the Jovian System and probe Jupiter's atmosphere. This will be followed by the Venus Radar Mission. In addition, a solid foundation for the future has been established through the efforts of the Solar System Exploration Committee, which has envisioned a comprehensive and fiscally responsible series of missions to follow up on the decade of the 1970's. The first of a new class of missions is the Mars Geochemical Orbiter, which is expected to be launched in the early 1990's. In addition to the upcoming HST launch in the 1980's, a detailed survey of the IR universe was conducted in 1983 by a joint NASA, Dutch, U.K. project called the Infrared Astronomy Satellite (IRAS). A detailed study of the gamma ray region of the spectrum will be conducted with the Gamma Ray Observatory (GRO) in the late 1980's. As with the Solar System Exploration Program, an aggressive program for the next decade has been defined by the Astronomy Survey Committee of the National Research Council, which at intervals of 10 years develops a comprehensive program for both ground-based and space-based astronomy. The emergence of space astronomy, as a major element of astronomy, is reflected in the top priority given the Advanced X-Ray Astrophysics Facility as the most important major project to be started in the next few years. Of all the research areas in space, life sciences has benefitted most from the return of manned flight in the U.S. with the numerous shuttle missions, and stands to make major advances when the space station is launched in the early 1990's.

The four chapters contained in this part give witness to the accomplishments, the status, and the excitement of space science. The chapters are divided by major areas of research for convenience. Space science is a mature, multidisciplinary, and interdisciplinary field of research and scientific interest that does not lend itself to simple arbitrary boundaries. For example, the sun as a star is of interest to astrophysics, while the sun as a source of radiation is of importance in understanding sun–earth relationships. The studies of particles and fields is essential to understanding the sun, the earth, the planets, the stars, and the many enigmatic objects in the universe. Space science encompasses everything from the center of the earth to the edge of the universe! The past 25 years have been a Golden Age for space science investigations. Our views of the universe, the solar system, the earth, and our place in the total system have been changed forever. We are deeply indebted to those who made it possible . . . even more so to those who made it happen.

2

SUN-EARTH RELATIONS

L. J. Lanzerotti

INTRODUCTION

On September 1, 1859, sketching sun spots in the course of his studies of solar phenomena, the British scientist Richard Carrington suddenly observed an intense brightening in the region of one of the spots. He was so taken by this occurrence that he quickly called his associates to the telescope to witness the event. Within a day, violent fluctuations in the earth's magnetic field and intense auroral displays were observed on earth, with reports of aurora observations as far south as Honolulu. Carrington was very intrigued about a possible link between his "white light" flare and the subsequent aurora. Nevertheless, he urged caution in connecting the two, commenting "one swallow does not a summer make."

During the several days of significant auroral displays, telegraph systems throughout Europe and in eastern North America suffered severe impairments in operation, or even complete disruptions of service. The new technology of telegraphy had never experienced such widespread impacts on service, occurrences which were attributed by telegraphic engineers to currents flowing in the earth, somehow produced by the auroral displays. Indeed, during some particularly intense aurora, the telegraph lines running from Boston to points both north and south could be operated for several hours without benefit of the battery supplies. Even in daylight some telegraph lines were disrupted, indicating to the engineers that the "aurora" must still be present, even though it could not be observed because of sunlight. During a nighttime episode, the extraneous voltage on a Boston telegraph line was observed to vary with an approximately 30 s periodicity that was simultaneously seen in the auroral display.

The occurrences of 1859 were a significant spur to action by telegraph engineers, with considerable professional activity in Europe devoted for many years to the studies of "earth currents." While the engineers often discussed the relation of these currents to enhanced fluctuations in the earth's magnetic field (magnetic storms) and possibly to disturbances on the sun, natural scientists of the time were less likely to see such a cosmic causal connection. The British scientist E. Maunder seriously discussed such a connection, but the words of the eminent authority Lord Kelvin, in discussions of a particular magnetic storm (June 25, 1885), were more in tune with the times when he concluded that ". . . it . . . is absolutely conclusive . . . that terrestrial magnetic storms are [not] due to magnetic action of the Sun. . . ." He further went on to conclude that ". . . the supposed connection between magnetic storms and sunspots is unreal, and that the seeming agreement between the periods has been a mere coincidence."

The first half of the twentieth century, which saw the implementation of transatlantic radio broadcasts and the discovery of the ionosphere, gave rise to more considered—and quantitative—discussions of the sun-earth connection by such individuals as S. Chapman, V. C. A. Ferraro, and H. Alfvén. Such considerations were warranted, not only scientifically, but for practical reasons as well. For example, transatlantic telephone traffic via low-frequency radio was often disrupted during magnetic storm occurrences, a situation which prompted many scientific and engineering experiments and publications in the 1930's, as well as the laying of the first transatlantic telephone cable in the 1950's.

Nearly 100 years after Carrington's discovery, at the threshold of the space age, the first transatlantic telecommunications cable was severely affected on February 11, 1958 by a magnetic storm which followed closely on the heels of a large solar flare. Toronto suffered an electrical blackout during the February 1958 solar disturbances. In 1972, the year after the last astronaut had gone to the moon, a series of large solar flares in early August resulted in such large magnetic disturbances that there was a severe disruption of a transcontinental communications cable in the United States. The energetic particles from these flares, primarily protons and electrons, would have been lethal to an explorer on the moon's surface or, probably, in a spacecraft in transit there.

By the time of the February 1958 solar events, there was no longer any doubt, of course, that the sun and the earth were intimately linked. The advent of scientific spacecraft demonstrated conclusively that the linkages occur not only by means of the optical emissions of the nearest star, but also by the "invisible" tenuous mixture of ions, electrons, and magnetic fields which we now call the solar wind, as well as by much more energetic solar particles. The space age has provided the means to dispatch unmanned instruments into the earth's environment to measure and study this medium which can produce both disruptions to technology and the dramatic, visual spectacle of the aurora.

The advances made by the use of spacecraft in understanding the solar terrestrial environment have been enormous. However, even as the basic morphology of this environment has been unraveled it has become clear that there are physical processes occurring in it that yet defy complete understanding. Thus, predictions as to the state of the environment and its possible effect on the earth remain very rudimentary. The physics of the processes in the medium is complicated. However, the increasing sophistication of experiments (ground-based, rocket, and spacecraft), as well as theoretical and computational capabilities, have begun to provide some insights into the fundamental processes in the solar-terrestrial system.

SOLAR PROCESSES: THE SOURCE FOR SUN-EARTH COUPLINGS

The linkage of the sun to the earth by its visible radiation provides the heat and light which has enabled life to evolve and thrive on the third planet. Because of the central role which the sun plays in the well-being of earth, the mythologies of many ancient civilizations attributed god-like qualities to this nearest star. The constant, unvarying nature of the source of the life-sustaining light and heat was the most important of these. This myth was shattered when Galileo's telescope revealed that the sun was not "perfect": spots were observed on the solar surface which varied with time and with location. Since Galileo's time, the sun has been found to have a number of other changeable features, many of which affect the earth in important ways. The existence of a cyclic variation in the appearance of sunspots was confirmed by a German druggist and amateur astronomer, S. H. Schwabe, in the mid-nineteenth century.

Sunspots are the best known manifestation of solar activity. The cyclic variation in sunspot numbers, with an approximate 11 year period, is shown in Fig. 2-1. However, for a significant interval after Galileo discovered them, there was a long hiatus when the number of spots was very small, or even zero. This so-called Maunder minimum, named after the aforementioned British scientist who first drew attention to the occurrence, is a significant enigma in terms of basic understanding of solar variability.

While the spot cycle is 11 years, the fundamental cycle, based upon the magnetic polarity of the sun, is approximately 22 years. At the beginning of a sunspot cycle, the spottedness increases first in the mid-latitude regions on both sides of the solar equator. As the cycle continues to develop and the number of spots increases, they migrate slowly towards the equator in both hemispheres until, near the end of the cycle, the equator-most spots begin to disappear. During and following the disappearance, spots again begin to appear in the middle latitude regions.

A reasonably current view of the solar machine, which produces the spots and other solar phenomena, is shown in Fig. 2-2. The temperatures and densities are illustrated as a function of depth in the sun. We will touch upon a

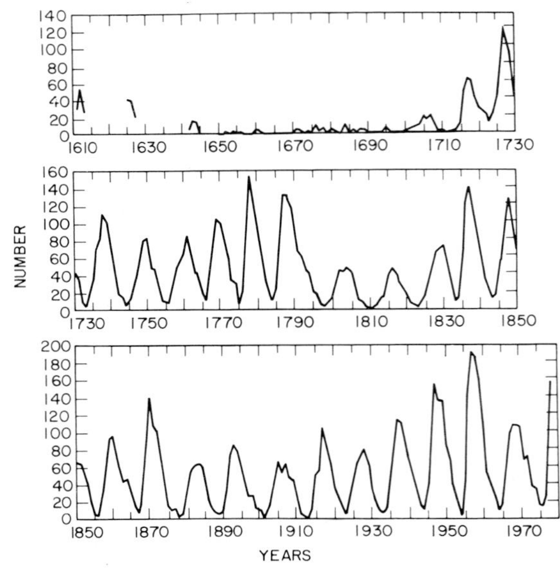

Fig. 2-1: Sunspot cycle, 1610–1979 A.D., as represented by the annual mean sunspot number for those years. (Courtesy of J.A. Eddy, High Altitude Observatory, NCAR.) The interval of about seven decades beginning in 1645 when sunspot numbers were unusually low is known as the "Maunder Minimum," which coincided with a significant drop in global temperatures on earth. The sunspot records between 1610, when Galileo used a telescope, and 1640 are mostly too scanty to reconstruct with considerable confidence.

number of key questions regarding the sun and its overall structure.

The precise nature of the nuclear burning cycle in the center of the sun has been called into some question because neutrinos (key elementary particles with no mass which should result from the burning) have not been observed in the expected abundances in an experiment designed to measure them. Whether this lack of detection can be simply explained in terms of minor uncertainties

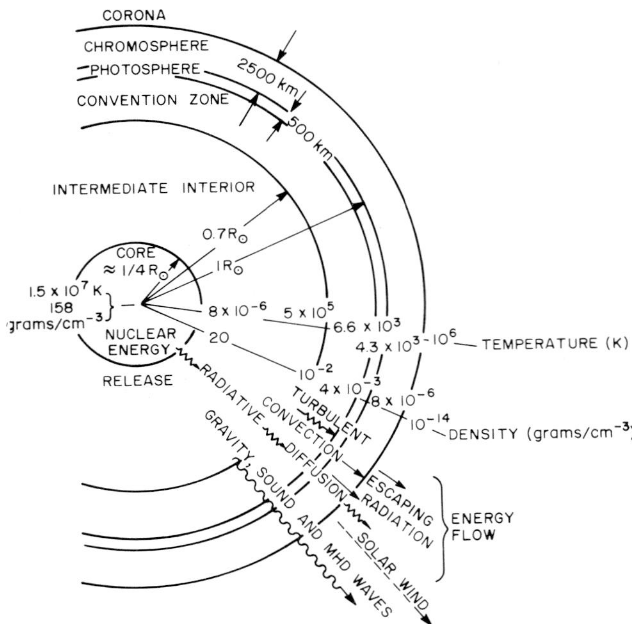

Fig. 2-2: The solar machine, showing conditions as a function of depth in the sun.

in the nuclear burning processes or whether the experimental results reflect a more profound ignorance of nature (of the sun and of the neutrinos themselves) yet remains to be understood.

The process by which the energy from the interior of the sun is transported to its surface, and how this convective transport of heat interacts with the internal magnetic field of the sun, carrying the magnetic field through the surface in some regions, is incompletely understood. What is known from several recent measurements of the solar surface is that the sun apparently oscillates with a wide range of discrete frequencies. Analysis of these oscillations in terms of the observed frequencies and the spatial scale sizes across the solar surface shows very distinct, discrete patterns in the frequency-spatial scale diagram, as illustrated in Fig. 2-3 from data taken at the South Pole, Antarctica. Such results, used with sophisticated models of the possible modes of oscillation of the sun, can help deduce the depth of the convective zone (see Fig. 2-2). The longer the period of the oscillation, the deeper the sun is effectively being "probed." The situation, in analogy to seismological studies of the earth's interior, is often called "solar seismology." Long-period (many hour) oscillations are difficult to discern because the rotation of the earth moves a telescope at a mid-latitude observatory out of direct observation of the sun after approximately 6–8 h of daylight. Thus, measurements have recently begun at the South Pole, where the continuous daylight existing during austral summer provides the length of continuous sunlight needed for detailed analyses of long-period oscillations.

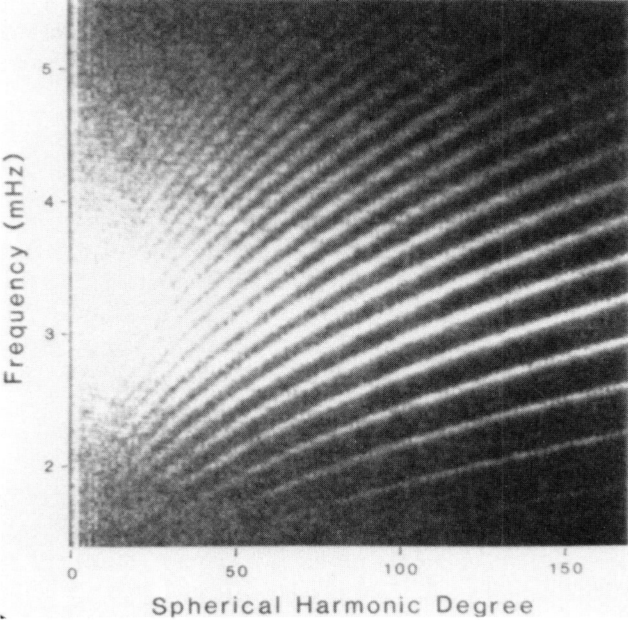

Fig. 2-3: Diagram of oscillations of the solar surface. The bright ridges indicate the spatial frequency (*k*, horizontal axis) and the temporal frequency (*w*, vertical axis) of the higher amplitude oscillations. The oscillations result from the natural resonance of acoustic or pressure waves trapped beneath the solar surface. The spatial axis ranges from 0 to about 300 cycles per solar diameter while the temporal axis ranges from 0 to about ½ cycle/min. (Based on observations by Duvall and Harvey at Kitt Peak National Observatory.)

Analysis of these oscillations indicates that the depth of the convective zone is approximately 30 percent of the solar interior beneath the surface, almost a factor of two greater than that incorporated in most solar models only a few years ago.

The convective zone transports the interior heat to the surface, providing an approximately constant output of the sun in the visible range, where most of the energy is contained. For more than a half a century, stimulated largely by the work of C. G. Abbott of the Smithsonian Institution, various observations have been attempted on earth in order to ascertain whether the solar output, the so-called "solar constant," is truly constant. Very recently, sophisticated measurements on the Solar Maximum Mission spacecraft have finally disclosed that the "solar constant" is not a constant at all, but that the total energy output of the sun is variable and can change by order of a fraction of a percent from day to day. This remarkable result is poorly understood. In particular, it is still a matter of considerable scientific debate as to what happens to the "excess" energy when the "solar constant" decreases in value: is the energy stored in the sun, only to be released at some other time? Exactly what takes place in the convective zone during such intervals? Is a decrease in the solar constant related to the occurrence of sunspots, long known to be regions of cooler temperatures on the sun, or are there other factors involved as well in producing a decreased solar output? Are such short-term changes in the solar constant relevant for weather and climate considerations? There also appears to be a longer term (many months) trend in the "solar constant" value measured thus far, but what such a trend implies in terms of cyclic behavior is not known.

Sunspots (Fig. 2-4) are not just regions of cooler gases on the solar surface. They are also regions where the solar magnetic fields from the interior penetrate the surface, producing very large, of the order of several thousand gauss, intensities (as compared to the earth's magnetic field at the surface of approximately ½ G). Sunspots occur primarily in groups of two, with opposite magnetic polarities, plus and minus. Magnetic field lines arch from one spot to the other. Under conditions which are poorly understood, the magnetic fields of the sunspot structures can become very unstable, ultimately producing a solar flare: intense brightening of a region of the solar surface abruptly occurs as atomic ions and electrons in the solar atmosphere are suddenly energized. The energy content of a solar storm is 10^{32}–10^{33} ergs and is probably derived from conversion of energy stored in the magnetic fields. Some particles impact the solar atmosphere, producing light, while the remainder propagate rapidly into interplanetary space. Intense radio bursts, easily detectable at earth, accompany the particle energization.

Following a flare, the sunspot region is altered in appearance and can produce additional activity. Solar flares, as evidence of violent processes occurring in the magnetized plasmas of the solar atmosphere, are likely

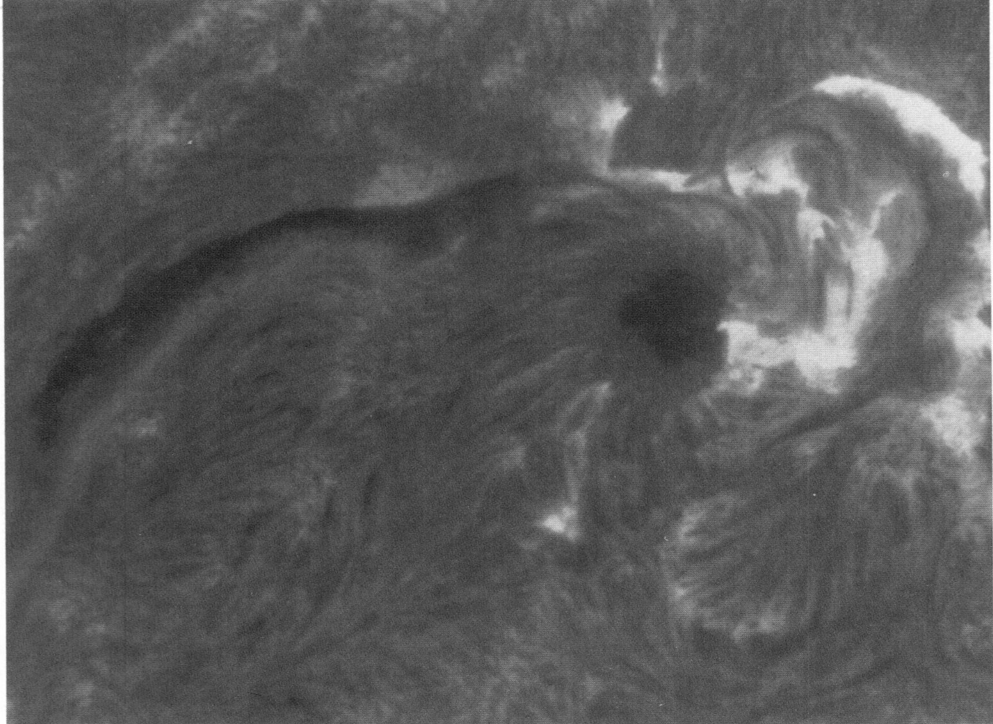

Fig. 2-4: Solar flare on the sun taken at Big Bear Observatory, California, in the hydrogen alpha line spectral region.

to involve plasma physical processes similar to those associated with the aurora on earth, as will be discussed further below. There may also be some close relationships to the unstable conditions that are found to occur in laboratory plasma fusion devices, albeit on a substantially smaller scale. Many facets of solar activity were studied by spacecraft in the Orbiting Solar Observatory (OSO) series. International programs such as the International Quiet Sun Year and the Solar Maximum Year have been important in coordinating research on solar activity.

In addition to solar flares as a violent form of solar activity, large injections into the interplanetary medium of solar coronal material, approximately 10^{12} kg or more, sometimes occurs. The true nature of such "coronal transients" was identified from analyses of data acquired during the Skylab era of solar research in the mid-1970's. While such transients can apparently produce disturbed interplanetary conditions, the mode of propagation and dissipation of the ejected mass and energy through the solar system, including their potential impacts upon the earth's plasma environment, is largely unknown at present.

INTERPLANETARY MEDIUM: THE MECHANISM FOR SUN-EARTH PLASMA COUPLINGS

Early atomic spectroscopic measurements of the solar atmosphere established that the outermost region—the solar corona—has a much lower density, and a much higher temperature than the underlying surface (the photosphere), where solar activity occurs. The precise mechanism(s) for heating this high upper solar atmosphere is (are) unknown. Possible causes include heating by acoustic waves generated by the convection of heat from the solar interior and/or heating by the damping of Alfvén waves produced in the solar photosphere and which propagate outward, away from the sun. The structure of the solar corona is a complicated plasma environment, which often appears as radial streamers and arches in eclipse photographs (Fig. 2-5).

Independent of the precise mechanism by which the corona is heated, the end result finds the very hot outer atmosphere continually flowing away from the sun, forming the solar wind. The speed of this wind typically lies in the range 400–600 km/s, although much higher velocities associated with solar activity have been measured. At a distance of a few solar radii above the photosphere, the velocity of expansion of the corona becomes larger than the Alfvén velocity (Fig. 2-6). That is, the velocity of expansion becomes faster than the velocity with which information can be transferred in the highly conducting (ionized) gas. Beyond this point, the solar wind expands and flows throughout the solar system, dragging with it the magnetic field from the solar surface. The region of our galaxy which is influenced by the solar wind is called the heliosphere. It is the solar wind that provides the plasma link between the sun and the earth and which is the source of geomagnetic disturbances on earth. Lord Kelvin, in his categorical statements of nearly a century ago, could not have been aware of this crucial link.

The outward expansion of the solar wind combined with the rotation of the sun provides two important

Fig. 2-5: Eclipse photograph of the sun, made from the earth's surface, showing coronal streamers and coronal structure.

physical phenomena in the heliosphere. First, the solar plasma tends to be confined to a sheet-like region around the solar equator, forming a "plasma sheet" in the heliosphere, somewhat analogous to the plasma sheet in the earth's magnetotail (as will be discussed below). Second, because the solar magnetic field is firmly embedded in the radially outward-flowing solar plasma, and yet is tied firmly to the sun at the solar photosphere, the field in interplanetary space forms a spiral pattern, similar to that produced by a rotating garden water sprinkler. The heliosphere plasma sheet is not rigid, but rather assumes a warped, wavy pattern in interplanetary space (Fig. 2-7); at any given time the earth can be above or below (or within) the plasma sheet. At the orbit of earth, the interplanetary magnetic field makes an average angle to the radially outward direction of approximately 45°; at the orbit of Jupiter, the interplanetary field is nearly perpendicular to the sun–planet line.

The solar wind has been extensively studied by instruments of ever-increasing sophistication on interplanetary spacecraft, particularly in the Pioneer and IMP (Interplanetary Monitoring Platform) series. More recently, the ISEE-3 (International Sun–Earth Explorer) spacecraft, located at a gravitationally stable position at approximately 10 percent of the sun–earth distance in front of the magnetosphere, has been able to study the wind unimpeded by possible magnetospheric disturbances. The Pioneer and Voyager missions to Jupiter and Saturn have provided crucial solar wind data from the outer solar system.

Sunspot activity significantly disturbs the tranquility of the rotating, warped interplanetary plasma sheet. In addition to sending high energy particles into the interplanetary medium, solar flares also emit greatly enhanced solar wind streams. These streams have a higher velocity and greater particle number density than the normal solar wind. Shock waves thus are formed in the interplanetary medium between the boundaries of

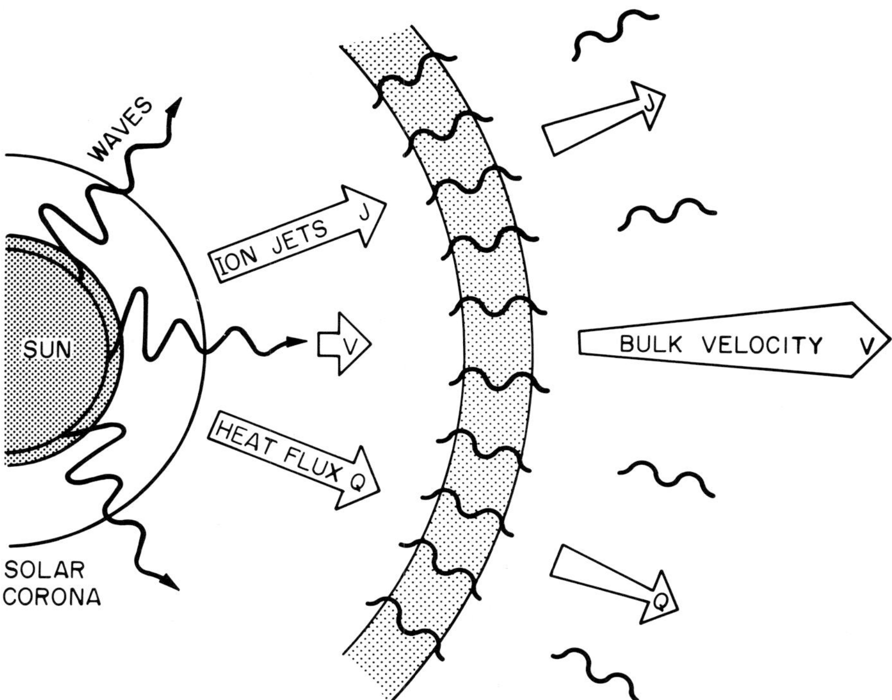

Fig. 2-6: Schematic illustration of the solar coronal expansion close to the sun. The details of the expansion, including the distance scales, remain uncertain. The coronal gas, heated by wave damping, is believed to drive an outward heat flux Q, a bulk heat convection V, and, possibly, transient ion jets J.

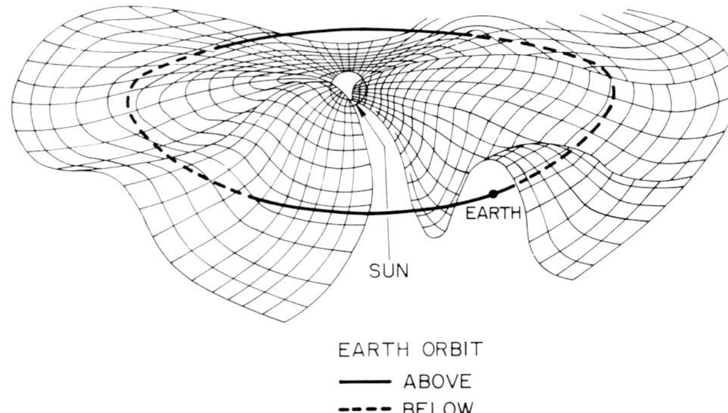

Fig. 2-7: Schematic representation of the solar-interplanetary current sheet. (Courtesy of S. -I. Akasofu, Geophysical Institute, University of Alaska.) The sun is the center of an extensive and warped disk-like sheet in which electric currents flow azimuthally, that is, around the sun. The average plane of the disk is approximately the plane of the equator of the sun's average dipole magnetic field, which may be tilted with respect to its equator of rotation. The sheet separates solar-interplanetary magnetic-field regimes of nearly opposite (or at least greatly different) average direction.

the faster and slower moving winds. These shock waves can themselves accelerate interplanetary particles to higher energies and can greatly agitate the earth's magnetosphere.

During solar minimum conditions the winds observed in interplanetary space appear to be emitted primarily from coronal "holes," especially polar coronal holes. These are regions of the solar corona where the solar magnetic fields tend to be more open, extending into interplanetary space. These open field regions are normally observed (from ultraviolet and X-ray measurements of the solar surface) to be confined near the solar pole regions. However, during solar minimum conditions the polar regions can extend, in areas limited in longitude, to near the solar equator. These coronal hole regions emit high velocity solar wind streams, streams which produce significant disturbances of the earth's plasma environment during the periods when solar flares and sunspots are nearly absent from the solar surface.

What happens to the solar wind at great distances from the sun? It is likely that the heliosphere, with the embedded sun and planets, forms a kind of magnetosphere in the local interstellar medium. In the direction in which the sun is moving relative to the nearby stars, there is likely to be a boundary established between the outward flowing solar wind and the interstellar plasmas and magnetic fields. Pioneer spacecraft measurements show that this boundary certainly does not exist within the orbit of Pluto, some 50 times the earth–sun distance. It is likely that the boundary will be found to occur between 100–150 earth–sun distances. The boundary will likely be a turbulent region, with perhaps a shock wave established in the interstellar medium, similar to the situation for the earth in the solar wind (see below).

At the orbit of earth, the solar wind carries an energy density of about 0.1 ergs/cm^2, very small compared to the energy in the visible and infrared wavelengths ($\sim 1.4 \times 10^6$ ergs/cm^2), a factor of about 10^7 difference. Yet, as will be seen in the next section of this chapter, this low energy density, when integrated over the earth's magnetosphere, can have profound effects on the earth's space environment.

Largely unknown at present are the plasma and energetic particle environments in the polar regions of the sun. They have never been measured directly by spacecraft. Whether the magnetic field configurations of these regions resemble the polar regions of the earth's magnetosphere as discussed below, or whether there is some other topology remains to be discovered by the flight of the Ulysses Mission spacecraft in 1986–1990.

EARTH'S MAGNETOSPHERE: THE EXTENSION OF THE EARTH'S MAGNETIC FIELD AND PLASMAS INTO SPACE

Overall Morphology

The earth announces its existence to the solar wind in a manner analogous to that of a supersonic airplane traversing the earth's atmosphere. Since the solar wind velocity is faster than the characteristic speed in the ionized gas (the Alfvén speed, which is a function of the magnetic field strength and the density of the plasma), a shock wave is established in front of the earth. The average location of the shock wave is some 10–14 earth radii on the sunward side. Behind the shock wave is a region of more turbulent solar wind flow. Finally, at an average altitude above the earth of some nine earth radii, the magnetopause exists as a thin current-carrying boundary separating regions of solar wind control of space from regions controlled by earth's magnetic field (Fig. 2-8). To a reasonable approximation, the solar wind flow around the earth at the magnetopause can be described by analogy to the atmosphere flow in a wind tunnel past an aerodynamic object. Also to a good approximation, the subsolar magnetopause location can be approximated by a balance in pressure between the flowing solar wind and the earth's magnetic field.

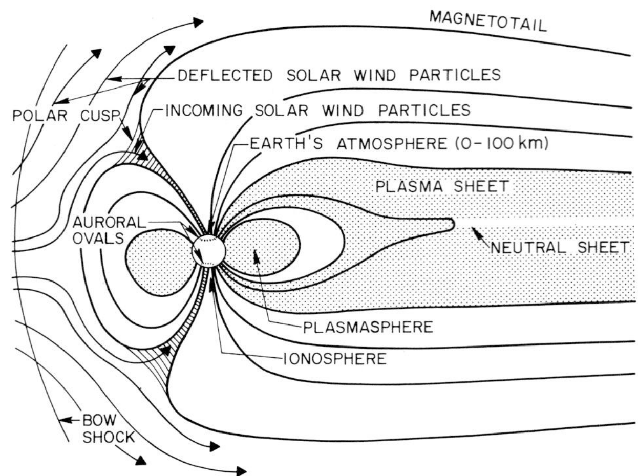

Fig. 2-8: Schematic illustration of the earth's magnetosphere, the nearby plasma environment formed by the interactions of the solar wind with the geomagnetic field.

Two different NASA scientific satellite series, beginning with the Explorer series, have been especially important in delineating and investigating in detail many aspects of the magnetosphere and its interaction with the solar wind. The other of these series included several spacecraft of the Orbiting Geophysical Observatory (OGO) type. The IMP series, numbered in the Explorer sequence and primarily devoted to interplanetary investigations, also contributed importantly to studies of the magnetotail, the magnetopause, and the bow shock. The three spacecraft ISEE mission (a joint program between NASA and the European Space Agency) was designed for detailed studies of the solar wind interaction with the magnetosphere, using one spacecraft permanently located in the solar wind and two orbiting the earth in tandem to give detailed spatial information about the magnetopause and bow shock. Spacecraft from other NASA programs and other agencies have frequently carried instrumentation that has contributed significant data for developing concepts of the plasma physics of the magnetosphere. These have included the ATS (Applications Technology Satellite) series from NASA, the Tiros and GOES series from NOAA, the DMSP (Defense Meteorological Satellite Program), the Space Test Program, and navigation programs (e.g., Triad) from DOD, the Vela program of DOE, and university spacecraft, such as the Hawkeye series from the University of Iowa. The International Magnetosphere Study was a major program in coordinating ground-based, rocket, and spacecraft research on the earth's magnetosphere.

The physics describing the actual processes forming the configuration of the earth's magnetosphere, while approximated by these simple concepts, is much more involved in detail. The formation of the shock wave in front of the magnetopause is more complicated than the simple airplane analogy. For one thing, the solar wind contains an imbedded magnetic field, which is intimately associated with the formation of the detailed characteristics of the shock. Furthermore, the shock wave is a collisionless shock—that is, the solar wind particles do not form the shock by collisions among themselves. Rather, the distribution of plasma particles becomes unstable because of the magnetic field and wave conditions produced by the presence of the obstacle. Waves generated by the shock wave can propagate upstream in the solar wind, making the solar wind environment in front of the shock quite complicated and very turbulent. Solar wind particles encountering the shock wave can be accelerated and sent back upstream, toward the sun, confined to flow along the magnetic field.

Similarly, the earth's magnetopause is a complicated physical regime. While the earth's magnetosphere has a vast extent, the magnetopause boundary itself is a very thin region where currents flow in the plasma. These currents define a thin "cellular" boundary separating the two quite different plasma regimes. This cellular-like boundary is not completely impenetrable and its location is variable, dependent upon solar wind conditions. Its location can move inward, towards the earth, under higher solar wind velocities, and can move outward during periods of lower solar wind speed conditions. Nevertheless, the two plasma regimes can be treated quite separately in analyzing many problems involving internal processes in each system. The interplanetary magnetic field also plays an important role in the formation of the boundary. It is likely that some type of "reconnection" of the earth's internal magnetic field and the interplanetary field occurs sporadically at the boundary, allowing plasma to escape from the magnetosphere and heliosphere plasma to become entrapped in the earth's magnetosphere.

The solar wind, by a viscous interaction with the magnetosphere plasma and magnetic fields, forms the magnetosphere into a long, comet-like (but invisible) object. A large, dawn to dusk, electric field is imposed across the magnetosphere. The energy transfer rate from the solar wind into the magnetotail is about 10^{19} ergs/s. The magnetotail may extend to more than a thousand earth radii (some 6×10^6 km) in the anti-sunward direction. In the center of the magnetotail, a "sheet" of highly conducting plasma separates the magnetic fields which originate in the southern hemisphere from those which terminate in the northern hemisphere. A current flows from dawn to dusk through the plasma sheet. The earthward extensions of this plasma sheet protrude along magnetic field lines which connect with the nightside auroral zone (ionosphere currents) in both hemispheres. The power dissipated in these currents is about 10^{18} ergs/s (10^{11} W). On the front side of the magnetosphere, the separations in each hemisphere between magnetic field lines closing on the dayside and those extending into the magnetotail form a "cusp"-like region through which solar wind plasma can reach the upper atmosphere, forming auroral emissions and currents in the ionosphere. The topology of the auroral zone, shown in a picture from a spacecraft during winter (dark) conditions in the northern hemisphere in Fig. 2-9, provides a qualitative, visible measure of the geometry of the magnetosphere and magnetotail. The aurora can also be seen under daylight conditions if measured in the ultraviolet emissions, as shown in the spacecraft picture of Fig. 2-10.

During intervals of geomagnetic storms (very disturbed magnetosphere conditions), the auroral zone can extend to much lower latitudes, and the currents flowing in the ionosphere can be significantly enhanced. The energy dissipated during a geomagnetic storm, which may last 2–3 h, is about 10^{19} ergs/s. These magnetic disturbance conditions seem to occur from the energization of the plasma sheet, often triggered by changed solar wind conditions. This enhanced plasma is then transported into the auroral zones. Additional energization seems to occur along magnetic field lines above the aurora. The energization in the magnetotail plasma sheet is believed by many to occur by the conversion of magnetic field energy into plasma particle

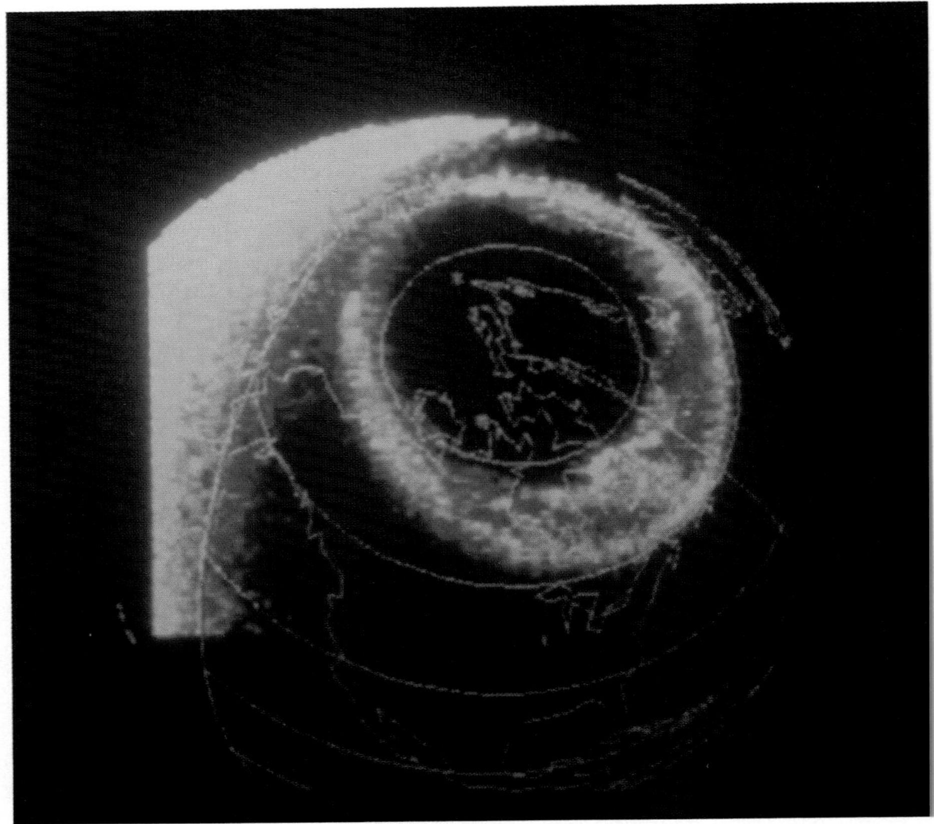

Fig. 2-9: Auroral zone of the earth measured in ultraviolet light under nighttime conditions by the Dynamics Explorer satellite.

kinetic energy through a process of reconnection of magnetic field lines across the plasma sheet.

This overall concept of the earth's magnetosphere was rapid in developing once spacecraft were sent above the earth's atmosphere. Soon after the discovery of the radiation belts by Van Allen and his students on Explorer 1 in 1958, the magnetosphere boundary and shock were detected. In the last few years, with the discovery and detailed measurements of the characteristics of the Jovian and Saturnian magnetospheres by Pioneer and Voyager spacecraft, as well as the mini-magnetosphere around Mercury (measured by the Mariner 10 spacecraft) and the magnetosphere formed by the solar wind interaction with the ionized upper atmosphere of Venus (which has particularly been elucidated by the Pioneer Venus orbiter spacecraft), the concept of a magnetosphere has become quite general in cosmic plasma physics. Indeed, many exotic astrophysical objects, such as pulsars and some radio galaxies, are now commonly discussed in magnetosphere terms. The capabilities for measuring by spacecraft the plasma properties of solar system magnetospheres have proven quite important in extending the concepts learned to other astrophysical magnetospheres which can only be studied remotely by detection of their emitted radio waves.

Plasma Processes

There are many localized and/or small-scale plasma processes throughout the global magnetosphere system which provide the underlying mechanisms that determine the state of the system. A large variety of plasma waves exist in the magnetosphere (Fig. 2-11). In the magnetosphere these waves can be electromagnetic, that is having both electric and magnetic properties, or they can be electrostatic, resulting from the separation of oppositely charged plasmas. Lightning storms within the earth's atmosphere create radio frequency waves which not only produce static in radios, but also result in waves which propagate, with a whistling tone, along geomagnetic field lines extending into the outer magnetosphere. Other waves, with wavelengths of the order of the length of a magnetic field line (several earth radii or more), are known as Alfvén waves. These waves are generated by the flow of the solar wind past the earth's magnetic field and by instabilities that can occur within the magnetospheric plasma. Still other radio waves are generated above the auroral regions (within perhaps one earth radius altitude) where electrons and protons are accelerated into and out of the earth's ionosphere. Detected on an earth-orbiting or interplanetary satellite, these radio waves from the polar regions would have many resemblances to the radio waves detected on earth from Jupiter's and Saturn's magnetospheres. The other plasma waves are internal to the magnetosphere, and would not be observable outside the magnetopause.

A most interesting aspect of the magnetosphere, as well as of solar flares, is the possible existence of regions where magnetic field energy is converted directly to plasma particle energy. These regions are believed to be those where magnetic field lines of opposite polarities

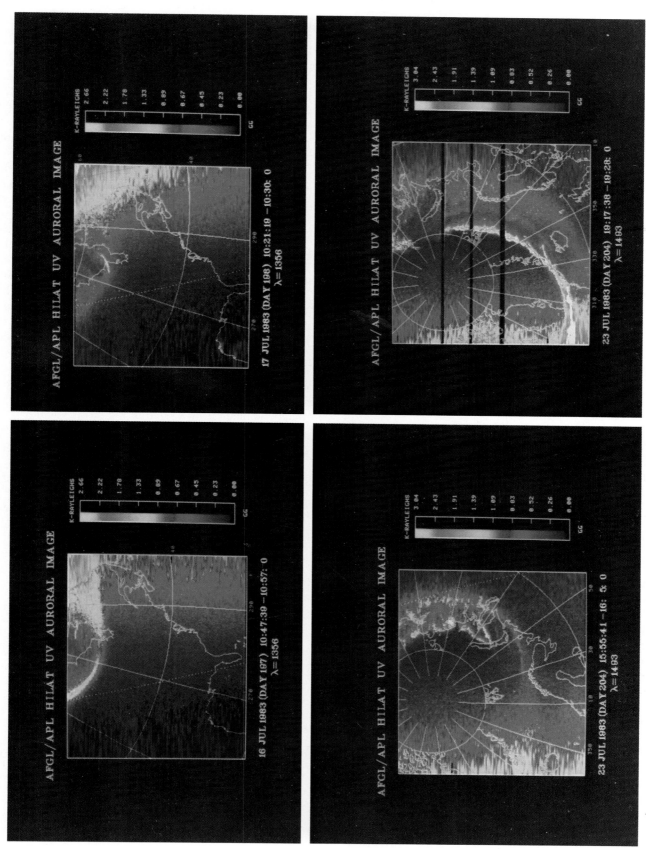

Fig. 2-10: Auroral activity measured in the ultraviolet under daylight conditions by the HILAT satellite.

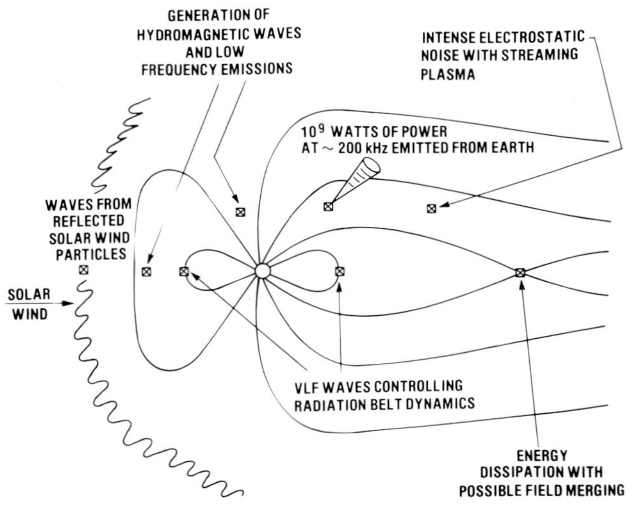

Fig. 2-11: Schematic view of significant plasma wave regions within the earth's magnetosphere (noon midnight meridian profile).

can suddenly reconnect and, in the process of reconfiguring, give up energy to the embedded plasma. As noted above, such regions can exist at the front side of the earth's magnetosphere, possibly in the cusp regions, and in the earth's magnetotail. In the magnetotail, the magnetic field lines in the plasma sheet are always oppositely directed. If the plasma sheet conditions should change in such a manner that the plasma became less conducting, reconnection of the magnetic field lines might occur spontaneously. Alternatively, the magnetic field lines could be pushed together by an external force (such as the solar wind interaction), which would slowly compress the plasma sheet. As the plasma in the center of the region becomes more dense, collisions among plasma particles or certain plasma instabilities could alter the conductivity, providing an environment for reconnection to occur. Many theoretical and computer simulation considerations of the reconnection process have been carried out and it is one of the most active areas of basic magnetospheric plasma research at present.

The earth's ionosphere is an intriguing plasma environment, with both neutral and ionized gases threaded by the geomagnetic field. NASA satellites that have contributed importantly to studies of the ionosphere and upper atmosphere include several OGO's, the Atmospheric Explorer (AE) series, and the dual (high and low altitude) Dynamics Explorer mission. At times, the ionization layers become unstable, causing the ionization to be patchy, with different ionization densities. This can be particularly prevalant in the auroral zone currents. In the equatorial regions of the ionosphere, where the magnetic field lines are parallel to the ionization layers (as well as to the surface of the earth) the ionosphere layers can become unstable; plasma bubbles can form and rise through the ionosphere to the upper levels. Studies of the basic plasma physics of such bubbles and ionization patches have led to significant new insights to cosmic plasma processes. Furthermore, joint theoretical and observational treatments of the plasma conditions in the equatorial regions of the earth's environment have probably attained the greatest amount of self-consistent understanding of any cosmic plasma phenomena studied to date.

The background plasma density in the magnetosphere varies from several thousand particles per cubic centimeter within the first few earth radii altitude to only a few per cubic centimeter at higher altitudes. There is ordinarily a rather sharp discontinuity between the two plasma regimes, and the boundary is called the plasmapause (Fig. 2-8). The boundary is formed at the location where there is a balance between the electric fields produced by the rotation of the geomagnetic field and the large-scale electric field imposed across the magnetosphere by the solar wind flow. This discontinuity in the plasma distribution is the source of magnetosphere plasma waves and significantly affects the propagation of Alfvén waves. The background plasma density inside the plasmapause results from cold ionosphere plasma diffusing into the magnetosphere during local daytime conditions. Outside the plasmapause, the cold plasma from the ionosphere is swept (convected) out of the magnetosphere by the cross-magnetosphere electric field.

The earth's radiation belts are populated by solar wind ions and electrons and by particles accelerated out of the upper atmosphere/ionosphere. The motions of radiation belt ions and electrons are controlled by the earth's magnetic field, as illustrated in Fig. 2-12. The inner Van Allen belt of electrons is located earthward of the plasmapause while the outer belt is outside the plasmapause. The relative importance ascribed to these two sources of radiation belt particles has varied during the history of magnetosphere research, with much emphasis being placed at present on the ionospheric source. The ions and electrons can be accelerated to their radiation belt energies by internal plasma instabilities and by large-scale compressions and expansions of the magnetosphere under action of the variable solar wind. In the innermost part of the magnetosphere, the decay of neutrons, produced by high-energy cosmic ray impacts on the upper atmosphere, yields electrons and protons to the trapped radiation belts.

An abbreviated sketch of major current systems in the

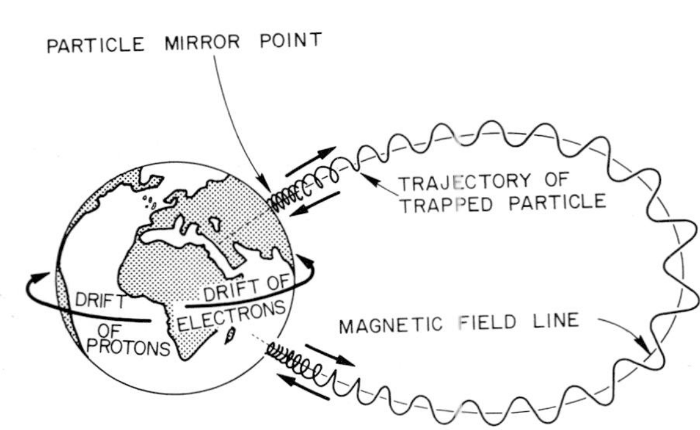

Fig. 2-12: Motion of charged particles in the earth's dipole magnetic field.

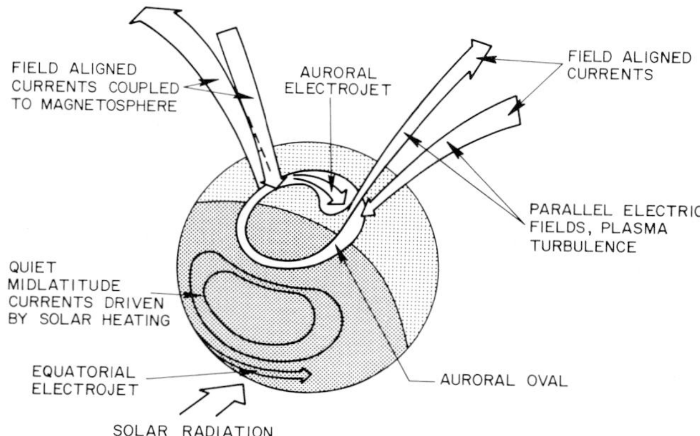

Fig. 2-13: Simplified picture of ionospheric current systems, and the connection of the auroral current systems to the magnetosphere via flowing currents along geomagnetic field lines.

magnetosphere/ionosphere system is shown in Fig. 2-13. The auroral current system is linked to the magnetosphere and the plasma sheet via the currents which flow along geomagnetic field lines. Solar radiation, in the visible and ultraviolet wavelength ranges, produces ionospheric currents on the dayside of the earth, including an intense current in the equatorial regions.

The auroral current systems can heat the upper, neutral atmosphere of earth, altering the overall circulation patterns in these regions. Without auroral heating, circulation would tend to be from the equatorial, hotter regions, toward the poles, colder regions. As illustrated in Fig. 2-14, which shows the results of theoretical calculations of the effect, increasing the intensity of the

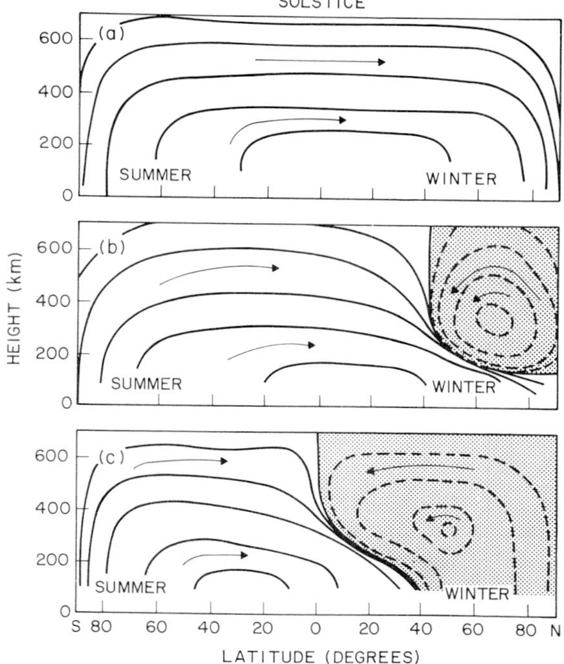

Fig. 2-14: Diagrams of the mean meridional atmospheric circulation in the earth's thermoshere at the time of the winter solstice in the northern hemisphere, for three levels of auroral (geomagnetic) activity. As auroral heating of the upper thermosphere increases, the influences of the heat are felt at lower latitudes.

auroral currents (and hence the amount of heating in the ionosphere) causes the upper atmosphere circulation patterns to reverse. The latitude of reversal depends upon the intensity of the heating.

THE SUN-EARTH CONNECTION: IMPLICATIONS FOR HUMAN ACTIVITY

A number of implications of the sun-earth plasma connections—the sun linked to the earth through the solar wind and solar flare particles—were mentioned in the introduction to this chapter. Most of these examples involved technologies that employ long conductors, such as telephone and telegraph lines, and electric power grids. These impacts of the sun–earth connection probably occur because the time-changing geomagnetic field, produced by the variable solar wind, causes currents to flow in the earth. These earth currents enter the long conductors and can produce deleterious effects on the connected electronics. The building of the Alaskan oil pipeline, directly through the auroral zone, meant that considerable effort was expended in studying the problem of the induction by auroral currents of earth currents in the pipe. Procedures had to be adopted to avoid potential difficulties. While earth currents have been much studied for over a century, their description remains largely empirical for two major reasons. First, it is difficult, if not impossible as yet, to predict the exact spatial scale of the disturbed, time-varying magnetic field. Second, the inhomogeneities in the conductivity structure of the earth are poorly known, so that regions of expected higher current flow for a given geomagnetic field variation cannot easily be predicted.

The extension of technology into space brought a new set of concerns. When communications satellites were originally proposed, it was not realized that the environment they would fly in would be anything but benign. However, we now know that the magnetosphere plasmas can significantly alter the properties of solar cells and spacecraft thermal control blankets. The energetic components of the plasmas can cause damage to semi-

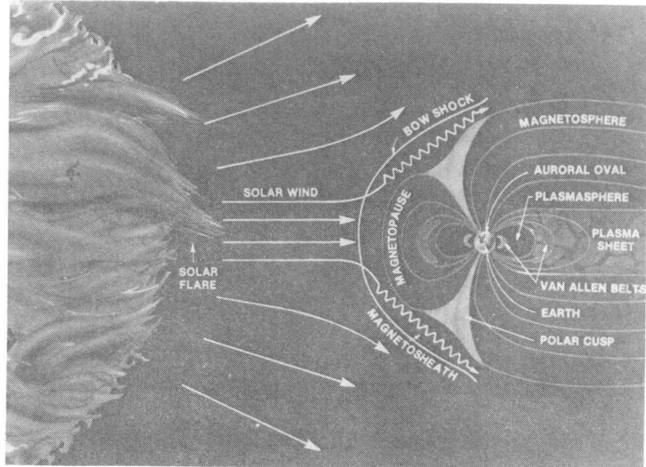

Fig. 2-15: The sun–earth plasma connection.

conductor components and anomalies in memory devices. Spacecraft have been known to suddenly go to a noncontrolled state because of radiational effects on such components.

The space shuttle, flying at about 200–300 km above the earth's surface, has been found to have a constant, visible glow around it. This glow is produced by the interaction of the shuttle with the space environment at these altitudes—mostly oxygen ions. However, the precise physical and chemical processes involved are not understood as yet. Such a glow, if present around orbiting telescopes, both earth-sensing and astronomical, could seriously affect the sensitivity and operational conditions of these instruments.

Prospecting for mineral resources by air and satellite surveys often uses magnetic sensing to detect such deposits. Geomagnetic field fluctuations can seriously impair the interpretation of surveys, and must be removed by various techniques in order to resolve geologically interesting features. It is particularly difficult to perform subtractions of magnetic disturbances in the auroral zone, and ground and aero survey parties are scheduled, insofar as feasible, to operate during geomagnetically quiet intervals. The accuracy for predicting such intervals is of considerable economic importance to prospecting firms.

The influences of the variable aspects of the sun—the solar constant, solar activity, solar wind—on the earth's lower atmosphere, the troposphere, where weather patterns are established, remains a great enigma. Statistically, certain elements of climate (and weather) in some regions appear to be related to solar variability, particularly the long-interval solar cycles. The elements of solar variability may all be interrelated in complex ways, which thus far have obscured some of the statistical results. A variability in the solar constant can be fit into many modern atmosphere circulation models in order to test for cause and effect relations. On the other hand, relating variability in solar activity and the solar wind to weather and climate (if these quantities are uncoupled from the solar constant) presents the significant problem of the driving mechanism involved. If such a mechanism—or mechanisms—exists to couple the solar particle and magnetic field flows to the troposphere, it remains to be discovered in the complex atmosphere–ionosphere–magnetosphere system.

THE FUTURE

Major future advances in understanding sun-earth relations will likely occur under the auspices of the International Solar-Terrestrial Physics (ISTP) Program and the Upper Atmosphere Research Satellite (UARS) Program. Both of these efforts are likely to be important elements of the International Geosphere and Biosphere Program (IGBP), presently under consideration by the International Council of Scientific Unions (ICSU). The ISTP program, beginning with the launch of an interplanetary spacecraft in 1990, is designed to address, in a coordinated and unified manner, energy flow throughout the solar–terrestrial system and plasma physical processes occurring in specific regions of the environment. It is designed as a major international collaborative venture, with the principal participants being NASA, the Institute for Space and Aeronautical Science (Japan), and ESA (Fig. 2-16). Opportunity will be provided (and encouraged) for participation by national space efforts as well as by ground-based researchers throughout the world. Active and passive science investigations in low-earth orbit from the shuttle can also be incorporated as required. The success of the coordinated effort will result from the implementation not only of the individual spacecraft components of the ISTP, but, just as importantly, from a sophisticated database system for ready exchange and correlation of diverse data sets from the many individual program elements.

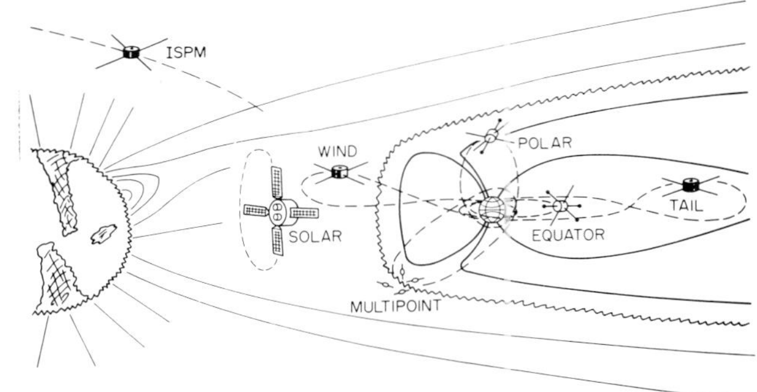

Fig. 2-16: Spacecraft configuration in the ISTP Program. The Wind, Equator, and Polar spacecraft will be sponsored by NASA and will concentrate on studies of the energy flow in the solar wind (Wind), the energy content of the magnetosphere plasma population (Equator), and the dynamics of the high latitude magnetosphere (Polar). The Tail spacecraft (Japan) will study energy flow in the geomagnetic tail and the mid-magnetosphere. The Solar and Multipoint (ESA) spacecraft will study solar interior/surface/coronal dynamics and macroscopic plasma turbulence processes, respectively. An important aspect of the ISTP is the International Solar Polar Mission spacecraft, a joint ESA–NASA mission that will orbit the poles of the sun in 1989–1990.

The UARS program, beginning development in 1985, is designed to gather comprehensive, long-term data on the upper atmosphere by the use of sophisticated remote sensing instruments. These instruments will measure ozone, other chemical species, solar radiation, incoming magnetosphere particles, and the temperature and dynamics of the upper atmosphere with high precision on a global scale. The UARS program will involve a central data handling facility that will process data into atmospheric physical quantities for use by the investigator teams. Researchers at their home laboratories will be able to study and analyze the data sets using remote computers.

These two major initiatives in studies of the sun-earth system will yield, in the next decade, large advances in knowledge of the solar–terrestrial environment and its impact on human activities.

3

SOLAR SYSTEM EXPLORATION

Geoffrey A. Briggs and William L. Quaide

The U.S. program of solar system exploration, now just over 20 years old, has as its scientific motivation the seeking of answers to some of the most fundamental questions that mankind has puzzled over throughout history: how did the solar system come into being and how did the earth and other planets evolve to their present states?

The question of solar system origin underlies all of our other questions about the planets and it also provides a unifying scientific theme, tying together astrophysics and planetary sciences. We address this question partly through the direct route of studying primitive materials (presently limited to meteorites but, in time, to include the comets and asteroids) and partly by the reconstruction of the histories of each of the planetary bodies in the solar system. This detective process is one which, additionally, allows us to substantially improve our understanding of our own planet earth, another central theme of our solar system exploration program. We also directly address another fundamental scientific goal—an understanding of the origin of life in the solar system; this enquiry requires a deep understanding of the chemical history of the solar system and, therefore, is complementary to the planetological thrusts of the program.

When the planetary exploration program was started in the early 1960's these goals seemed almost idealistic. Today, the challenge remains, but our progress has been dramatic and the goals are clearly realizable.

Planets are cool, nonluminous objects not readily accessible to detailed telescopic study (other than precise determination of ephemerides) in spite of their relative proximity to the earth. As a result, therefore, once the Newtonian principles of gravitation had been enunciated in the seventeenth century—thereby satisfying a long standing quest to explain planetary motions—the astronomical study of the solar system lapsed into a quiescent state punctuated by the exciting discoveries of the outermost planets. Around the turn of this century interest in the planets revived with the popularization by Percival Lowell of his ideas about extant civilizations on Mars, ideas supported by his interpretations of telescopic images of that planet—the now notorious "canals." The fallacy of the Lowellian view was soon demonstrated by other more rigorous astronomers and, perhaps as a result, planetary astronomy again fell from favor. A serious revival of interest in the telescopic study of the solar system had to wait until Gerard Kuiper arrived on the scene in the 1940's. Kuiper, with enormous energy and professionalism, began again to seriously investigate the planets, using the latest spectroscopic techniques. These techniques enable us to analyze the composition, temperatures, and pressures of planetary atmospheres, thereby immediately placing important constraints on the original composition and the evolutionary history of these objects. Coupled with the continuously improving astrophysical understanding of stellar formation processes (of which planetary system formation is thought to be a by-product), these new insights into the widely differing characteristics of solar system bodies with atmospheres, from Venus out to Titan and beyond, brought a whole new field of science into being. With the developments of radio and radar astronomy and with the availability of new detectors able to make sensitive measurements in the infrared—where the cool planets are best observed—planetary astronomy has prospered over the last few decades. In parallel with the astronomical advances, 20 years of study of meteorites, pieces of asteroids delivered to earth, have given us great scientific insights about the composition of the early solar nebula and the evolution of the accumulating planetesimals. All these pieces of information by themselves, in the absence of spaceflight, would have allowed us to come a long way in understanding our local neighborhood in the universe.

It has been the dawn of the space age, however, that has led to the most spectacular opening up of the solar system to man's exploration. Not much more than a single generation has separated Kuiper's identification, in 1944, of methane in the atmosphere of Titan from the historic flyby of that Saturian moon by NASA's Voyager 1 spacecraft. Recently another NASA spacecraft—Pioneer 10—left behind all the planets as it departed the solar system toward interstellar space. By 1989, when Voyager 2 passes Neptune, we will have visited all of the planets except Pluto. In the 1990's, a major emphasis of NASA's solar system exploration program will be the investigation of the last virgin territory—the comets and minor planets (asteroids).

In this short review, the accomplishments over the last two decades of NASA's planetary exploration program will be examined together with a brief look into the near future. One of the first things to point out is that solar system exploration only recently became a systematic, carefully planned enterprise. In the beginning, technological limitations and international politics played a far greater role than scientific strategy in shaping NASA's planetary activities. Naturally enough, the first spacecraft launches to the moon and to the planets were made using modified military launch vehicles, with the Atlas rocket as the basic launch system. Lacking a worldwide tracking system, it was necessary for NASA to establish a Deep Space Network (DSN) to allow us to command and receive data back from these spacecraft. Initially, the available launch vehicles and DSN receivers, together with the uncertain reliability of spacecraft systems, constrained us to investigations of the moon, Venus, and Mars. Such a constraint was, in fact, not too hard to live with since these three planetary bodies will always be among the most important to us. (In our odyssey to the planets, we often find ourselves applying our new knowledge to our understanding of the earth; the moon, Venus, and Mars are the most instructive planets from such a point of view.)

Before Mariner 2 blasted off the launch pad at Cape Canaveral toward Venus on August 27, 1962, the U.S. had already committed itself to the Apollo project, whose goal was to land a human on the moon, and return him safely to earth, before the end of the decade. As a result, several unmanned projects were already underway to survey the moon in preparation for the selection of the Apollo landing sites. Caltech's Jet Propulson Laboratory (JPL) had initiated the Ranger project to crash instrumented probes into the lunar surface, the emphasis being on the return of high-resolution images to assist geologists in understanding the nature of that surface. As a result, there were many subsystems in common between Ranger and the Mariner family of deep space vehicles. The Mariners, however, had a much broader scientific mission than the Rangers—they were instrumented to probe the atmospheres of Venus and Mars (the first nine Mariners were launched to these bodies, not always successfully) at infrared, ultraviolet, and radio wavelengths, to study surface temperature characteristics, to measure gravity and magnetic fields, as well as to transmit back pictures. The first successful Mariner, Mariner 2, did not in fact carry a camera because Venus was known from telescopic observations to be completely cloud-enshrouded (with no visible contrast in the clouds except at UV wavelengths where we did not as yet have an imaging capability). So when Mariner 2 flew by Venus at a distance of 21 680 mi on the historic date of December 14, 1962, the principal excitement lay in the measurement of very high surface temperatures (800° F) and the absence of a discernable magnetic field. At the time it was a sufficient achievement that Mariner 2 had made it all the way to another planet; Mariner 1 had been lost during the launch and Mariner 2 had more than one emergency in flight. But a successful start had been made—the scientific objectives of the mission had been achieved—and the human race had passed another milestone in the history of its civilization.

In contrast to the Mariner spacecraft sent to Venus, all of the Mariners launched to Mars were equipped with imaging systems, with the parodoxical result that the first missions were disappointing. Although by the early 1960's even the lay public had long since ceased to believe in advanced Martian civilizations desperately sustaining a dying planet (supposedly by seasonally transporting water in canals from the polar regions to lower latitudes), there was still good reason to expect that Mars would be an exciting planet for geological study because it is a moderately large terrestrial planet and has an atmosphere. Size is a key characteristic in determining the evolution of the rocky inner planets. Geologic and atmospheric evolution are linked to the thermal history of the planetary interior: the large ratio of surface area to internal volume of small planets allows them to lose their inner heat rapidly and reach early senescence, like our own moon. The heat engines of larger planets like the earth remain active much longer so that high mantle temperatures are sustained—which can lead not only to widespread volcanism but even to crustal plate tectonics. Mars—a body intermediate between the moon and the earth in size—was expected to show evidence of a quite complex evolution. The high resolution surface coverage obtained by the imaging systems of Mariners 4, 6, and 7 showed us a landscape that most resembled the lunar highlands—a brillant success for the engineering teams, but an undoubted disappointment to the scientists.

Fortunately, the momentum of NASA's planetary exploration program was such that another Mars mission was already underway—a project that would add sufficient retropropulsion capability to the Mariner spacecraft to enable orbit insertion about Mars. High resolution studies of the planet could take place over an extended period rather than for the few hours that characterized flyby missions. The Mariner 1971 project in question had a lot of inheritance from Mariner 1969 (Mariners 6 and 7) in terms of spacecraft systems and instrumentation. The big new development was in the area of mission operations—how to operate two complex spacecraft carrying many instruments on a programmable scan platform while in orbit for months. The planning for the mission revealed just how complicated this would be, because each of the science teams was competing vigorously for the limited resources of observing time and data bits. Many new organizational problems had to be addressed and a large system of computer hardware and software had to be brought into being to ensure that the ground system could keep pace with the two spacecraft once they were active in orbit. Skeptics said that the project would never be ready with a system able to do the job. We will never know because the first of the two spacecraft was launched to a watery grave in the Atlantic Ocean. There are still some who believe that this was a deliberate act by desperate project engineers unable to contemplate further the enormous confusion that would ensue when Mariners 8 and 9 arrived in Mars orbit together. It must be said, however, that these same cynics admire the confidence of a project manager who would choose to "deep six" the first-launched spacecraft rather than the second. In any event, this first planetary orbiter mission (Mariner 9 was

Fig. 3-1: Mosaic of 102 individual photos of Mars taken in February 1980, illustrating the grandeur of Valles Marineris, a chasm stretching across the center of the mosaic and having a length equivalent to that of the United States, ocean to ocean. Three great volcanoes of the Tharsis Ridge are also prominent, to the left of the chasm.

inserted into Mars orbit on November 13, 1971) turned out to be a spectacular success. After waiting in orbit for two months in order for a global scale dust storm to settle out, Mariner 9 systematically mapped Mars from pole to pole, over a period of a year during which a whole suite of other surface and atmospheric measurements were made. We learned that the earlier Mariners had sampled only part of Mars and that the most spectacular—and scientifically rewarding—regions had still been awaiting discovery. The pre-exploration view that Mars would be an evolved body, unconfirmed by early missions, was proven correct; Mars has indeed experienced a complex history manifested in terms of global scale tectonism (the Valles Marineris canyon system would lose the Grand Canyon in one of its minor branches) and enormous volcanic constructs that dwarf their terrestrial equivalents. Most extraordinary of all was the discovery of massive arroyo-like features—channels arising full-bore out of characteristically broken up depressions in the broad Martian plains. After more than a decade of study effort, we are still unable to offer explanations for these features that do not invoke the presence of liquid water. The features appear to have been created in the quite distant past (perhaps 2–3 billion years ago). Orbital observations from the Viking mission of 1976–1980 have augmented this picture of a planet that must have been dramatically different early in its

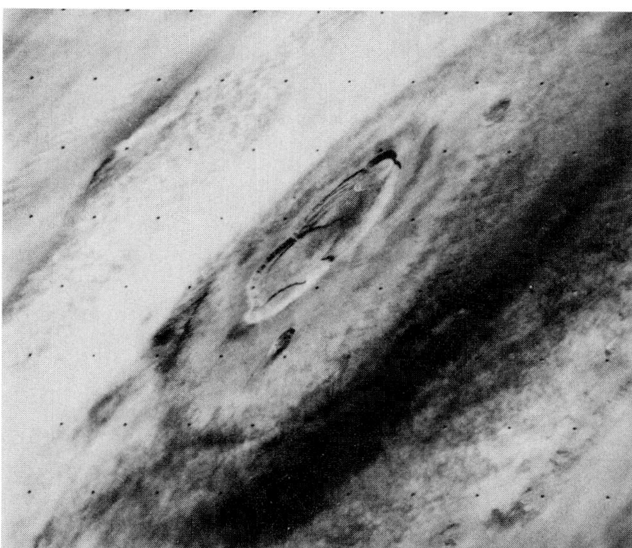

Fig. 3-2: The summit caldera of the Martian volcano Olympus Mons, seen surrounded by water clouds in this photograph taken by the Viking Orbiter.

history. The Viking data show complex dendritic networks of smaller channels spread out over much of the planet.

These possible drainage networks inevitably raise the question of whether Mars might once have had a clement climate. The Mariner 9 mission also produced evidence of relatively recent climate change, though not necessarily change in which water was an issue. The

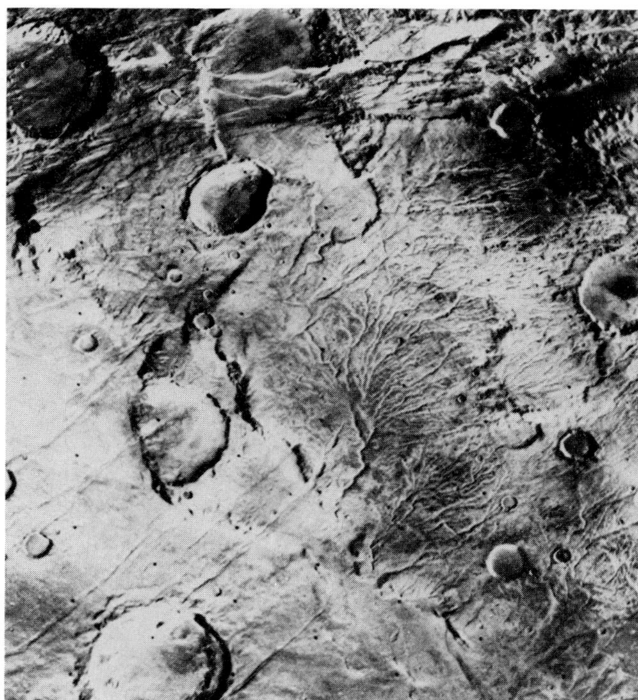

Fig. 3-3: Dense, dendritic drainage networks in the ancient Southern Highlands of Mars. The dendritic nature of these channels argues against their origin by wind or lava erosion and for an origin by water delivered to the surface by rainfall. The long, straight markings in the lower left are north trending fault valleys apparently created by flexing of the crust of this region of Mars.

observations in question are of layered sediments near both poles of Mars—layers some tens of meters in thickness in deposits that are a kilometer or more in total depth. The layers are thought to be sediments of dust and ice (water and/or carbon dioxide) formed by the deposition of dust raised in annual global dust storms. The argument is made that the global dust storms are modulated by variations in the orbital characteristics of Mars—cycles in the eccentricity and obliquity of the Martian orbital elements. Important objectives for future Mars exploration include the unraveling of the climatic history of that planet, a history that may also throw light on the earth's past.

The Viking landings on Mars took place in the summer of 1976 with the principal goal of searching for evidence of extant life. One of the fundamental objectives of planetary science is to understand the chemical history of the solar system in sufficient depth that biologists can answer the question of how life arose on our own planet. This understanding is expected to be derived from the synthesis of a comprehensive body of chemical and physical data being assembled for all the planets and for the small bodies in the solar system. The Viking mission leap-frogged this systematic approach to the problem and went straight to the task of looking for living organisms. Although we have little understanding of what conditions are required to allow life to evolve, there is little doubt that, outside of the earth, Mars is the most promising habitat for life in the solar system. Carbon, nitrogen, and liquid water are essential ingredients to life on earth. Mars, as a neighboring terrestrial planet, offers the most similar environment to our own; we know that is has carbon, nitrogen, and water ice available. No mission short of soft-landing sophisticated laboratories on the surface of Mars could be identified to tackle the life issue directly and, given the compelling interest in the issue, NASA undertook the development of this extraordinarily challenging mission.

Viking made all of the preceding unmanned NASA missions look simple by comparison. In addition to the development of miniaturized biology laboratories (with all the problems attendant to carrying out wet chemistry experiments remotely), the Viking Project had to develop the planetary entry vehicles, and the landers themselves with their sampling systems. Two orbiters able to seek out and "certify" two suitable landing sites were needed; they had the additional task of relaying the lander data to earth. The mission operations system involved hundreds of scientists and engineers all linked together by a complex and unforgiving sequence of mission events—the "time line." By comparison, the historic Mariner 1971 effort appeared easy. This national effort called for the dedicated efforts of two NASA Centers—the Langley Research Center which managed the project and the Jet Propulsion Laboratory which built the orbiters and provided the mission operations facility. As in all planetary missions, numerous industrial contractors contributed to the effort. The Martin Marietta Corporation—the prime contractor for the landers—had the center-stage role.

Fig. 3-4: The rock strewn nature of Mars' windblown surface is illustrated by the two photographs taken in September of 1976 by the camera of Viking 2. The scene is to the northeast of the spacecraft, looking to the horizon some 3 km distant. The rock in the lower right corner is about 25 cm across. The largest rock near the center of the photo is about 60 cm long and 30 cm high. The slope of the horizon is due to the 8° tilt of the spacecraft.

The Viking landings, the lander science, and the orbital operations were carried out virtually flawlessly. The results from the biology experiment have demonstrated that the soil samples at the two widely separated landing sites do not contain living organisms nor, on the basis of a mass spectrometer/gas chromatograph analysis, do they contain any organic material to the parts per billion level. It is evident that the solar ultraviolet flux at the Martian surface is inimical to the stability of any organic chemistry. (Meteorites are a constant, if low level, source of abiologic organics and some evidence of meteoritical carbon was expected.)

The further study of the question of life processes on Mars is likely to center on the search for fossil evidence in returned surface samples, rather than on the search for exotic locales on the Martian surface that sustain life at the present time.

With the demise in late 1982 of the last surviving Viking spacecraft—the Viking 1 Lander, which had by then been renamed the Thomas Mutch Memorial Station—our current Mars research activity has been based on the continuing analysis of the substantial body of data acquired by the two orbiters and landers. Mars has had a complex geologic history which is gradually being better understood by detailed analysis of the stratigraphy and processes revealed in the high-quality orbiter imaging coverage. Looking to the future, it is recognized that we must add an additional dimension to our knowledge base in the form of global chemistry maps (both elemental and mineral chemistry), we must make direct measurements of the internal structure of Mars by means of implanted seismometers, and we must return to earth documented surface samples from representative geologic units to make the same kind of physical, chemical, and chronological analyses performed on the Apollo samples. This agenda for the 1990's is planned to begin with a simple remote sensing orbiter to be launched in 1990. Carrying a payload of geoscience instruments, the orbiter will acquire the crustal geochemical information and will also directly address questions related to the current and past climate on Mars: the sources and sinks of volatiles; the processes of seasonal and long-term transfer of volatiles between different sinks; the general circulation of the atmosphere; and the photochemical processes that maintain the composition of the atmosphere in the apparent disequilibrium we observe today.

The kind of global geochemistry and geophysical information that we need for Mars is still, remarkably, lacking for the moon also. Through the scientific fallout of the Apollo project (which returned several hundred kilograms of lunar samples in an unprecedented bonanza for planetary science) we now know more about the moon's history than any other planetary body beside the earth (and we know the moon's early history far better than our own). However, because of the rapidity with which the Apollo project moved forward and the lack of a dominant scientific motivation for that project, we now find ourselves lacking basic global lunar data to provide the context in which to understand the Apollo results. The Apollo data do tell us clearly the sequence of major events that shaped the lunar surface—the early chemical separation of the crust, followed some half billion years later by the giant impacts that produced the mare basins, their subsequent flooding by lava, and the tailing off of major geologic activity as volcanism ceased about three billion years ago. In contrast to the earth's present state, the Apollo data tell us that the moon now has a deep lithosphere—1000 km—and little internal seismic activity. The moon is a small planet which has long since been in the final stage of its evolution.

Our continued high level of interest in the moon comes from a desire to better understand the earliest phase in its history, and its origin—no less elusive to us now than before Apollo. The chemical signature of the lunar highlands—the original lunar crust—was first interpreted to imply an extraordinary episode of global melting, presumably as a result of meteoritic infall. This exciting "magma ocean" concept has not been entirely abandoned in the light of more recent analyses which reveal a wider variety of igneous rocks than can be produced from a simple global ocean of magma. Current modeling of the early moon seeks to include greater complexity, but is limited by lack of global data. In particular, we need better global maps of surface composition, gravity field, and magnetic patterns to provide the clues and insights necessary to elaborate upon our present simplistic models of lunar history. There is at present no compelling need to bring back more lunar samples even though the Apollo and Soviet Luna landings sampled only a few sites that were heavily biased toward safe mare regions. The impact processes that spread lunar material all the way around the moon ensured that we already have samples of all the major geologic units. Additional samples would add greatly to our knowledge base but, given our present priorities, we are prepared to wait for the next wave of manned lunar exploration rather than plan unmanned sample acquisitions.

In spite of the several early Mariner missions to Venus and the additional Mariner Venus Mercury flyby mission of 1973–74 (which flew by Venus and then Mercury, giving us our first close-up views of the lunar-like surface of this innermost planet, and which also measured its dipolar magnetic field), exploration of Venus was dominated by the Soviet Union until quite recently. Beginning with the Venera 4 mission in 1967, which returned data from an atmospheric entry probe, the Soviet Union has returned to Venus with atmospheric entry and lander vehicles at almost every launch opportunity. Atmospheric chemistry and physical state have been probed by the Soviets; surface composition has been measured in several locations and, from four spacecraft, surface panoramas have been returned. In 1978 the U.S. returned to Venus with the Pioneer Venus mission consisting of four entry probes on a carrier bus and a long-lived orbiter which is still active in returning atmospheric and ionospheric data. The entry probes provided new data about the relative abundance of the inert rare gases in the atmosphere—information that has been compared with the corresponding abundances on earth and Mars (measured by Viking) and in meteorites. The Venus pattern, which represents a complex remnant signature of the chemical composition of Venus when it was formed, is quite unlike that of the earth, a twin of Venus in many respects. This unexpected finding has cast considerable doubt on our ability to interpret such signatures (which we attempt to do through various models of planetary evolution). Since there is also some doubt about the validity of the measurements them-

Fig. 3-5: Scientist-astronaut Harrison H. Schmitt photographed standing near a giant lunar boulder during one of the Apollo 17 extravehicular activities at the Taurus-Littrow landing site. Evidence of Schmitt's sampling of the debris on the rock can be seen on the near surface. The photograph was taken by Astronaut Eugene A. Cernan.

selves, we plan to go back to Venus to repeat them with higher accuracy and reliability.

The various Mariner, Venera, and Pioneer measurements at Venus have painted a consistent picture of a planet that has followed a dramatically different evolutionary path from the earth. Today the Venus atmosphere is made up largely of carbon dioxide in an amount calculated to be roughly equivalent to that outgassed by the earth. However, while most of the terrestrial CO_2 is now locked up in rocks, the Venusian CO_2 remains in the atmosphere, creating a massive greenhouse effect that has raised the surface temperature to an extraordinary 700°K. Also contributing to the "greenhouse" is the planet-wide blanket of cloud and haze, thought to be made up of sulphur particles and other aerosols. Wind profiles measured by the probes show a tremendous wind shear in the clouds and an almost stagnant atmosphere near the surface. UV observations of the Venus clouds show that the winds in the upper atmosphere are blowing at sustained velocities of hundreds of miles per hour, leading to an apparent four-day rotation of the upper atmosphere, in contrast to the 240-day retrograde rotation of the planet itself. Given the deceptively simple physics of the Venus atmosphere, it is sobering to find that we still have only the most modest understanding of how the Venus general circulation "works."

One of the most exciting measurements of the Pioneer Venus entry probes determined the ratio of the hydrogen isotopes in the atmosphere. The observed unexpectedly high abundance of heavy hydrogen (deuterium) has been interpreted to imply that Venus must have lost a significant amount of hydrogen to space over its history (the heavy isotope would be preferentially retained). In turn, this indicates that Venus must have outgassed a significant quantity of water (the only reasonable source of hydrogen); presumably this water would have formed a shallow ocean early in the history of the planet. It is supposed that Venus experienced a "runaway greenhouse" when sufficient CO_2 and water vapor had accumulated in the atmosphere, leading to the catastrophic heating up of the atmosphere and boiling off of the ocean. Evidence of such an event may be contained in fossil ocean shores that will be looked for in the radar maps acquired on orbital missions in the future.

The still operating Pioneer Venus orbiter is, in fact, equipped with a simple radar altimeter which has provided us with our first global view of the Venus

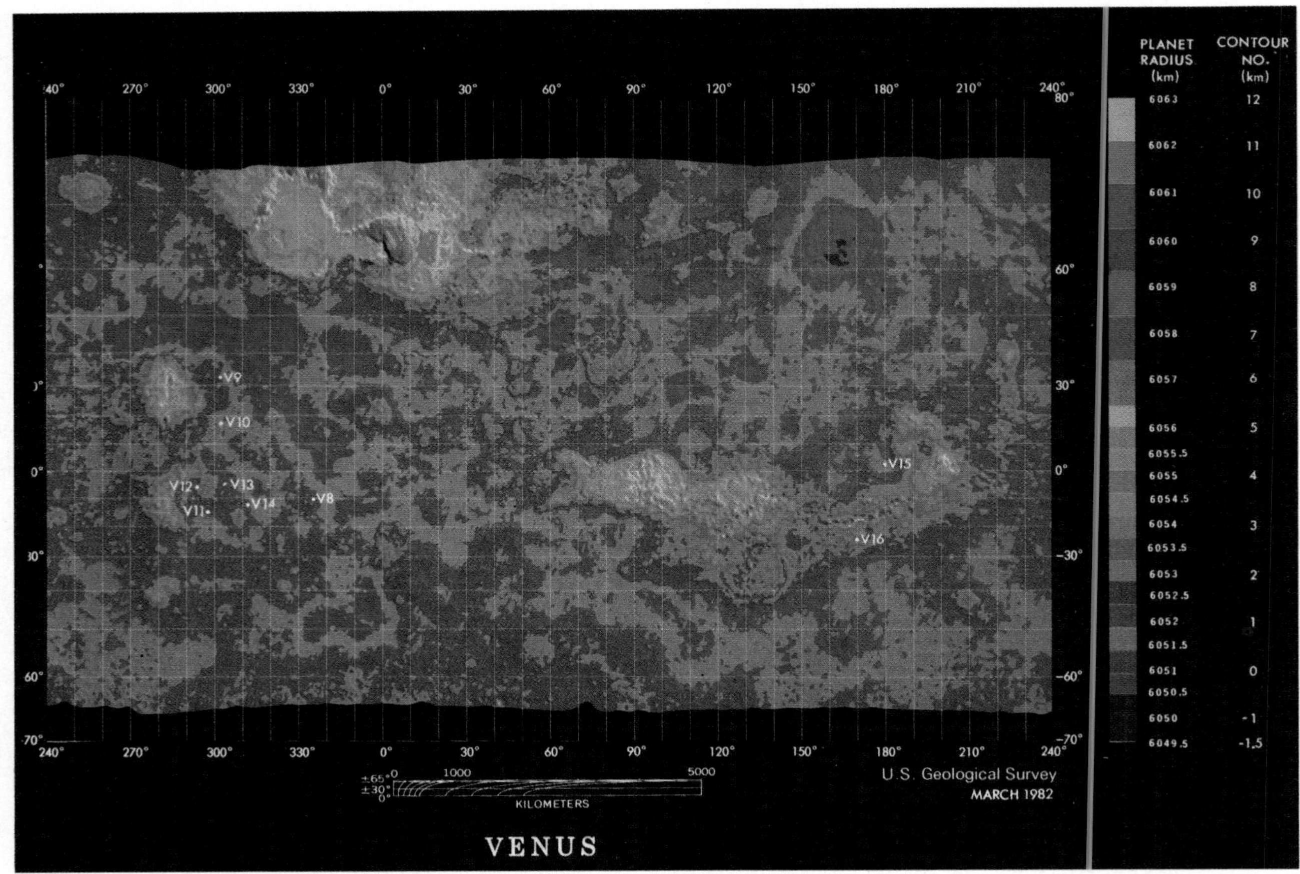

Fig. 3-6: The topography of Venus as determined from altimetry records obtained by the Pioneer Venus spacecraft. The continent size masses of Ishtar Terra, in the north, and Aphrodite Terra, near the equator, are the most prominent features of the planet. The highest point, in Maxwell Montes of Ishtar Terra, rises some 11 km above the mean surface. The mountains Beta Regio and Phoebe Regio at 280° longitude are probably volcanic in origin. It is around these mountains that the Venera probes landed.

surface. Venus, we find, is remarkably spherical in shape with the larger part of the surface in the form of gently rolling plains at elevations not far distant from the mean. Unlike the earth, where the continents and the ocean basins lead to a bimodal distribution of surface elevations, Venus has a unimodal distribution, suggesting a fundamental difference in the way in which the crusts of the two planets have evolved. There are half a dozen upland regions that are continental in scale, the most prominent being the equatorial Aphrodite region and the northerly Ishtar. Two other highland regions, Alpha and Beta, are also of note; most of the Venera landings have occurred in this general area. It will require high-resolution radar coverage to interpret the nature of these uplands, but we can already speculate that Alpha and Beta are of volcanic origin on the basis of the Venera composition measurements and the panoramas which appear to show lava flows. Adding to the weight of the evidence are the Arecibo radar telescope data which have been reconstructed as a radar image of the area in question and which show surface features whose morphology clearly resembles a volcanic construct.

The several mountain ranges that, together with a broad elevated plateau, make up Ishtar, have aroused considerable excitement among geologists because the high-resolution radar images show the ranges to be characterized by a series of linear bands that are generally parallel to their long axes. The mountain ranges are themselves up to a thousand kilometers in length and hundreds in width. The highest mountains—the Maxwell Montes—reach 11 km above the mean. The size scales of the linear bands are hundreds of kilometers in length and 10–20 km in width. A variety of explanations for the observations have been proposed: wind activity, mass wasting, volcanism, and tectonics. In the view of some, the latter hypothesis—in which the linear bands would be analogous with the Appalachian fold mountains or with the Basin and Range province in the western U.S.—is most likely. Such an origin would imply that Venus is a geologically evolved planet with much to teach us about the evolutionary possibilities of terrestrial planets. The currently available radar resolution and coverage are inadequate to do more than point to what are potentially the most exciting regional targets for future high-resolution radar mapping—mapping that can only be obtained from Venus orbit. In late 1983, the Soviet Union deployed a radar imaging system at Venus, with results yet to be fully disclosed. Meanwhile, NASA began full scale development of a synthetic aperture radar mission in 1983 for launch in 1988, a mission that will open a new scientific window on the inner solar system with promise of untold scientific discoveries.

Missions to the outer planets—where most of the planetary mass of the solar systems resides—call for a new level of technological capability: in launch vehicles, in tracking, in the supply of power and of thermal control, and in reliability. For the first outer solar system spacecraft, Pioneers 10 and 11, there was also a sense that a high degree of luck would be needed to cross the asteroid belt without damage caused by orbiting dust particle impacts. We have found from these pathfinding missions that, in fact, the region between Mars and Jupiter—the region of the mainbelt asteroids—is quite benign and not an important factor in mission planning. The Jovian radiation belts, however, have proven to be potentially devastating, so that the Voyager spacecraft that were to follow the Pioneers had to use new electronics designs and shielding to withstand the environment. With the 1979 Voyager measurements of the oxygen/sulphur cloud in the orbit of Jupiter's moon Io, it was realized that the Galileo spacecraft—under development for launch in 1986—was also in serious jeopardy because the evolving microminiaturization trend in electronics has made modern spacecraft electronics highly susceptible to the kind of high energy particle environment near Jupiter. As a result, radiation hardening has become a challenging technology frontier for planetary spacecraft headed to Jupiter.

Scientifically, the outer planets visited by Voyager—Jupiter in 1979 and Saturn in 1980 and 1981—have been breathtaking. These planetary giants, with their retinue of moons and rings and with their complex, dynamic magnetospheres, each present an array of atmospheric, geologic, and plasma physics questions that demand a fresh new perspective. Jupiter and Saturn, though both formed predominantly of hydrogen and helium, have evolved in ways that are different in important respects. The atmospheric chemistry and the general circulations (which are evidently dominated by the flow of heat outwards from the interior rather than by differential latitudinal solar heating) are substantially different, with Jupiter exhibiting the greater complexity. We remain at a relatively primitive stage in our understanding of these subjects. A major advance will be made by the Galileo probe in 1988 and, at Saturn, by a similiar probe sometime in the 1990's.

The systems of moons—some of the size scale of Mercury and our moon—are fascinating both because of their individuality and because they reveal systematic variability in density (and hence composition); this argues for their origin as a by-product of the planetary primary's formation, in analogy with the formation of the solar system itself. Most of these moons are ancient bodies composed principally of ices and marked by cratered surfaces that were formed billions of years ago. Among the Jovian moons, however, both Io and Europa have young surfaces characterized by intense volcanism in the former case and by ice tectonism in the latter. Nowhere do we see impact craters, even though occasional large impacts must still occur. Io, we observe, is in the process of resurfacing itself continuously through the volcanic activity that evidently is the result of tidal flexing of the entire body as it orbits Jupiter (Io is tugged gravitationally by Europa, which occupies an orbit whose period is resonantly synchronized with Io's). It appears that, long ago, Io lost its original supply of icy

Fig. 3-7: Jupiter and four of its planet-size moons, the Galilean satellites, as photographed in March 1979, and assembled into the composite picture. The satellites, in order of decreasing distance from Jupiter, are Io (upper left, orange in color), Europa (center), Ganymede, and Callisto (lower right). Nine other smaller satellites are also known.

volatiles, and present-day volcanism is driven by sulphus—a cosmically abundant element that, on earth, largely resides in the core. Europa, by contrast, still has an icy crust, perhaps a few tens of kilometers deep. Its surface too is evidently being reworked constantly by tectonic processes whose origin remains obscure to us. This moon, too, is subject to tidal gravitational effects which, plausibly, may have led to its unique aspect.

In the Saturnian system, the giant moon Titan (Mercurian in size), long an object of close study because of its atmosphere, has grown even more in scientific importance following the Voyager flybys, even though these encounters failed to observe its haze shrouded surface. The atmosphere of Titan is the key to the excitement which now surrounds the study of this moon. Composed primarily of nitrogen (perhaps derived from an original ammonia atmosphere) and with some argon, Titan's atmosphere also contains a significant proportion of methane and other more complex organic molecules—these latter species are evidently created by photochemical reactions in the upper atmosphere—stimulated by solar UV energy, and are the principal cause of the haze that envelops the moon. Clouds of methane ice at lower altitudes probably also contribute to the obscuration of the surface. Calculations suggest that the organic aerosols will precipitate out and that we may expect a surface "rind," perhaps hundreds of meters thick, of organic tars. Add to this the possibility of an ocean of liquid ethane, and Titan assumes a quite bizarre character. However, such extraordinary phenomena are not the principal basis of our eagerness to return to study Titan by means of an atmospheric probe and a remote sensing radar mapper. Rather, it is the prospect of being able to investigate the complexities of the photochemical

Fig. 3-8: A montage of the Saturnian system assembled from images taken by the Voyager 1 spacecraft during its encounter in November 1980. The satellite Dione is in the foreground, obscuring part of Saturn. Enceledus and Rhea are to the left of Saturn, and Tethys and Mimas are to the right. Titan appears most distant at the top right.

processes that are occurring in this natural laboratory, processes that may throw light on the early history of the earth's primitive atmosphere. It seems likely that we will learn more about the chemical evolution of the solar system and its relationship to the formation of life by investigating Titan in depth than in any other way.

Planning is already underway to probe the atmosphere of Titan using a vehicle similar to the Galileo probe in conjunction with a Saturn flyby spacecraft or, preferably, a Saturn orbiter; the mission is to be launched in the early 1990's.

This new mission to Saturn will also take the next step in studying the complex ring system and magnetosphere of that planet. Voyager provided us with our first comprehensive close-up view of the rings, thereby showing them to be even more glorious than we had expected and also revealing several new rings and many characteristics over which we still puzzle. One of the newly discovered rings—the F ring—lies outside the bright outer A ring that is clearly visible from earth. The F ring is, in fact, a family of several individual strands that are notable because they are not exactly circular and

Fig. 3-9: Io's southern hemisphere, photographed in 1979 by a wide-angle camera on Voyager 1. Calderas (volcanic collapse craters) are numerous, some surrounded by diffuse sulfurous markings and others by eruptive flows. The mountain in the lower left is probably an outcrop of the silicate crust of Io.

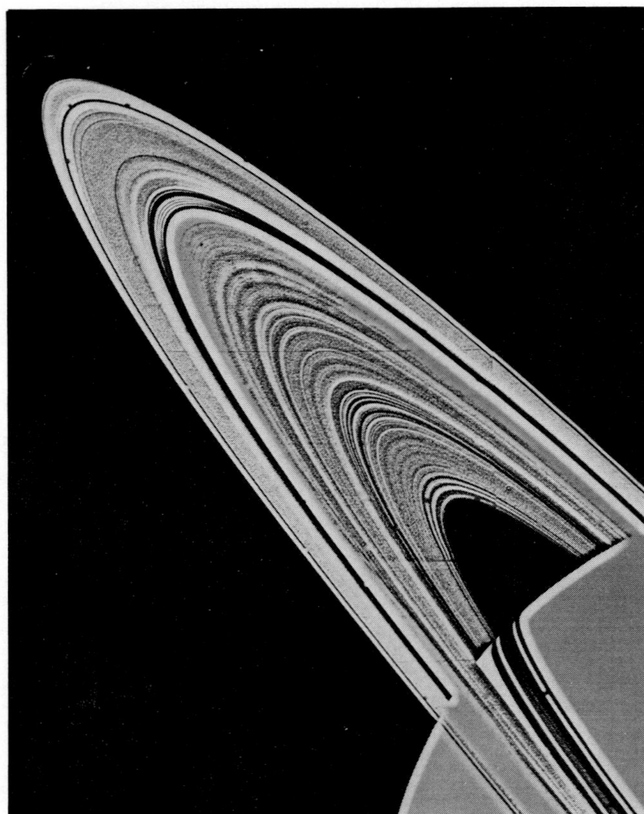

Fig. 3-10: Saturn's rings as viewed from 8 million km by Voyager 1. The image shown has become computer-processed so that the large-scale brightness differences between the A, B, and C rings have been suppressed and finer detail emphasized. Note the four rings inside the Cassini Division and the faintly visible, thin F ring, lower left, beyond the edge of the A ring.

because they appear to be "braided," one about another. While this bizarre phenomenon defies explanation at present, a Voyager discovery of two relatively small moons, one inside and one outside the F ring, appears to explain the isolation of this ring through resonant gravitational "shepherding" of the individual ring particles. Some of the complex structure within the major ring divisions may be explainable by the presence within the rings of "moonlets," but these have yet to be discovered. Other mechanisms governing the form of the rings are certainly at work, including the resonant gravitational effects of Saturn's moons (influencing ring particles whose period of rotation is a simple fraction of a moon's period); the 50 or so strongest resonances all have identifiable ring features—clearings, spiral "density" waves, and embedded opaque "ringlets." Other features in the rings are being modeled on the basis of treating the rings as a viscous medium.

Voyager's ring discoveries have not been limited to Saturn. We now know that Jupiter also has a ring, one quite different in kind from that of Saturn (which is apparently made up of icy particles whose size distribution has been inferred from the Voyager measurements and includes objects as large as a house). The Jovian ring may very well be made of silicate material rather than ice, and apparently is composed of very tiny particles, based on their light scattering properties. Moreover, the Jovian ring shows little structure, apparently being limited in its outer extent by a small shepherding moon and diffusing inwards towards the planet's cloud tops.

At present, Voyager 2 is en route for a January 24, 1986 encounter with Uranus, a planet little known because of its great distance from the earth. We do know enough about Uranus from astronomic studies, however, to be able to predict another historic encounter of great scientific and popular excitement: Uranus is unique in its polar orientation direction (pointing almost at the sun at the present epoch) and in the organization of its nine faint, eccentric rings. Further, Uranus has another family of moons to observe and, most probably, a distinctly different magnetospheric structure, given the unique polar orientation. By August of 1989, if all goes well, we should have in hand a fascinating database with which to compare all four of the giant gaseous outer planets, including Neptune. The 1989 Neptune encounter will be the last planetary event of the Voyager mission, which will then join Pioneers 10 and 11 in their probing of the heliospheric boundary and their search for the gravitational signature of a possible tenth planet lying beyond Pluto. Pluto itself will not be the target of any mission in the foreseeable future because of the great trip time and launch performance requirement to carry out a capable enough mission to justify a 15-year journey.

In the next decade, an important focus of planetary exploration will be the comets and asteroids, objects that have thus far been neglected because of the great propulsive performance required of any spacecraft that would study these tiny objects closely. The U.S. strategy for exploring the small bodies of the solar system has placed stress upon being able to match orbits with the target in order to spend an extended period in rendezvous. As a result, probably mistakenly, the U.S. has let pass opportunities for reconnaissance-level exploration of these objects in a simple flyby mode. By leaving the reconnaissance phase to telescopic observations, the U.S. finds itself sitting on the sidelines while other spacefaring nations—the Soviet Union, Europe, and

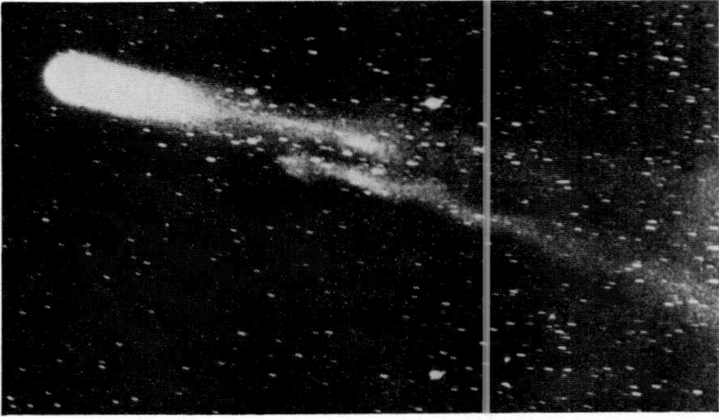

Fig. 3-11: Halley's comet against the star background. This photograph was taken from Hawaii on June 6, 1910. It shows the plasma tail of the comet, extending in the anti-sunward direction, and a disconnection of that tail which had occurred on the previous day. The disconnected portion was observed to recede from the coma, the bright head of the comet, at an average rate of 57 km/s.

Japan—send spacecraft to intercept Halley's comet in 1986. The U.S. will, however, be participating in the international cometfest in 1985–86 by directing a solar wind monitoring spacecraft—the third International Sun Earth Explorer (ISEE-3), now renamed International Cometary Explorer (ICE)—to intercept the tail of Comet Giacobini-Zinner in September 1985. ICE will return unique data from Comet Giacobini-Zinner because the Halley intercepts are all targetted toward the coma or sunward side of that comet. ICE will also be remembered for the most incredibly complex series of lunar swingby maneuvers required to achieve the escape trajectory. Telescopic observations from the shuttle in earth orbit and from the ground as part of the International Halley Watch will comprise the remainder of the U.S. activity during the first Halley appearance of the space age.

Scientific interest in comets is extremely high, not because of the spectacular appearance of these objects when they are near the sun, but, rather, because comets are probably the most primitive objects to which we can ever have access. Their primitiveness derives from their small size and, hence, their ability to radiate away the internal heat (from radioactive decay) that causes planetary scale objects to evolve and lose most of the evidence of their origin. It is the origin of the solar system that forms the central preoccupation of planetary science. The tiny comets, kept in deep freeze on the outermost edge of the solar system, promise to provide us with the most direct insight into the nature of the solar nebula from which the sun and all the planets formed. Eventually we plan to return samples from comets and asteroids to allow the most scientifically sophisticated laboratory techniques to be brought to bear on the analysis of the primitive material. The first dedicated cometary mission undertaken by the U.S. will be an orbit-matching rendezvous mission to a short period comet (such as Kopff or Honda-Mrkas-Pajdusakova) to be launched in 1990. A shuttle/Centaur launch will provide enough energy to reach the comet on a ballistic trajectory after a few years of flight time. A chemical propulsion unit will then be used at encounter to synchronize the orbits of spacecraft and comet. At the time of rendezvous the comet will be distant from the sun and still inactive; initial studies will focus on the elemental and mineral chemistry of the comet and measurements of the comet's morphology and basic physical characteristics. Later, ices and dust will be liberated from the comet as it heats up, so that direct measurements of these components can be made on board the spacecraft. Maneuvering around the active comet, the spacecraft will be able to observe the local variations of sublimation activity that are known to characterize comets and will study the interaction of the solar wind with the comet's tenuous atmosphere. One or two passes through the comet's ion and dust tails will provide us with measurements of these phenomena, also helping to complete our picture of these most complex celestial visitors. Later in the 1990's, following a similar rendezvous, an automated lander will drill a core from the crust of another short period comet and return it to the earth.

This mission is, however, one for which we do not yet have all the required technology; a low thrust solar electric propulsion system will be needed for this mission. As such, unlike the comet rendezvous mission, the comet sample return mission belongs to the augmentation missions recommended for the 1990's rather than the recommended core program (NASA's planetary program for the rest of the century is based on the recommendations of the NASA advisory council's Solar System Exploration Committee. Another scientifically compelling mission that requires enabling technology development, and is therefore an augmentation mission, is one to place mobile laboratories on the surface of Mars in order to make *in situ* analyses and to collect samples for return to earth. Such a mission will probably take full advantage of an earth-orbiting space station to launch the spacecraft(s) to Mars and to receive the samples on return. The mission will require an array of new technology developments and will be of a complexity to challenge NASA at least as much as the Viking mission. The landings on Mars could take place as early as 1996 and the samples could be back on earth before the end of the century, to climax what will by then be almost four decades of exploratory excitement unequaled since the opening up of the New World in the fifteenth and sixteenth centuries.

4

ASTROPHYSICS

David A. Gilman and Charles J. Pellerin, Jr.

INTRODUCTION

When the first X-ray observatory was launched on December 12, 1970, astronomers already knew that the X-ray source Cygnus X-1 was different from the other bright X-ray sources. It changed intensity dramatically in seconds, and its spectrum did not follow a purely thermal curve. Because of its peculiar nature, Cygnus X-1 became an important object for the first orbiting X-ray observatory, designated the first Small Astronomy Satellite by NASA, and it was observed many times. SAS-1 was renamed "Uhuru" by Ricardo Giacconi and the other scientists that developed its instruments, and with the Uhuru observations and suborbital rocket flights, the position was determined to fall within an area that had only two reasonably bright stars. In May 1971, Braes and Miley and Hjellming and Wade discovered a weak radio source that was quickly identified with the blue star, HDE 226868. But was this the optical and radio counterpart of Cygnus X-1?

The answer came in the affirmative from Tannabaum and his associates when they compared the Uhuru and radio observations. Cyg X-1 had been observed frequently at both wavelengths in hopes that the object would do something to reveal its true nature, and it turned out that the failures in early 1971 to find a radio source were very significant. Prior to May, there was no radio source, but Cyg X-1 was very strong in X-rays. In late April, over a period of less than two weeks, the X-ray source shifted intensity to a new, lower level, and at the same time, the radio source appeared. The identification was made, and by 1974, it was found that movements in its visible light spectra showed that HDE 226868 was part of a binary system with an invisible star having a mass at least six times that of our sun.

This fits well with the picture Uhuru had brought us that the bright X-ray sources in our galaxy were mostly binary star systems with an invisible, compact companion orbiting a more normal star. Neutron stars were likely candidates in most cases, but the invisible companion to HDE 226868 was much too massive for that. The resistance of neutrons to compression can only support a neutron star with a mass less than about 2.5 times the mass of the sun. Beyond that, gravitation wins and all traces of the star other than its gravity vanish forever. The companion to HDE 226868 was a black hole.

The 29-year history of space astronomy is full of stories of discovery like the one for Cyg X-1. The discoveries come at a rate than makes clear the richness of the universe and the certainty of reward for every increase in sensitivity we can achieve.

The objective of astronomy is to determine the physical character of the universe and its components. We want to know the temperature, density, magnetic fields, and structure, and we want to understand how energy is generated and transported. We want to know how the universe was formed, what can make quasars so luminous that they are visible on the edge of the universe, how stars are born and planets formed, and what laws of physics apply when matter is compressed to nuclear densities and beyond. So, this chapter is about space astrophysics: an endeavor which is a mix of the ancient study of astronomy, modern physics (e.g., quantum mechanics and general relativity), and the current high technology of space experimentation.

First, our overview provides an historical perspective with a few examples from recent events. The second section of this chapter, "Observing the Universe," begins by discussing the role of gravity in controlling the universe and continues by describing the channels of information (electromagnetic waves, cosmic rays, gravitational waves, and neutrinos). This section closes by summarizing the measurement parameters of observational astrophysics. In the third section, "Examples of Current Problems," several current problems in astrophysics are discussed in some detail. These illustrate the current state of knowledge. The next section describes the tools to address these problems, the "Observatories."

What separates space astronomy from ground-based astronomy is the interference (opacity, distortion, and background emission) of the atmosphere to most forms of electromagnetic radiation. With spaceborne instruments we can access the entire electromagnetic spectrum with high sensitive and high angular resolution.

Space astronomy had its beginnings on the top of a V-2 rocket in 1946 when a team led by R. Tousey obtained the first spectrum of the sun in the ultraviolet beyond the atmospheric cutoff. In 1948, T. R. Burnight used a V-2 to carry photographic films covered by thin metal plates to detect X-rays from the sun. Since then, the parallel advances in launch capability and detection capability make space astronomy one of the most rapidly evolving fields of scientific research today. Yesterday it was small telescopes carried on sounding (suborbital) rockets and small satellites, today it is long-lived observatories in space launched and serviced by the space shuttle, and tomorrow it will be telescopes and

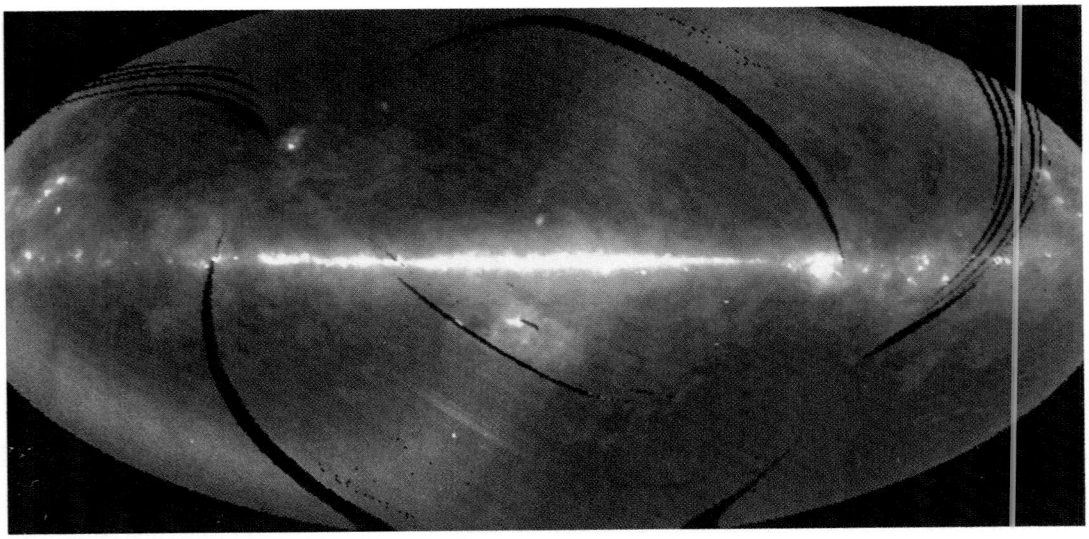

(a)

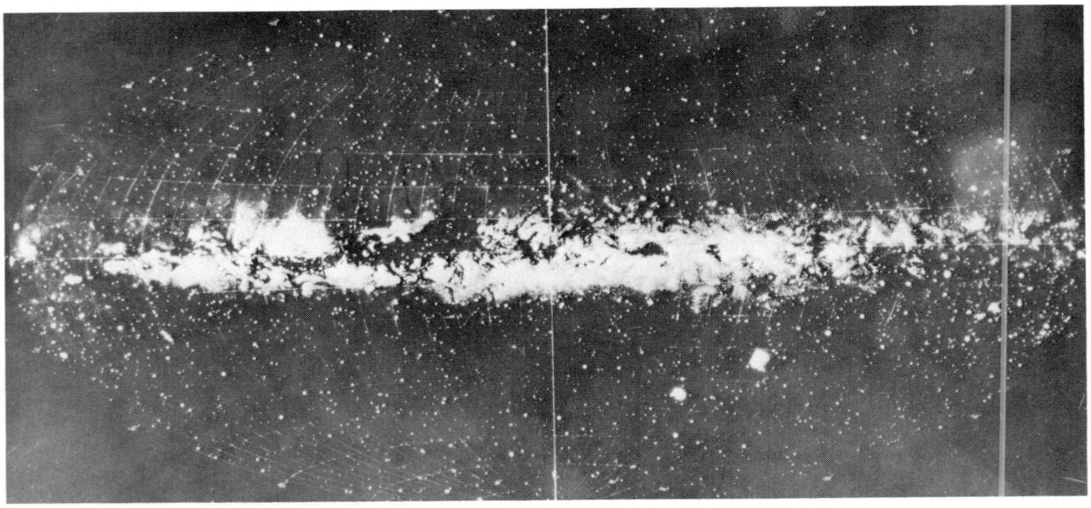

(b)

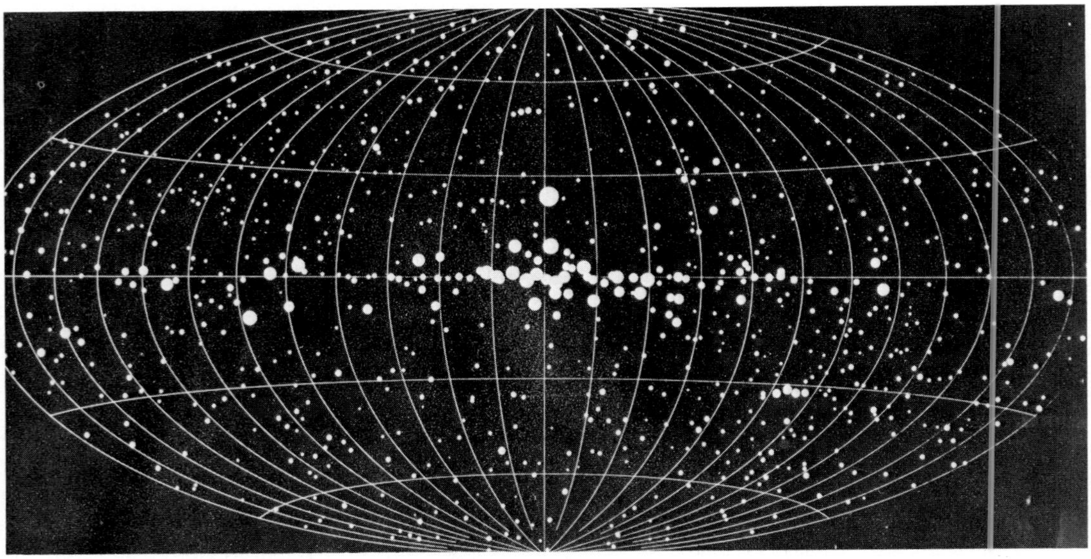

(c)

Fig. 4-1: The sky in (a) infrared, (b) visible, (c) X ray, and (d) gamma ray (next page) wavelengths. The infrared observations are from IRAS, with blue indicating hot objects (young stars, for example) while the red indicates cold objects (interstellar dust). The visible light picture comes from the Hale Observatory and shows primarily stars. The X-ray map comes from the survey conducted by HEAO-1, and the gamma ray map comes from the COS-B mission.

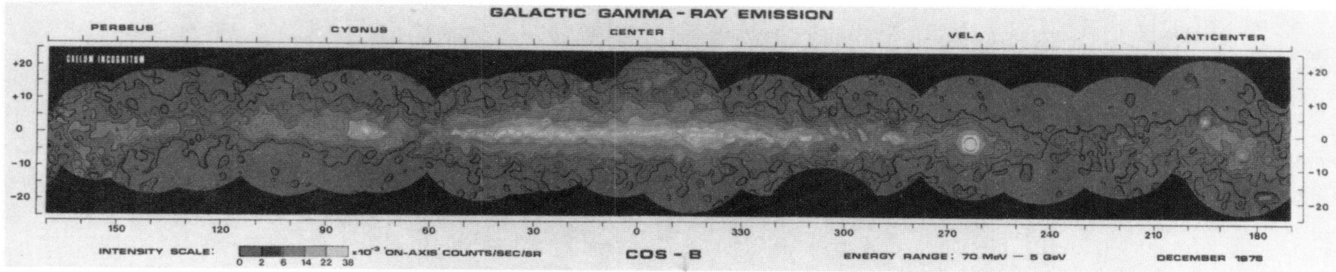

(d)
Fig. 4-1 (*Continued.*)

facilities so large that they have to be assembled in orbit at a manned space station.

The first orbiting space observatory was the Navy's Solrad satellite which carried a monitor of solar X-rays starting in June 1960. This was followed in 1962 by the first of NASA's Orbiting Solar Observatories, a series of missions which investigated the rapidly varying ultraviolet, extreme ultraviolet, X-rays, and gamma rays coming from the sun. The thread of solar physics investigations in space continued in 1973, when high resolution images of solar ultraviolet light and X-rays were obtained from the first American space station, Skylab. Today, investigations of solar activity are being conducted by the Solar Maximum Mission (SMM). With these missions, we discovered that the structure of the corona is dominated by magnetic fields, we determined the sequence of events in solar flares, and we found the regions of the sun that give rise to the solar wind.

These same missions have also left a legacy of unanswered questions. How are the sun's magnetic fields generated, how are solar flares triggered, how is the solar wind accelerated, and how is the corona heated? Much of this missing information is contained in structures too small for us to observe with our present instruments, and we are designing high resolution instruments to investigate the physics of our nearest star. The first will be the Solar Optical Telescope (SOT) which will make its early observations and discoveries from the bay of the space shuttle and later will contain the core of instruments around which the Advanced Solar Observatory (ASO) will be developed in the space station system.

Having discussed some of the space missions in solar physics, let us address astronomy. After ultraviolet telescopes made their first exploratory observations from sounding rockets, it was clear that if astronomers could have regular access to the portions of the spectrum that are absorbed by the earth's atmosphere, it would be possible to investigate the nature of stars significantly more massive than the sun and also the coronae of other stars. This possibility spurred the development of the Orbiting Astronomical Observatories (OAO).

The last of the OAO series was named "Copernicus" by the astronomers who used it and was especially significant because it observed into the far ultraviolet—a region where photons have sufficient energy to excite interstellar hydrogen and be absorbed. This ability to sensitively probe the interstellar medium at short wavelengths revealed the surprising result that the medium is far from uniform. Copernicus, and the International Ultraviolet Explorer (IUE) that followed, found that the sun was in a small cloud of warm gas in an enormous cavity of low-density, very hot gas surrounded by great walls of denser, cooler gas of what we used to think was the interstellar medium. Thus, in less than ten years, we have come to understand that the interstellar medium is shaped by forces we had not predicted—violent, high temperature forces—and that we will not understand the formation of stars until we know the sources, processes, and structure of the hottest components of the gas between the stars.

The fact that there is a cavity with very little absorbing gas out to distances of at least 200 light years in some directions has driven astronomers to seek a survey of sources which are very hot and emit primarily in the extreme ultraviolet: radiation that can actually ionize the interstellar medium. We have just begun development of the Extreme Ultraviolet Explorer to make that survey.

There are enormous advantages in going to space for even those few regions of the spectrum where the atmosphere is relatively transparent, such as the visible. The richness of the field of visible light astronomy is well established from ground-based observatories. Enormous collecting areas have been achieved, but investigations are principally limited by angular resolution that is never much better than 1 arcsec except in unusual circumstances. To push significantly beyond this limit to investigate the planets, stars, structure of nebulae, and galaxies near and far, and to measure the size of the universe itself, we are developing the first mature observatory in space, the Edwin P. Hubble Space Telescope. Like ASO, it combines the best technical developments in scientific instrumentation with the best in spacecraft technology to produce a powerful, flexible capability for investigating the physical nature of the universe. The space shuttle will be used to bring it to orbit, and it will be serviced from the shuttle and later from the space station to make sure that the spacecraft continues to perform and that the detectors in the focal plane take advantage of the state of the art.

In the X-ray region, the Uhuru survey of the sky discovered a violent, energetic universe of X-ray sources. These range from black holes, neutron stars, and remnants of supernova explosions in our galaxy to active radio galaxies, clusters of galaxies, and quasars. Detailed investigations were conducted by the third

Small Astronomy Satellite and the first High Energy Astronomy Observatory (HEAO-1). A major breakthrough came in 1978 with the launch of the second HEAO. In the telescope carried by HEAO-2, cosmic X-rays were reflected by mirrors at very shallow, "grazing" angles to form an image on detectors at the focal plane. This dramatic increase in angular resolution and sensitivity led to the discovery that nearly everything in the universe produces detectable amounts of X-rays. Jupiter was found to emit X-rays from its auroral zones, stars in the Hyades cluster that otherwise appear identical to our sun were found to have coronae hundreds of times more luminous in X-rays than our sun, galaxies were found to have hundreds of times the mass previously derived from starlight alone, and the structure of gas between galaxies in giant clusters was revealed.

Astrophysical research to discern the physical nature of the universe clearly requires a mature observatory for X-ray astronomy, and we are planning for the Advanced X-ray Astrophysics Facility (AXAF) to be built around an imaging X-ray telescope a hundred times more sensitive than HEAO-2. It will be launched by the space shuttle and serviced both from the shuttle and the space station.

Finally, the highest energy photons are characterized as gamma rays. Gamma ray astronomy got its start at the top of the atmosphere using balloons. These giants of plastic film and helium serve an important role in the early development of instruments for space astronomy. Because gamma rays accompany nuclear reactions, the earliest gamma ray instruments in orbit were carried on military spacecraft like the Environmental Research Satellite series and later the Vela series. Low energy gamma ray telescopes for the first surveys were carried on the first and third HEAOS. The discoveries at low energies include extremely rapid and transient gamma ray sources that outshine the entire sky for a few seconds and then vanish, detection of anti-electrons annihilating in collisions with normal electrons at the center of our galaxy, and radioactive aluminum from nova and supernova explosions distributed in the plane of the Milky Way.

Orbital high-energy gamma ray astronomy started with a small telescope on the second Orbiting Solar Observatory, but the first sensitive survey was conducted by the second Small Astronomy Satellite (SAS-2). SAS-2 and the European Space Agency's gamma ray telescope COS-B discovered that there are many places in our galaxy, and beyond, that produce copious amounts of electromagnetic radiation far more energetic than any developed in terrestrial laboratories. Some are associated with the most enigmatic objects in the universe—quasars and pulsars, as well as cosmic rays interacting with atoms in a giant molecular cloud. The rest, however, have not been identified and their true nature remains a mystery. NASA is developing the Gamma Ray Observatory to investigate the nature of cosmic sources of low- and high-energy gamma rays, hoping to solve the puzzles left by earlier missions and discover what other secrets there may be at the highest energies in the universe.

Because of the high technology involved with cryogenic telescopes, infrared astronomy is just coming out of its infancy. Only in 1983 was the first survey of the infrared sky performed from a satellite, the Infrared Astronomical Satellite (IRAS). It is already clear, however, that another revolution in astronomy is beginning. IRAS recorded data on the formation of stars throughout our galaxy, discovered five new comets, and found a new asteroid (possibly a dead comet) that passes closer to the sun than any planet or known asteroid. One of the most exciting discoveries has been the system of particles around the star Vega that must be related to the formation of planets there. Other stars show similar features, and it is clear that a new, important opportunity to investigate the formation of planets has appeared. IRAS also discovered clouds of dust above and below the sun in the plane of the galaxy and not just one, but three giant dust shells asymmetrically placed around the star Betelgeuse. IRAS has also found bright infrared galaxies that can barely be discerned in the most sensitive photographic atlases. These are just the beginning; analysis of the IRAS data has not yet been begun in earnest.

IRAS was designed to be especially sensitive to sources smaller than about 1 arcmin across, but it found great clouds of warm dust above and below the sun in the plane of the galaxy. These clouds are in patches that sometimes extend for many degrees. IRAS did not observe, however, in the band of wavelengths between 0.2 and 1.0 mm where local, extended sources of radiation like these clouds are overwhelmed by the cosmic microwave background. We now think that this background radiation comes from a time when the universe was less than a million years old; investigating the submillimeter spectrum and structure across the sky will give us the structure of the universe at the earliest observable time. This is the objective of the Cosmic Background Explorer (COBE), now being built at the Goddard Space Flight Center: to observe the structure of the Big Bang.

The IRAS telescope was kept at a few degrees above absolute zero so that thermal emission from the telescope itself would not overwhelm the signal from cosmic infrared sources. Cryogenic cooling is required for the next generation telescope that will follow IRAS and investigate its discoveries, the Space Infrared Telescope Facility (SIRTF) now planned by NASA for the early 1990's. SIRTF will be an observatory like the Hubble Space Telescope, serviced from the shuttle and the space station system and resupplied with cryogens. As a result, SIRTF will be able to conduct investigations of cosmic infrared radiation into the next century; it will take that long just to scratch the surface of the hundreds of thousands of sources discovered by IRAS. Who knows what SIRTF will discover along the way?

Infrared observation without cryogenic cooling can be

done with very high spectral resolution, very large telescopes, and high light-gathering power because then the thermal emission from the telescope is divided into very narrow bands. Beginning at a few centimeters of wavelength and going down to the far infrared at about 50 μm, interstellar molecules produce narrow spectral features, and the information in the features can tell us a great deal about the atmospheres of the outer planets and about the formation of stars and planets in our galaxy. A telescope for this spectral range would have to have a diameter of 10–30 m to match the performance of ground-based instruments operating at wavelengths longer than 1 mm, and NASA and ESA are each studying telescopes that have to be deployed automatically or assembled with the help of men to investigate cosmic submillimeter radiation. The NASA telescope is called the Large Deployable Reflector.

It is our plan in this chapter to show how rich the universe has been discovered to be and describe the instruments that will enable our continued scientific analysis of the universe and show us our place in it. It appears that gravitation is the force that organizes the universe, and we begin with a discussion of its role. We will follow this with a section on the channels of astrophysical information that can be exploited, emphasizing electromagnetic radiation and what it can tell us. Next we summarize a few of the outstanding problems of astrophysical research today, and then we show how the program of future missions is aimed at solving these problems.

OBSERVING THE UNIVERSE

The Unique Role of the Gravity Force

Gravity is the force that directly or indirectly shapes everything in the universe. Gravity dominates everywhere, reaching all the way to the edge of the universe where its pull even now may be so great that it will eventually halt the universe's expansion and ultimately crush all matter and energy back into the inferno from which it arose 20 billion years ago. Equilibrium between the force of gravity and the kinetic energy of matter in motion is revealed in the orbits of moons around planets, planets around stars, stars around galaxies, and galaxies within clusters and superclusters of galaxies. Equilibrium is indicated when we observe the roundness of planets and stars.

The other forces of nature are made apparant as a consequence of the action of gravitation. When the solid-state structure of condensed matter provides the dominant resistance to compression, irregular shapes can appear as in interstellar grains only a few micrometers across up to objects the size of major asteroids. As long as atomic structure is important for resisting the force of gravity, the object is a round, planet-sized object.

The point is reached, as we consider more and more massive objects, when atomic structure gives way under

(a)

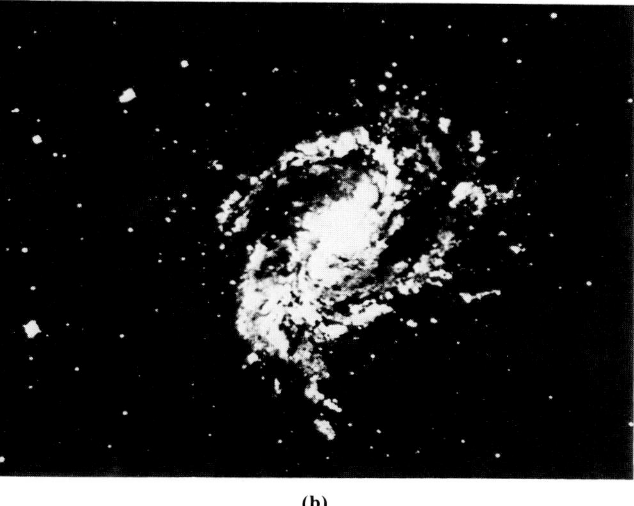

(b)

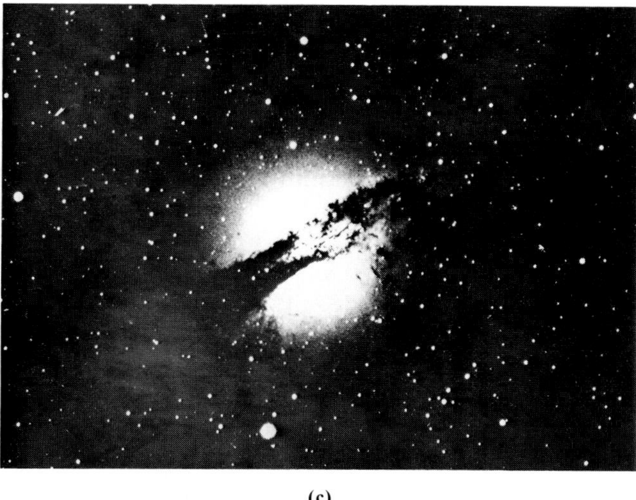

(c)

Fig. 4-2: Examples of equilibrium between gravitation and kinetic energy: (a) a globular cluster of stars; (b) a spiral galaxy; and (c) the nearest strong radio galaxy, Centaurus A, an elliptical galaxy with an unusual dust lane. Most elliptical galaxies are gigantic versions of globular clusters, having very little gas and dust.

the pressure of overlying mass, and the object becomes a star. Gravity will do work on the star, compressing it and causing it to shine until the temperature and density at the core is sufficient to ignite nuclear reactions. Then it becomes a true star with the inevitable crush of gravitation halted as nuclear energy is liberated to keep the star shining.

There are many exciting episodes in the life of a star, but there will always come a time when the nuclear fuel is exhausted and the collapse resumes. When a final equilibrium is achieved at last, the star may be in one of three states, depending on its mass. As long as the final mass is less than about 1.2 times the mass of the sun, the star will become a white dwarf, with the resistance of electrons to compression balancing gravitation. Our sun will ultimately be about the size of the earth when this happens. If the star finishes its life with mass between 1.2 and about 2.5 times the mass of the sun, the force on incompressible electrons becomes so great that gravity can do the work to form neutrons out of electrons and protons. In normal matter, this is energetically unfavorable; free neutrons spontaneously decay into an electron, a proton, and a neutrino. In the formation of a neutron star, however, getting rid of electrons this way allows the star to shrink until finally the resistance of neutrons to compression halts the collapse. The collapse is very rapid, and the liberation of gravitational energy as the star falls from an object the size of the earth to an object the size of Washington, DC, is staggering. It is the source of energy thought to power one major type of supernova explosion, producing more radiation than the star ever did over its whole normal life and leaving most of its outer layers expanding into space-like interstellar shrapnel. The third possible, final state for a star is one in which gravitation alone persists. As we mentioned in the introduction, when the mass of the star is greater than about 2.5 times the mass of the sun, not even the incompressibility of nuclear matter can halt the compression, and the star inevitably becomes a black hole.

Ten years ago, there were still significant arguments about whether there really were black holes, whether there was a third star in the Cygnus X-1 system that could account for the motions of the companion star, and whether another form of matter might arise at extreme compression and prevent the ultimate collapse. These arguments have largely gone away. Over 10 years of observing HDE 226868, it has now been firmly established that the unseen companion has greater than six solar masses. Another answer comes from the fact that a neutron star is only a factor of ten times larger than a black hole of its same mass. Prof. Hans Bethe was awarded the Nobel prize in physics in part for showing that we know enough about nuclear physics to see that there are no stable states of matter that lie between the incompressibility of neutrons and a black hole when the mass of the final star is greater than about 2.5 solar masses. We also realize that something must happen when the escape velocity of an object exceeds the speed of light. The bending of starlight in a gravitational field has been established, and every theory of gravity that takes relativity into account comes to the same conclusion that when light cannot escape, neither does anything else.

One usually imagines that black holes have unthinkable densities. After all, a black hole a few kilometers across has a density several times that of nuclear matter. But what about the density of a truly enormous black hole, say one the size of the universe? The radius of a black hole is proportional to its mass, so we expect the density to be very much lower for a black hole so large. We can estimate the size of the universe from its expansion which appears to be exactly like an explosion from a point. In such an explosion, the particles disperse at rates given by their velocities: the fastest moving particles at the edge and the slowest moving particles still close to the site of the explosion. From the point of view of a particle in the explosion, everything about it is moving away—the ones ahead because they are faster, the ones behind because they are slower—and the velocity is proportional to the distance. This is just how the universe appears. Within a factor of 2, the galaxies we see are moving away from us with a velocity of 25 km/s for every million light years away each galaxy is found.

Because we have learned that nothing travels faster than the speed of light in a vacuum, a rough estimate of the size of the universe is the boundary at which this expansion would reach the speed of light—about 1.2×10^{10} light years. The mass of a black hole having a radius the size of the universe would be 6×10^{56} g. This turns out to be 10^{11} solar masses (about 1 galaxy) per cubic million light years—very nearly the density we observe. It is a major goal of space astronomy to observe the motion of matter in the universe, observe the effect of gravity so far on the expansion, and determine whether the expansion exceeds escape velocity. We want to know whether or not we are inside a black hole.

Besides determining the ultimate fate of the universe, gravity also makes it wonderfully observable. Gravitational energy is released as matter is drawn together by its pull. This is the ultimate source of power in everything we see in the universe (except perhaps the Big Bang), but the problem is to work out how it happens. It is easy to see that matter flowing onto a neutron star or black hole in a binary system could be heated to several million degrees, but what is the source of friction and how efficient is it? How does gravity, driving the convection in our sun, manage to heat the corona to nearly the same temperature as the gas funneling into a black hole? Clearly the energy for a supernova explosion comes from the explosive release of gravitational energy, but how is it transferred to the infalling matter? By neutrinos? What becomes of the stellar debris ejected at 10 000 km/s into interstellar space? That energy is somehow responsible for many of the phenomena we have discovered but which elude our understanding:

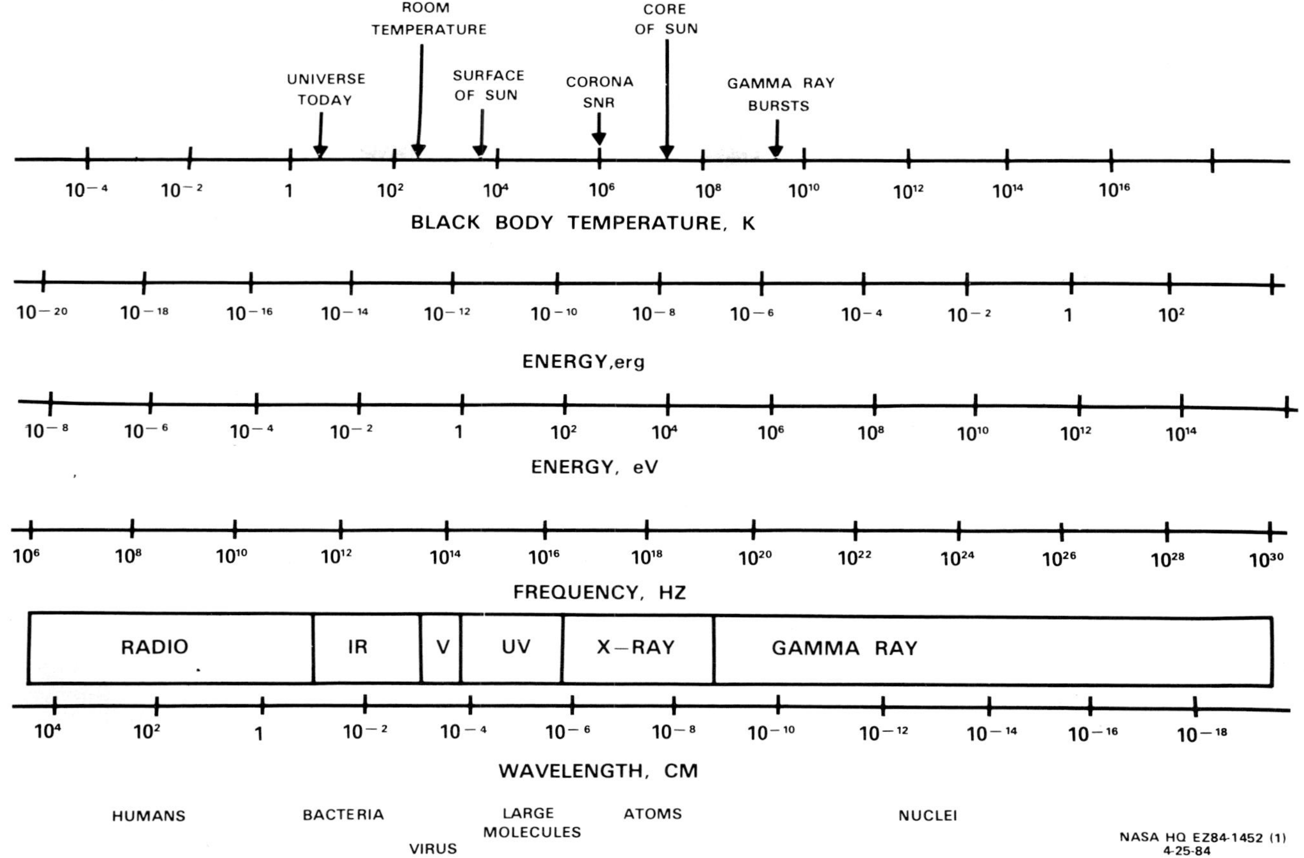

Fig. 4-3: The relationship among the parameters used to characterize the frequency of electromagnetic radiation, including wavelength and energy per photon. Also given is the temperature of a perfect radiator giving its maximum power at a given frequency.

supernova remnants, cavities of superheated and very rare gas in the interstellar medium, and perhaps even the acceleration of cosmic rays. We know that gravity holds neutron stars together, but what makes them into pulsars?

Channels of Information

From our laws of physics, we expect to receive information about the universe through four channels: electromagnetic radiation, gravitational radiation, neutrinos, and cosmic rays. Electromagnetic radiation is the most familiar, and we will spend most of our time discussing astronomy from this channel. Cosmic rays have also brought us information about our galaxy, and it is fruitful to think of them as a sample of material from which we try to derive the history of the galaxy much as one would from a sample of the moon. Unlike lunar samples, however, cosmic rays have had their electrons completely stripped off, and they have been accelerated to very close to the speed of light. How this happens is the oldest question in high-energy astrophysics. Neutrinos are a promising carrier of information about nuclear processes as they are occurring, but we have not yet positively detected even our own sun. The most promising, unopened channel is through gravitational radiation, predicted to travel at the speed of light as a consequence of rapid accelerations of great masses.

Electromagnetic radiation

Except for the design of radio telescopes, cosmic electromagnetic radiation is most usefully discussed in quantum mechanical terms rather than through Maxwell's equations. In quantum mechanics, electromagnetic radiation carries energy in quanta called photons, each carrying an energy proportional to the frequency of the radiation. The constant of proportionality is Planck's constant. Photons are massless and move at the speed of light, and therefore have a characteristic wavelength inversely proportional to the frequency. Depending on the domain of discussion, one refers alternately to the frequency, the wavelength, or the energy of the photons. Fig. 4-3 shows the relationship among some of these parameters.

Also shown is the temperature associated with each energy. When an astronomical body is optically thick, that is, when the mean free path of a photon is small compared to the size of the object, the radiation can be in equilibrium with the object and a smooth, characteristic spectrum of intensity with frequency is produced. This is the Planck spectrum, and it depends only on the temperature. Stars always have a basic spectrum with the energy at each frequency in accordance with Planck's spectral shape. Material can also be optically thin, that is, nearly transparent. When this happens, the spectrum depends strongly on the temperature and composition, and strong lines of absorption or emission are observed.

In astronomy, electromagnetic radiation is divided into six bands: radio (including microwave and submillimeter astronomy), far-infrared, optical (including near-infrared and near-ultraviolet), ultraviolet (including far-ultraviolet and extreme ultraviolet), X-rays (including soft X-rays and hard X-rays), and gamma rays (from low energy to high energy). Each has special observational techniques and sources, and each major band has its own way of specifying the radiation it observes. The bands and their characteristics are summarized below.

Radio: This is one of two regions where excellent results are obtained from the ground. The typical range of wavelengths is from 1 mm to 30 m, bounded by the transparency of the atmosphere at short wavelengths and the ionosphere at long. In the customary units for this region, the range is 10 MHz to 300 GHz.

Radio photons are detected with large antennas which operate in arrays to achieve very high angular resolution (~1 m·arcsec) using coherent, aperture-synthesis techniques. Orbiting radio telescopes in space are now under study because of the extremely long baselines and correspondingly high angular resolution possible. A typical astronomical source for radio astronomy investigation might be a jet of relativistic electrons extending from the core of a radio galaxy into intergalactic space. These electrons radiate by the synchrotron mechanism, a relativistic effect that boosts the electron gyro frequency in the weak intergalactic magnetic fields into observable radio frequencies. Radio telescope systems are excellent for studying sources of relativistic electrons and also regions of ionized hydrogen around newly formed stars.

Millimeter and submillimeter waves (not accessible from the ground) are also the preferred bands for radiation from excited molecules. These transitions are relatively rare at wavelengths shorter than 50 μm, and a space submillimeter telescope (1 mm–50 μm) with very high spectral resolution will be a powerful tool for investigating the dynamics of molecular clouds.

Far-Infrared: Although there are a few "windows" where IR can come through the atmosphere, IR observations with high sensitivity can be accomplished only from space for most of the wavelengths between 5 and 200 μm. This band overlaps the submillimeter band in wavelength, but there is a great difference in technique and in the kind of information we try to obtain. This is not the last overlap you will find in this summary.

Currently, IR astronomy is conducted with normal-incidence mirror systems with solid-state detectors at the focal plane. The IRAS telescope was a Cassegrain system with a 60-cm primary mirror cooled to about 2 K. Objects with temperatures between 5 and 1000 K emit primarily in the infrared.

Because IR wavelengths are long compared to the size of a typical grain of interstellar dust, 100 μm IR from the center of our galaxy is attenuated by only about 30 percent, while visible light photons are attenuated by about 10^{12}. IR astronomy is, thus, used for the study of cold sources such as dust, primordial planetary systems, and regions of star formation. Also, hot objects enshrouded in dust are studied in IR as well as distant hot objects which have their radiation shifted into the IR by cosmological recession. We also expect to see IR from "brown dwarf" stars—the cinders of burned out stars.

Optical: This band is defined by the second transparent window in the earth's atmosphere between 3000 Å (10^{10} Å/m, or 10 000 Å/μm) and 5 μm. Recent developments in the technology of image-intensification and solid-state detectors systems have made ground-based optical astronomy increasingly more sensitive and as vital a part of future astronomical research as it has been in the past.

In this band, atmospheric turbulence acts to limit the quality of an image to about 1 arcsec (unless great pains are taken to extract higher resolution information from the spots in the general blur.) Also, dust and aerosols in the atmosphere attenuate the light and reflect city lights, creating undesirable background. For a significant improvement in sensitivity and in angular resolution, optical astronomy has to go into space. High angular resolution means that the angular extent of a pixel is smaller, and with photon-counting detectors, the sensitivity is increased as the background is reduced.

Stars are excellent sources of visible light, as are galaxies. Some galaxies have bright nuclei that are unresolved and have spectra that are smooth but completely unlike starlight. These are the Seyfert galaxies and N galaxies. The shape of these spectra suggest that synchrotron radiation is causing emission all the way to the near ultraviolet. The relativistic boost of the electrons to make this possible is many orders of magnitude higher than for radio emission, and the source of their power is a great mystery. Quasars appear to be extreme examples of this kind of emission.

Ultraviolet: At wavelengths shorter than 3000 Å, ozone in the atmosphere absorbs the ultraviolet, and we have to go to space again to observe. The ultraviolet extends from about 100 to 3000 Å and is especially sensitive to gas along the line of sight from the telescope to the source of radiation. Detectors range from photographic film for sounding rocket flights and shuttle missions to electronic detectors that use microchannel plates over multi-anode array readouts. Image intensifiers and CCD arrays are also used. Mirror coatings become very important in the ultraviolet. In the extreme UV at wavelengths shorter than about 1000 Å, normal-incidence mirrors become too inefficient, and grazing-incidence systems are necessary. This can make a telescope very long, especially if high-dispersion gratings are used for taking spectra.

The ultraviolet is especially important for studying ionized gas. Hydrogen is the most abundant element, accounting for more than 90 percent of the mass of the universe. When electromagnetic radiation has wavelength shorter than about 900 Å, it can ionize hydrogen and produce highly ionized states of other atoms as well. Gas can also be ionized by energetic charged particles like cosmic rays. The ions give off and absorb ultraviolet light in narrow lines that contain a great deal of information on the physical state. Thus, astronomers infer the current production rate of ionizing radiation, the density structure and the dynamic structure of the ionization region, its probable history, and the flow of matter and energy in the ionization region.

X-Rays: At wavelengths shorter than 100 Å, radiation is penetrating enough to reach through the interstellar medium even from the other side of our galaxy. These are the X-rays and gamma rays. These photons carry so much energy, that they are usually described by the amount of energy they carry. The units are electron volts of energy, measured by the thousands in X-rays (keV) and by the millions (MeV) and billions (GeV) in gamma rays. A photon with 1 keV energy has a wavelength of 12.4 Å.

X-ray detectors generally operate by sensing the amount of ionization that is caused by X-ray photons, either in proportional counters, microchannel plates, solid-state detectors, or scintillation counters. Proportional counters are sensitive descendents of the old geiger counters; the size of the output pulse is proportional to the amount of ionization. Microchannel plates accelerate electrons released by the X-ray and produce cascades of electrons which are sensed to determine the arrival of an X-ray photon. No ionization information is obtained in microchannel plates, only time and location. Solid-state detectors are diode equivalents of the proportional counter; electrons produced by ionization are collected at the outputs under high voltage. A low-voltage version of this has great potential for high-resolution imaging and good spectroscopy using CCD arrays. Crystal scintillation counters are important at higher energies when the penetrating power of the X-rays requires greater stopping power. Also gas scintillation counters are being developed which look very promising for detection and spectroscopy at energies around 1 keV.

X-rays are important for observing 10 and 100 million degree gas we did not know existed until 1962, when a sounding rocket attempting to observe solar X-rays fluorescing from the moon discovered instead the first cosmic X-ray source, Scorpius X-1. X-ray spectra reveal the temperature of hot gases (several million degrees) and also components that are not in equilibrium, such as gas from a stellar wind that is swept up by the gravitational field of an orbiting neutron star. Magnetic fields on neutron stars have been measured by observing the hard X-rays corresponding to the cyclotron frequency of an electron in magnetic fields as high as 10^{12} G. There is nothing like an X-ray telescope for investigating high energy phenomena and objects such as black holes and neutron stars.

Gamma Rays: "Gamma radiation" was originally used to designate electromagnetic radiation that came from nuclear reactions. While the wavelengths of X-rays are typically the size of atoms, the wavelengths of gamma rays are typically the size of nuclei. Two modifications have been made to the definition. First, it now extends to include all electromagnetic radiation that is detected with a gamma ray detector whether it came from a nucleus or not, and second, it is extended to any electromagnetic radiation with more energy than a typical nuclear gamma ray. Although there is overlap with hard X-rays at energies between 100 and 300 keV, any photon with energy above 300 keV is called a gamma ray.

Because the energy range is open ended, not just one kind of detector will work for the whole range. At low energy, large scintillation crystals and large solid-state detectors are used. These are typically shielded from gamma rays by massive collimators on all sides except for the aperture in order to obtain directional information. At high energies, spark chambers are used. In these, photons are converted to a pair consisting of an electron and an anti-electron which carry virtually all the momentum and energy of the photon. By tracking the pairs and measuring their energies, the direction and energy of the photon can be obtained. In between low-energy and high-energy gamma ray detectors are the Compton telescopes. These operate in the range between about 1 and 30 MeV and are, in a way, the gamma ray equivalent of an imaging telescope.

Narrow gamma ray lines come from nuclear decays in the interstellar medium, and also from the annihilation of electrons and anti-electrons. Both types of radiation have been detected, the first from the decay of radioactive aluminum in the interstellar medium and the second from a mysterious source of anti-electrons at the center of our galaxy. Gamma rays have also been observed from pulsars and quasars. Gamma ray telescopes observe directly some of the most energetic phenomena in the universe.

Cosmic Rays (Particle Astrophysics)

While electromagnetic radiation is carried by the massless, chargeless photon, cosmic rays are electrons (positive and negative) and nuclei (mostly protons with some helium nuclei and traces of more massive nuclei).

Like X-rays and gamma rays, cosmic rays can be detected by the ionization left in detectors, but because they are charged, relativistic cosmic rays will produce Cerenkov light in appropriate detector media. This is very important for modern cosmic ray telescopes. Cerenkov light is like a sonic boom of light that occurs when a charged particle travels through a medium faster than the speed of light in the medium. Depending on the charge and energy range of interest, cosmic ray telescopes are made from combinations of Cerenkov detectors, scintillators, trajectory measuring device-like multi-wire proportional counters, ionization chambers, and calorimeters for measuring total energy.

Cosmic rays are a unique source of astrophysical information. While all the other possible carriers are neutral particles, cosmic rays are charged. Thus, cosmic rays are accelerated, guided, and stirred by electromagnetic processes, principally the magnetic fields. The number of cosmic rays reaching the earth is "modulated" by the solar wind plasma that fills the heliosphere. During times of solar activity, fewer cosmic rays can penetrate deep into the heliosphere. Cosmic ray spectra follow "power laws," showing that some nonthermal process is responsible for their acceleration. How cosmic rays achieve their extremely high energies (extensive air showers have shown cosmic ray particles reaching the earth with a joule of energy) is one of the oldest outstanding questions in high-energy astrophysics.

Cosmic rays are a sample of matter from beyond our solar system; they carry the signature of their source and the scars of all their travels. As cosmic rays pass through the interstellar medium, they occasionally collide with atoms in their path. Because cosmic rays have such high energies, these collisions often smash the cosmic ray nucleus, producing a shower of secondary stable and unstable nuclei. Examining the composition of cosmic rays, the signature of the secondaries can be read, particularly if one can identify the individual nuclear isotopes.

Beryllium is a million times more abundant in cosmic rays than in normal galactic material and is undoubtably the product of collisions of cosmic ray carbon, nitrogen, and oxygen with the interstellar medium. In the trans-relativistic regime (100–600 MeV/nucleon), it has been possible to identify the isotopes and observe that there are about 10 times more beryllium nuclei of the stable isotope (5 neutrons and 4 protons) than the unstable isotope (6 neutrons and 4 protons). From this ratio and the half-life of unstable beryllium (3 million years), it is easy to see that on the average, trans-relativistic cosmic rays are kept in the galaxy about 10 million years before escaping into intergalactic space. This also means that they have been created recently by cosmic standards, and an equilibrium between acceleration and loss has been established.

Using the nuclear collision probabilities for creating beryllium from cosmic ray carbon, nitrogen, and oxygen, it can be calculated that the trans-relativistic cosmic rays have traveled through about 7 g of matter per square centimeter. Because they travel at about half the speed of light, this means that they have been in interstellar space with an average density of 1 atom per cubic centimeter. The gas around the sun has less than 1/10 this density, but higher density regions are less than 100 light years away in most directions.

Gravitational Waves

Because no information can propagate faster than the speed of light, all modern theories of gravitation include traveling waves of gravitational energy as part of their description of gravity far from rapidly accelerating masses. Dipole radiation is not allowed because mass comes with only one sign (i.e., there is no negative mass). As a result, the least complicated form is described by quadrupole terms. So far, no gravitation radiation has been directly detected, although there are indirect observations showing energy loss from a binary star system that may be due to gravitational radiation.

The best developed and certainly most widely accepted theory of relativistic gravity is Einstein's theory of general relativity. In general relativity, gravitation is described by bending in space and time that determine how freely falling matter will travel, and it differs from Newtonian gravitation with effects that appear as the velocities of bodies and test particles approach the speed of light. A classic example is the bending of starlight in the gravitational field of the sun. In this case, the test particles, photons of starlight, are moving at exactly the speed of light.

All our tests of relativistic gravity have been of this kind: test bodies moving in a gravitational field and exhibiting deviations from Newtonian gravity because of their velocities. With accurate ranging to our interplanetary spacecraft, tests have been carried out with high enough precision that we can even say that the gravitational constant G changes less than 1 part in 10^{12} per year [6]. This is very significant. Because we think the universe is less than 10^{11} years old, this result can be taken to show that the gravitational constant does not depend on the age of the universe and is likely to be truly constant.

Another kind of test for general relativity is to measure deviations from Newtonian gravity coming from a moving source of gravitation. This effect is analagous to the creation of magnetic fields by moving electric charges. In order to be detectable, the source must be massive and the velocities should not be utterly insignificant compared to the speed of light. It turns out that a perfect gyroscope in a polar orbit should be torqued 0.05 arcsec in a year due to the relativistic correction in the earth's gravitational field due to the earth's rotation. The effect is called "frame dragging"; the rotation of the earth under the satellite causes the satellite to sense a "gravitomagnetic" force. The technical ability to build gyroscopes of sufficient precision to measure this extremely small torque is just now in hand after 20 years of development. We call the mission Gravity Probe B. (Gravity Probe A confirmed the prediction of general relativity for the slowing of a maser clock on the ground—deep in the gravitational well of the earth's field—compared to a maser clock in space.)

Neutrinos

Neutrinos are weakly interacting particles with very little mass and in most theories have no mass at all. Recently, electromagnetism has been "unified" with the weak interaction that governs neutrino interactions. ("Unification" is one of the major goals of contemporary physics.) Neutrinos are created in the nuclear reactions that heat stars and should leave normal stars without any further interactions. There is currently a dilemma because experiments have failed to detect neutrinos from our sun. Neutrinos can be very important when a neutron star forms, and the high-energy neutrinos may be responsible for converting the gravitational energy of collapse into the kinetic energy of a supernova explosion. One day astronomers will need to observe the neutrino signature of the death of a star.

This subject has largely been mentioned for completeness. Unless there is some major technical breakthrough, there is little hope of measuring neutrinos from any source other than the sun or possibly a supernova explosion in the galaxy.

Observing in Space Astronomy

Leaving the other channels of information, we now want to concentrate on astronomy with electromagnetic radiation. Because we have summarized the attributes of each frequency band above, we will concentrate here on other parameters. All that there is to know about electromagnetic radiation is characterized by its quantum numbers: frequency, direction of propagation, location of arrival, time of arrival, and polarization. For astronomical observations, the most useful description is frequency, direction, time, intensity, and the Stokes' parameters of polarization. We will use these to discuss what it takes to get more information from astronomical observations. Each parameter will be illustrated with examples from recent solar physics.

Angular Resolution

Angular resolution describes how well one can determine the direction of travel of the incoming photons. Fig. 4-8 shows the sun with just about the highest angular resolution ever achieved. The spot that looks like a fine water spot just above the center of the picture is a feature on the surface of the sun called filigree. Fine structure on the surface of the sun has been briefly seen with resolution better than about 0.1 arcsec. Finer resolution is of limited use because there are limits on how small physically important structure on the surface can be. This size is about 100 km, the mean free path for a photon at the photosphere. All visible light is physically averaged over this scale. However, until the Solar Optical Telescope (which is designed to measure many physical parameters on this scale) is developed and observes from space, all other observations are necessarily averages over even larger scales.

The improvement in knowledge about the surface of the sun that we expect with SOT is illustrated by recalling the problem of determining how oxygen is carried by the blood. Experiments had shown that the blood carried the oxygen, and changes in color could be observed, but no one knew how it worked. When the microscope was invented, and blood cells were observed for the first time, the solution was obvious. This example shows a lesson that is often repeated—breakthroughs in understanding frequently follow advances in observational technology. We expect that understanding the structure and dynamics of the surface of the sun will be in hand when the sun is finally observed at 100 km resolution by SOT since the solar magnetic fields appear to organize the surface into cells of this size.

Spectral Resolution

By breaking light into its components (i.e., spectroscopy) one measures the abundances of atoms and their internal and kinetic energies. Thus, increases in spectral resolution are analogous to the advantages of increases in angular resolution. Fig. 4-9 shows how the sun appears in the light of a narrow spectral line, H-alpha. It has a quite different appearance from white light and shows only gas above 100 000°—right at the bottom of the corona. Thus, with high spectral resolution, we see a very different physical situation.

Temporal Resolution

Ancient astronomers thought the sky was perfect except for a few wandering lights and occasional transient stars. Perfection meant that it would never change. Modern astronomers feel just the opposite; change means that another dimension, time, is available for discovering the physical nature of a source. Many of the most interesting phenomena change quickly, like a solar flare. In a flare, energy in tangled magnetic fields above sunspots is rapidly converted into the kinetic energy of electrons and ions which are accelerated down toward the sunspots. As the electrons reach the thick lower layers, an impulsive burst of hard X-rays is released, and heated gas comes back up the field lines to produce X-ray emission from the arch. This picture has been developed by making X-ray observations with high time resolution; the impulsive part of a flare often lasts only a few seconds. In order to understand how solar flares are triggered, it will be necessary to observe solar flares with time resolution of milliseconds in order to record the collapse of magnetic fields that initiates the flare.

Polarization

The polarization of electromagnetic radiation also carries important information. Most often, the polarization is caused by magnetic fields, and magnetograms of the sun are derived from polarization data. Polarization is observed in radio waves from radio galaxies, in starlight (the dust grains of the interstellar medium are aligned over a galactic scale), and in X-rays from the Crab pulsar. Even gamma rays are certainly polarized from quasars and pulsars.

Spectral Coverage

In order to understand the sequence of energy transport within many sources, observations have to be made at wavelengths from radio through gamma rays simultaneously. This is true, for example, in understanding the white light from solar flares. Observations in hard X-rays and UV light made by the Solar Maximum Mission are combined in Fig. 4-12 with a white light picture taken from the ground to show how energy coming from the electrons accelerated in the flare are stopped in the lower atmosphere producing hard X-rays. The gas where the electrons are stopped is heated by the bombardment, giving the ultraviolet light. The X-rays and UV light provide heat to gas even lower in the atmosphere which radiates it away as a white light flare. Similarly, without knowing the radio, IR, visible, UV, X-ray, and gamma ray spectrum from a quasar, one cannot determine the strength of the magnetic field, the energy density of relativistic electrons, or even the total power in the quasar.

Bounds

What is the range of things we can observe? Until 28 years ago, our knowledge was limited to theories and the observations that could be obtained from visible light. The discovery of quasars as the optical counterparts of some cosmic radio sources first showed us that the universe in visible light may be only a very small fraction of all the phenomena accessible through the other portions of the electromagnetic spectrum. Space has opened up these new channels, but even in space there are limits. Cosmic background is a vital source of information on the scale of the universe as a whole, but it is also a background against which other kinds of observations have to be made. Interstellar dust is a major and important component of the interstellar medium, but it also absorbs so much visible and ultraviolet light that we may never be able to view starlight from most of our own galaxy.

There are even limits in the range of accessible frequency. The low frequency cutoff due to the plasma frequency of the interstellar medium is 10^5 Hz. The high frequency cutoff is 10^{29} Hz, corresponding to 400 TeV, the energy above which a gamma ray will be absorbed in a collision with photons of the microwave background. Furthermore, the quantum mechanical principle of uncertainty limits our spectral resolution by the amount of time we have available to make the observations. Angular resolution is limited by the sizes of the telescopes we have available or can synthesize from arrays of telescopes.

A clear explanation of the limits of astronomical observation can be found in Martin Harwit's book, *Cosmic Discovery*. This thought-provoking investigation into the process by which astronomy comes to find the major forms of matter and energy in the universe shows just how rich the universe is. Boundaries on what we can observe and how big the universe is leads to the startling conclusion that there may be as few as 150 kinds of objects as unique as quasars and pulsars. We live in a time when our major observatories in space and on the ground give us the opportunity to discover the major pieces of the universe the way the explorers of the sixteenth and seventeenth centuries discovered the continents of the world.

EXAMPLES OF CURRENT PROBLEMS

Cosmic Backgrounds

Besides being an unavoidable source of noise in astronomical observations, cosmic background radiation carries a wealth of unique information. The cosmic microwave background and the cosmic X-ray background are two that dominate the sky at their wavelengths and are so smooth and uniform that they are clearly related to the universe as a whole.

The cosmic microwave background was originally proposed by Gamow [9], Alpher and Herman [10], and later developed theoretically by Dicke, Peebles, Roll, and Wilkinson [11] to come from the "Big Bang" at the creation of the universe. Their calculations showed that the universe must have been opaque and dominated by radiation until it cooled to the point that hydrogen ions could become hydrogen atoms. Then the universe must have become transparent to the radiation, which would have the spectrum of a black body at about 4000°. This radiation should permeate the universe and be visible today with a characteristic temperature of a few degrees Kelvin.

The celebrated discovery of the cosmic microwave background came independently of these predictions with work by Wilson and Penzias [12]. When its significance was realized, the microwave background became the main reason most astronomers subscribed to the "Big Bang" theory over "steady-state" theories of the universe; there had been a successful prediction.

The microwave background is very isotropic and has a temperature of almost exactly 3 K. Departures from perfect isotropy contain information on the shape of the universe and conditions during the Big Bang. Departures from a perfect black body spectrum contain information on the evolution of the universe at the time matter "decoupled" from radiation and major evolutionary events since then. Just as starlight probes the material along the line of sight, the cosmic microwave background probes the state of matter since the Big Bang. Most matter is seen in emission above the background, but fast electrons will scatter the microwave photons to higher energies and reduce the microwave radiation coming to us through rich clusters of galaxies. The number of fast electrons can be determined by observing how much the background radiation appears to be reduced. The number of fast electrons could also be determined from X-ray observations if only we knew the distance to the cluster, but turning this problem around, we can find the distance to the cluster if we can perform sensitive microwave and X-ray observations. This independent and sensitive method of determining distances on a cosmological scale is yet to be exploited.

The cosmic microwave background is not the only isotropic background that space astronomy has to investigate; an X-ray background was discovered with the first observations of the X-ray sky. It appears to be just as smooth as the microwave background, implying that it comes to the earth from cosmological distances. Unlike the microwave background, however, the X-ray background was not predicted in advance. It could be composed of so many X-ray emitting galaxies that it appears smooth except at very high angular resolution. Deep observations made by HEAO-2 suggested that quasars might be responsible, but the spectrum of the diffuse X-ray background extends into gamma ray energies of more than 100 MeV and does not resemble the spectra of quasars that are known to be X-ray emitters. It may also be truly diffuse: a gas at a temperature of 5×10^8 degrees with a nonthermal component and a significant fraction of the mass of the universe. Deeper observations with higher angular resolution and higher sensitivity to bring up the contrast will tell us whether or not distant X-ray galaxies make up the background or whether it is an intergalactic gas. In either case, it is more evidence for a remarkably parallel evolution of the universe in all directions.

Distance Scale

Establishing the distance to objects in the universe is a fundamental problem not only for obtaining a physical understanding of the objects, but also for determining the nature of the universe itself. We can tell if the universe will expand forever or not by observing the effects of gravity on rate of expansion out to near the edge of the universe. Measuring velocities is the easy part; measuring distances is the hard part.

It will surprise someone who is not an astronomer to find out how poorly distances are known even for the nearest stars. The very closest star, Proxima Centauri, a member of the triple star system known as Alpha Centauri, has a parallax of scarcely 2 arcsec as the earth goes around the sun. One appreciates the problem of measuring the distance even to Alpha Centauri when one remembers that twinkling in the atmosphere has kept angular resolution to little better than 1 arcsec. With thousands of observations, this can be beaten down statistically to improve the knowledge of Proxima Centauri by two orders of magnitude, but stars are so far away that parallax measurements quickly become very difficult.

Luckily, the Hyades cluster of stars in the constellation Taurus is nearby and moving away from us quickly enough to allow astronomers to estimate its distance by the rate at which the cluster appears to have shrunk since the earliest photographs. The Hyades cluster is quite rich, and knowing the distance of about 140 light years has allowed astronomers to estimate the distance to other stars from their apparent brightness and color.

The method of relating color and apparent brightness is most accurate when it is applied to a large number of stars that are at the same distance. This is possible in clusters of stars, and by this means the distances to the nearest Cepheid variables has been estimated. Cepheid variables are bright, yellow stars that change their brightness significantly with regular periods on the order of several days. The most important fact about these variables, however, is that within the several classes of Cepheids, their periods are a function of their luminosities, with brighter stars having longer periods. It was by observing Cepheid variables in the Andromeda nebula that Edwin P. Hubble proved that the Andromeda nebula was at least 1.5 million light years away and therefore a completely separate galaxy from our own. It can be said by this measurement, he discovered the universe.

Although Cepheids and other variables are bright, they have not been observed much further away than Andromeda, and other distance indicators have been developed in a sequence of one indicator calibrating another out to greater and greater distances. After a hundred million light years, the expansion of the universe becomes more important than the random motions of galaxies, and redshift is often used, but in the current state of affairs, this method of measuring distance is many steps removed from the parallaxes to nearby stars. Most recently, it has been discovered that the gravitational pull of galaxies in the Virgo cluster, about 60 million light years away, has reduced the expansion of galaxies near our own to about half the rate found well away from Virgo.

Quasars and Pulsars

Two problems that have been with us ever since their discovery in the 1960's are quasars and pulsars. "Quasar" is short for quasi-stellar radio source while "pulsar" is short for pulsating star. Quasars are extremely luminous sources of radio, infrared, visible, X, and gamma radiation found billions of light years from our galaxy. They are "quasi-stellar" because they are compact enough that ground-based optical telescopes cannot resolve them. In contrast, pulsars are rotating, magnetized neutron stars in our own galaxy that primarily give off pulses of radio waves, but in some of the most interesting and revealing cases they also give off visible, X, and gamma radiation.

Quasars were discovered in 1963 when the radio source 3C 273, identified with a remarkable "star" and jet in 1962 by Cyril Hazard, was found by Maarten Schmidt to have a redshift (the ratio of measured to original wavelength) of 0.16. Since then, quasars have been found out to a redshift of more than 3.5, making them by far the most distant objects observed. Their intrinsic luminosities can be as high as 10^{41} W, and some have been seen to change their brightness significantly in only a few hours.

The fundamental problem of quasars is to find out how the luminosity of a galaxy can appear to be coming from a volume the size of our solar system. A power of

Fig. 4-4: An optical image of the first quasar discovered, 3C 273 with its jet. The central part of the quasar is unresolved and resembles the star next to it except that by being brighter, it appears bigger in this image.

10^{41} W corresponds to more power than would be liberated by the complete conversion of all the matter in a sun into energy once a month. As bizarre as it may seem, the most conservative theory is that quasars are powered by supermassive black holes in the cores of very distant galaxies. In normal nuclear burning, less than 1 percent of the matter is converted to energy, but in accretion onto a black hole, the gravitational energy available is on the order of the Einsteinian mass-energy equivalence of Mc^2.

The luminosity that can be achieved by accretion is limited by the mass of the gravitational body that drives the process. Radiation from accretion scatters off the electrons of the infalling material, reducing the effective force of gravity, and when radiation pressure on an electron just balances the gravitational pull on a proton (electrical forces are much stronger than gravitational forces, and charge neutrality is always maintained), further accretion is impossible. The limiting luminosity of the system is called the Eddington limit.

Many of the X-ray sources in the galaxy have luminosities very close to the Eddington limit for masses typical of a neutron star. If we suppose that the quasar is radiating at its limit of luminosity allowed for accretion, the calculation shows that the mass must be at least 10^{10} times the mass of the sun!

Pulsars radiate by a completely different mechanism. The current best theory describes a pulsar as a rapidly rotating, highly magnetized neutron star. The magnetic field is on the order of 10^{12} G, and rotation generates electric fields that supports a complex magnetosphere. Typically, pulsar periods range from milliseconds to several seconds, and at the magnetic poles, the electric fields are so strong that the vacuum breaks down in a shower of electrons and anti-electrons which are then accelerated to very close to the speed of light. The electrons are forced to follow the magnetic field lines, and as their trajectory bends, radio emission is generated. This theory satisfies much about the radio characteristics of a pulsar, but there are also many problems that remain even in this application.

A more profound problem is to determine how pulsars radiate visible, X, and gamma radiation. One class of theories proposes that this radiation comes from the light cylinder—the surface around the rotation axis where corotation reaches the speed of light. Another class proposes that the radiation comes from close to the neutron star. What is maddening is that the pulsars are giving us clue after clue and we cannot figure them out.

A comparison of two famous pulsars illustrates this point. Pulses from the Crab occur twice in every 33 ms period. The pulses are exactly the same phase for every wavelength we have observed from radio through

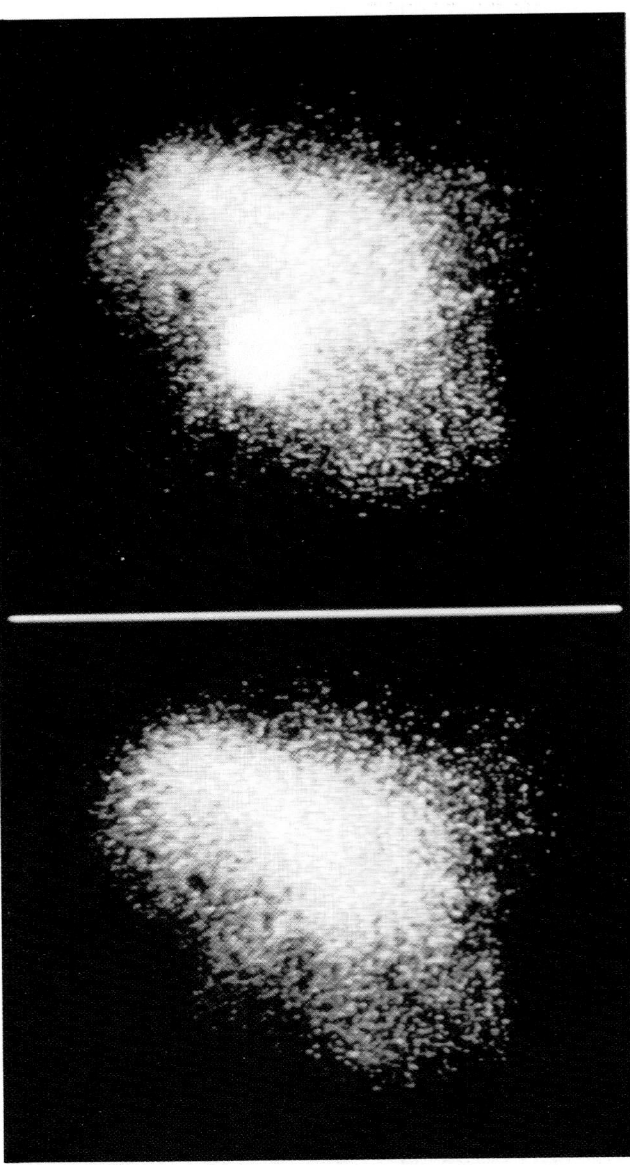

Fig. 4-5: The pulsar in the Crab Nebula seen in X-rays. This image was taken by HEAO-2 by folding images taken when the pulsar was off and when it was on. The brightest part of the X-ray nebula is seen in both images.

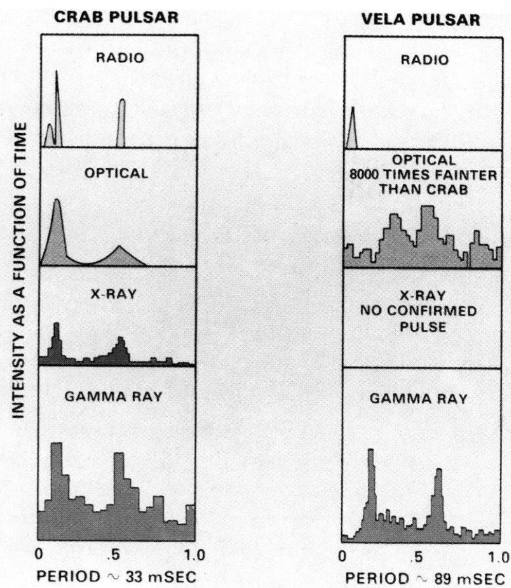

Fig. 4-6: Two famous pulsars: The Crab and Vela.

gamma rays. The Vela pulsar, in contrast, has only one radio pulse in each 89 ms, while it has two at completely different phases in both visible light and gamma rays. Even more puzzling, there are no X-ray pulsations—just a steady, very compact X-ray source. Fig. 4-6 illustrates the situation. The most recent clue in the enigma is that the second gamma ray pulse in the Crab seems to be fading. This must be very significant, but we don't know what it means.

The Formation of Stars and Planets

The most fundamental problem for the astrophysics of the interstellar medium is how gas and dust can be swept up into giant, self-gravitating clouds, collapse to form stars and planets, and then be dispersed after the formation. The main action occurs deep within these clouds, and all visible and ultraviolet light is blocked even though these are essential components of the dynamics within the star-forming region. What we know comes primarily from radio and microwave observations, near infrared, and some from X-ray and gamma ray observations. Only just now with the IRAS results do we have significant data from the primary open channel, far-infrared radiation.

The discovery of very hot expanding components in the interstellar medium has shown us forces that are very effective in moving the cooler gas around. These components are heated by supernova explosions and violent winds coming from blue (one might say ultraviolet) extremely massive stars. In addition, white dwarf stars—a fast-moving population giving off highly ionizing, extreme ultraviolet radiation—pass through the interstellar medium like buckshot, leaving a wake of hot, ionized gas behind. Because the youngest, bluest, most active stars are embedded in the gas and dust that was their nursery, the most dynamic parts of stellar birth are best observed in the shortest wavelengths that can penetrate the surrounding material: the infrared.

The effect of this activity is to apply pressure within the cloud. In some cases, globs of dense gas and dust are ejected at high velocity, as in the case of Herbig-Haro objects. In other cases, the molecules are excited far out of thermal equilibrium, and interstellar masers of ammonia, water, and hydroxyl molecules switch on. These forces and their effect on the cold matter determine whether the material is blown away or compressed to form other new stars. Some regions of star formation, like the one around Eta Carina, have a great abundance of very massive stars. It is estimated that Eta Carina itself has a mass greater than 200 times that of the sun and is shedding into the gas and dust around it between 0.1 to 0.001 solar masses per year [7].

The vibration and rotation states of interstellar molecules produce very narrow lines, and with millimeter and submillimeter spectroscopy at high angular resolution, the dynamics of the dense gas can be observed. Furthermore, supernova explosions, bright young stars, and strong stellar winds can all be observed by the X-rays they produce, no matter how deep they are in the cloud. The ionization from cosmic rays can also be assessed from the gamma rays given off by the energetic collisions of cosmic rays with gas in the molecular clouds. It is clear that a multispectral attack on the problem of star formation, led by infrared and submillimeter observations, can let us see into the darkest regions of our galaxy.

Heating the Corona

Many problems in solar physics were used to illustrate the information gained by improvements in polarization sensitivity and in angular, spectral, and time resolution. We have not observed the sun with the angular resolution to even know the structure of the surface, nor have we observed flares with the spectral coverage, angular and time resolution to know what triggers solar flares.

One major problem that has resisted our understanding is the transport of energy and mass to the corona and the solar wind. Ten years ago there was a good theory that showed the potential for acoustic waves generated at the surface of the sun to supply the upper atmosphere of the sun with energy. As the waves ascended, they would steepen into shocks and finally dissipate as heat at the bottom of the corona or at the base of the solar wind. It was a good theory with much to calculate and also observe. The waves propagating in the chromosphere between the surface and the corona were found to produce a recognizable signature in the ultraviolet, and experiments were put on the last two Orbiting Solar Observatories (OSO-7 and OSO-8) to confirm the theory.

Unfortunately, the theory was incorrect. The acoustic waves dissipate long before the base of the corona is reached, and we no longer know how the heat is delivered. This has more than academic interest. The corona is an important and highly variable source of ionizing radiation for our ionosphere, and the solar wind has a direct dynamic effect on the earth's magnetosphere. As man's life becomes more complex, we become more sensitive to the electric and magnetic environment of earth. Enormous electric potentials develop in the Alaskan pipeline, for example, when there is a strong magnetic storm, and some earth-orbiting satellites have failed completely just when energetic particles from the sun have arrived at earth. Our interest in the sun is to understand the physical processes on the nearest star in hopes that some day we may be able to predict the behavior that has such direct influence on our daily lives.

THE OBSERVATORIES

Discoveries in astronomy inevitably emerge from two parallel developments—advances in technology of observational techniques and advances in physical knowledge. The advances in technology are exemplified by the new ability to put astronomical instruments in orbit, but technology of the instruments themselves has also improved dramatically. The physics of quantum mechanics, general relativity, and theories of elementary particles are being applied to equilibrium and non-equilibrium environments to help us interpret our observations. As a result, we find ourselves in a universe populated with not only stars but also white dwarf stars, neutron stars, black holes, pulsars, and quasars with creation, birth, death, and destruction happening on vast scales of temperature, distance, and time.

A Major Observatory for Visible and Ultraviolet Light

The premiere space observatory for optical and UV astronomy will be the Hubble Space Telescope (HST), named for Edwin P. Hubble, the astronomer who is credited with discovering the awesome size of the universe. NASA, through the Marshall Space Flight Center, has the HST well into development. The observatory will be operated by the Goddard Space Flight Center in a scientific program managed by the Space Telescope Science Institute at Johns Hopkins University. The telescope has a Ritchey-Cretien design (Cassegrain with hyperbolic primary and secondary mirrors). The aperture is 2.4 m across, and the focal ratio is 24. The mirrors are coated with Al and MgF_2, giving the telescope high reflectivity at all wavelengths longer than 1200 Å. The telescope will be held stable to 0.007 arcsec to take full advantage of its diffraction-limited optical system.

Besides the Find Guidance Sensor, there will be five scientific instruments sharing the focal plane of the telescope: the Wide Field Camera, the Faint Object Camera, the Faint Object Spectrograph, the High Resolution Spectrograph, and the High Speed Photometer. The telescope and its instruments have been described by C. R. O'Dell [2] and in more detail by Hall [3].

It is significant that the size of the telescope was set by the desire to observe Cepheid variables in the Virgo cluster; it was Hubble's observation of Cepheids in the Andromeda nebula that established it as a galaxy beyond our own. With the HST, there will be a recalibration of the distance scale of the universe that

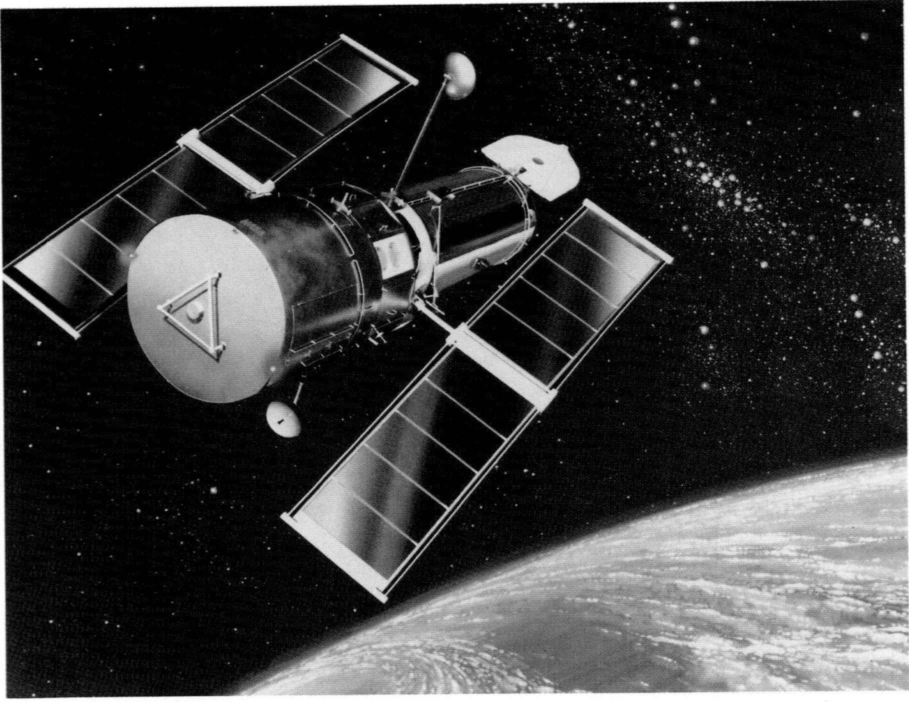

Fig. 4-7: The Edwin P. Hubble Space Telescope will be the first major observatory in space for visible and ultraviolet light astronomy.

will allow us to determine the law of expansion with considerable certainty. This law of expansion is crucial to all our physical understanding of extragalactic objects; the law is used to find out how far away extragalactic objects are and is a key part in determining the age of the universe. The law is also named for Edwin Hubble.

The other part of determining the age of the universe is to measure the "deceleration." Deceleration measures the mass of the universe by observing the effect of gravitation in slowing the expansion. If gravitation has been effective in slowing the expansion, then except for relativistic effects, the further back we look in time, the faster the Hubble expansion should appear. We will be able to measure the deceleration of the universe when we have established new distance indicators to very great distances. The indicators will have to be independent of redshift and probably will have to take evolution of the universe into account. The deceleration will tell us not only how old the universe is, but also its ultimate fate. Will it expand forever, or will it ultimately collapse? The HST has been designed to answer this fundamental question.

The range of wavelengths that will be observed using the HST go from 1200 Å to 10 000 Å (1 μm). The HST will be about 50 times more sensitive in visible light than the 200-in Hale telescope on Mount Palomar and hundreds of thousands of times more sensitive in UV light than our present capability (IUE and telescopes carried above the atmosphere in sounding rockets). Another advantage of the HST is its high angular resolution that comes from the diffraction-limited optical system operating above the distorting atmosphere.

The improvement in sensitivity will be so great that we may confidently predict another major revolution in our understanding of the universe will come from its operation. Spiral galaxies are quite different from ellipticals, resembling long-playing phonograph records more than footballs; yet with ground-based telescopes, we cannot tell the difference between an elliptical and a spiral more than a few percent of the way across the universe. With the HST, the difference will be apparent halfway across.

The HST will be launched by the space shuttle into an orbit 500 km high and inclined 28.5°. This is a good orbit for astronomy because the cosmic radiation is low and the orbit is easily accessible by the space shuttle. The scientific instruments will be upgraded during the 15-year mission as the telescope is serviced by later flights of the shuttle; and when the space station becomes operational in the early 1990's, it will open up a new level of servicing capability that may make it unnecessary for the HST ever to be returned to earth for maintenance or repair. Servicing, maintenance, and repair in orbit will become an increasingly important part of space astronomy, beginning with the HST mission.

An Observatory for Gamma Ray Astronomy

Electromagnetic radiation with the shortest wavelengths are gamma rays, and gamma ray astronomy observes a universe of extremely energetic phenomena—antimatter annihilation, cosmic ray interactions, quasars, pulsars, and some sources that have yet to be identified. NASA, through its Goddard Space Flight Center, is

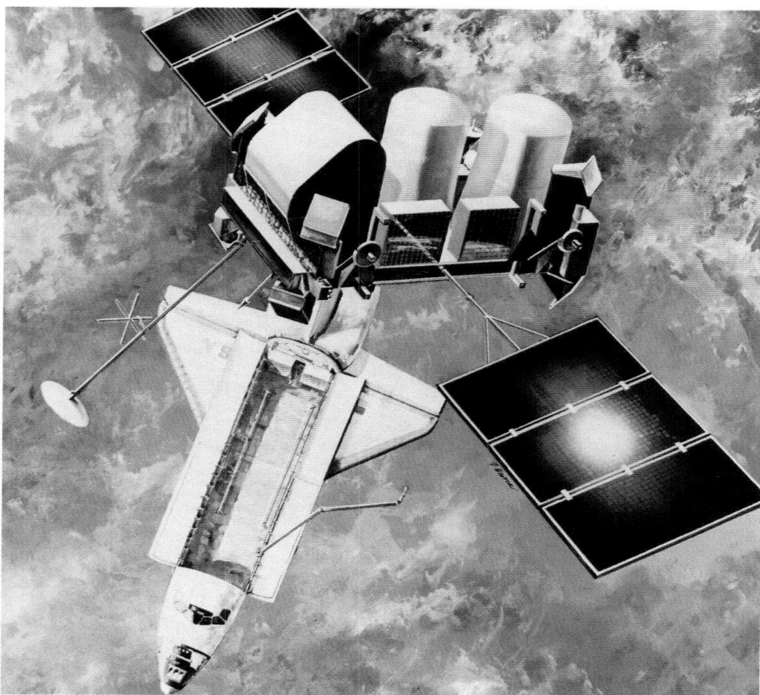

Fig. 4-8: The Gamma Ray Observatory with four instruments covering the spectrum from 0.1 MeV to 30 GeV.

currently developing the Gamma Ray Observatory (GRO) to be the first observatory to carry the complement of instruments necessary to completely cover the wavelength band from 0.1 MeV to over 30 GeV.

Perhaps the most exciting discovery in low-energy gamma ray astronomy (0.1–10 MeV) was the discovery by gamma ray telescopes on balloons that anti-electrons (positrons) were being annihilated at the center of our galaxy. This source appears to be unique; the general absence of annihilation radiation shows that over scales of at least 100 million light years, the universe around our galaxy is made of normal matter with antimatter appearing only as a consequence of nuclear and subatomic processes.

The discovery of antimatter annihilation at the center of our galaxy was followed by a pair of observations with HEAO-3 which found that the rate of annihilation had declined by more than a factor of 3 in the six months between the two observations. Several important conclusions follow from this result. Not only did the creation of anti-electrons cease, but the anti-electrons that had been created were mostly annihilated in the six months. This allows us to estimate the density as greater than $10^5/cm^3$, and from the narrowness of the annihilation line, we conclude that the temperature must be less than 10^5 degrees. In all, 10^4 times the solar luminosity must be converted into electron and anti-electron pairs to produce the luminosity observed in annihilation radiation by the balloon-borne and satellite gamma ray telescopes.

In order to obtain detailed information about the physical conditions in the source of these anti-electrons and about other sources in the band between 0.1 and 10 MeV, the Oriented Scintillation Spectrometer Experiment (OSSE) on the GRO will rotate to view each source, then view a background region. It is being developed by the Naval Research Laboratory, and it will have enough sensitivity to extend the search for gamma ray line emission begun by the high-resolution gamma ray spectrometer on HEAO-3.

In the low-energy gamma ray range, mysterious sources occur that flare for just a few seconds and then disappear (gamma ray bursts). During the time that they are producing gamma rays, they can outshine the entire universe. Small gamma ray detectors in satellite networks have established that the usual gamma ray events are evenly distributed in the sky, indicating that they come from bodies that are typically less than 300 light years from the earth. The strongest and most unusual event was found to have come from the direction of a supernova remnant in the Large Magellanic Cloud more than 150 000 light years away. If it was at that distance, it must have been extremely luminous. Furthermore, the fluctuations in intensity were periodic, giving support to the theory that gamma ray bursts come from the collision of comets or asteroids with neutron stars. In order to observe fainter and more distant gamma ray bursts, the GRO includes the Burst and Transient Source Experiment (BATSE). If gamma ray burst sources are from within 300 light years of the sun, BATSE will be able to extend the range well into the rest of the galaxy, giving a distribution concentrated into the plane of our galaxy rather than uniformly distributed. BATSE is being developed by investigators at the Marshall Space Flight Center.

While X-ray telescopes can reflect X-rays at shallow angles and bring them to a focus, gamma ray telescopes with energy below 20 MeV generally rely on massive shielding to collimate their detectors and define the field of view. Gamma rays do scatter, however, and in the Imaging Compton Telescope (ICT) on GRO, this scattering is used to obtain some imaging capability for the energy range 1–30 MeV. Knowing where the gamma ray scattered, how much energy it lost, which way it went, and how much energy it carried away from the scattering defines a cone of possible directions that the gamma ray can have come from. The ICT is being developed by an international collaboration under V. Shonfelder of the Federal Republic of Germany. Between 5 and 75 MeV, the sky has never been surveyed, and the ICT will give us a first sensitive look at the bottom half of this band.

Above 30 MeV, GRO will observe the universe with the Energetic Gamma-Ray Experiment Telescope (EGRET), being developed at the Goddard Space Flight Center. A primary objective of this instrument will be to locate sources of energetic gamma rays discovered by the SAS-2 and COS-B missions to within 0.1°. Of the 25 sources in the COS-B catalog, only four have been identified: two pulsars, one molecular cloud, and a quasar. There is also a source in a compact binary system familiar from X-ray astronomy, Cygnus X-3, which was detected by SAS-2 but which has not been detected by COS-B. It is probably variable. No one class of object stands out, and it is very important to discover what the remaining unidentified sources might be. Gamma rays of this energy must be coming from extremely energetic astronomical events. The molecular cloud, for example, may be radiating as a result of cosmic rays colliding with the gas and dust of the cloud.

Major Observatory for X-Ray Astronomy

X-ray astronomy is now ready for its mature observatory, the Advanced X-ray Astrophysics Facility (AXAF). AXAF will double the energy range previously accessible to imaging X-ray telescopes, improve the best angular resolution by a factor of almost 10, and in the energy range where it overlaps previous X-ray telescopes, it will be more than 100 times more sensitive. Unlike any X-ray telescope before, it will have a lifetime measured in decades and instruments that can be replaced in orbit. AXAF is the new facility given the highest priority by the Astronomy Survey Committee of the National Academy of Sciences in its report *Astronomy and Astrophysics for the 1980's*.

Some aspects of the universe observable by AXAF include measurement of masses, temperatures, densities,

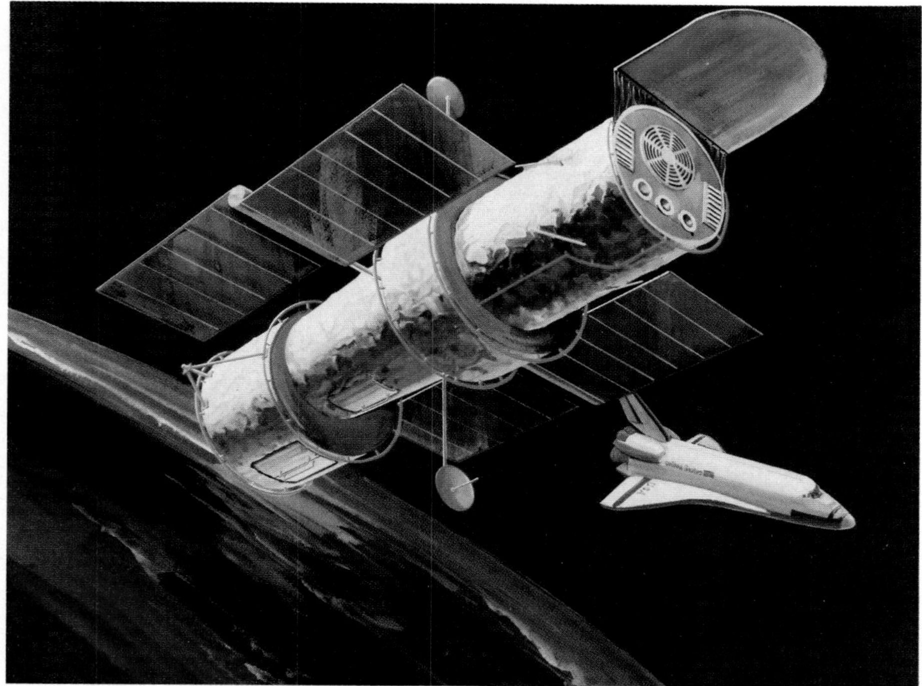

Fig. 4-9: The Advanced X-ray Astrophysics Facility will be as great an advance over HEAO-2 as the Hubble Space Telescope will be over ground-based telescopes.

magnetic fields, composition, evolution, velocities, and structure in the most energetic environments in the universe. With its high angular resolution (0.5 arcsec), AXAF will make deep exposures of the X-ray background to determine if the source is truly diffuse and constitutes a major fraction of the mass of the universe or if the source is really the superposition of very many distant X-ray galaxies. With its high sensitivity, AXAF will observe bright X-ray binaries in many nearby galaxies to help establish an extragalactic distance scale. High sensitivity will also allow high spectral resolution in observation of the composition of supernova remnants, the aurorae of Jupiter, and perhaps the composition of the surface of Io.

One of the most important new capabilities in AXAF will be its ability to image X-rays with energies up to 8 keV. This covers the important lines of iron near 7 keV, which AXAF can use as a probe of the evolution of matter in clusters of galaxies. Iron should not be abundant until the second generation of stars has exploded, and the enrichment of intracluster media should be quite noticeable in AXAF observations which will be able to reach a redshift of 3.

Finally there is the importance of the long life of AXAF. Martin Harwit, in his analysis of astronomical discovery, lists 43 discoveries that figure most prominently in the literature and at scientific meetings. The list goes back to antiquity, and although each astronomer would probably make a slightly different list, the basic conclusions of Harwit are difficult to dispute. From this list, one sees that 29 out of the 43 discoveries have been made in visible light. This has occurred because of the long time we have been able to investigate objects by the visible light they produce. Furthermore, there has been enough time to find out which objects in a category are the most likely to show us their conditions and processes. The second-oldest branch of astronomy, radio astronomy, accounts for 8 of the remaining 14 discoveries. The advent of long-term high-sensitivity X-ray observations will inevitably lead to profound new discoveries.

During all of HEAO-2, in contrast, there was time to observe only one cluster of galaxies with emission from its intracluster gas weak enough to leave some of the galaxies visible above the background. Seven galaxies could be observed, but there were an additional three objects that appeared to be galaxies in X-rays but which could not be identified with galaxies in visible light. We are left to wonder what they are, but if we suppose that they are galaxies with dust so thick that we cannot see their visible light, then anemic clusters (low in density of galaxies and intracluster gas) may be fruitful objects to investigate for X-ray galaxies. Having a mission long enough to find the most revealing objects and then to investigate their physical conditions is a vital, new aspect of space astronomy.

AXAF will use six grazing-incidence, X-ray, Wolter type I mirror pairs with a focal length of 10 m. The outer mirrors will have a grazing angle of 50 arcmin while the inner mirror will have a grazing angle of about 25 arcmin, permitting images at energies as high as 8 keV. The diameter of the outer mirror will be 1.2 m, and the effective on-axis area will run from 1700 cm^2 below 1 keV to 1100 cm^2 at 2 keV, 600 cm^2 at 4 keV, and 200 cm^2 at 7 keV.

Scientific instruments for AXAF have recently been selected. To take advantage of the high imaging quality of AXAF (0.5 arcsec), they include a high resolution camera that will cover 32 arcminutes of the AXAF field of view and a CCD imaging spectrometer which will cover a smaller portion of the field but will give good spectral resolution and increased sensitivity at higher energies. Much of the spectral information astrophysicists need requires higher resolution than the CCD's can give, and three instruments were selected to obtain this resolution. For good sensitivity and simultaneous coverage of a wide energy band, high throughput transmission gratings were selected that will disperse the X-rays into several thousand spectral channels across one of the two imaging detectors described above. For high sensitivity and high spectral resolution (10 eV) over the entire 0.1 to 8 keV range, an X-ray calorimeter was selected, while for the very highest spectral resolution for extended sources, a Bragg crystal spectrometer will be developed. Each of these instruments is particularly suited for some special measurement, optimizing one or more parameters such as field of view, energy band, sensitivity, angular resolution, or spectral resolution for a particular application.

AXAF will be launched by the space shuttle into an orbit 500 km high and inclined 28.5° to the equator. The first revisit to replace instruments and service the spacecraft is expected to occur about three years after launch. Although AXAF is being designed to be serviced from the shuttle, we expect that servicing will normally occur at the space station. The nominal lifetime for the observatory will be 15 years, and there will be provisions for returning the spacecraft to earth for major repairs.

Infrared and Submillimeter Astronomy

The discoveries of the first far-infrared survey of the sky are just beginning to wash across the astrophysical community, and they are showing us things never before seen. The far infrared favors radiation from temperatures between ten and a few hundred degrees Kelvin, and the galaxy is transparent to the radiation. Fig. 4-10 shows the central 50° of our galaxy in false colors that indicate temperature; the warmest material, including dust within our solar system, is blue while cooler material is red. The yellow and green knots scattered along the plane of the galaxy are giant clouds of interstellar gas and dust heated by nearby stars. One is struck by how thin the region of star formation is in our galaxy compared to the stars of the Milky Way seen at night in visible light. Like other spiral galaxies, most of the stars in our galaxy are confined to a disk with the proportions of a long-playing phonograph record. Our view at night is limited to the nearest few hundred light years; if there were no dust, we would be dazzled by the light from stars up to a thousand times farther away, nearly all confined to the narrow plane revealed in the infrared.

Among other discoveries of IRAS was a ring of dust and debris about the star Vega that is likely to be the by-product of the formation of planets. Vega was observed many times as a calibration for the IRAS detectors, and in analyzing the calibration data, the IRAS astronomy team found that what should have been a point source unresolved by the telescope turned out to be slightly extended and have a component of emission with a temperature of 90 K in addition to the component near 10 000 K from the star itself. Small dust particles of a few hundred micrometers in size would be

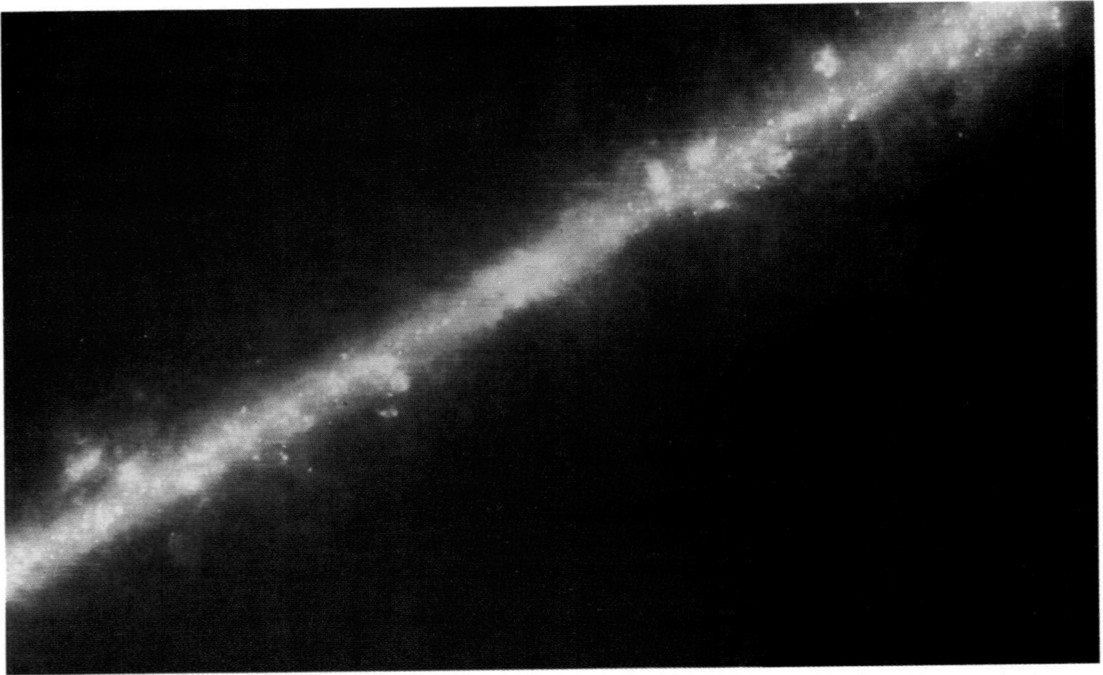

Fig. 4-10: The central 50° of our galaxy in the far infrared observed by IRAS. The puffy, blue foreground objects seen crowding together in the distance are regions of star formation. The center of the galaxy is at the center of the picture.

the most efficient radiators of far-infrared emission, but the lifetime of these particles against the drag of radiation pressure is so short that we can confidently say that the particles seen around Vega must be at least 1 mm across. Since these are not such efficient radiators, the mass in the cloud of debris must be at least 10^{30} g, or enough to form several times the number of planets in our solar system. The larger the body, the less efficiently it radiates infrared per unit mass, and certainly no planets outside our solar system have been directly detected by IRAS.

There may be very many such stars and debris systems in our galaxy. IRAS appears to have recorded several dozen and perhaps many more, but the best instrument for a sensitive search will be the Space Infrared Telescope Facility (SIRTF). It is not enough to see that a star has an infrared excess; the extent of the debris must be resolved to confirm its nature. SIRTF will have at least a factor of 10 times better angular resolution than IRAS, and by using several hundred times the number of detectors in IRAS, SIRTF will provide detailed infrared pictures of even the faintest IRAS sources.

Because it operates in space and its telescope is cooled to below 4 K, IRAS had a sensitivity far beyond any previous instrument, and it will not be possible to view the objects in the IRAS survey catalog again until cryogenically cooled telescopes return to space. SIRTF will be the most sensitive of all the telescopes planned for the 1990's. Its wavelength range will go beyond the 8 to 120 μm coverage of IRAS to the entire infrared band from 2 to 700 μm. Where they overlap, SIRTF will be 100–1000 times more sensitive than IRAS. This will come from two major differences. The first is that IRAS was primarily a survey instrument which swept rapidly across the sky, while SIRTF will be a true observatory, carrying a variety of focal plane instruments and capable of extensive observations of a single target. The second is that there have been dramatic increases in the sensitivity of infrared detectors, even over the last several years. Scientific instruments for SIRTF have been selected, including an infrared array camera for observations between 2 and 30 μm using the entire SIRTF field of view, a multiband imaging photometer which will use arrays of several types of detectors to cover the entire SIRTF band from 3 to 700 μm with high photometric precision, and a moderate resolution spectrometer which can divide the infrared light into between 300 and 1000 channels per octave over the 4 to 200 μm range.

Between 100 and 1000 μm, there are very many spectral lines of molecular rotation and vibration that can in principle be used to obtain chemical, temperature, velocity, and other physical characteristics in the many regions of star formation discovered by IRAS. At high spectral resolution, the background from a warm telescope is less important than having a large signal from the source because the noise will come primarily from the detectors. It is therefore reasonable to develop large antennas not much cooler than room temperature to observe the molecules in the atmospheres of the planets, the interstellar medium, and regions of star formation in this and other galaxies. In the United States, we are embarking on a program to develop the technology for the Large Deployable Reflector (LDR) needed for extremely high resolution spectroscopy of interstellar molecules. Spectral resolution of one part in a million in frequency is needed in some cases to resolve the velocity structure in very cold molecular clouds.

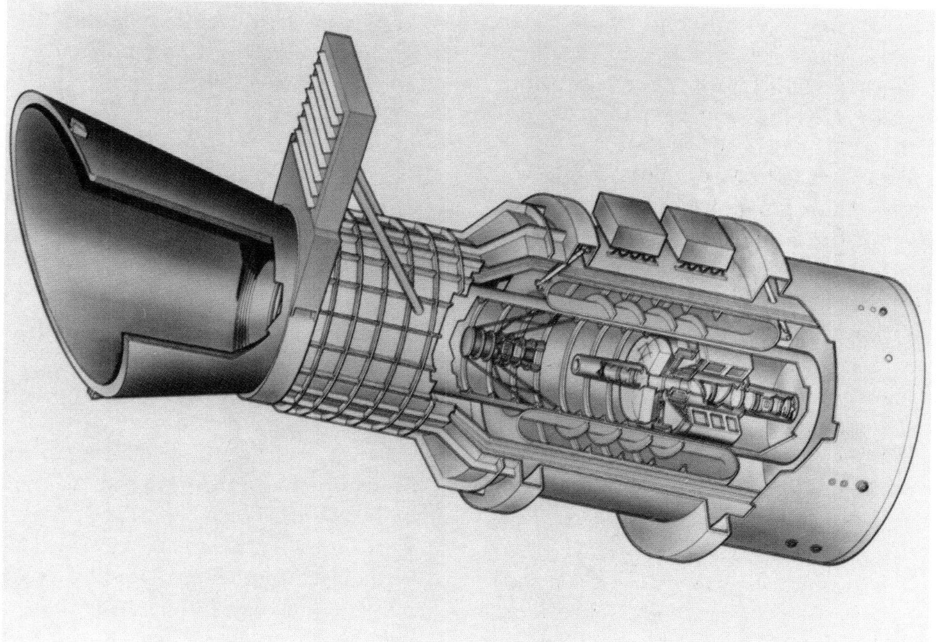

Fig. 4-11: Cut-away picture of the Space Infrared Telescope Facility (SIRTF) showing the telescope and focal plane instrument package in its superfluid helium dewar.

Fig. 4-12: Artist's concept of how the Large Deployable Reflector (LDR) might be assembled from the space shuttle. The assembly of LDR would be much easier to accomplish at a manned space station.

Major Observatory for Solar Physics

The brightest example of gravitation making the universe visible is our nearest star, the sun. Although most of the sun's radiation can penetrate easily through the earth's atmosphere, atmospheric conditions completely prevent us from observing the dynamics of the surface of the sun on a scale comparable to the fundamental averaging length—the mean free path of a photon on the surface of the sun. Studies of the sun have preceded all studies of other stars, and there has developed a symbiotic relationship between stellar astronomy and solar physics. This symbiotic relationship has continued through the development of space astronomy and is likely to be with us forever.

Having observed flares on the sun, other stars were examined for flaring behavior and found in some cases to have flares millions of times more energetic than on the sun. The newest discoveries about the sun, that it vibrates with a spectrum at periods near 300 s, will undoubtedly lead astronomers to observe other stars with the long times needed for high frequency resolution on stellar vibration. Vibration carries information on the internal structure and dynamics of a star, allowing us to observe the interior. Observations of X-rays from stars otherwise identical to our sun make us wonder about the past and future state of the sun's corona.

There are many fundamental lengths involved in observing the sun. There is the length over which the triggering of a solar flare occurs; there is the thickness of the transition regions between the cool chromosphere and the hot corona of the sun. We do not know any more about what initiates the conversion of magnetic energy to kinetic energy in a flare than we do about what heats the corona and accelerates the solar wind. A major observatory in space for solar physics will allow us to observe each phenomenon at the appropriate scale. This is the fundamental principle behind the Advanced Solar Observatory (ASO).

We plan for ASO to grow from the Solar Optical Telescope (SOT) and the Pinhole/Occulter Facility to include:

1) a High Resolution Cluster of instruments for studies of the interior dynamics, the solar photosphere, chromosphere, and low corona at optical, ultraviolet, extreme ultraviolet, and soft X-ray wavelengths;

2) a Pinhole/Occulter Facility with sensors in the visible spectrum to observe the corona close to the sun, in hard X-rays to image fast energetic events in flares;

3) a Solar High Energy Facility for spectroscopic observations of transient phenomena in hard X-rays and low-energy gamma rays;

4) a Solar Low Frequency Radio Facility for studying particle acceleration and propagation in the corona and inner heliosphere.

The space station system, through its co-orbiting scientific space platform or possibly even attached to the manned base itself, provides an excellent new opportunity for implementing the ASO. The capabilities now being studied would allow the ASO to take space astronomy from the era of long life through space servicing into an era of observatory evolution in space as instruments are added year by year to produce the mature observatory. The success of Skylab suggests that there may be advantages in operating the observatory itself from the space station's manned base.

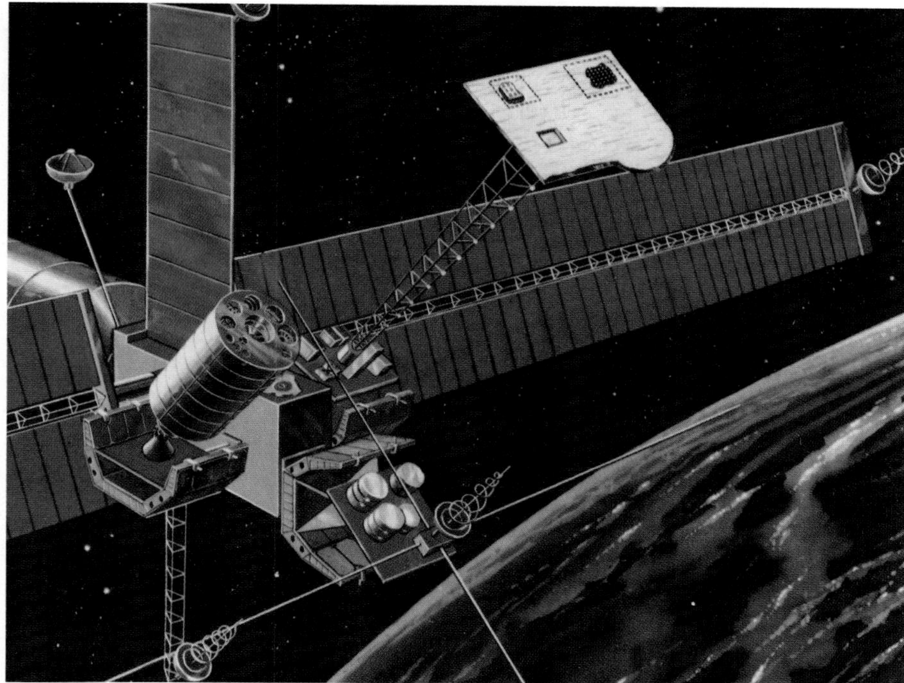

Fig. 4-13: The Advanced Solar Observatory showing its High Resolution Cluster, the Pinhole/Occulter Facility, the Solar High Energy Cluster, and the Solar Low Frequency Radio Facility.

CONCLUSION

Following on the discoveries of the last three decades, the next half century will surely be known as the Golden Age of Astronomy. Although prophecy is risky, it is reasonable to expect that we will succeed in solving many of the problems that drive us to our planned program of space astronomy missions. If we succeed, we will understand the physics of our sun's surface on the distance scale over which the gas and radiation do the averaging, and we will know its structure down to the core. We will know how quasars are powered and how they convert their power into luminous energy. We will know the shape of the universe a million years after its creation, and we will understand how galaxies condensed out of the primordial gas. We may even know how the universe was created. We will have discovered perhaps a hundred new kinds of cosmic phenomena, many of which will resist our comprehension for as long as quasars have, but by the end of the next 50 years, more will have been understood in quantitative detail than not.

Nearly all the discoveries and understanding will come from astronomy and astrophysics conducted in space. This can be said without any reduction in our high expectations for the achievements of ground-based radio, infrared, visible, and ultraviolet astronomy. It is said because only five decades of the electromagnetic spectrum can penetrate the earth's atmosphere out of the 24 decades that reach the top. Even for these five there are many advantages for going into space. In visible light, the atmosphere is transparent but refractive. This leads to scintillation, or twinkling, that makes the best angular resolution for optical telescopes on the ground about 1 arcsec. Furthermore, dust in the atmosphere and our growing cities make the night sky glow, and compared to visible light astronomy from space, ground-based telescopes are looking through a "dirty basement window." In radio waves, space offers the possibilities of increased angular resolution with baselines many times the size of the earth and of increased sensitivity as we eventually take advantage of the protected radio silence of the far side of the moon.

Our surveys have shown that the three richest bands in space astronomy are X-rays, UV/visible, and IR/sub-millimeter. Nearly everything we have discovered—from planets to quasars—produces radiation in each of these three bands that reveals an important aspect of what is happening in these objects. The richness of astronomical information available in these bands is further indicated by the large number of nations with plans for developing instruments to observe in these bands. Japan has a series of X-ray satellites in operation and under development, including the Hackucho in 1979, Hintori in 1981, Tenma in 1983, and Astro-C in 1987. The European Space Agency is operating Exosat, launched in 1983, and the Federal Republic of Germany is developing the Roentgen Satellit (Rosat) for launch in 1987. Both Exosat and Rosat have grazing-incidence optical systems approximating the HEAO-2 system. The European Space Agency is a major partner in the development of the Hubble Space Telescope, and it is performing preliminary design for an Infrared Space Observatory that will have many of the capabilities of SIRTF. These space astronomy missions complement the radio astronomy and visible light astronomy that can be done from the ground, and they are complemented in turn by gamma ray telescopes, very low-frequency radio telescopes, and other specialized missions in space astronomy.

As we look deeper into the universe, we will be able to see more of the kinds of objects that make up the universe. If Martin Harwit is right and the number of significant components of the universe is finite, then one of the most important jobs of space astronomy will be to search for the remaining, undiscovered categories of objects in the universe. Most of them will be within our reach. Except for extremely luminous objects like quasars, astronomy has not observed further away than a few percent of the way across the universe. Not limited by the atmosphere, our major observatories in space, the Hubble Space Telescope, the Gamma Ray Observatory, the Space Infrared Telescope Facility, the Advanced X-ray Astrophysics Facility, and the Advanced Solar Observatory will make a major fraction of the universe accessible to observation.

REFERENCES

[1] M. Harwit, *Cosmic Discovery*. New York: Basic Books, 1981.
[2] C. R. O'Dell, "The space telescope," in *Telescopes for the 1980's*, G. Burbidge and A. Hewitt, Eds. Palo Alto, CA: Annual Reviews, 1981.
[3] D. N. B. Hall, Ed., "The space telescope observatory," NASA Publ. CP-2244, 1982.
[4] R. Weiss, "Measurements of the cosmic background radiation," *Ann. Rev. Astron. Astrophys.*, vol. 18, p. 489, 1980.
[5] R. A. Sunyaev and Ya. B. Zel'dovich, "Microwave background radiation as a probe of the contemporary structure and history of the universe," *Ann. Rev. Astron. Astrophys.*, vol. 18, p. 537, 1980.
[6] R.W. Hellings *et al.*, "Experimental test of the variability of G using Viking Lander ranging data," *Phys. Rev. Lett.*, vol. 51, p. 1609, 1983.
[7] R.M. Humphreys and K. Davidson, *Science*, vol. 223, pp. 243–324, 1984.
[8] Astronomy Survey Committee, *Astronomy and Astrophysics for the 1980's*. Washington, DC: Nat. Acad. Press, 1982.
[9] Gamow, *Phys. Rev.*, vol. 70, p. 572, 1946.
[10] Alpher and Herman, *Phys. Rev.*, vol. 75, p. 1089, 1949.
[11] Dicke, Peebles, Roll, and Wilkinson, *Astrophys. J.*, vol. 142, p. 414, 1965.
[12] Wilson and Penzias, *Astrophys. J.*, vol. 142, p. 419, 1965.

5

NASA'S LIFE SCIENCES PROGRAM

Gerald A. Soffen

HISTORY

NASA's interest in the life sciences was initially in response to the early needs of the manned missions, Mercury, Gemini, and Apollo. In those pioneering times with little experience, there was a wide range of speculation about how humans would endure the new environment of space. The crews of those early missions were the subjects of many tests and procedures that by today's standards might appear unnecessary or unduly conservative. Those entrusted with the crew health realized that it was wiser to risk criticism than to risk a serious mistake by overlooking possible hazards. Their philosophy, which was consistent with contemporary medical practices, was well matched by the engineering teams who built and flew the spacecraft.

The biomedical requirements for the Mercury astronauts were established by NASA life scientists and an advisory group of prominent physicians. Facilities and personnel of the Department of Defense were used to conduct psychological and stress testing of the candidates. The screening and test procedures of the Mercury astronauts were later used for the selection of Gemini and Apollo astronauts.

The medical success of Mercury demonstrated that humans could exist in the space environment for short periods. These important results showed that there was no loss in pilot performance capability. All of the measured physiological functions remained within normal tolerances. There was no evidence of abnormal psychological response. The radiation dose received for the short flights was considered medically insignificant. There were some changes in heart rate and blood pressure that persisted briefly after the landing, but normal physiological function was shortly resumed.

During Gemini, the flights were extended to 14 days and more extensive observations were made of the stress of spaceflight. In that period, the medical results showed that even in the longer periods there was no performance decrement. The main physiological change was a postflight decrease in the response of the cardiovascular system to rapid postural changes. Standing up quickly caused a light-headedness which lasted for a couple of days. Also, the first reports of bone mineral loss were made, and a small decrease in red blood cell mass was noted. These studies were the medical basis for the planned Apollo missions.

In the Apollo flights there was a wide range of medical events during the missions, but the only significant finding was the high metabolic cost of performing extravehicular activity. During Apollo, the ground crew surgeons observed common medical events such as viral upper respiratory infections, dehydration, urinary tract infections, cardiac arrhythmia, fatigue, space motion sickness, and nausea. There were extensive postflight measurements made on the cardiovascular system, and nutritional and immunological studies were done. Some concern emerged about the possible vestibular problems of motion sickness. There was no medical show stopper.

During this early period of measurement and observation, there was an opportunity to survey the range of physiological and anatomical changes that take place during human flight. On the Skylab missions, some of these changes were dealt with quantitatively. The first U.S. physician (Dr. Joseph Kerwin) was flown on the Skylab 2 mission. The large interior volume of Skylab provided an opportunity for extensive testing, and we saw the initial focusing of scientific theories and questions about the data. The crew was used for performing biomedical tests, especially investigations dealing with the brain function (neurophysiology) and the heart (cardiovascular). The physiological data obtained suggested ideas of a more general response by biological organisms.

While the biomedical answers were the central theme of life sciences, there was also a parallel effort by the medics to work with the engineers in designing the life support system and related equipment. The life scientists were heavily involved in setting standards and requirements for the cabin atmosphere and the limits of toxic agents. They prescribed the diets and monitored the preparation of the food. They set standards for cleanliness and toilet care. They helped to design the first space suits and developed a human factors effort that guided the engineers in the man–machine interface.

In 1975, the Apollo–Soyuz mission was conducted as a joint venture between the U.S. and the U.S.S.R. The program was a demonstration of the ability to transfer crew from one space vehicle to another. During the mission there were several life sciences experiments performed dealing with microbiology, immunology, and dosimetry. During the re-entry phase, the crew was exposed briefly to toxic nitrogen tetroxide, but this was handled expeditiously and there was no medical consequence.

Animals were used on short duration rocket flights as a precursor to human beings. The Soviets preferred the dog because of their extensive experience with those animals. The U.S. has preferred the monkey because of its close physiological relation to humans, especially for

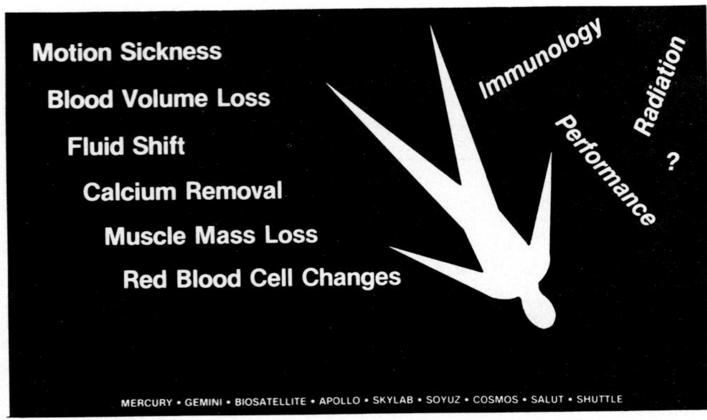

Fig. 5-1: The human in space experiences a variety of physiological changes. NASA is working towards understanding how these affect the ability to work in space and is developing effective medical countermeasures.

cardiovascular responses. Monkeys were used as early as 1948 in the Blossom missions. Aerobee 1, 2, and 3 carried monkeys in 1951 and 1952, and in 1958 Able and Baker were flown on Army medical sounding rockets. Ham, a celebrated chimpanzee, flew on the Mercury capsule, and was followed later by Enos who went for three revolutions. The animals helped to secure the path for our successful manned missions.

Free-flying biosatellite missions in 1967 and 1968 carried a variety of organisms from simple bacteria, to insects, invertebrates, seeds, and some simple preparations like bacteriophages. On a third biosatellite mission in 1969, a nonhuman primate (macaque monkey) was flown with its own life support system, and automated measurements were made on the brain, blood, and urine. This mission proved to be very difficult, but a good deal was learned about developing a complex automated life support system. During this period we learned that the two elements of space affecting biological organisms are weightlessness and radiation, and that organisms react with different tolerance.

Recently, two squirrel monkeys and a dozen rats were flown on Spacelab to verify the performance of the animal life support systems. All animals survived the flight in excellent condition.

Another aspect of life science was introduced through the lunar and planetary program. This program initially involved the search for organic molecules on the moon and extraterrestrial life on Mars. The samples of the moon were proven to be absent of organic chemistry; scientists clamored then to investigate Mars. As a challenging national goal, this enigmatic problem became one of the most fascinating and difficult technical feats by NASA: how to detect life for which there was little theory on a distant planet about which there was little information? Many schemes were suggested which culminated in a complex set of interlocking experiments on board the Viking missions, involving chemistry, microbiology, and a photosynthetic investigation. The results, while not conclusive, have led most biologists to believe that Mars is uninhabited now. We are left with the questions of why not and what is the frequency of life in the cosmos?

The life sciences program has been characterized by extraordinary breadth and very specialized interests of the individual scientists. The medical responsibility has required great prudence and judgment, while the exobiological goals have required great flights of the imagination.

TODAY'S DIRECTIONS

Currently, NASA's life sciences program has three major thrusts: one is to support man in space, the second is to understand the effects of microgravity on biological organisms, and the third is to deal with the possibility of extraterrestrial life and the origin and distribution of earth's biota. Life scientists have one common theme—understanding the nature of life either for practical needs, such as in medicine, or for esoteric reasons, such as in space biology or biogenesis (origin of life). Generally speaking, contemporary biology and medicine are in their most fruitful period. The recent discoveries in biochemistry, genetics, physiology, microanatomy, and plant physiology enjoy a very large research audience and command a significant part of today's scientific literature. Unlike some of NASA's other scientific endeavors, such as planetology or aeronomy where NASA plays a major role in determining a whole field, NASA's life sciences program is very specialized, and deals with only a limited segment of the life sciences community. However, with the expanding opportunity to come from the space shuttle and space station, there will very soon be a strong interest by the large life science community to use space for its research needs. NASA's golden age of life sciences lies ahead.

The program will be discussed in six areas: space medicine, space biology, life support systems, exobiology, global biology, and flight investigations.

SPACE MEDICINE AND PHYSIOLOGY

The achievements in manned spaceflight have been magnificent. Both the U.S. and the Soviet Union have demonstrated the human's adaptability to the microgravity of the space environment for long periods, and for a wide variety of people. The most dramatic short-term influences on humans are the malaise and illness experienced by astronauts in the first few days of spaceflight, resulting from motion sickness, and the changes in a cardiovascular system that is no longer experiencing the constant *g* force. The problems of long-term flight are largely bone demineralization, radiation, and unpredictable psychological problems. Presently, we are entering a phase of extended human travel at micro *g* levels with only brief periods of hyper *g* during launch and landing. We frequently use the term "weightlessness." However, this really denotes the subjective sensation. The movements of the crew members produce some small *g* force on themselves. On an extensive mission in

TIME COURSE OF PHYSIOLOGICAL ADAPTATION TO WEIGHTLESSNESS

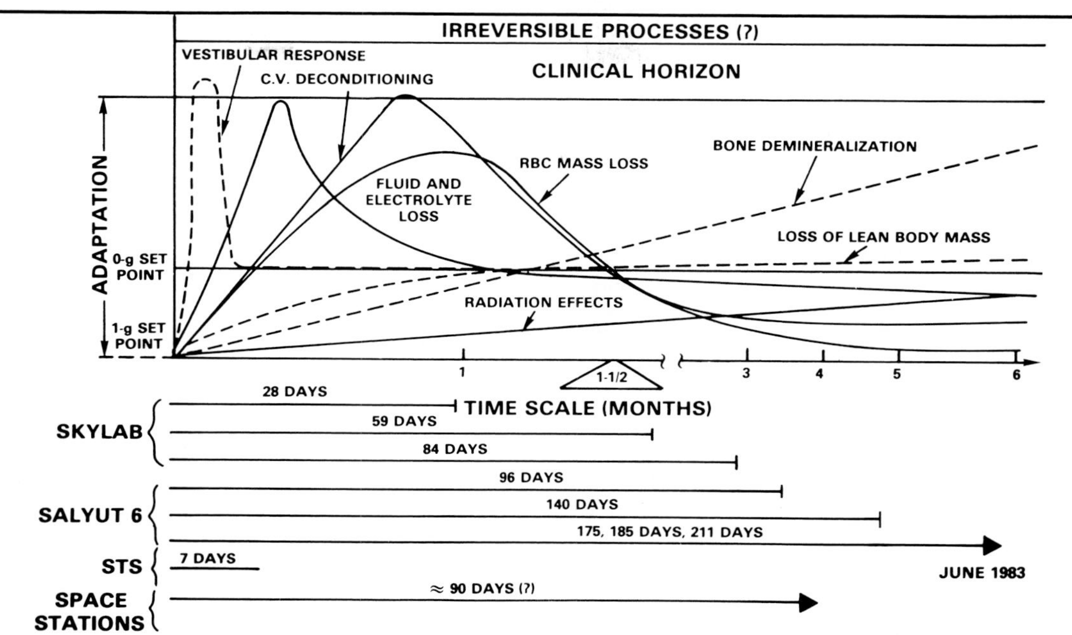

Fig. 5-2: The human physiological changes during spaceflight begin the first few days. Most changes are compensated for within the first month or so, but some such as bone demineralization or radiation effects are continuous.

low-earth orbit, such as the space station, the crew will be exposed mostly to hypo *g* unless the habitability module is "spun-up."

The Soviets have recently reported results of a 211-day Salyut earth-orbiting mission with no serious irreversible change in general physiology. They have considered a manned Mars mission which is viewed by experienced space physiologists as difficult, but plausible. In the next century, we will see humans travel throughout the solar system, possibly at prolonged acceleration above one *g*. This will have its special associated problems and their solution will be different from those of today.

In the first few days of spaceflight, many of the astronauts and cosmonauts report a group of distressing symptoms that have come to be called space adaptation syndrome. Between one-third and one-half of all space travelers experience some degree of distress. It consists of a malaise and can be accompanied by headache, cold sweat, pallor, stomach awareness, nausea, and vomiting. It differs in its time progression and intensity from individual to individual, and prediction of its occurrence has been poor, both by ourselves and by the Soviet physiologists. This phenomenon is frequently associated with rapid head movement and optical disorientation.

Current physiological opinion focuses on a theory of sensory conflict. Human spatial orientation is normally achieved by interpreting signals from several sensory modalities. The eyes give visual cues, the proprioceptor pressure sensors in the skin (bottom of the feet or seat of the pants) give cues, and stretch muscles in the neck contribute to the signal, but the main organ is the small vestibular organ in the inner ear. Acting as a small inertial sensor, the organ can detect small changes in pitch, yaw, and roll, which then go to the brain where they are interpreted so as to allow correction for all ongoing motor activity. There is also a small linear detector organ called the otolith, for detecting forward motion. The current theory is that in the lower *g* environment the gravity vector is removed from this organ, and that this acts as a less damped and more sensitive system. Sensitivity threshold is believed to be changed. What is not clear is how this triggers the symptoms of "motion sickness." The expressed symptoms are common to several other human maladies including, gastroenteritis, anxiety, influenza, pregnancy, and also other forms of motion disequilibrium. The body is probably using some archaic form of rejection to this hostile condition, and what we are seeing is a pattern of avoidance. Neurophysiology is still unable to explain this. Space sickness lasts only a few days. By the fifth day all is back to normal; all crew members of the shuttle flights have returned without symptoms of distress. Our concern is that during some flight the illness will prevent a crew member from performing his task. NASA has a very active research program in this area.

The solution to this problem lies in several lines of work. One is to develop better predictive capabilities when enough space passengers have been measured. We already have some statistics. There may be some common test that could reveal who are the more susceptible individuals. Our laboratory tests that expose crew members to rotating chairs, other motion-produc-

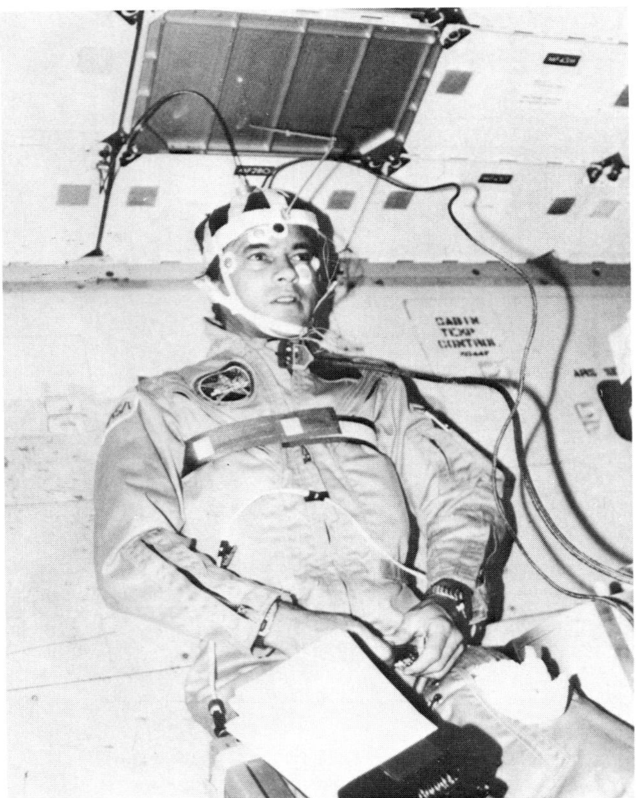

Fig. 5-3: November 1982: shuttle flight number 5. Astronaut Joseph P. Allen, IV, one of two mission specialist astronauts for STS-5, participated in a biomedical test in the mid-deck area of the earth-orbiting space shuttle Columbia. A series of electrodes was connected to his face for monitoring of his response in zero gravity. He was assisted in the test by Astronaut William B. Lenoir, the flight's other mission specialist. Dr. Allen is wearing the multi-pieced constant wear garment for space shuttle astronauts.

ing devices, or free-fall on a descending aircraft have not yielded strong correlation between who becomes ill on the ground and who becomes ill in space. If we could find a reliable test, those crew members could be screened or at least assigned duties that are not needed until late in the mission. Physiological adaptation to the symptoms might be accomplished through training or previous exposure. Currently the crew members are trained in a KC 135 free-fall aircraft for short periods. Also, using aircraft maneuvers (aileron rolls) they are able to experience short-term exposures to weightlessness.

Biofeedback, a physiological conditioning of the sympathetic nervous system, is a new and promising technique. During several training sessions the subject is taught to obtain control over several involuntary reactions such as heartbeat and blood flow. This is then used to obtain control over biological functions, and finally to retard or suppress the motion sickness symptoms. Biofeedback training has been shown by the U.S. Air Force to be useful for pilots suffering from air sickness. This will be tried on upcoming shuttle flights. Lastly, the use of pharmacologicals is believed by many clinical physiologists to hold the most promise. The current use of a mixture of scapalomine and dexidrine is controversial. Scapalomine is known to suppress the symptoms of nausea, and also to induce sleepiness and lack of appetite. The dexidrine is added as a stimulant. At the levels used, so as to maintain alertness of the crew, there is some question of its value as a preventative drug or as a counter-measure. Some crew members have reported benefits. Ultimately, the physiological basis of the problem must be understood for any pharmacological approach to be effective. Then we might be able to tailor the right substance and the right dose.

A second physiological change that occurs with human flight concerns the cardiovascular system. With the lowered *g* vector, the blood that is normally in the lower extremities becomes pooled in the upper torso. The body, sensing this anomalous increase through baroreceptors, reacts by lowering the blood volume through normal kidney control of fluid filtration. (More urine is produced.) This results in a reduction of blood volume and conservation of the whole blood elements. There is simply a decrease in fluid for the new physiological state. This causes no acute problem until re-entry.

During the period of re-entry when the crew is re-exposed to gravity, the blood is redistributed throughout the body, and in the normal *g* condition is again drawn down to the lower extremities with a consequent reduction of the blood in the brain. The body compensates, but there is some net decrease in perfusion resulting in anoxia to the brain (sometimes seen as dizziness or semiconsciousness). As a counter-measure, we prescribe the pilots and commanders to ingest a large quantity of isotonic saline (salt and 1–2 liters of water), just prior to landing. This has the effect of increasing the blood volume temporarily while the re-entry is taking place. We have seen no adverse effects of this practice. The normal physiological readjustment begins immediately after landing, and appears to be adequately restored within the next day or so. Despite the success of this technique, some cardiovascular physiologists are concerned because we have so little quantitative data. We do not know the margin of safety of this short-term volume decrease. The critically important alert state of the commander during the last few moments of landing the spacecraft requires that we operate from a better quantitative basis than is presently in hand.

As acknowledged by experts in space medicine, the most serious physiological concern of long-term spaceflight is the loss of calcium from the bones. There is a slow removal of a percent or so each month in space of the bone's mineral matrix. Measurements are made by X-ray densitomety or tomography. Studies on board the Skylab showed that the calcium is released from the bone to the blood and removed by urine and feces. This phenomenon is similarly seen in bedridden patients, and also in people with osteoporosis. The mechanism of mineral release is not understood. In normal bone metabolism, the mineral is constantly being replaced by new material. We do not know if the "breakdown"

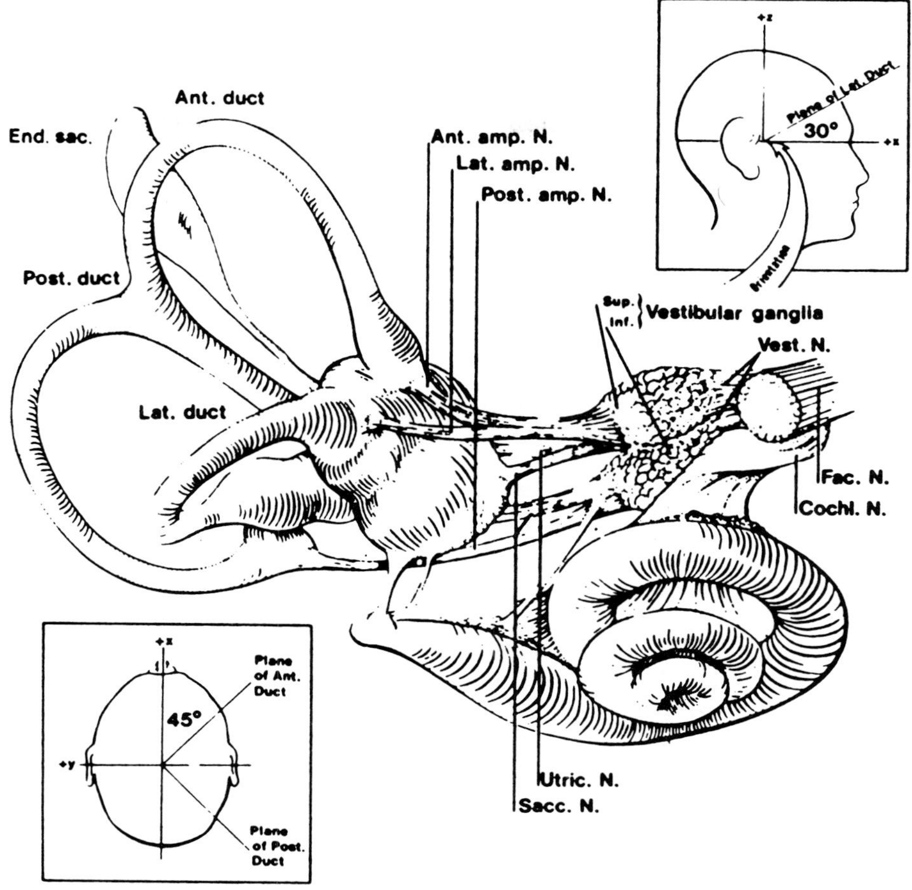

Fig. 5-4: Space motion sickness, also called space adaptation syndrome, is the major problem of short-duration flights. It appears to be related to sensory conflicts in signals coming from physiological sensors, the main one being the vestibular system of the inner ear.

process is speeded up, or if the "build-up" process is slowed. The loss appears to continue throughout the flight, although some recent Soviet data suggest that mineral loss may be slowed during very extended flights. The use of exercise is believed to be an effective countermeasure for long-term flights. During the post-flight period the recovery is reversible. The period for full recovery appears to be about the same as that of the exposure. Since mineral metabolism is under complex endocrine control, it is not surprising that merely increasing dietary calcium does not solve this problem. In fact, the increased intake of calcium is considered a very adverse physiological practice, since it invites the possibility of precipitating a kidney stone. Such an event would be dire on board a spacecraft without adequate facility for handling this serious clinical problem. Many physiologists are also concerned about other salts and electrolytes that are affected by the massive calcium release. The physiological ionic balance is critical for maintaining normal health. While the body has control mechanisms for readjustments to changes, we do not know what systems are being changed or the consequences of these changes.

The real permanent problem for long-term travel is probably radiation exposure. Radiation damage is cumulative for most biological organisms and tissues. While there is considerable variation from tissue to tissue, organisms are intolerant of more than a few tens of rads. The largest exposure reported by the U.S. to date was the 84-day Skylab mission, with a total exposure of under 10 rad. Shielding of most radiation is reasonable; however, high cross-sectional particles of high energy such as iron will present a problem. We have not seen any flight during solar flares or passes through the concentrated regions of the radiation belts. The most serious problems are likely to occur during an extensive planetary mission if there is any unanticipated solar activity. We currently have a good deal of radiation dosimetry data from past flights.

Some other physiological problems of lesser importance are being studied. They include fluid and electrolyte shifts that occur in conjunction with the blood volume changes, and muscle mass reduction associated with lessened physical activity. This may be handled by an adequate exercise regime. There are red blood cell changes that remain unexplained, although this does not appear to cause a problem. Some physiologists anticipate changes in cardiac muscle that may occur on long-term flights due to changes in the work load on the heart. Immunological changes involving the white blood cells that have recently been reported go unexplained.

One of our more perplexing areas to deal with is that of behavior and psychology. This is not so much a problem related to the spaceflight as it is one of prolonged isolation in a small environment. On the present short-term very busy missions, this has not been an issue. The Soviets have expressed concern that during the very long-term missions there could be a serious group interaction problem that is difficult to diagnose. In their extended missions, they place great importance on establishing a high-fidelity communication link to give their cosmonauts a feeling of attachment to their close family members and friends on the ground. They place great emphasis on this area. They bring in important Soviet officials, celebrities, and entertainers to "keep up the spirits" during those long missions.

SPACE BIOLOGY: GRAVITATIONAL PHYSIOLOGY

The applied physiology deals essentially with the health of the human crew members or their food and nutrition in spaceflight. To illustrate how the fundamental and applied problems are interlinked, we might consider a unique but not impractical problem: a trip to visit Jupiter or its moons. Over a hundred people have flown in space at various levels of micro *g*. Subjects on large-scale centrifuges have performed thousands of experiments at hyper *g* for short periods of time. The human is not fragile. He is well-engineered to endure spaceflight, and adjusts to the changes and the fluid shifts. The central nervous system accommodates, the metabolism adjusts, and perhaps the body accomplishes mineral loss suppression. The cosmonauts of Soviet missions have experienced relatively long periods at low *g* levels, but we have only observed high *g* for short periods. Our systems cannot adjust to high *g* forces for long periods without extensive countermeasures. However, given today's poor state of knowledge, it is clear that a human traveling near the planet Jupiter at hyper *g* for long periods could be in serious trouble. There is a practical need to understand gravitational physiology. There is also an interest in this science in its own right.

One interesting theoretical question raised by scientists working in this area of increasing *g* is the limit to the size of organisms that can exist as related to the *g* of their particular planet. There are redwood trees that grow to almost 100 m. Are they limited in their size by the *g* of the earth? What are the biological or structural properties of animals limited by the force of *g*? We certainly know of very large aquatic animals, and have evidence of the giant land-dwelling reptiles of the past, the dinosaurs. The size of these creatures was limited mostly by temperature and food supply. But, what are the limitations imposed by gravity in permitting an animal to bear its weight and retain its mobility and nutrition? This area of evolutionary biology as determined by gravity has been a fertile area of study, enhanced by field observations, but its solution is limited; conventional laboratory experimentation is not very useful.

In the experiments and observations at micro *g* that have been made in plants and animals we know that below the level of the whole organism there are considerable changes. In plants the major change is in the growth cycle. Some changes in leaf epinasty (movement due to growth) have been observed in weightlessness, but not sufficient to suggest interference with the normal function of the plant's physiology. During development there are extraordinary changes in the formation of roots, stem length and diameter, and internodal elongation. Leaf size, flower, and fruit dimensions are all likely to be influenced by a change in *g* during their growth periods. Experiments have been done that suggest some differences among the plant species, but not enough to understand the underlying mechanisms.

Recently, the U.S. has flown some plants on the space shuttle. Pine seedlings, oats, and mung bean seedlings were flown for several days to observe lignification (wood formation). Generally, the experiment showed that plants can be grown in space, but with some effects on their morphology. There was a reduction in lignin content (the woody part). The role of lignin in protein synthesis may be important if plants are to become an important amino acid source of nutrition for extended flight. Two enzymes employed in lignin biosynthesis were reduced. Fortunately, the centrifuge offers a way of simulating the hyper *g* condition for extending our range, and also will be used as a control experiment to obtain 0–1 *g* during flight.

An experiment called the HEFLEX (Helianthus Flight Experiment) is planned for an upcoming flight. An engineering test was performed on the shuttle in the mid-deck locker. This will be used to study the reduction of circumnutation during weightlessness. Circumnutation is the wobble in growth following the earth's off-axis rotation. The study is to validate the laboratory clinostat as a tool for this kind of work.

In the laboratory, we have learned some of the dynamics of the processes of response to the *g* stimulus. The biochemical processes of plants do not act independent of one another. Normally, plants respond to changes in the direction of the gravity vector; root tissues grow down and stems grow up. Recently, one investigator learned that this process is light dependent. Plants grown in the dark do not respond to changes in the direction of *g*; however, brief flashes of light will trigger the bending process.

The mediator of gravity sensitivity is becoming more evident. We have been studying the difference between the effects of various growth hormones. Some are gravity sensitive while others are insensitive. There are asymmetric gradients of protons and calcium ions that parallel the growth patterns of response to gravity, known as geotropism. This implication of calcium gradient in the root tip as the gravity sensing mediator is an important idea, and suggests new methods of simulating weightlessness. It has been shown that the blade of the leaf of some plants is the part that senses gravity. The hormone auxin regulates and affects the pulvinus of the leaf by regulating the ethylene production. Auxin-induced ethylene synthesis appears to be the link between blade perception and leaf angle response to gravity.

Small inclusions called amyloplasts, which exist in special sensitive cells, have been studied by Dr. Arthur Galston of Yale University. Dr. Galston has succeeded in isolating the organelles of cells in a viable condition and studied their fall as vertically mounted under the microscope. The amyloplasts have a net negative surface charge. Using ion techniques and fluorescent indicators, it was determined that the calcium is bound to the charge shell around the amyloplasts. Here again, calcium is linked as the mediating ion. He has theorized that the settling of the amyloplasts redistributes the calcium which may be the basic sensing mechanism.

Recently, one scientist has succeeded in a seed-to-seed demonstration of a plant (arabidopsis) grown on a rotating clinostat. We have learned that CO_2 and moisture control are critical for the success of this experiment. Viability of the seeds is currently being tested. At a recent conference the Soviets reported their success in performing this seed-to-seed full growth cycle on board the Salyut.

Other investigators have shown that in addition to the *g* force, other stimulators are needed for successful natural plant physiology. Friction and flexure are mechanical stresses that also can change the pattern of growth in stems; this may be due to a change in ethylene production. One theory of gravity stimulation of cells is that there are deformational shifts in the macromolecules that control permeability of the membrane to the calcium ion. This would result in calcium migration into the cell against an actively controlled concentration gradient. As the calcium concentration rises, it binds to a modulator protein called calmodulin. This stimulates ethylene production of the surrounding cells. The result is an activation of growth hormone production at one place, and a suppression at another place, which produces the asymmetry and therefore the bending.

In the animal kingdom, we have been examining a few lower order organisms, such as spiders and insects, for sensitivity to gravity, but most work is being done on mammals. One scientist is studying the lyriform organ on the leg of the spider which appears to be gravity sensitive. This investigator is studying changes in the heart rate following centrifugation. Current theory suggests that the spider's gravity response system operates by sensing the strain in its external skeleton. It is suggested that there are reflex mechanisms of the heart which respond to the *g* stimulus, producing a blood pressure change in the legs. This allows the spider to maintain postural control. It is not surprising to find special gravity-sensing structures that have evolved in different animals for their special problems of sensing *g*.

Fruit flies have been studied on a Cosmos flight, and recently bees were flown in the shuttle. A curious conclusion by one scientist comparing the flight of the two different insects suggested that the bees were unable to fly in weightlessness, whereas the fruit flies attempted to fly but through their poorly controlled flight they suffered injury due to impact against the experiment's housing unit. These enigmatic results need to be repeated under more controlled conditions.

Several investigators are using a rat suspension model developed by Emily Morey-Holten of Ames Research Center. In this preparation the forepaws of a restrained rat are permitted to bear weight but not the hind paws. The animal is suspended by various devices, and certain organ systems, bone, muscle, blood vascular, etc. are studied. There is some dispute over the rat as a suitable model for the study of change in humans. There are anatomical differences that may cause the systems to react differently. Nevertheless, since animals will be flown in space for other experimental purposes, it is important to establish their baseline, independent of those results that will be gathered on humans.

Elaborate studies using this model have shown that in rodents calcium loss is somewhat controllable by dietary intake. This has been corroborated by several scientists following the time course of bone loss for several weeks and comparing these data from suspended animals to controlled organisms. Using this same preparation, and studying the histology, we have determined that the bone cells on the surface are the sites most affected by the nonweight-bearing condition. Electron micrographs reveal a reduction in communicating junctions between adjacent cells. This could mean that variations in mechanical stress affect the entire musculoskeletal system.

This same rat model has been used to study haemopoeisis (fabrication of new blood cells), and we have found transient changes in haemoglobin, reduced production of red blood cells, decrease in the average size of the cells, reduced blood volume, anemia, and a postsuspension increase in the number of white blood cells. These changes are all compatible with data from human spaceflight. It appears that this immobilized rat may be a useful tool in which to study the hematological effects of weightlessness.

Dr. Pace and Dr. Smith of the University of California at Berkeley have devoted years to better understanding the scaling laws that govern metabolic rates. How does the change in size of an animal relate to the rate of its metabolic processes and is this affected by gravity? They have studied a variety of organisms including rats, guinea pigs, hamsters, and rabbits and established that gravitational loading very likely influences the scaling of these metabolic rates. They postulate that in weightlessness this will be considerably less than the classical 0.75 power for normal scaling of the metabolic processes.

Dr. Muriel Ross of the University of Michigan has been studying the vestibular system in the rat, particularly the otoconia of the saccule and utricle. These are microstructures of the inner ear. The classical view is that dense crystals simply add mass to the otoconial membrane, making the sensory cells more sensitive to linear acceleration. Dr. Ross' eloquent ultrahigh resolution micrographs have shown a very elaborate structure in the otoconia, far more complex than previously considered. There appear to be highly ordered composites of organic and inorganic materials quite different from pure calcite. The crystals may exhibit a piezoelectric effect by altering the adjacent field during linear acceleration. If this is the case, they may not actually move over the underlying receptor area as classical theory explains.

We are beginning to see some results in the study of gravitational effects on embryology. One of our investigators who is well-versed in the individual control timed-events of the developing rodent embryo has been studying embryos at hyper *g*, especially the developing central nervous system. He anticipates that at less than one *g* the alterations in the cardiovascular system are likely to affect the formation of the peripheral and central integrating elements of the vestibular system. In a recent Cosmos flight this experiment was performed and is being examined.

What We Need to Know About Gravitational Physiology

Starting from the most basic question we need to know the answers to these general questions. As is always true, each question leads to another question:

- What are the fundamental sensing systems of gravity in biological organisms, e.g., hair cells, starch grains, membrane deflection, piezoelectric effects?
- How do these sensors work?
- How do they mediate their signals to the rest of the organism?
- How do these systems develop embryologically?
- How did they arise evolutionarily?
- What are the physiological (or anatomical) consequences of changes in *g*, on what time scale, and to what magnitude?
- How does the whole organism compensate for the changes to hypo and hyper *g*?
- Are the changes reversible?
- What countermeasures can be employed or developed for the safety of human spaceflight?
- Are there changes in *g* that can be beneficial to living organisms? (We tend to think in terms of the negative aspects. Where can we use this low *g* to our advantage? There have been several suggestions for special medical treatment in low *g*. One such idea is in the treatment of massive burns where the pressure on the body interferes with the healing process.)

From my perspective, most of NASA's biomedical program is linked to gravitational physiology. Most of our research work has been done in ground-based laboratories. In recent years, with access to spaceflight limited only to short-duration shuttle flights and the Soviet Salyut and Cosmos missions, we have used the time to develop ground-based laboratories, theories, and techniques. Models and methods have been strongly supported. Now, with the planning of extensive flights on a space station, there is a great opportunity to develop a permanent clinical, physiological laboratory in space to make extensive human measurements, and also to develop a special space laboratory for highly controlled experiments in space biology to answer some of these questions.

LIFE SUPPORT SYSTEMS

One of the more interesting technical challenges of keeping humans in space is providing them with an adequate life support system. What are the needs and what are the problems? Besides the obvious needs for air, food, and water, there are some very basic requirements that are levied on the spacecraft designers. Adequate sanitation is of utmost concern and has been criticized by crew members. Even for the short duration missions, provision for toilet care is an obvious necessity. Other requirements include concern for toxic or microbiological contaminants, adequate removal of unpleasant vapors, temperature control, noise limits less than 80 dB, light levels for vision, and human factor design for ease of operation.

The current missions are stocked with foods that are basically the choice of the crew members and based on sound dietetic practice. They are essentially dehydrated foods. We are running the mission somewhat like a camping trip, taking up food and returning with the waste. Potable water is supplied on board the shuttle as a by-product of the fuel cell. Oxygen is supplied by fresh makeup gas from on-board gas tanks. The respired carbon dioxide is removed by passing the air over a bed of lithium dioxide where it is converted into the carbonate, thereby trapping the carbon dioxide. Crew members change the lithium dioxide canisters several times during the mission.

For more extensive missions it will become desirable to have a larger supply of water and air. In our current programs we are exploring the feasibility of reconstituting the water by flash evaporation, or by distilling it at very high pressures and temperature where the physical solubility characteristics of water change, and distilled water is produced. For carbon dioxide removal, it is possible to reduce the CO_2 to free carbon and recirculate the O_2. In another scheme, we have studied the reduction of CO_2 to carbon and converting it to methane which could be vented from the space craft. Details of this physical chemistry have been worked out, and breadboards being built will soon be tested.

For missions of considerable duration and large

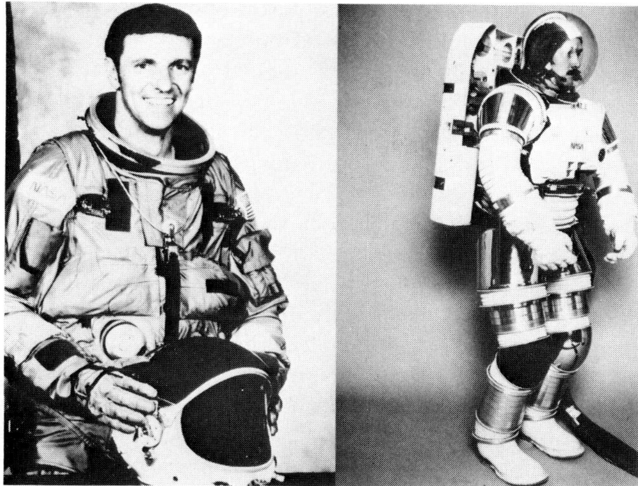

Fig. 5-5: Extra vehicular activity will become an important part of the space station. NASA is developing an 8 psia hard suit, which has complete mobility, is easily put on, and has interchangeable sizing for different individuals.

numbers of crew members, it is reasonable that we will supply some of the nutrition from synthesis on board the spacecraft. This may be done using plants for photosynthesis, or if we can make some chemical breakthrough this might be done by chemosynthesis. Our current program is called CELSS (Controlled Ecological Life Support Systems). The goal is to develop the knowledge to design and test such a semi-closed system. The blackboard equations for converting wastes back to food are rather elementary. CO_2 is reduced to carbohydrates by using plants and sunlight, thereby producing food; and the human wastes are used as fertilizer to restore the nitrogen and other essential trace elements. This constitutes a small mini-ecosystem with man as the CO_2 producer and the plants with sunlight as the CO_2 user and oxygen producer. But converting this to an active system is no simple task. We know of only one successful long-term biological operating system. That is the earth as a whole. It is not just a matter of miniaturizing the earth because the scaling does not work; sources and sinks are all different. There are no oceans and large volumes of atmosphere. Indeed, we are not certain that such an artificial system can be made to work practically for any long period of time. In such a closed (or semi-closed) system, a major engineering problem is the control subsystem, the component that governs the rates of the process and the flow characteristics of the system. The necessary measurements and feedback loops to maintain this dynamic system are as complex as any spacecraft by itself, and will require painstaking details and numerous tests.

Another technically interesting area involves the operation of EVA (extravehicular activity), known as the space walk. This excursion into a vacuum is not a trivial exercise of jumping into a space suit and into space. However, these EVA's are very important for upcoming missions and for handling construction and repairs of spacecraft. The current suit employed operates at 4.7 psia. This pressure is a compromise between using a higher pressure inside of the soft suit which tends to expand the suit, making it too rigid in which to work, and using a lower pressure which results in "physiological bends." The "bends" are caused by nitrogen coming out of solution of the blood due to a change in solubility at lower pressures. This produces bubbles in the blood stream and great pain in the joints. We are sufficiently close to the margin of safety so that the current procedure calls for the crew member to breathe pure oxygen for 90 min with the suit on, prior to donning the helmet. This procedure washes much of the dissolved nitrogen out of the tissues and gives several hours of protection against the "bends." The process is cumbersome and time consuming. Another procedure being tried is to lower the cabin pressure to 10.2 psia prior to suiting-up.

Currently under development is an 8 psia hard suit with very flexible joints that will avoid the "bends" problem by holding a high pressure within the suit. This suit looks very promising. One problem still unsolved is that of an adequate space glove. Working in space may frequently require manual dexterity and handling beyond what a mechanized hand can do. High pressure gloves are difficult to work against, low pressure gloves tend to restrict circulation; the glove problem may be around for a while.

EXOBIOLOGY: THE ORIGIN OF LIFE ON EARTH

For two decades, NASA has supported and developed a program called "Exobiology." This name is a holdover from the past when the program was stimulated by a desire to search for extraterrestrial life, hence the name exo and biology. In parallel with several lunar and planetary missions—Apollo, Viking, and Voyager—there emerged a laboratory program whose object was to understand the origin of terrestrial life. This has been extended to include the evolution and distribution of life and life-related molecules on the earth and throughout the universe. It is intended to find out how the direction of chemical and biological evolution was affected by solar, planetary, and astrophysical phenomena.

This question of the origin and prehistory of terrestrial life is one of the great scientific questions of our age. There is extreme difficulty in reconstructing the chemical events that lead to the origin of life. We have assembled several pieces of the puzzle.

Life on earth (a self-replicating system) appeared 3½–4 billion years ago, in the earth's earliest history. Current evidence suggests that in this solar system the only planet to have developed life is the earth. The two Viking missions to Mars had as a primary goal the search for life. There were three active biological experiments performed on the Mars soil sample and one chemical test for organic compounds. The results of the two spacecraft at different locations on the Mars surface did not give us a conclusive answer; however, all of the experiments worked. What we found was that there appears to be no organic material on the Mars surface. Life without organic chemistry is not very plausible. The biological experiments all gave some kind of signals, but the results were readily explained in simple chemical terms based on what we found in the Mars soil. It appears that the Mars surface has a good deal of iron in a clay component in the form of an oxidizing substance. This oxidizing state is probably due to UV radiation of the very dry condition of the soil. Such a condition would make life difficult for the survival of microorganisms and would account for the bizarre "biological results" and the absence of organic chemistry on the surface. Many biologists were disappointed that either there was no life or that the results were not more unequivocal, but nature is known for its mysteries.

The chemistry of all terrestrial life from its origin is believed to be similar to modern biochemistry, based on proteins made of polymers of amino acids for structural and functional units, and based on nucleic acids for the information bearing units. Recent laboratory studies suggest that the nucleic acids can also serve as catalysts in certain reactions. This uniformitarian principle of how the chemistry of life works is becoming the foundation of how life first started on the earth, and while we are getting closer to the source of the origin, we are still not there. There is still no detailed theory of the chemical details that led to the actual first self-replicating units that might have been called life!

The chemistry of the building blocks of life is relatively easy to synthesize in a test tube. The conditions require an adequate source of reduced carbon, some water, and a form of energy (light, electrical, or even heat will work). Some complex molecules have also been synthesized in the laboratory. Going from laboratory results to what actually happened in nature is misleading. In the laboratory, some scientists use water as a solvent, some use dry conditions, some hot, and others need cold conditions. Our problem is that we do not know what the earth was like 3½–4 billion years ago when life was being formed, so that we can reconstruct those conditions in the laboratory.

All of the known living organisms have similar chemistry, which leads us to believe that the origin of life on earth was a single chain of events that happened only once. This common chemical basis is striking in light of the vast possible variation offered by the world of organic chemistry. An example of this is the universality of a molecule called adenosine triphosphate. This is a relatively simple molecule capable of exchanging energy. The last phosphate bond is a high-energy bond that can be used to do work. This same molecule is involved in such diverse physiological processes as muscle contraction, vision, moving water up in plants against the pull of gravity, the light of the firefly, and the flagella which move sperm. Adenine triphosphate is present in all living organisms. It appears that once nature found this kind of "rechargeable battery" it no longer needed to solve the problem again. (The teleological expression is only intended for simplistic illustration.)

Recent laboratory work suggests that inorganic minerals played a key role in early organic chemistry. The most recent work suggests the importance of clay minerals in the early pre-biological synthesis. Clay offers a magnificently complex structure capable of providing a wide variety of chemical catalytic surfaces and states of solubility. Its lattice can hold large or small quantities of water with only simple changes in the local conditions of pH, temperature, and particular ions. Some scientists believe that the earliest functional polymers were attached to the walls of primitive clay cells, concentrating the chemistry of the newly emerging self-replicating molecules.

The origin of the genetic code was probably the most striking event giving rise to a self-replicating system. The genetic code is a universal transcription of information from one individual to another. This is the very heartbeat of contemporary molecular biology. Dr. Leslie Orgel of Salk Institute has recently been able to demonstrate very short segments of nucleic acid (five units long) that are able to self-replicate. The rate of the chemical reaction is a strong function of the ionic character of the medium.

No other life has been found in the solar system, but the relatively high organic content of carbonaceous chondrites and the discovery of interstellar organic molecules suggests that organic chemistry is a common occurrence in the solar system, and in the universe.

Life has been found in unusual and extreme terrestrial environments, e.g., in rocks, near hot geysers, in radiation cooling waters, and at extremely low temperatures. This suggests the very high adaptability of life to its environment. This is not to suggest that life can survive under any conditions. Rather, it suggests that once life is started under some relatively stable conditions, it can keep up with slow changes of those conditions and be carried to some rather extreme examples. When the changes are either too abrupt or too severe, it is likely that the particular life ceases in that changed condition.

Recent studies of terrestrial events that coincide with dinosaur extinction suggest that large-scale planetary phenomena can interfere with the orderly evolutionary process. These catastrophes are important to the overall process in determining the resultant biota.

Many scientists believe that other stars have planetary systems and that the probability that living systems started elsewhere is very high. With this assumption it is reasonable that intelligent life is sufficiently common in the cosmos to warrant a search for coherent signals among the incoming radiation. It is assumed that an intelligent civilization more advanced than ours will be heavily endowed with the technology for carrying out high-level communications over long distances. Technology has reached the stage where this is possible, and NASA is supporting the development of the technology for a program called SETI (Search for Extraterrestrial

Intelligence). This is basically an eavesdropping project in which the incoming radiant energy is searched for coherent signals. This is planned as a part of the listening aspects of on-going radioastronomy at places like the DSN (Deep Space Net) antennas and at the Arecibo facility in Puerto Rico.

GLOBAL BIOLOGY

The study of the earth's biosphere is one of the major scientific challenges of the next several decades. The use of remote sensing, new methods of chemical analytical techniques, and large capacity for handling data all make this possible. Global biology is the study of the interaction of the earth's biota on the physical and chemical makeup of the earth, and conversely, the effects of the earth in determining the existing biota. It is a contemporary study which begins with today's conditions, and will eventually translate in time back to the beginning of life and the planetary environment that made life possible. This science of the biosphere includes aspects of the radiation environment, the atmosphere, the oceans, and the land mass including polar regions, lakes, and rivers. In this sense it is a highly interdisciplinary science. Its ultimate goal is to understand the interconnections between the living and the nonliving world and to determine the vulnerability of living systems.

The program is new and most work has been to establish plans, goals, and priorities. Conceptual modeling will play a vital role in determining the measurements that need to be made. Of special interest are the biogeochemical cycles of nitrogen, carbon, phosphorus, and sulfur which cycle through both the living and nonliving systems. Biogeophysical processes determine some of the climate parameters which, in turn, determine the support of the biology. Some biologists have postulated a control mechanism on the part of the biology in helping to fix certain conditions which promote life-supporting processes. This is known as the Gaia hypothesis.

FLIGHT EXPERIMENT PROGRAM

Central to any NASA scientific program is its flight experiments. In life sciences there is a variety of activity.

U.S./U.S.S.R. Cooperation

The Soviet program is based on two sources of data, biomedical data on the cosmonauts from the Salyut missions and dedicated biomedical flights of the Cosmos class. Through our joint exchange program U.S. scientists have become investigators and co-investigators on these Cosmos series. Until recently, there has been an annual exchange of data at these formal joint meetings (halted as part of the Administration sanctions). Both the U.S. and the Soviets have benefited from these direct exchanges. Biomedical data from the Soviets' long-term missions are very useful for our planning a space station. In return, NASA's excellent university affiliations allow the Soviets direct access to a broad range of physiologists across this country. In the opinions of all of the scientists on both sides it was felt there was a great deal derived by each side from these mutual exchanges. When politically possible, they should be resumed.

Shuttle

This is by far the most important database we have. We have flown almost 50 crew members, and in the next several years this number will grow to several hundred. Until recently, the low number of individuals exposed has made statistical treatment impossible. Now we will begin to analyze the quantitative nature of the results and to deal with the wide variations of physiological response due to differences in age, sex, size, constitution, and genetic type. The data from crew members of the shuttle flights are being assembled as a longitudinal study. This is a very long-term biomedical record of a large sample that is used to observe statistical changes of a cohort group of individuals. We are interested to see if there are any important irreversible changes that take place as a result of either long periods of spaceflight or repeated exposures. A question that remains open is whether there are synergistic effects by several changes that occur concomitantly. Does the sensitivity to radiation change during the time that calcium is being released from the bone? Does a disturbance of the cardiovascular system affect any long-term capability of the immune system? There is no evidence to suggest the synergism, but biomedical systems are so tightly coupled that many physiologists have wondered if this may be happening at a very low level of activity. When we are able to observe an aging population of astronauts we may possibly see changes in life expectancy, rates of heart disease, oncology, osteoporosis, or some indicators of early permanent physiological changes due to flight. We may also find that spaceflight is no more of a hazard than air flight has been to commercial pilots.

Spacelab

The European Spacelab is being outfitted as a complete biomedical laboratory for experiments planned in the next few years. This pressurized tank is ideal to house the human subjects, animals, and plants that are a part of the planned investigations. These missions are dedicated missions that introduce complete studies of the selected biomedical subjects, as compared with the abbreviated tests that we are able to do on the mid-deck during a normal shuttle flight. In the first of these dedicated missions, about 20 investigator teams will be involved. The experiments selected are of eight areas as shown below. They are done on humans, or where that is not possible because of the invasive nature of the experiment, the experiments are performed on laboratory animals: the white rat or a small monkey. Also, there will be experiments on some plants and one dealing with fertilized frog eggs.

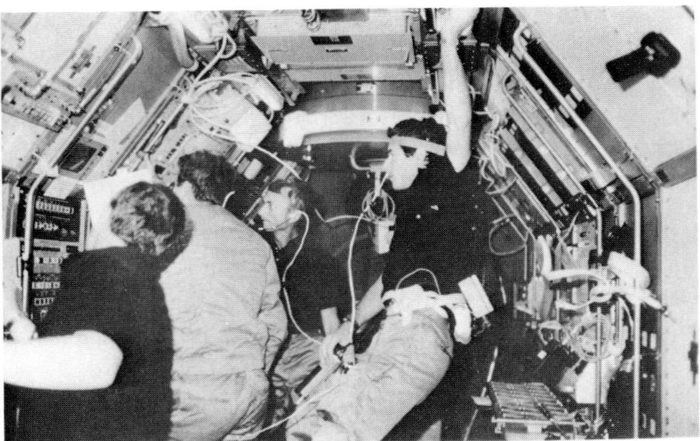

Fig. 5-6: The crew of Spacelab I (left to right): Robert A. R. Parker, Byron K. Lichtenberg, Owen K. Garriott, and Ulf Merbold. Spacelab IV includes a dedicated biomedical laboratory to carry out medical and biological experiments on board the shuttle. The new laboratory is being constructed at the Ames Research Center and Johnson Space Center. Payload specialists will carry out experiments on themselves, laboratory animals, and other biological specimens such as plants and amphibian eggs.

The primary objectives of this first mission are to determine cardiovascular adaptations, vestibular physiology, and to look at the acute fluid shifts in humans. Our secondary objectives are observing reduction in red blood cell mass, negative nitrogen and calcium balance, suppression of the immune system, geotropism in plants, and fertilization and early development of the amphibian egg. The short-duration flights of the space shuttle limit observing long-term biological effects such as calcium loss or radiation damage. For these shuttle missions we decided to concentrate primarily on the short-term biomedical problems of human spaceflight, and to select a few important and exciting biological areas. One of the primary goals of the first mission is to validate the animal as a surrogate for studying human physiology in space. While the final knowledge is to benefit the human, experimental animals provide an ideal source of data. By selection they are a more controlled group. Their genetic background is less random than that of humans. They are all of a similar size and have a common physiological background. We can control their diet and behavior more than we can with human beings; but most important is that we already have a large database of the "normal limits" for these animals.

The recommended investigations for Spacelab are presented in Table 5-1.

Space Station

For the space station, the demands on life sciences will change. There will be two central themes, one of maintaining health care on board, and second, examining the long-term effects of flight on biological systems. The consideration of health maintenance will influence any space station concept because of its contribution to sustaining the life and well-being of the crew. A space station will bring NASA into a new era when "industrial activities," long stay times, frequent extravehicular activity, and scientific and flight test activities will be commonplace. The individuals sent into orbit will have a wide range of physical attributes, educational backgrounds, and occupations. The varied activities of such a diverse space station crew will increase the potential for industrial accidents, such as trauma, burns, infections, and psychological problems. Any of these may complicate the physiological changes already associated with microgravity. Infectious diseases will also be of concern in the closed environment of the space station, and methods to diagnose, treat, and possibly quarantine individuals must be developed. Current thinking does not envisage a rapid medical emergency return to earth. Treatment of several weeks duration should be accommodated. If a crew member becomes ill, that person will be treated and cared for in the space station facility—and that care will place demands on the rest of the crew.

Several types of physiological deconditioning in weightlessness have been identified. For this reason, health care must involve not only the diagnosis and treatment of accidents and diseases, but also the prevention of chronic problems and the maintenance of a healthy state. Thus, the health care program should include a physiological/psychological monitoring system, a crew health database for tracking routine exams, recommendations for individual exercise, diet, and rest/work schedules, the monitoring of environmental parameters, the assessment of human performance, and the development of standards for habitability factors.

Health maintenance facilities will be developed in several stages, beginning with the already proven Shuttle Orbiter Medical System (a sophisticated first-aid kit) augmented to provide emergency care. As the number of individuals, lengths of stay, and size of the station increase, the area dedicated to health care and maintenance will expand to include a first-aid station, an exercise area, the diagnostic and treatment facilities of a physician's office, and eventually the equivalent of a two-bed hospital. With some modification, these facilities can be used to accommodate clinical studies that support human biomedical research and health care issues.

Studies that employ a wide range of subjects will be conducted to characterize the flight problems, determine their underlying mechanisms, assess the longer term effects, and suggest countermeasures and/or solutions. Several foreseeable problems in long-term habitation in space include increased propensity for bone fractures and anemia; impaired immune response; cardiovascular intolerance; and complications resulting from space sickness. In addition to these known biomedical problem areas, other potential problem areas have been identified and need careful assessment. These include investigations of microgravity effects on basic metabolic activity, gastrointestinal function, central nervous activity biorhythms, the aging process, the development of chronic

TABLE 5-1

Discipline	Subject of Experiment	Investigator	Species
Cardiovascular/Pulmonary	Cardiovascular adaptation to zero *g*	Blomqvist	Human
	Inflight study of cardiovascular deconditioning	Fahri	Human
	Autonomic cardiovascular control in zero *g*	Eckberg	Human
	Pulmonary function in zero *g*	West	Human
	Cardiovascular adaptation to zero *g*	Popvic	Rat
	Microcirculation in zero *g*	Hutchins	Rat
Vestibular	Vestibular experiments in Spacelab	Young	Human
	Autogenic-feedback training to prevent space motion sickness	Cowings	Human
	Zero *g* effect on gravity receptors	Ross	Rat
Renal/Endocrine	Fluid-electrolyte regulation in zero *g*	Leach	Human
	Mechanisms of fluid-electrolyte homeostasis in zero *g*	Moore-Ede	Monkey
	Thermoregulation of primates in zero *g*	Fuller	Monkey
Hematology	Erythrokinetics in zero *g*	Dunn	Human
	Blood volume regulation in zero *g*	Johnson	Rat
	Erythropoiesis in zero *g*	Dunn	Rat
Muscle	Protein metabolism in zero *g*	Stein	Human
	Skeletal myosin isoenzymes in zero *g*	Hoh	Rat
	Biochemistry of skeletal muscle in zero *g*	Baldwin	Rat
	Histology and biochemistry skeletal muscle	Ellis	Rat
Bone	Pathophysiology of mineral loss in zero *g*	Arnaud	Human
Immunology	Lymphocyte proliferation in zero *g*	Cogoli	Human
General	Gravitropic response in zero *g*	Brown	Plant
Biology	Post-illumination onset of nutation in zero *g*	Heathcote	Plant
	Effects of zero *g* on development of amphibian eggs	Tremor	Frog eggs
	Bone, calcium, and zero *g*	Holton	Rat

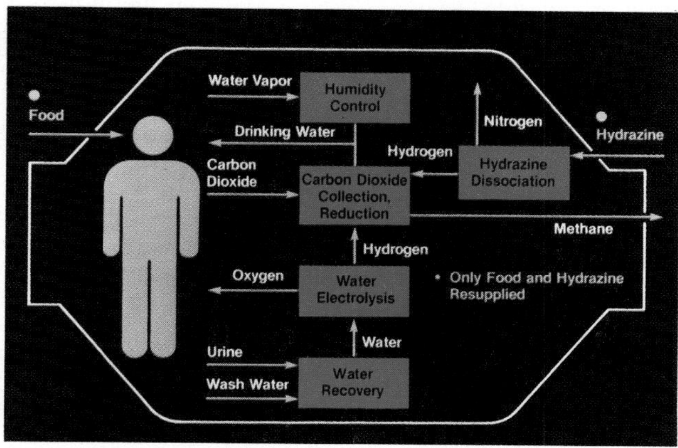

Fig. 5-7: The space station will need a regenerable life support system in which water and air are cycled internally. A scheme like this requires only food and a supply of hydrazine.

degenerative diseases, the neurophysiology of vestibular system control, and long-term radiation effects.

At present, there is no adequate characterization of in-flight trauma to bone and skin tissues, which are most susceptible to injury during spaceflight. Bleeding problems, clotting, inflammation reactions, and healing rates need to be examined, characterized, and analyzed to determine their appropriate treatment in space. The efficacy of drugs in disease therapy must also be reassessed because microgravity effects have the potential to alter certain aspects of rates of reactions such as gut absorption, distribution in body fluids, and excretion. Finally, the development of techniques and instrumentation necessary for a major surgical facility (e.g., surgical tools, containment of bleeding, anesthesia, and fluid handling) need to be assessed. A variety of plant and animal species will be flown in space. One of the most interesting problems identified for future study in space concerns animal development. Some issues in this area concern the effect of gravity on tissues whose function on earth is skeletal support or circulation of blood and fluids against gravity. On earth, the genetic expression of the bone tissue of each developing vertebrate may be adapted, in form and in quantity, to that animal's needs in a one-*g* environment. In space, the absence of gravity may alter the normal development of bone. The mechanisms by which higher organisms discern direction and orient themselves in a one-*g* field develop soon after conception. It is not known if gravity is necessary to the embryonic development of those mechanisms. On earth, phenomena such as the parental rotation of bird eggs, the self-generated rotation of developing frog eggs soon after fertilization, and the orientation of embryos in the rat uterus, support the notion that gravity is a significant factor in the production of viable offspring. Among the other major problems requiring future study in gravitational biology are those that concern the influences of gravity on reproduction, biorhythms, plant biology, and radiation biology.

To carry out research using animals and plants, to study the physiological effects of zero *g* and the general role of gravity on living systems, a vivarium is required. The vivarium must provide air, water, food, and waste treatment to keep the animals and plants healthy and stable. It is projected that plants, ranging in size from algae to leafy vegetables, and animals, ranging in size from mice to large primates, will need to be accommodated. The vivarium will also contain facilities to support "active" experiment requirements, such as tissue analysis, microscopic examinations, chemical preparation and storage, surgery, and dissection. Some experiments will be "passive" and simply require that specimens be maintained in the vivarium for long periods of time prior to study.

The vivarium should be designed to automate as many life support and experiment functions as are reasonable to minimize impact on crew time. In some cases, however, it will be necessary for crew members to collect metabolic wastes and blood samples, perform invasive measurements at discrete time intervals, prepare chemicals, analyze specimen samples, perform experiments, and monitor the subjects' condition.

To determine gravity effects on genetics, reproduction, embryological development, maturation, and population dynamics, many generations of space-born plant and animal species need to be studied. It will also be desirable to study the combined effects of microgravity and other stresses, such as exercise, light cycles, heat, radiation, and diet on these organisms.

In addition, a centrifuge facility is needed to provide a one-*g* control for the zero-*g* experiments, and to study graded responses between zero *g* and one *g*. The centrifuge area could represent a one-*g* holding facility in which a limited number of plants and animals would remain until needed for zero-*g* experimentation.

SUMMARY

It is clear that the life sciences are in the dawn of development within NASA. With our knowledge of the relative safety of spaceflight we are now able to pursue the interesting research questions that have been exposed. The opportunity for in-depth study provided by Spacelab and a space station will identify human requirements, determine basic biological responses to microgravity, and provide data on zero-*g* adaptation. Similarly, basic research will assist biomedical research to understand the underlying mechanisms of observed physiological responses to zero *g*. In turn, studies in biomedical research feed data back to basic research by providing inter-species correlation of results. Carrying out these elements in parallel will provide a comprehensive life sciences experimental flight program that contains an interactive analytical approach among investigations, minimum redundance, and maximum use of the exciting opportunities offered by a space station.

In parallel will be the continued thrust to understand our own origins and how life operates at the global level. With extensive use of satellites we will have a view of the changing earth as man has never before seen. The pressing needs to care for the six billion inhabitants of the earth during the next century will place high demands on the earth as a life support system. Already we are seeing limitations on soil capability, changes in air quality, pollution of large natural water reservoirs, and possible climate changes, either natural or man induced. The release of carbon dioxide and methane into the atmosphere may result in global changes that permanently affect our life support system. It is time to use the knowledge of NASA's life sciences program to help understand the earth, man's ability to injure it, and the natural fluctuations that govern our limitations to life.

PART II:
REMOTE SENSING FROM SPACE

6

REMOTE SENSING FROM SPACE: AN OVERVIEW

Ralph Bernstein

Remote sensing is much older than commonly thought, and the application areas are currently extensive and growing. This part provides a brief historical perspective to remote sensing, and discusses application areas that are both active and productive at the present time and have significant potential for the future. We have attempted to illustrate the applications with pictorial data, as this is an efficient and convincing way to demonstrate the potential and performance of remote sensing from space.

HISTORICAL PERSPECTIVE TO REMOTE SENSING FROM SPACE

Remote sensing, as its name implies, is the acquisition of physical data relating to an object or feature in a manner which does not involve direct contact in any way. In most cases, the term refers primarily to the sensing of the electromagnetic field. In the past, due to technology limitations, remotely sensed data were derived primarily from the visible portion of the electromagnetic spectrum using our eyes, and later cameras, for sensors. It is interesting to note that the development of cameras stemmed from experiments over 2300 years ago by Aristotle with a "camera obscura," which led to attempts to record images so formed [1]. The photographic process of Daguerre and Nièpce was applied by Tournachon (later known as Nadar), and a camera was first used to view the earth from a balloon in the 1850's. Laussedat experimented with a glass plate camera carried aloft by a balloon over Paris in 1858, and succeeded in capturing a number of images. Nadar's technique was applied to an earth observation experiment in the U.S. Boston was photographed from a balloon platform in 1860 (See Fig. 6-1). Nadar later developed a mathematical analysis for converting overlapping perspective views into orthophotographic projections. These results were impressive at the time, and began the observation of the earth from aerospace and space platforms.

REMOTE SENSING TECHNOLOGY

Fig. 6.2 summarizes the flow of earth observation data from acquisition through information extraction and understanding. Most development to date has been on sensor development and preprocessing. Future emphasis will be on information extraction, understanding, and modeling.

There has been impressive progress in sensors, platforms for elevating the sensors to suitable altitudes, and systems for communicating and processing the remotely densed data. New sensors have been developed for supporting such observations. The Landsat 1–3 Multispectral Scanner (MSS) with 79 m resolution data was first launched into earth orbit in 1972. The Landsat 4 and 5 Thematic Mapper (TM) provides data in the visible and infrared portion of the spectrum with 30 m resolution data and with a greater dynamic range.

The spectral bands selected for the TM are shown on Table 6-1, and provide the rational for band selection. Scanning sensors use an oscillating mirror to scan the earth and detectors to convert the radiance data into voltages which are digitized within the sensor system. Sample TM images are shown in Figs. 6-3 and 6-4, illustrating the capability of this technology. The enlarged view of Washington, D.C., [2] obtained from the TM allows for a clear delineation of cultural features such as the Washington monument and its shadow, various governmental buildings and other monuments, and cultural and natural features. Earth observation from space provides the potential for large-area frequent coverage and low-cost data acquisition.

Scanning sensors are increasingly being used to provide multispectral image data over a wide spectral region. The data can be used to augment, or even replace conventional map products, with the benefit that they

Fig. 6-1: Aerial photo of Boston, MA, taken from a captive balloon in 1860.

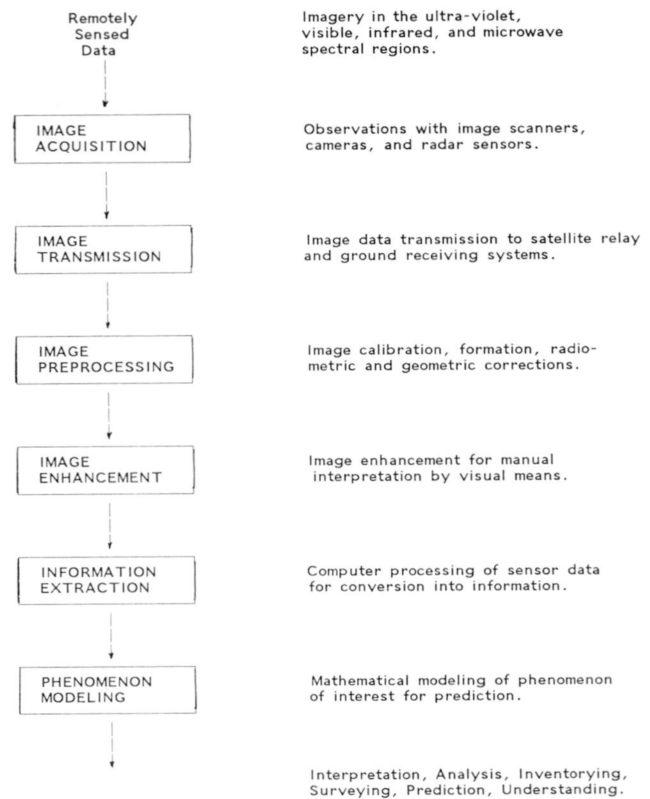
Remotely Sensed Data
Imagery in the ultra-violet, visible, infrared, and microwave spectral regions.
IMAGE ACQUISITION
Observations with image scanners, cameras, and radar sensors.
IMAGE TRANSMISSION
Image data transmission to satellite relay and ground receiving systems.
IMAGE PREPROCESSING
Image calibration, formation, radiometric and geometric corrections.
IMAGE ENHANCEMENT
Image enhancement for manual interpretation by visual means.
INFORMATION EXTRACTION
Computer processing of sensor data for conversion into information.
PHENOMENON MODELING
Mathematical modeling of phenomenon of interest for prediction.
Interpretation, Analysis, Inventorying, Surveying, Prediction, Understanding.

Fig. 6-2: Processing flow of remotely sensed data.

TABLE 6-1
LANDSAT 4 AND 5 THEMATIC MAPPER BAND SELECTION

Band	Wavelength / Applications	
Band 1	0.45–0.52 μm	(Blue)
	Bathymetry in less turbid waters	
	Soil/vegetation differences	
	Deciduous/coniferous differentiation	
	Soil type discrimination	
Band 2	0.52–0.60 μm	(Green)
	Indicator of growth rate and vegetation vigor	
	Sediment concentration estimation	
	Bathymetry in turbid waters	
Band 3	0.63–0.69 μm	(Red)
	Chlorophyll absorption/species differentiation	
	One of best bands for crop classification	
	Discriminate vegetation cover and density	
	Good for ferric iron detection	
	Ice and snow mapping	
Band 4	0.76–0.90 μm	(Near-IR)
	Water body delineation	
	Sensitive to biomass and stress variation	
Band 5	1.55–1.75 μm	(Mid-IR)
	Vegetation moisture conditions and stress	
	Snow/cloud differentiation	
	May aid in defining intrusive of different iron mineral content	
Band 7	2.08–2.35 μm	(Mid-IR)
	Distinguish hydrothermally altered zones	
	Mineral exploration	
	Soil type discrimination	
Band 6	10.4–12.5 μm	(Thermal-IR)
	Surface temperature measurement	
	Urban versus non-urban land use separation	
	Burned areas from water bodies	

provide a current view of the observed area and provide surprising detail. Figs. 6-5 to 6-7 show a Landsat 4 TM image of the San Francisco area [2] both with and without a map overlay. It is apparent that the sensor data, suitably annotated, can serve as a new type of a map product. New linear array technology sensors are soon to become operational, such as the French SPOT satellite, scheduled for launch in 1985 (10 m resolution) with the potential for providing data from space with even greater detail.

Communications technology is keeping pace with data transmission needs. The NASA Tracking and Data

Fig. 6-3: Sub-image of the Washington, D.C. area.

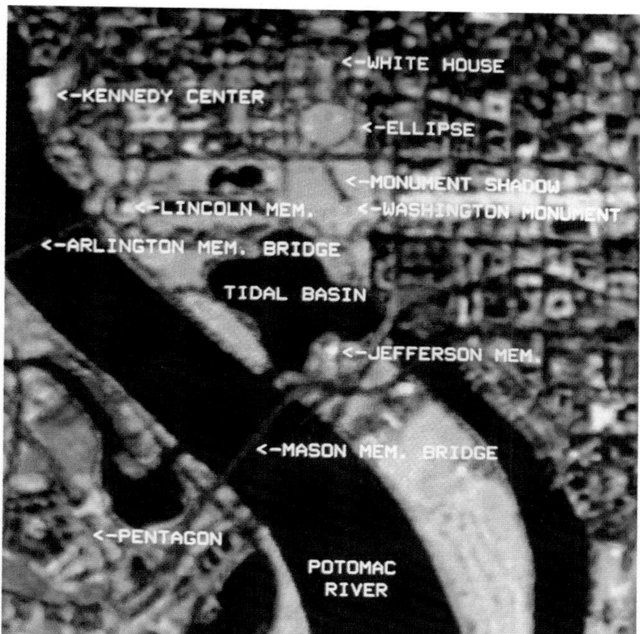

Fig. 6-4: Enlarged sub-image of the Washington, D.C. area.

Relay Satellite (TDRS) relays the data from the sensor platform to the ground at data acquisition rates. For example, the Landsat 4 and 5 TM data is relayed to ground receiving stations at the rate of 85 million bits/s.

The last two decades have seen an explosive growth in digital processing technology. This technology had brought about the beginning of a second industrial revolution—one that is multiplying man's mental energy, whereas the earlier industrial revolution multiplied man's physical energy. This technology, when applied to scientific objectives, and in particular to image processing, has supported the achievement of remarkable scientific results that would not have been possible with earlier available technology. This has been particularly true in the earth, lunar, planetary, space physics, and astronomical observation programs. Concurrently, there has been impressive progress in image technology, transitioning from photography to multiband two-dimensional detector array sensors. It is anticipated that future imaging systems will merge the imaging sensors with the digital processing technology.

TECHNOLOGY DIRECTIONS

It is interesting to view the remote sensing and image processing activities on a historical basis and to project future technologies and applications. This provides a basis for understanding limitations and trends.

Fig. 6-8 summarizes the technological approaches for past, present, and future terrestrial observation and processing activities. The past and many current activities involve the use of camera systems with photo-optical processing and manual information extraction. Although this will continue to some extent into the future, the trend appears to be towards the use of more advanced multispectral sensors and sophisticated image processing operations to reduce the labor-intensive nature of manual image analysis and processing. The present approach involves the use of multispectral scanners that provide data in numerous spectral bands, including many outside the spectral response of the human eye. The data, in a digital form, are transmitted to the ground, where high-speed digital processors are used to correct the data and convert them into information products for distribution. Initially, general-purpose computers were programmed to experimentally implement the image processing algorithms. Some processors are being specifically designed to implement frequently used algorithms,

Fig. 6-5: Landsat image of the San Francisco Bay area.

Fig. 6-6: Landsat sub-image of the San Francisco Bay area.

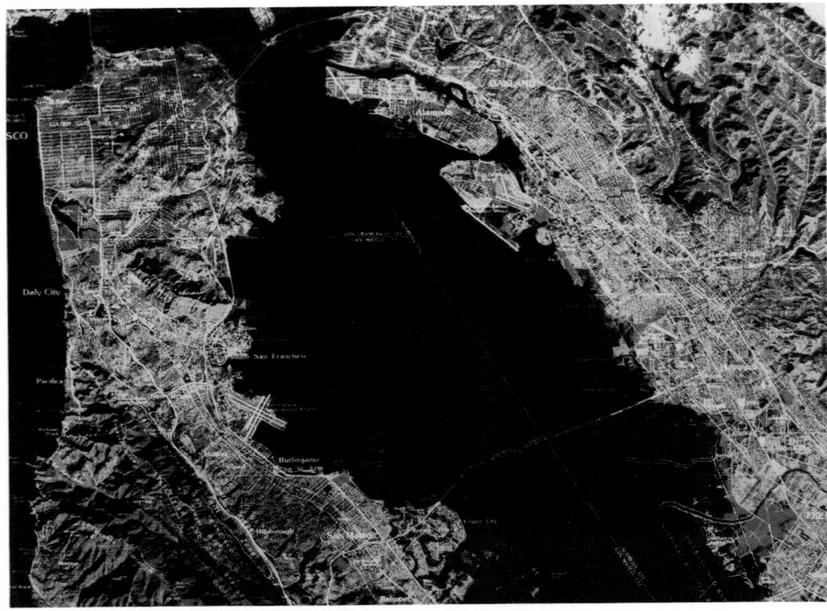

Fig. 6-7: Landsat sub-image of the Bay area with a map overlay.

improving the performance and reducing the cost of image processing for some applications. Systems using personal computers are also finding application in remote sensing [2], [3].

The very recent and important development of techniques for the extraction of information by computers, as opposed to human operators, is impressive. This development, which is still in its infancy, has resulted in a significant reduction of time and costs and an improvement in accuracy in a number of important applications. The algorithms used in some cases mimic the operation of the human; in other cases they are developed from physical laws and processes. In general, machine processing exploits the fact that computers can examine each image data sample, can discriminate subtle intensity differences, and can perform sophisticated statistical tests on anything from individual samples to large aggregates of data.

The future approach may go in the direction shown in Fig. 6-7. Sensors in all spectral bands, including the ultraviolet and microwave, will be used to acquire remotely sensed data. Many narrow spectral bands may

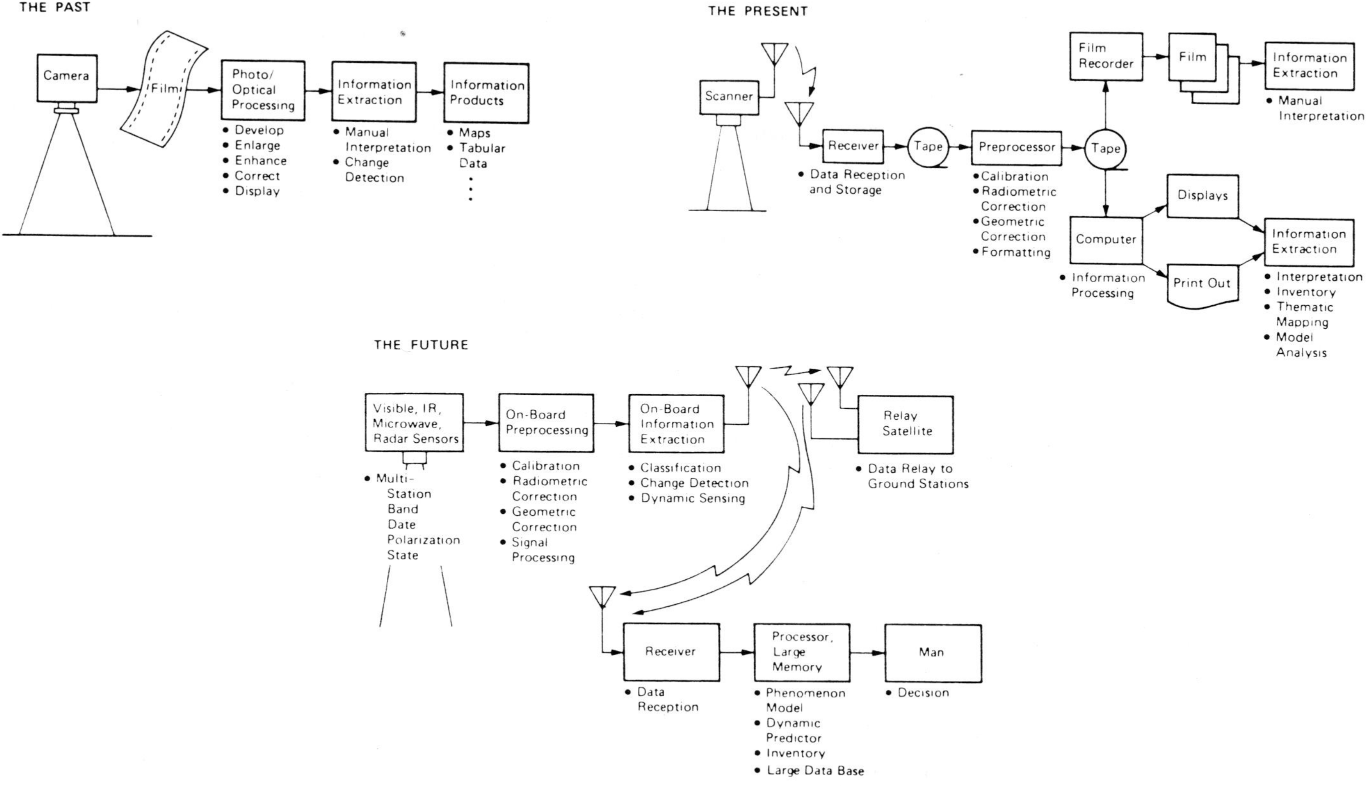

Fig. 6-8: Remote sensing and image processing: the past, present, and future.

TABLE 6-2
SUMMARY OF APPLICATIONS OF LANDSAT DATA IN THE VARIOUS EARTH RESOURCES DISCIPLINES[a]

Agriculture, Forestry, and Range Resources	Land Use and Mapping	Geology	Water Resources	Oceanography and Marine Resources	Environment
Discrimination of vegetative types: Crop types Timber types Range vegetation Measurement of crop acreage by species Determination of range readiness and biomass Determination of vegetation vigor Determination of vegetation stress Determination of soil conditions Determination of soil associations Assessment of grass and forest fire damage	Classification of land uses Cartographic mapping and map updating Categorization of land capability Separation of urban and rural categories Regional planning Mapping of transportation networks Mapping of land-water boundaries Mapping of wetlands	Recognition of rock types Mapping of major geologic units Revising geologic maps Delineation of unconsolidated rock and soils Mapping igneous intrusions Mapping recent volcanic surface deposits Mapping land forms Search for surface guides to mineralization Determination of regional structures Mapping linears (fractures)	Determination of water boundaries and surface water area and volume Mapping of floods and flow plains Determination of aerial extent of show and snow boundaries Measurement of glacial features Measurement of sediment and turbidity patterns Determination of water depth Delineation of irrigated fields Inventory of lakes	Detection of living marine organisms Determination of turbidity patterns and circulation Mapping shoreline changes Mapping of shoals and shallow areas Mapping of ice for shipping Study of eddies and waves	Monitoring surface mining and reclamation Mapping and monitoring of water pollution Detection of air pollution and its effects Determination of effects of natural disasters Monitoring environmental effects of man's activities (lake autrophication, defoliation, etc.)

[a]From N. M. Short, P. D. Lowman, Jr., and S. C. Freden, *Mission to Earth: Landsat Views the World,* NASA Scientific and Technical Office, 1976.

be used for particular applications. It will be possible in the next two decades to implement selected preprocessing and information extraction algorithms with onboard digital image processing systems and relay the processed information directly to the user via communications satellites. This will provide near real-time distribution of information to users so that dynamic events can be detected and monitored in real time. The space shuttle and the planned space station provide important sensor platforms for earth observation of this type.

Clearly, an integrated data collection approach which will provide data acquired at various altitudes from different platforms (space, aircraft, ground) at different viewing angles and times must be and will be designed. A large database, containing spatial, spectral, and temporal data and information acquired over all geographical areas of interest will develop. It is anticipated that a database for terrestrial applications alone could be global in nature and contain a vast amount of data [4]. Physical phenomenon models can be structured, which, when provided with past and present data, will allow the prediction of future events more reliably. Thus, the role of man in future systems will be elevated to a higher level and will consist of an interesting interaction with a large database and phenomenon models, as opposed to the conventional visual image processing of the past.

SUMMARY OF APPLICATION AREAS

Remote sensing has been applied to many applications. They include biomedical, industrial, surveillance, and earth and space applications. This part of the book deals with results for earth observation applications.

Earth observation applications have been extensively developed, and are summarized in Table 6-2. It is apparent from this table that the use of remotely sensed data coupled with innovative data processing and information interpretation results in many useful applications that benefit man and his environment.

The next four chapters of this book provide detail in important earth application areas. The organization has been structured to include land, ocean, and atmospheric measurements and results. The earth observation has been categorized in terms of surface (land) measurements and geophysical measurements in separate chapters. Experts in the field have provided these inputs, and their contributions are greatly appreciated.

REFERENCES

[1] *Manual of Remote Sensing*, vol. 1, *Theory, Instruments, and Techniques*, American Society of Photogrammetry, 1975.

[2] R. Bernstein, J.B. Lotspiech, H.J. Myers, H.G. Kolsky, and R.D. Lees, "Analysis and processing of Landsat-4 sensor data using advanced image processing techniques and technologies," *IEEE Trans. Geosci. Remote Sensing*, vol. GE-22, pp. 192–222, May 1984.

[3] H.J. Myers and R. Bernstein, "Image processing on the IBM personal computer," *Proc. IEEE*, vol. 73, June 1985.

[4] R. Bernstein, "Data base techniques for remote sensing and image processing applications," in *Lecture Notes in Computer Science, Proc. Data Base Techniques for Pictorial Applications*, A. Blaser, Ed. New York: Springer-Verlag, 1980.

7

LAND APPLICATIONS FOR REMOTE SENSING FROM SPACE

Robert N. Colwell

INTRODUCTION

The face of the land looks to the sky. It follows that the sky is the most favorable vantage point from which to look at the face of the land. This is a major reason why balloonists speak ecstatically of their "bird's eye view" and why astronauts exclaim "What a ride!" as they command what has been termed the "God's eye view" of the earth.

But there is a far more serious purpose to our viewing, from above, the land and its resources. That purpose has its origin with God's commands, given in the book of Genesis, for man to "exercise dominion over all the earth—and to subdue it." Realizing that it would be a bit much for one man, Adam, to do this by himself, God further directed man to "be fruitful and multiply." Especially in recent years, however, man's vigorous response seems to have gotten seriously out of hand. His clumsy efforts to subdue the land, whether with the bulldozer or chain saw, the concrete mixer or the match—the better to satisfy the needs and desires of the rapidly multiplying population that God had ordained—must surely be prompting the Divine Creator to exclaim, "Enough already!"

In any event, the most important reason for our viewing the earth from above is simply this: by such means we can better inventory, monitor, and manage the earth's land resources, i.e., its timber, forage, agricultural crops, soils, water, minerals, fossil fuels, domesticated animals, fish, and wildlife, as well as its recreational, aesthetic, and environmental resources. The better we perform the inventory and monitoring tasks, the better we will be able to comprehend what man has done and is doing to the land and its resources and, even more importantly, the better we will be able to determine what needs to be done if these resources are to be managed wisely in the future.

Since shortly after the dawning of the space age a quarter century ago, man has been given the opportunity for the first time to view the planet earth in its entirety. Some important information about the earth and its resources has been obtained merely through direct visual observations by astronauts as they looked down from an altitude of 100 mi or more and watched the terrain go by at the rate of 5 mi/s. But of greater significance, by far, are the records of the earth's surface that have been obtained by cameras and other remote sensing devices, mounted in either manned or unmanned earth-orbiting satellites. It is not surprising that these highly detailed and permanent records, when variously enhanced and subjected to painstaking examination by teams of expertly trained earth scientists, have yielded far more information about the earth's land resources than could possibly have been obtained merely from the fleeting glance that is afforded to an earth-orbiting astronaut. As some of the photographs in this chapter will show, the old Chinese axiom that "a picture is worth ten thousand words" certainly is true when the picture is in the form of a remotely sensed image of the earth, and the words (which could not be supplied except from a careful study of the picture) pertain to the amount, distribution, and condition of the earth's land resources.

RATIONALE FOR REMOTE SENSING OF THE LAND FROM SPACE

We can express in a simple, five-part statement the rationale that governs most of our efforts to remotely sense the land and its resources from space:

1) Whether viewed on a local, regional, national, or global basis, the supply of most of these land resources is rapidly dwindling and/or the quality of them is rapidly deteriorating.

2) This process is occurring most dramatically at the very time when the demand for such resources is rapidly increasing.[1] (See Fig. 7-1.)

3) The combination of these factors creates a compelling need for the wisest possible management of the earth's land resources.

4) An important step leading to such management is that of obtaining accurate resource inventories at suitably frequent intervals, so that the resource manager (i.e., the farmer, forester, range manager, etc.) will know at all times both the quantity and condition of each kind of land resource that is present in each part of the area for which he has management responsibility.

5) As will be illustrated and otherwise documented in this chapter, the required inventories almost invariably are best made through the analysis of data acquired by cameras and related remote sensing devices as they look at the earth's surface from space.

[1] *There is a multiplication factor with respect to this increased demand: not only is the world's population rapidly increasing, but so is the per capita demand for its natural resources, in both developed and developing countries, occasioned by man's insatiable desire for an ever higher standard of living.*

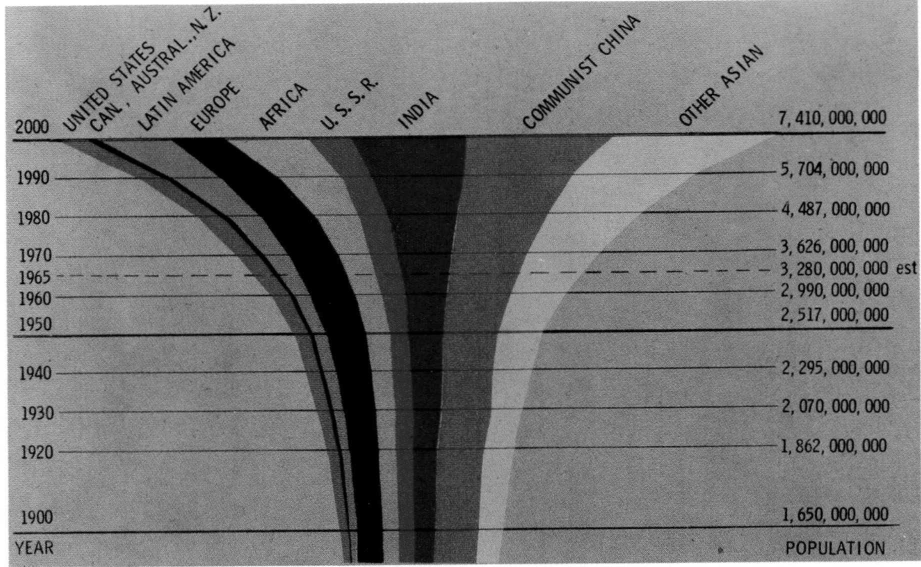

Fig. 7-1: The growth rate of earth resources consumers during the 20th century, both globally and as a function of countries or continents, can be readily appreciated from this "inverted cornucopia". If one is to appreciate from this figure the tremendous increase in demand for the earth's land resources, an additional factor also must be considered, i.e., the increasing *per capita* demand for such resources in both the developed and developing countries. (Courtesy of NASA.)

A broader and far more eloquent expression of the rationale for remote sensing of the land from space was provided by Abraham Lincoln when he said, although in a quite different context than the present one: "If we better knew where we are and whither we are trending, we would better know what to do and how to do it." More than a century later, another president of the United States, Lyndon Johnson, provided logical amplification of Lincoln's statement, but again in a different context, when he said: "it is easy for a responsible official to do the right thing, once he knows what is the right thing to do." Despite the differences in context, those two statements are highly germane as we seek to devise and implement better measures for the inventory, monitoring, and management of earth's land resources.

HOW PHOTO-LIKE IMAGES ARE ACQUIRED FROM SPACE

As alluded to elsewhere in this volume, there are two major means by which photo-like images are acquired from space for use by ground-based image analysts as they seek to inventory and monitor the earth's land resources:

1) by means of a conventional camera, operated by an astronaut and returned with him to earth, complete with its exposed photographic film, which is then processed in the conventional manner to produce photographic prints and/or transparencies; and

2) by means of an "electro-optical scanning device" (such as the Landsat Multispectral Scanner or Thematic Mapper) which observes, for each small picture element or "pixel" of the terrain that is being scanned, the "scene brightness" in certain wavelengths of the electromagnetic spectrum to which that particular sensor has been designed to respond. By the time this scanning device has observed and recorded scene brightness for the last pixel in a given scan line, the vehicle (e.g., the Landsat spacecraft) has advanced to a point from which scanning of the next line of terrain should and does begin.

Usually, when a scanning device is used, each observation of scene brightness is converted by the space-borne sensor system into a digital value; these digital values are then transmitted, pixel by pixel, from the spacecraft to a receiving station on earth where the signals are recorded and stored on magnetic tape. Later, as the taped record is played back, pixel by pixel and scan line by scan line, a corresponding amount of electrical current is made to actuate a small light bulb known as a "photo diode." This very small spot of light glows brightly or dimly during the playback process, as it responds to each digital value in turn, while traveling across a piece of previously unexposed photographic film. In this way it correspondingly exposes the film, pixel by pixel and scan line by scan line. Subsequent processing of the film produces a continuous photo-like black-and-white image for that portion of the earth's surface that has been observed by the sensor system. If a photo-like *color* image is desired as is usually the case, this can be obtained by color-coding and optically combining the black-and-white images that have been obtained, as described above, from two or more spectral bands.[2]

[2] *In addition to the direct visual analysis of these space-acquired photographs, an ever-increasing amount of success is being achieved through computer-aided analysis of the "scene brightness" digits, themselves. Such analysis exploits the fact that, just as the multiband photographic record tends to provide a unique tone signature for each category of earth resource that is to be identified from the space photo, so the corresponding multiband digital record tends to provide a unique digital signature which, by proper computer programming, can likewise lead to identification of the feature.*

HOW LAND RESOURCE INVENTORIES ARE FACILITATED BY SPACE PHOTOGRAPHY

In order to appreciate both the usefulness and the limitations that are associated with the use of space photography for the inventory and monitoring of land resources, two terms must be employed: discrimination and identification. As applied to the mapping of resources within a land area that has been remotely sensed from a spacecraft, the term discrimination pertains to the act of determining where differences exist in one component or another of the total resource complex. Usually discrimination also includes the drawing of "stratification" boundaries between one condition and the others that surround it, even though it usually is not possible to determine, merely from an examination of the space photography, the exact identity of features on either side of the stratification boundary itself. On the other hand, "identification" pertains to the act of determining what the specific attributes of the resource complex actually are within a given portion of the area (e.g., within the confines of each stratification boundary). (See Figs. 7-2 through 7-4 of the Alice Springs, Australia area.)

On space photographs the spatial resolution is such that, when combined with the spectral or color differences exhibited by such photographs, one usually can draw remarkably accurate and highly useful stratification boundaries. Once these boundaries have been drawn, however, it usually is necessary to acquire, within each category of area that has been outlined by the discrimination process, limited amounts of aerial photography for sample areas which, from a study of the space-acquired imagery, have been selected as being representative of that category. In keeping with this concept, the generic term "aerospace photography" has been coined to embrace all photographic observations made from either an aircraft or a spacecraft. Furthermore, in keeping with the concept of "multistage sampling," still more detailed observations with respect to the land and its resources than can be obtained from the aerospace photography must commonly be made by means of direct on-the-ground inspection. The making of such observations on the ground is a costly and time-consuming process. Fortunately, however, these observations need to be made only within subsample areas that are selected from a study of the aerial photographs as being representative of the sample areas to which they apply, respec-

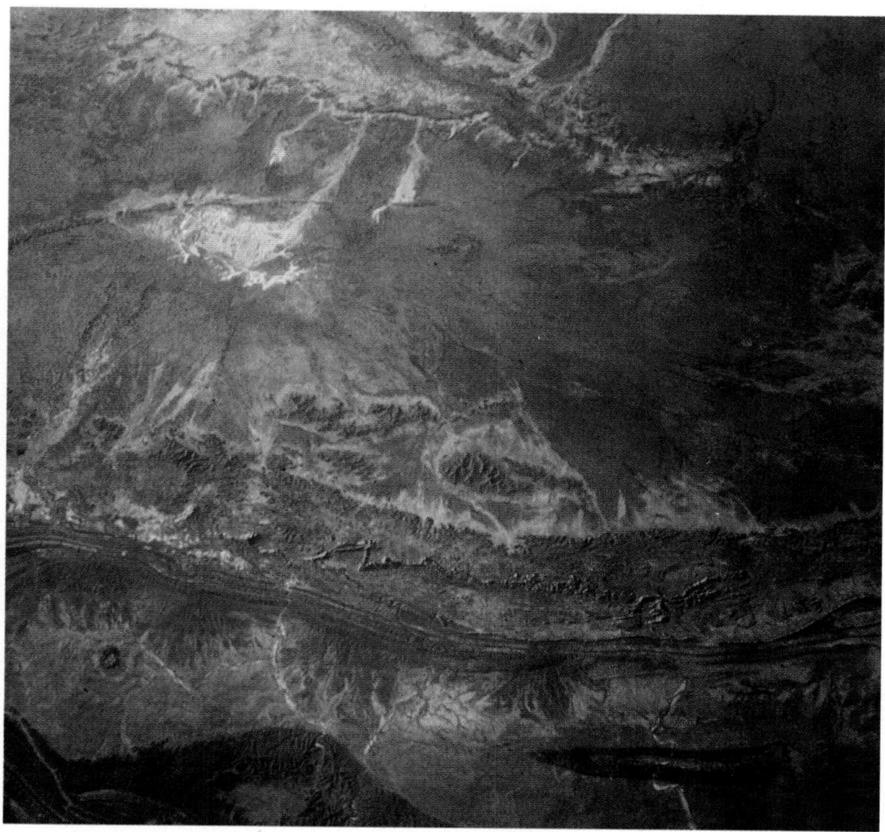

Fig. 7-2: Ektachrome space photography of a portion of central Australia taken from an altitude of 120 miles by the Gemini 5 astronauts. As seen in Fig. 7-3, which pertains to the same area, most of the tone or color differences that can be discerned on this photo are closely correlated with significant differences in vegetation, geology, soils, and land use. As explained in the accompanying text, those who seek to use modern remote sensing technology as an aid to the inventory, monitoring, and management of land resources find space photography such as this to be very useful for the drawing of resource boundaries, but of much less use for determining the exact nature of the resource complex that is present within any given boundary. (Photo courtesy of NASA.)

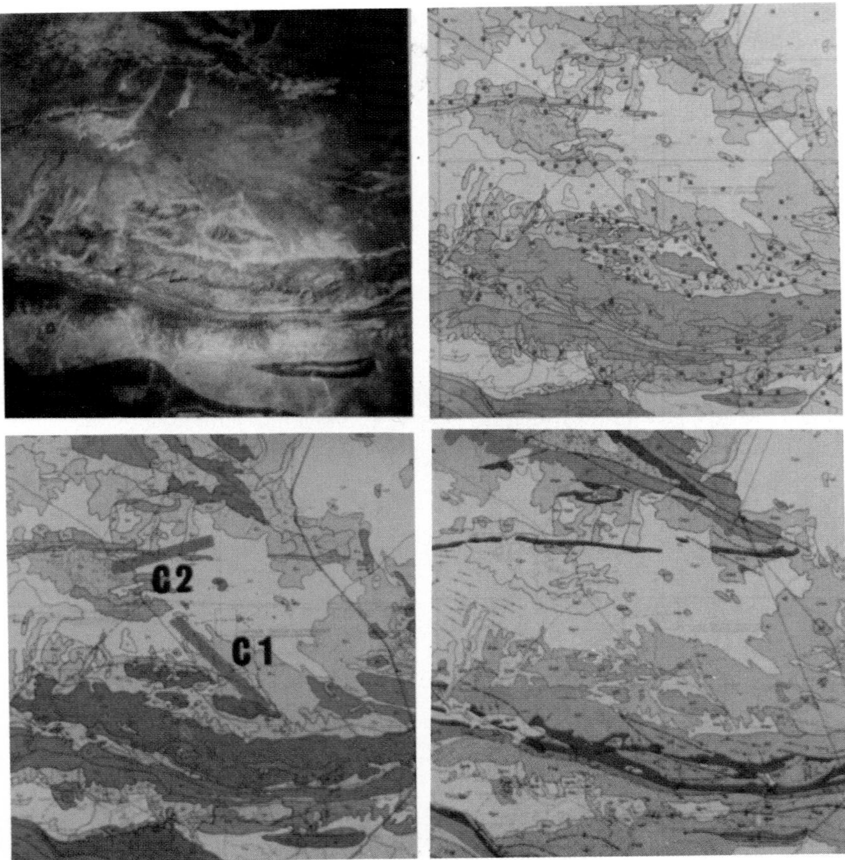

Fig. 7-3: The same space photo as in Fig. 7-2 is shown here (top left) but at a greatly reduced size to facilitate its being compared with "ground truth" maps for the same area that were compiled even before the dawning of the space age. Reading clockwise from the upper right, these maps pertain to vegetation types, geology/soils types, and land-use types, respectively. Note the remarkable extent to which boundaries coincide on each of the three maps and on the space photograph. (Top left photo, courtesy of NASA; other photos courtesy of CSIRO, Australia.)

Fig. 7-4: The tremendous land area that is covered by a single space photograph (typically about 12 000 square miles) can be appreciated when one realizes the following: the above photo, which was taken from an altitude of 1500 ft above the terrain, shows less than 10 percent of the area covered by the space photo of Fig. 7-2, i.e., the portion in the upper right quadrant of Fig. 7-2. Many land resource boundaries, including vegetation boundaries, can be accurately delineated on a space photo, and often even more accurately than on an oblique photo of much higher spatial resolution, such as the one shown here. For further evidence of this important fact, as applied to this and other geographic areas, see [8].

tively. Examples of the ground observations that need to be made are those that will permit determination of the following land resource attributes per unit area (e.g., per acre): volume of timber by species and size class of the trees; animal carrying capacity by species and forage class of the rangeland vegetation; and predicted yield, by crop type, of the agricultural crops.

Once the space photos and aerial photos have been interpreted and the on-the-ground observations made, the resource inventory of the entire area of interest is obtained by first determining photogrammetrically the total area (acreage) of land that is within each stratum category and its constituent sample and subsample area. Given this information it is merely a matter of simple arithmetic to 1) expand each set of subsample ground observations to the corresponding sample area for which it is representative and 2) similarly, to expand each set of aerial photographic observations to the corresponding stratum, i.e., to the entire area for which it is representative.

APPLICATION AREAS FOR LAND REMOTE SENSING FROM SPACE

Basic Considerations

We have already stated that the primary application for remote sensing from space is the inventory and monitoring of the land's resources, so that these resources can be managed more intelligently. But even the smallest "management unit" of the land is likely to contain a great many different kinds of resources. Hence, for each small part of the land, the resource manager needs information as to the quantity and quality of its entire "complex" of natural resources.

More specifically, the resource manager needs information, for each area and sub-area, that pertains to 1) its *vegetation* attributes, consisting primarily of its timber, forage, and agricultural crops; 2) its *geologic* attributes, especially those pertaining to its minerals and fossil fuels; 3) its soil attributes, primarily as they pertain to soil texture and soil fertility; 4) its *hydrologic* attributes, often with particular reference to the availability of water for use by plants and animals or for the generation of hydroelectric power; and 5) its *geographic* attributes, with particular reference to the spatial distribution of each of the various resource components.

Specific Informational Requirements

In Table 7-1 we have listed the kinds of information sought in various scientific disciplines (and hopefully obtainable by means of remote sensing) by those who seek to inventory, monitor, and manage the earth's land resources. We have listed, under each discipline area, both the *basic* and *applied* information sought with

TABLE 7-1

BASIC AND APPLIED INFORMATION SOUGHT THROUGH REMOTE SENSING BY WORKERS IN VARIOUS DISCIPLINES THAT DEAL WITH EARTH'S LAND RESOURCES

1) FORESTERS, RANGE MANAGERS, AND AGRICULTURALISTS
   a) Basic
      i) Amount and distribution of the "biomass"
      ii) Nature, extent, and function of important "ecosystems"
      iii) Amount and nature of energy exchange phenomena
   b) Applied
      i) The species composition of vegetation in each area studied
      ii) Vigor of the vegetation
      iii) Where vegetation lacks vigor, the causal agent
      iv) Probable yield per unit area and total yield in each vegetation type and vigor class
      v) Information similar to the above on the dynamics of livestock, wildlife, and fish populations
      vi) Changes resulting from past practices
2) GEOLOGISTS
   a) Basic
      Worldwide distribution of geomorphic features
      i) Energy exchanges associated with earthquakes and volcanic eruptions
      ii) The nature of geomorphic and mineralization processes
   b) Applied
      i) Location of certain or probable mineral deposits
      ii) Location of certain or probable petroleum deposits
      iii) Location of areas in which mineral and petroleum deposits of economic importance probably are lacking
3) HYDROLOGISTS
   a) Basic
      i) Quantitative data on factors involved in the hydrologic cycle (vegetation, snow cover, evaporation, transpiration, and energy balance)
      ii) Quantitative data on factors governing climate (weather patterns, diurnal and seasonal cycles in weather-related phenomena)
   b) Applied
      i) The location of developable aquifers and target areas for ground water exploration
      ii) The location of suitable sites for impounding water
      iii) The location of suitable routes for water transport
      iv) The moisture content of soil and vegetation
      v) Systems of enhancing ground water recharge
4) GEOGRAPHERS
   a) Basic
      i) Global, regional, subregional, and local land-use patterns
      ii) The nature and extent of changes in vegetation, animal populations, weather, and human settlement throughout the world
   b) Applied
      i) The exact location, at any given time, of facilities for transportation and communication
      ii) The interplay of climate, topography, vegetation, animal life, and human inhabitants in specific areas
      iii) The levels of economic activity and the purchasing habits of inhabitants in specific areas
      iv) Geographic distribution and dynamics of socio-economic and political factors influencing the production and use of earth resources
      v) Land cover and characteristics related to land-use potential

respect to that discipline. The logic in our making this two-way subdivision stems from the fact that it is through the performance of both basic and applied research that workers in these various disciplines have been able to determine what categories of information are needed within their respective disciplines.

As applied to any given discipline, *basic research* by definition seeks to understand the fundamentals on which that discipline rests; the types of data derived from basic research are therefore called *basic data* and, when analyzed, lead to the types of *basic* information listed in Table 7-1. On the other hand, *applied research* seeks to solve specific problems in an applied or practical manner. The types of data derived from applied research (and subsequently from operational systems and procedures based on such research) are therefore termed *applied data* and, when analyzed, lead to the types of *applied* information listed in Table 7-1.

For example, we have indicated in Table 7-1 that foresters, range managers, and agriculturists wish to acquire basic information on "the nature and distribution of the earth's biomass," on "the amount and nature of energy exchange phenomena" which involve vegetation, and on the functioning of the ecosystems of which they are a part. University scientists in the disciplines of forestry, range management, and agriculture are, indeed, justifiably interested in such basic information, and in its long-term significance. This basic information may even be of some small interest to the forester, range manager, or farmer in his capacity as a citizen of the world. However, he finds little in this kind of information that tells him how better to grow timber, livestock, or agricultural crops on the parcel of land for which he has management responsibility, nor even whether, during the period in question, there is likely to be an overproduction or an underproduction of the type of forest, range, or agricultural product which he is in the business of producing. Instead, he needs information of the applied type as to the vigor, for example, of the vegetation in each part of the forest, rangeland, or farm area which he is attempting to manage. Furthermore, in those places where the vegetation is suffering from a vigor loss, he needs to know the identity of the causal organism or agent so that he can take the necessary remedial action. All of the above considerations are reflected in Table 7-1 under the heading "Foresters, Range Managers, and Agriculturalists."

The remaining portions of Table 7-1 give distinctions between the basic and applied information needs of other potential users of modern remote sensing technology as applied to the earth's land resources including geologists, hydrologists, and geographers. In considering more specifically the informational requirements of scientists in any of the disciplines listed in Table 7-1, we may find it helpful to prepare a second, more detailed table. For example, Table 7-2 shows more specifically

TABLE 7-2

USER REQUIREMENTS FOR VEGETATION AND LAND RESOURCE DATA

TYPE OF INFORMATION DESIRED

For Agricultural Crops	For Timber Stands	For Rangeland	For Land-Use Decisions
Type of agricultural system and crop type (species and variety).	Timber type (species composition).	Vegetation type (species composition).	Land uses and cover types.
Present crop vigor and state of maturity.	Present tree and stand vigor by species and size class.	Seasonal forage development, vegetation cover or density, condition and "Range Readiness" (for grazing by domestic or wild animals).	Present land-use patterns, infrastructure, and interrelationships; cover types giving clues to kind and quality of use on the "natural, undeveloped" landscape.
Prevalence of crop-damaging agents by type.	Prevalence of tree-damaging agents by type.	Prevalence of forage-damaging agents (weeds, insects, rodents, diseases, abnormal growing conditions, etc.) by type.	Land-use conflicts and potential conflicts; land-use needs under present and alternative economic growth policies.
Prediction of time of maturity and eventual crop yield per acre by crop type and vigor class.	Present timber volume by species capacity and merchantability classes per acre.	Present forage production and apparent trends in probable future productive capacity per acre by vegetation type and range condition class.	Land resource availability, lands (by present use classes and covertypes) with multiple use potential; for each land use and covertype area, determining capability for each potential use.
Total acreage within each crop type and vigor class.	Total acreage within each stand type and vigor class.	Acreage within each vegetation type and condition class.	Acreages by land-use and cover type classes.
Total present yield by crop type.	Total present and probable future yield by species and size class; alternative and secondary values of the area by timber type.	Total present and probable future animal carrying capacity (domestic and wild) and alternative or secondary values of the area by vegetation type.	Human carrying capacity of the land under present and alternative economic growth policies.

the informational requirements of those who are concerned with the management of *vegetation* resources and of the land which supports them. Starting with the left-hand column of that table, we see that, by and large, the users of agricultural crop data need only the following categories of information: that pertaining to crop type, crop vigor, crop-damaging agents, crop yield per acre by type, total crop acreage by type, and total yield (more properly called total "production"). Proceeding to columns two and three, we note that essentially these same categories of information likewise are the ones sought by the managers of timberlands and rangelands, respectively. Furthermore, when we compare column four with the preceding three columns, we see that there is a very strong parallel between the kinds of information required by land-use planners and vegetation managers. These informational requirements fall into the same general categories of basic inventory, negative production or decision influences, quality or productivity, acreage and, finally, carrying capacity.

With the present population pressure on vegetation resources and the land, the three categories of vegetational resource managers shown in Table 7-2 are inextricably tied to the process of land-use planning and decision making because of the ever growing tendency of urban and industrially oriented man to usurp the best of the agricultural, rangeland, and timberland for these other "higher economic uses." Thus, it is important to put the land-use decision process in proper perspective whenever one discusses information needs and processes in land resource management. The land-use decision tends always to take place in a highly political atmosphere, often without regard to biological facts, further needs and pressures, and the wisdom of long-term economics. The only hope for improving this situation lies in providing a better quantitative basis for the land-use decision process and a good understanding of the alternatives, potentials, and consequences of land-use change in agricultural, forested, and rangeland areas. Thus, while the land-use decision process requires some unique kinds of information as indicated in column four of Table 7-2, it also requires an intimate awareness of the factors in each of the preceding three columns according to dominant vegetational resource.

Next, let us consider the question dealt with in Table 7-3, i.e., "how quickly and how frequently do the users need the information?" Obviously, this matter will be of great relevance when consideration is given, in a later section of this chapter, to man's ability to derive this information through the remote sensing of land areas from space. It will be noted in Table 7-3 that the same headings as appeared in Table 7-2 have been used for the four vertical columns. At the risk of some oversimplification, this table lists six time intervals that are indicative of the frequency with which various kinds of information about vegetation and land resources are needed (10–20 minutes; 10–20 hours; 10–20 days; 10–20 months; 10–20 years; and 20–100 years).

As we consider relationships between the frequency with which data relative to the earth's land resources should be collected and the rapidity with which the collected data should be processed, we may find it helpful to employ the term "half-life" in much the same way as it has been employed by radiologists and atomic physicists. The shorter the isotope's half-life, the more quickly a scientist must work with it once a supply has been issued to him. One half-life after he has acquired the radioactive material, only half of the original amount is still useful; two half-lives after acquisition, only one quarter of the original amount is useful, etc.

By coincidence or otherwise, this half-life concept seems to apply remarkably well to nearly every item listed in Table 7-3. Specifically, if the desired frequency of acquisition of any given type of information, as listed in that table, is divided by two, a figure is obtained indicating the maximum time after data acquisition by which that particular item of information should have been extracted from the data and put to use. It is true that some value will accrue even if that item of information does not become known to the resource manager until somewhat later. But the rate at which the value of the information "decays" is in remarkably close conformity to the half-life concept.

As previously indicated, there is a high degree of diversity in the user's requirements for information on the vegetation and land resources of an area. Consequently, the three tables which have just been presented probably would need major modification before they would accurately portray the information needs of any particular user. For example, many users think almost entirely in terms of protecting the vegetation resource from damaging agents, occasionally even through the extreme of non-use, and thus would view the problem somewhat more *narrowly* than we have viewed it here. On the other hand, there are those who think of the vegetation resource as merely one of the many items which comprise the total "resource complex" in a given land area. They are likely to view the problem more *broadly* than we have in these tables and, in so doing, they may point to the importance of such nonvegetational components as landforms, soils, water, minerals, wildlife, and recreational potential.

We could, of course, construct, for *nonvegetation* resources, tables similar to those that have been presented here for *vegetation* resources. It is probable that, upon completion of such tables, we would perceive a high degree of commonality with respect to informational requirements for all of the components which, collectively, comprise the "complex" of natural resources for any given area.

Table 7-4 is an attempt to set forth some of the characteristics of simply structured versus complexly structured areas in relation to the feasibility of making remote sensing based inventories. Such characteristics apparently have been given little consideration by remote sensing scientists up to the present time. As a result, there have been some seriously mistaken estimates made in the recent past as to the feasibility of using remote sensing techniques in various geographic areas.

As in Tables 7-1 to 7-3, the emphasis in Table 7-4 is

TABLE 7-3

USER REQUIREMENTS FOR VEGETATION AND LAND RESOURCE DATA
FREQUENCY WITH WHICH THE INFORMATION IS NEEDED (EXAMPLES ONLY). TO CONVERT THIS TABLE TO "RAPIDITY WITH WHICH INFORMATION IS NEEDED," USE "HALF-LIFE CONCEPT" (SEE TEXT)

For Agricultural Crops	For Timber Stands	For Rangelands	For Land-Use Decisions
10–20 min: Observe the advancing waterline in croplands during disastrous floods. Observe the start of locust flights in agricultural areas.	*10–20 min:* Detect the start of forest fires during periods when there is a high "fire danger rating."	*10–20 min:* Detect the start of rangeland and brushfield fires during periods when there is a high "fire danger rating."	*10–20 min:* Not applicable.
20–20 h: Map perimeter of ongoing floods and locust flights. Monitor the Wheat Belt for outbreaks of Black Stem Ruse due to spore showers.	*10–20 h:* Map perimeter of ongoing forest fires.	*10–20 h:* Map perimeter of ongoing rangeland and brushfield fires.	*10–20 h:* Not applicable.
10–20 days: Map progress of crops as an aid to crop identification using "crop calendars" and to estimating date to begin harvesting operations.	*10–20 days:* Detect start of insect outbreaks in timber stands.	*10–20 days:* Update information on "range readiness" for grazing, on forest use in critical periods, and also on times of flowering and pollen production in relation to the bee industry and to hay fever problems.	*10–20 days:* Monitoring compliance with certain codes and construction itself in critical areas of land-use change or during peak construction periods.
10–20 mos: Facilitate annual inspection of crop rotation and of compliance with federal requirements for benefit payments.	*10–20 mos:* Facilitate annual inspection of firebreaks.	*10–20 mos:* Facilitate annual inspection of firebreaks, range production, and range conditions.	*10–20 mos:* Monitoring development and land-use change in critical care areas for enforcement of codes and keeping valuations equitable and up to date.
10–20 yrs: Observe growth and mortality rates in orchards.	*10–20 yrs:* Observe growth and mortality rates in timber stands.	*10–20 yrs:* Observe signs of range improvement or deterioration, study the spread of noxious or poisonous weeds. Observe changes in "edge effect" of brushfields that affect suitability as wildlife habitat.	*10–20 yrs:* Reassess situation as per Table 7-2, col. 4 to fine-tune long-term land-use plan and reevaluate policies. Provide improved data for prediction models and trend analysis. Revise or set long-term economic development goals.
20–100 yrs: Observe shifting cultivation patterns.	*20–100 yrs:* Observe plant succession trends in the forest.	*20–100 yrs:* Observe major plant succession trends on rangelands and brushfields.	*20–100 yrs:* Document long-term changes in land use and monitor attainment of these development goals.

TABLE 7-4

CHARACTERISTICS OF SIMPLY STRUCTURED VERSUS COMPLEXLY STRUCTURED AREAS IN RELATION TO NATURAL RESOURCES

SIMPLY STRUCTURED AREAS	COMPLEXLY STRUCTURED AREAS
1) *Agricultural Vegetation*	1) *Agricultural Vegetation*
a) Fields large, regularly shaped, usually homogeneous with respect to crop condition.	a) Fields small, irregularly shaped, frequently heterogeneous with respect to crop condition.
b) Few competing crops and cultural practices.	b) Many competing crops and cultural practices.
c) Little interspersion of cropland with non-cropland.	c) Much interspersion of cropland with non-cropland.
d) All fields of a given crop planted on about the same date and hence developing in essentially the same seasonal pattern.	d) Fields of a given crop planted on many different dates and hence developing with many different seasonal patterns.
2) *Range and Forest Vegetation*	2) *Range and Forest Vegetation*
a) Blocks of rangeland and forestland are large and relatively homogeneous.	a) Blocks of rangeland and forestland are small and relatively heterogeneous.
b) Elevational range is low to moderate and hence vegetation of a given type tends to develop with essentially the same seasonal pattern.	b) Elevational range is high to very high and hence vegetation of a given type tends to develop with many different seasonal patterns.
c) Few vegetation types present, all adapted to the same elevational and climatic range. Topography flat to gently rolling so that few vegetational differences are the result of differences in slope and aspect.	c) Many vegetation types present, each adapted to a particular elevational and climatic range.
	d) Topography steep so that many vegetational differences are the result of differences in slope and aspect.
d) Cultural practices with respect to range and timber resources are few and uniform.	e) Cultural practices with respect to range and timber resources are many and varied.
3) *Geology, Soils, and Hydrology*	3) *Geology, Soils, and Hydrology*
a) Geologic, soil, and hydrologic formations are relatively large, simple, discrete, and homogeneous.	a) Geologic, soil, and hydrologic formations are relatively small, complex, intermingled, and heterogeneous.

placed primarily upon such renewable resources as agricultural crops, range vegetation, and forest vegetation; only limited treatment is given there to whether the geology, soils, and hydrologic resources of an area make it a simply or complexly structured one. Consistent with the so-called "land systems" concept, however (as developed, for example, by Christian and Stewart for use in Australia), an area that is of complex geologic structure is very likely to be complex also in terms of its soils and hydrologic attributes and therefore in its associated vegetative attributes.

The photo interpreter is likely to perceive these attributes in the reverse order because, in most areas, the vegetative attributes are the most photogenic ones and hence are the ones most easily perceived. Stating the matter in reverse order, therefore, it is highly probable that, when the vegetation attributes of an area are found to be complex, the geology, soils, and hydrologic attributes of that area are complex also. As a result, the feasibility of using remote sensing based techniques for the inventory of such an area's entire "resource complex" is likely to be considerably more limited, and the requirement for acquiring ancillary "ground truth" data much greater than if the area were more simply structured.

Generally speaking, there is need for a far greater realization by remote sensing scientists of the fact that there are fundamental differences, of the types suggested by Table 7-4, among various geographic areas. There also can be important differences in the specific kind of information desired, depending on who is the individual or user agency that needs the information. Table 7-5 provides a tabulation of some of the federal government agencies that are interested in using remote sensing derived information as an aid to the management of agricultural, forest, and range resources.

FACTORS GOVERNING THE USEFULNESS OF REMOTE SENSING

Table 7-6 seeks to list all of the major factors which govern the potential usefulness of any given type of aerial or space photography to those who wish to inventory, monitor, and manage an area's natural resources.

TABLE 7-5
USER REQUIREMENTS FOR VEGETATION RESOURCE DATA: AGENCIES AND GROUPS DESIRING THE INFORMATION

Agency or Group	Agricultural Crops	Timber Stands	Rangeland Forage	Brushland Vegetation
Federal agencies[a]	Agricultural Stabilization and Conservation Service; Cropland Conservation Program; Conservation Reserve Program; Agricultural Conservation Program; Emergency Conservation Measures Program; Commodity Credit Corp.; Agricultural Marketing Service; Statistical Reporting Service; Economic Research Service; Soil Conservation Service; Federal Crop Insurance Corp.; Farmers Home Administration; Rural Community Development Service; Foreign Agricultural Service; Famine Relief Program; Foreign Economic Assistance Program; Dept. of Commerce Agricultural Census Program	U.S. Forest Service; Bureau of Land Management; plus many Federal Agencies listed in col. 2	U.S. Forest Service; Bureau of Land Management; plus many Federal agencies listed in col. 2	Primarily U.S. Forest Service and Bureau of Land Management
State and county agencies	Agricultural Extension Service; State Tax Authority	Division of Forestry; Forestry Extension Service; State Tax Authority	Livestock Reporting Service; Range Extension Service; State Tax Authority	Division of Forestry; Division of Beaches and Parks; Water Resource Agency; State Tax Authority
Private groups	Producers of fertilizers and pesticides; crop harvesting industry; food processing and packing industry; transportation industry; food and fiber advertising and marketing industry	Producers of fertilizers and pesticides; logging industry; wood processing industry; transportation industry; wood and wood products advertising and marketing industry	Hunting and fishing clubs; public utilities commissions; local irrigation districts	

[a]The names of a few of these federal agencies may have changed since the time when the listing shown here was compiled.

TABLE 7-6

FACTORS WHICH GOVERN THE POTENTIAL USEFULNESS OF ANY GIVEN TYPE OF AERIAL OR SPACE PHOTOGRAPHY TO THOSE WHO WISH TO INVENTORY, DEVELOP, AND MANAGE NATURAL RESOURCES

Note: For each of the factors listed in this table, an entire "spectrum" of conditions is theoretically possible, ranging from very unfavorable to very favorable as regards its effects on the usefulness of the given type of photography in relation to the inventory, development, and management of natural resources. However, in any given instance, the applicable situation is likely to be well locatable and quantifiable. It follows that, in any given instance, the overall usefulness of this type of photography for the stated purpose will be determinable quantitatively by the aggregated effects of these various factors. Usually, however, a weight will need to be assigned to each factor, in proportion to its estimated importance; hence the aggregated value normally will reflect these individual weights.

1) Area to be analyzed is very *completely structured* in terms of the criteria appearing in Table 7-4.	1) Area to be analyzed is very *simply structured* in terms of the criteria appearing in Table 7-4.
2) *Only photos having a GRD of, say, 10 ft* are available for use. (GRD = Ground Resolvable Distance).	2) To the extent desired, *photos having a GRD of, say, 10 ft plus* any of all other forms of remote sensing can be used.
3) Clouds *usually* obscure the area that is to be analyzed.	3) Clouds *rarely* obscure the area that is to be analyzed.
4) Remote sensing can only be done on *one date* and at one *time of day*.	4) Remote sensing can be done on each of *many dates* and at *many times of day*.
5) There is a *very long delay* after the photos have been taken before they can be retrieved and placed in the hands of analysts.	5) There is only *a very short delay* after the photos (and other remote sensing data) have been obtained before they are retrieved and placed in the hands of the analysts.
6) Because of *rigid time constraints*, only a *"quick look" analysis* can be made.	6) For all practical purposes there are *no time constraints;* hence the making of a *complete data analysis* is feasible.
7) Only *one data analyst is available* and he is inexperienced, poorly trained, poorly funded, poorly equipped, little appreciated, and poorly motivated.	7) *An entire multidisciplinary team of analysts* is available and each of them is well experienced, well trained, well funded, well appreciated, well supported by consultants (when they are needed), and well motivated.
8) The analysis required is limited to only *one natural resource* and consists of a *one time inventory* of it in its static state.	8) The analysis required is one which will integrate all components of *the entire "resource complex,"* including renewable resources, and will make *repeated inventories* to monitor them in their dynamic state.
9) The resource classification scheme that is used is of *limited extensibility* because it is *locally specific*.	9) The resource classification scheme that is used has *great extensibility* because it comprises one component of an overall scheme that is *globally uniform*.
10) The derived inventory data must be *tightly held* because of sensibilities that relate to the economic or military security of the area under study.	10) The derived inventory data can be made *freely available* to all interested parties without fear of economic or military sensitivities.
11) The *sole purpose* of obtaining the inventory data is to facilitate *resource preservation*.	11) The *multifaceted purpose* of obtaining the inventory data includes the facilitating of *resource development*.
12) *Few funds are available* with which to implement decisions derived from a study of the resource information that has been acquired; furthermore, *the decisions themselves are suspect* because they were based on inadequate information as to the cost effectiveness of each of several resource management alternatives.	12) *Very substantial funds are available*, and with them the necessary equipment, engineering knowledge, and local political stability, to ensure that both short-term and long-term benefits will derive from implementation of the resource management decision; furthermore *the decisions themselves are sound* because they were based on reliable information as to the cost effectiveness of each of several resource management alternatives.
Note: To the extent that the factors listed in the above column pertain, there will be *minimum* benefit derived from the use of this type of photography in relation to the inventory, development, and management of natural resources.	*Note:* To the extent that the factors listed in the above column pertain, there will be *maximum* benefit derived from the use of this type of photography in relation to the inventory, development, and management of natural resources.

Consistent with the note appearing at the top of that table, there can be, for each of the listed factors, an entire "spectrum" of conditions, ranging from very favorable to very unfavorable, in relation to the usefulness of remote sensing. However, in any given instance, the relevant factors probably can be assessed reasonably well, and even quantified, if only we will make the effort to do so. If so, the overall usefulness of remote sensing will be determinable quantitatively by the aggregated effects of these various factors. In any specific instance, however, it probably will be necessary to assign a weight to each factor, in proportion to its estimated importance. Ideally, then, it will be this single aggregated value that will accurately indicate "remote sensing feasibility."

As emphasized by the note appearing at the bottom of Table 7-6, the statements appearing in the left column of that table are descriptive of highly *unfavorable* situations. Hence, to the extent that those descriptions apply in any given instance, there will be *minimum* benefit derived from the use of remote sensing techniques in relation to the inventory, monitoring, and management of natural resources. In contrast, the statements appearing in the right column of that table are descriptive of highly *favorable* situations. Hence, to the extent that those descriptions apply in any given instance, there will be *maximum* benefit derived from the use of remote sensing techniques.

While the statements appearing in Table 7-6 could be improved and expanded upon, they should suffice, even in their present form, to emphasize the fact that remote sensing scientists will need to give far more attention to such considerations in the future than they have in the past, and with the following beneficial result: there will be far less overselling of remote sensing techniques for situations where they are not likely to be successful, and there will be far more extensive and intelligent use of remote sensing techniques in situations where they have the potential for being highly successful.

It is the giving of proper consideration to statements and factors such as those listed in Table 7-6 that is at the very heart of remote sensing technology transfer and acceptance in the years to come. However, there also can be many other factors at play as will become apparent in the remaining sections of this chapter.

IMAGE ENHANCEMENT EQUIPMENT AND METHODS USED IN PRODUCING THIS CHAPTER'S ILLUSTRATIONS

The extraction of useful, resource-related information from an examination of space photographs is, at best, a difficult task. Consequently, there has been a great deal of interest in recent years in developing equipment and methods for producing "enhanced" images of various types that might facilitate the image analyst's task.

For the first one hundred years after the "discovery" of photography, virtually all of the images that were produced were either in monochromatic shades of gray, or (in the case of "sepia-tone" renditions) in shades of brown. All the while it was realized, however, that if a photograph could be produced in full color, the viewer would be able to extract much more information from it. It often was pointed out, in keeping with this thought, that the human eye is only able to perceive approximately 200 shades of gray under the very best of viewing conditions whereas, under comparable conditions, the human eye can perceive as many as 200 000 different colors (i.e., different combinations of hue, brightness, and saturation).

Development of Color by Means of an Optical Combiner

Shortly after the capability for black-and-white photography had been developed, serious consideration was given to means by which, through the simultaneous projection of three black-and-white photos and the use of "additive color" principles, photography in full color might also be obtained.

For example, as early as 1855 in a paper on color vision, James Clerk Maxwell gave a hypothetical example to clarify a discussion of Young's theory of vision. His pronouncement at that time has since been accepted as "the invention of *color* photography."

Speaking of Young's theory, Maxwell wrote:

> This theory of color may be illustrated by a supposed case taken from the art of photography. Let it be required to ascertain the colours of a landscape by means of impressions taken on a preparation equally sensitive to rays of every colour. Let a plate of red glass be placed before the camera, and an impression taken. The positive of this will be transparent wherever the red light has been abundant in the landscape, and opaque where it has been wanting. Let it now be placed in a magic lantern, along with the red glass, and a red picture will be thrown on the screen. Let this operation be repeated with a green and violet glass, and by means of three magic lanterns, let the three images be superimposed on the screen. The colour of any point on the screen will then depend on that of the corresponding point of the landscape; and by properly adjusting the intensities of light, etc., a complete copy of the landscape, as far as visible color is concerned, will be thrown on the screen.

A more lucid description could scarcely be written even today regarding the use of an "optical combiner" to exploit the principles of additive color, thereby producing a composite color image from black-and-white photos that have been taken with various film–filter combinations. As will later be illustrated, the same process can be used in combining multidate photos, all of which have been taken with the same film–filter combination.

Maxwell later demonstrated the above operation on May 17, 1861.

Toward the end of the 19th century, Louis Ducos du Hauron developed three-color separation to a fine art and made beautiful color prints using red, yellow, and blue pigments.

Development of Color Film

In 1924, Mannes and Godowsky patented their first work on a multiple-layer color film which eventually led to the development of Kodachrome. There has been emphasis on three-color systems ever since the three-layer color film came on the market in 1935. Almost immediately there developed an interest in the applications of color to aerial photography. In 1937, Colonel George Goddard made preliminary tests and was soon followed by Bradford Washburn and Walter Clark as they made aerial Kodachrome pictures of southeast Alaska in 1938.

The question soon arose of whether it was better to make color film which could be developed to a negative or to a positive. It was decided that "reversal positive film" was the type that should be made available to the military during World War II. This was done and the film was designated as Kodacolor Aero Reversal Film, which was the forerunner of Ektachrome. Principles governing the production of color images through use of the reversal process are described in the following section.

By the end of World War II, two main "reversal type" color films were in use. One of these was referred to as a "natural color" film (of which modern day Ektachrome film is an example) because there was a close match between the colors exhibited by various features in nature and those exhibited by their images when this film was used to phogograph them. The other was referred to as a "camouflage detection" film (of which modern day "color infrared" or "Infrared Ektachrome" film is a good example) because certain kinds of military camouflage materials were made to stand out in sharp contrast to their surroundings when photographed with this kind of film.

Present Day Infrared Ektachrome—A False-Color Reversal Type of Film

Figs. 7-5(a) and (b) comprise a matched pair of space photographs, the members of which were taken, respectively, with a false-color (Infrared Ektachrome) film and a natural color (Ektachrome) film. Both of these films employ three dyes (yellow, magenta, and cyan) and both are of the "reversal positive" type in that the color produced at any given point on such a film (when the film has been developed) is the result of a "subtractive" process. The means by which this is accomplished could be explained through the use of either member of the above-mentioned pair of photographs, together with information on the spectral reflectance of each of the various kinds of features that were being photographed. In providing that explanation, let us use Infrared Ektachrome as the example and, in so doing, let us attempt to answer the question: "Why does healthy green vegetation appear red when imaged on Infrared Ektachrome film?"

In answering this question we will find it helpful first to recognize that it is primarily the *foliage* (mass of leaves) that is exposed by plants to the overhead view, the better to absorb the sun's energy for use in photosynthesis (food manufacture). Hence, it is important to examine the spectral reflectance curve for a healthy green leaf (curve *a* in Fig. 7-6) and to have in mind the explanation of why that curve is shaped as it is. The explanation is provided by the annotations appearing in Fig. 7-7.

Given this information, we can provide the following explanation for the red color of healthy green vegetation, as imaged on Infrared Ektachrome film: this film, being of the subtractive reversal type, is one in which the dye responses, when the film is processed, are inversely proportional to the exposures received by the respective layers or wavelengths.

Specifically, in those places on the film where healthy green vegetation is imaged:

1) Only a small amount of cyan dye will be left in the processed film because:
   a) the cyan dye is coupled to the film's infrared sensitive layer;
   b) the film receives a great deal of radiation (reflectance) in the near-infrared part of the spectrum from healthy green vegetation (as indicated by curve *a* in Fig. 7-6); and
   c) at any given spot on the film, the amount of cyan dye left after processing will be *inversely* proportional to the amount of near-infrared energy to which that spot has been exposed.

2) Relatively large amounts of yellow and magenta dyes will be left in the processed film, however, because:
   a) the yellow and magenta dyes are coupled, respectively, to the film's green- and red-sensitive layers;
   b) relatively speaking, the film receives very little radiation (reflectance) in the green and red parts of the spectrum (as indicated by curve *a* in Fig. 7-6); and
   c) at any given spot on the film, the amount of yellow and magenta dyes left after processing will be inversely proportional to the amount of green and red wavelengths of energy, respectively, to which the spot has been exposed.

In view of the foregoing, any spot on the processed film where healthy green vegetation has been imaged will appear *red* (e.g., when the film is viewed in transparency form over a light table that has white light as the illuminant) for the following combination of reasons:

1) a great deal of *red* light will be transmitted through that spot to the observer because the spot contains very little cyan dye to interfere with the transmission of red light;
2) very little green light will be transmitted through that spot to the observer because the spot contains a great deal of yellow dye, which interferes with the transmission of green light; and
3) similarly, very little blue light will be transmitted through that spot to the observer because the spot contains a great deal of magenta dye, which interferes with the transmission of blue light.

Even on rereading the above explanation, the reader could easily get lost in a jumble of words and colors. Hence, Table 7-7 has been included to help clarify the explanation of why healthy green vegetation appears red on color infrared photography and also why it appears green on natural color photography. It is also apparent from a study of that table, together with curves *a*, *b*, and *c* of Fig. 7-6 (showing that as a plant loses vigor, a loss in infrared reflectance precedes any change in its visible light reflectance) why the loss of vigor in plants often can first be discerned by the blue-green appearance of such plants on Infrared Ektachrome photography. [One example of this can be seen by observing point *A* as imaged on Figs. 7-5(a) and (b).] It is for this reason that such a film is used to detect "previsual symptoms" of the loss of vigor in plants—a capability that often is of great importance if effective remedial measures are to be taken—e.g., by "nipping in the bud" any infestation caused in plants by insects or diseases, doing so before the infestation can spread to other plants. Also on Figs. 7-5(a) and (b) many examples can be seen of a more advanced stage of the loss of vigor in plants (corresponding to reflectance curve *C* of Fig. 7-7). One such example is labeled *B* on Figs. 7-5(a) and (b).

Table 7-7 facilitates an understanding both of the means by which a conventional Ektachrome film provides *true* colors in the processed image and by which an Infrared Ektachrome film provides certain *false* colors.

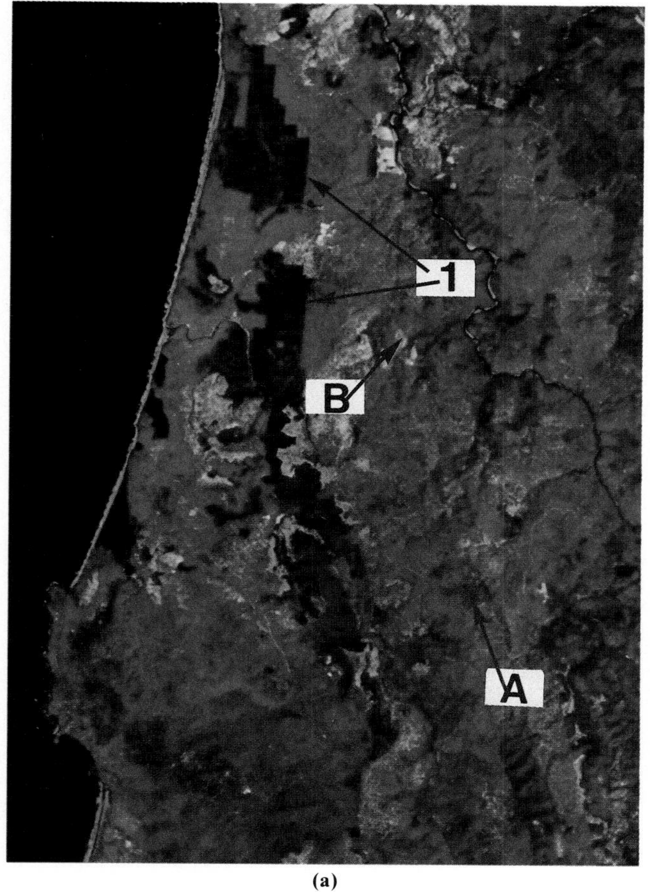

(a)

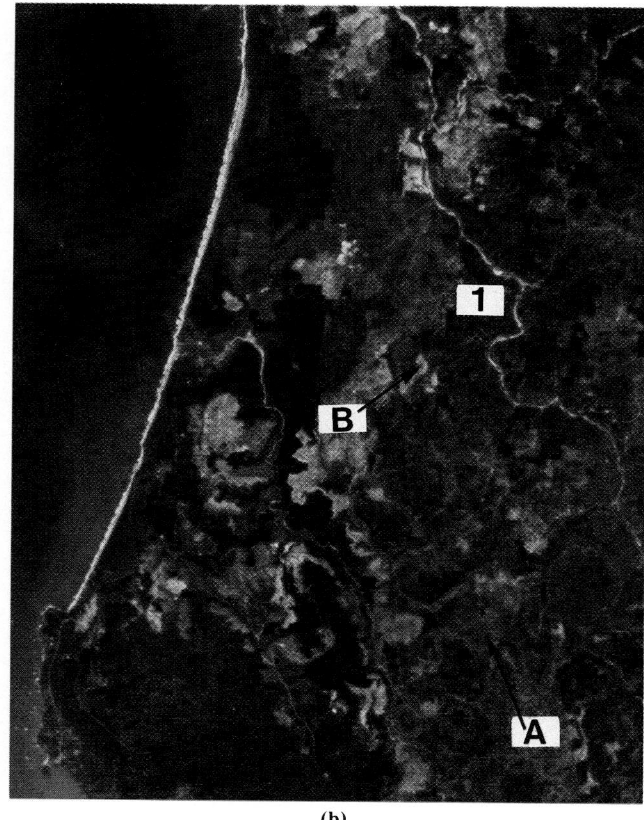

(b)

(c)

Fig. 7-5: (a) Color infrared space photo of Redwood National Park and its environs in northern California, taken by Skylab astronauts in June 1973, from an altitude of 270 mi and reproduced here at a scale of approximately 1:500 000. For reasons explained in the text, all of the virgin stands of redwood trees (*Sequoia sempervirens*) within this scene such as those labeled "1," appear in a very distinctive dark red color and hence are much more easily and accurately delineated here than on the matching frame of "natural color" (Ektachrome) photography shown in (b). Area *A*, as annotated above, is in an early stage of vigor loss and has spectral reflectance characteristics as diagrammed in curve *b* of Fig. 7-6. Hence it shows "previsual symptoms" of vigor loss on this photo but not on the matching color photo of (b). On the other hand, area *B* is in a more advanced stage of vigor loss and has spectral reflectance characteristics as diagrammed in curve *c* of Fig. 7-6. Hence it can be readily discerned on both of these space photographs. (Photo courtesy of NASA.) (b) Natural color (Ektachrome) space photo of Redwood National Park, taken by the same astronauts and at the same instant as (a) in order to permit a comparison to be made regarding the interpretability of land resource features as portrayed on the two kinds of photography. Note that the virgin stands of redwood trees appear in only a moderately distinctive dark green color here and that early vigor loss in plants at area *A* is almost indiscernible. (c) is an aerial photo of the right center portion of the area shown here. (Photo courtesy of NASA.) (c) This color infrared photograph shows a portion of Redwood National Park and its environs as imaged on October 6, 1972 from an altitude of 65 000 ft through the use of a U-2 aircraft and a precision aerial mapping camera having a 6-in focal length and a frame size of 9 × 9 in. This photography is very useful when employed as supplementary coverage of limited representative areas as selected from space photographs such as the one shown in (a).

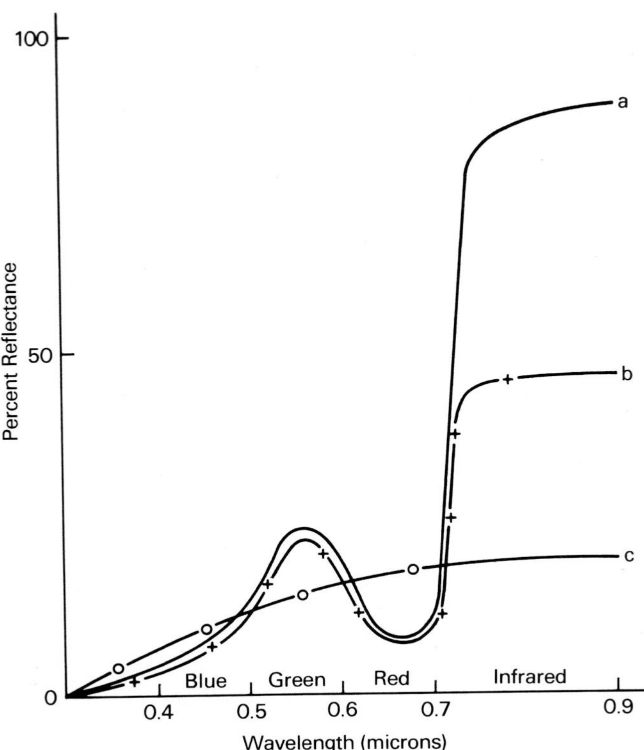

Fig. 7-6: Typical reflectance spectra for a leaf at each of three states of vigor. Curve a = a healthy green leaf; curve b = a leaf that is just beginning to suffer from a loss of vigor due, for example, to its having been invaded by some harmful insert or pathogen; and curve c = a leaf suffering from a more advanced stage of vigor loss. The significance of curve b in relation to the early detection of vigor loss on infrared photography (thereby permitting early remedial measures to be taken) is discussed in the text and an example of such detection of "previsual symptoms" is shown at area A as annotated on Fig. 7-5(a) and (b).

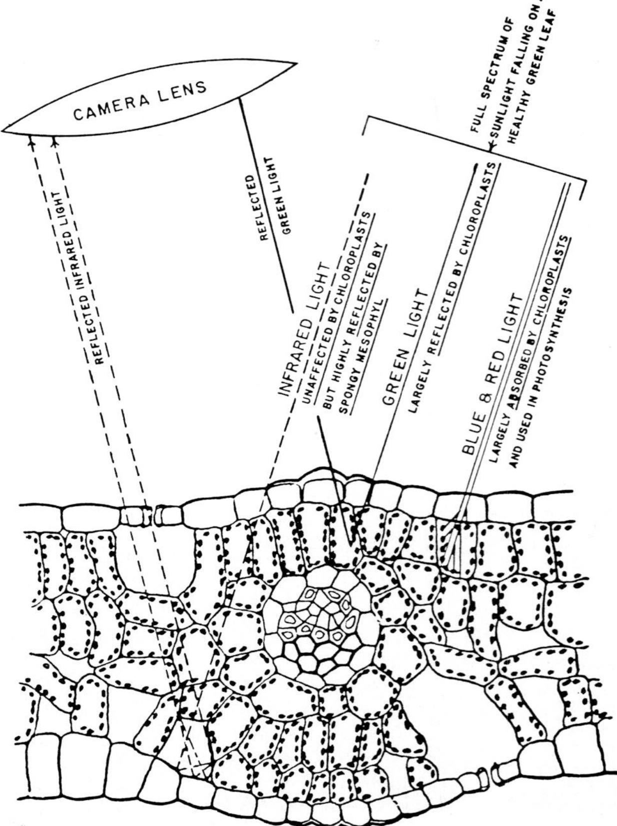

Fig. 7-7: Schematic drawing of cross-section of a healthy oats leaf. Note that certain wavelengths are largely absorbed while others are reflected to a high degree, either by the chloroplasts or by the spongy mesophyll ("middle-of-the leaf") tissue. Should this leaf start to lose vigor as a result of any of a large number of causes, such as attacks on the plant by various insects or fungi, its spongy mesophyll (which is now turgid and highly reflective of infrared light especially in the 0.7–0.9 μM region) would either collapse because of a restriction of the water supply to it or be plugged by hyphae of the fungus. This would happen long before the leaf's green color started to fade. Hence, the disease would be detectable much sooner on infrared than on panchromatic photography, since the latter, as its term implies, is sensitive to "all colors" of the visible spectrum but not to infrared radiation. Within the needles of coniferous trees, the spongy mesophyll, although present, is less well developed than on the broad leaves of hardwood trees or of shrubs and herbaceous vegetation, as diagrammed here. It is for this reason that in Fig. 5(a), for example, the coniferous trees (redwoods) are imaged in a very dark red to nearly black color. As a result, the redwood groves, by virtue of the increased color contrast between them and their surroundings, are much more distinguishable in Fig. 7-5(a) than 5(b) [7].

Image Enhancement Devices That Employ Color in Their Displays

For purposes of explaining how the various composite images that appear in the next section were produced, two kinds of image enhancement devices will be briefly described here. They are 1) optical combiners and 2) electrooptical devices.

A Representative Optical Combiner

Fig. 7-8 shows an unsophisticated but effective optical combiner designed in accordance with Maxwell's pre-

TABLE 7-7
SPECTRAL RESPONSES OF NORMAL COLOR AND INFRARED-SENSITIVE COLOR FILMS

Type of film	Spectral region			
	Blue	Green	Red	Infrared
Normal color film (e.g., Ektachrome):				
Normal color film sensitivities	Blue	Green	Red	
Color of dye layers	Yellow	Magenta	Cyan	
Resulting color in photographs	Blue	Green	Red	
Infrared sensitive color film (e.g., Infrared Ektachrome):				
Sensitivities with Wratten 15 yellow filter	([a])	Green	Red	Infrared
Color of dye layers		Yellow	Magenta	Cyan
Resulting color in photographs		Blue	Green	Red

[a]All three layers of Infrared Ektachrome film are sensitive to blue light, but the Wratten 15 yellow filter prohibits blue light from striking, and thus sensitizing, any of the three layers.

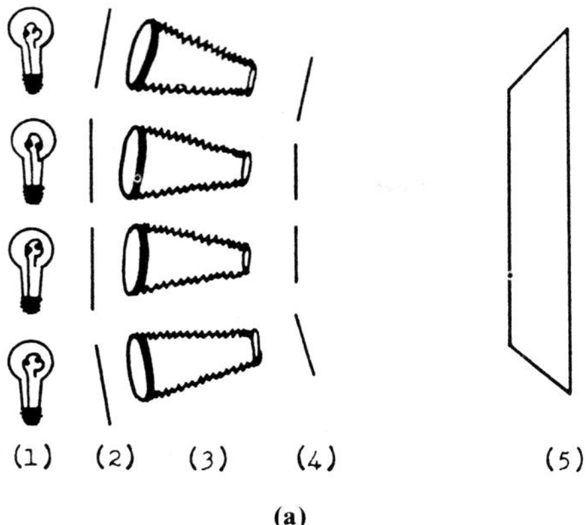

(a)

(b)

Fig. 7-8: An optical combiner similar to that described, in 1856, by Clerk Maxwell, as quoted in the text. (a) Schematic diagram in which the light sources are shown at (1), the multiband images at (2), the optical train for projection at (3), and the color filters at (4), through which the images are simultaneously projected in common register onto either a front-viewing or rear-viewing screen at (5). (b) A nest of four projectors comprising an optical combiner of the type diagrammed above. (Photo courtesy of the NASA/USDA Forestry Remote Sensing Laboratory, University of California, Berkeley, CA.)

viously discussed principles. The device shown in that figure is still used in one form or another even though this particular version was developed more than two decades ago by personnel of the NASA-USDA Forestry Remote Sensing Laboratory of the University of California at Berkeley. The diagram at the top of that figure indicates in a schematic fashion the structure and operation of that device when it is used to combine multiple images (four in this instance) and project them simultaneously in common register through selected color filters to produce a composite color-enhanced scene. The multiple images used as input may be either 1) those taken on a given date, but in multiple wavelength bands; 2) those taken in a given band, but on multiple dates; or 3) those that comprise any of several possible multiband, multidate combinations.

In all instances, a lantern slide must first be made of each of the black-and-white images that are to be combined. For example, the composite color photograph shown in Fig. 7-9, covering a portion of the Imperial Valley in California, was obtained by registering three black-and-white images (each acquired with panchromatic film and a Wratten 25 filter) taken on three separate dates, i.e., March 8, April 23, and May 21, of a typical growing season. The various colors exhibited by the fields in that scene are attributable to changes in tone that occurred in the black-and-white photos as recorded on those three dates within this single band (the 0.6–0.7 wavelength band). When the slides were simultaneously projected onto a screen in proper register, a filter was interposed in each instance between the slide and the screen. In this instance, blue, green, and red filters were used, respectively, for the three dates of imagery. The various multidate color codes that are seen, field by field in the color composite image on the screen, are attributable to the different vegetation states, and corresponding shades of gray, of the crops within those fields on those dates.

As previously discussed with respect to Figs. 7-5 to 7-7, healthy green vegetation tends to absorb to a high degree the *red* light that strikes it, i.e., light that has wavelengths in the 0.6–0.7 μ range. Hence, when only red light is being exposed for in the taking of a photograph (as when panchromatic film is being used with a Wratten 25 filter), vegetation appears black on the positive image. On the other hand, bare soil that exhibits the usual light gray to white color in nature, axiomatically reflects all wavelengths of visible light to a high degree, including red, and thus appears light in tone on a "panchromatic-25A" photographic positive. Given these facts, we are not surprised to find that

1) *black*, in the color composite image, denotes fields that were continuously vegetated throughout the period—in this case, they are sugar beets;

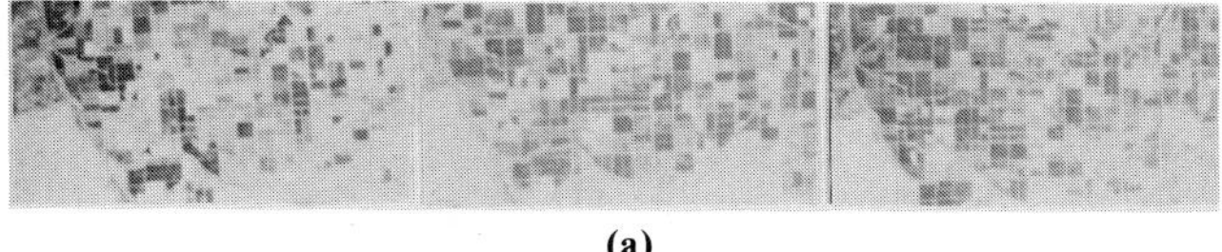

(a)

(b)

Fig. 7-9: (a) Multidate black-and-white simulated space photos taken from a U-2 aircraft flying at an altitude of 65 000 ft. Shown here are agricultural crops in California's Imperial Valley area as imaged in the months of April, May, and June of a given year with panchromatic film and a Wratten-25A (red) filter. (b) Compositer color photo made by projecting the top three photos simultaneously through blue, green, and red filters, respectively, by means of the "optical combiner" shown in Fig. 7-8. For an explanation of the resulting color code for various crop types, see text. (Photos in (a) courtesy of NASA.)

2) *green* denotes fields that had a crop present on date 1, but absent on date 2 and present again on date 3 (such as alfalfa that was mowed shortly before date 2); and

3) *red* denotes fields that were densely vegetated on dates 1 and 2, but from which the crops had been harvested by date 3 (such as barley).

A Representative Electrooptical Device for Enhancing Images

Various electrooptical devices have been developed in the past two or three decades that will permit multiband or multidate black-and-white photographs to be com-

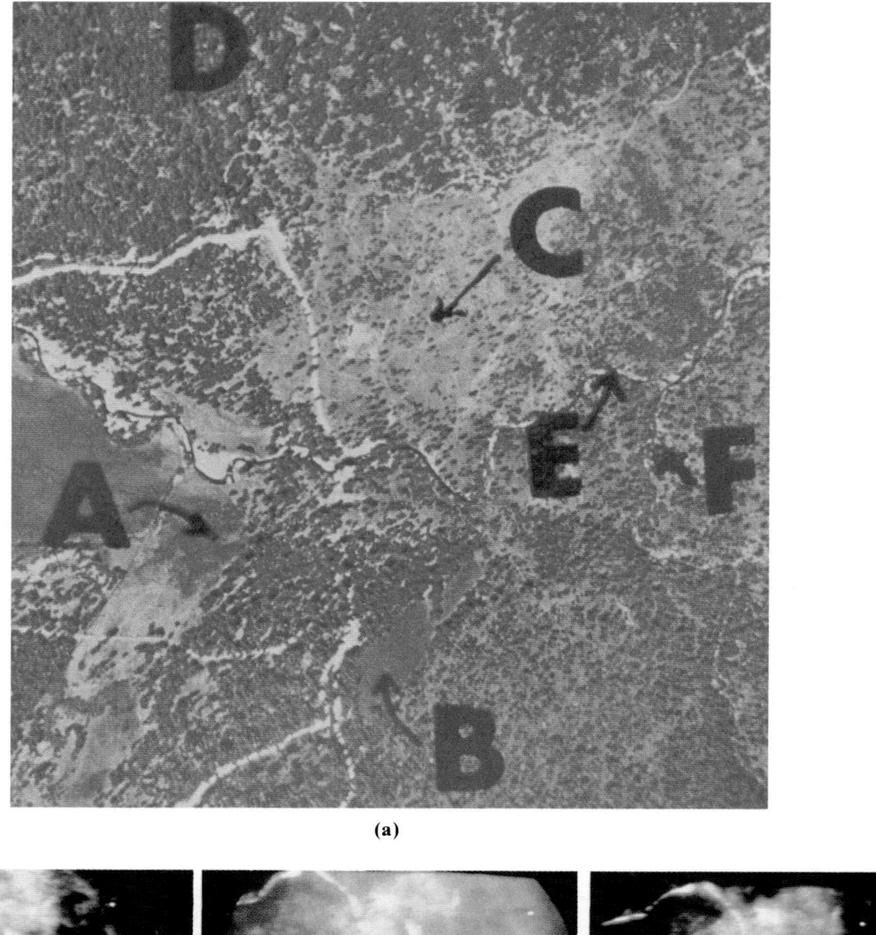

(a)

1 2 3

(b)

Fig. 7-10: (a) Despite the high quality of this color infrared aerial photo, there is a need to produce a series of color enhancements the better to differentiate among the low shrubs at *A*, the sedges and rushes at *B*, and the cordgrass just to the left of *A*. The use of an electrooptical device of the type described in the text has permitted each of these distinctions, in turn, to be made as illustrated in (b). (b) These image examples were produced through the use, as input to an electrooptical combiner, of three black-and-white photos taken in the green, red, and near-infrared wavelengths regions, respectively—the same regions as were used in taking the color infrared photo shown in (a), and covering the same geographic area. The advantage of this device was in permitting each of the three black-and-white images to be density sliced and color coded in order to produce any of a large number of enhanced images, including the ones shown here. Reading from left to right, these three color-enhanced images permit (1) the sedges and rushes at *B* of (a) to be differentiated from everything else; (2) the cordgrass to the left of *A* to be differentiated from everything else; and (3) the low shrubs at *A* from everything else. The dial settings of the electrooptical device required to produce such enhancements are established and recorded through the use of a "calibration area" such as this one, for which ground truth is known. Then, as successive frames of photography along the same flight line are to be inventoried for the presence of these same resource categories, each of these dial-setting combinations, in turn, is used to enhance the input photographs. This permits each resource category, in turn, to be differentiated from everything else, and mapped on each successive frame, thereby producing a resource map of the entire area. Such a map could not have been prepared from the color infrared photography itself, because of inadequate color differences exhibited on such photography among the resource categories to be mapped. This same technique can be successfully applied to space photography, as well. (Photos courtesy of George Dalke of the University of Kansas and John Estes of the University of California, Santa Barbara.)

bined into a single, color-coded, composite photograph, as displayed on the color monitor (i.e., the viewing screen) of a closed-circuit color television system. (See Fig. 7-10(a) and (b) for an image example.) The device that produced this color composite consists of a closed-circuit color television system that is equipped with three cameras. As with an optical combiner, the input for this system consists of three photographs of the same scene taken with three different combinations of photographic film and filter or else on three different dates, but with the same film–filter combination. Each camera scans one of the three photos and each is linked to one of the three phosphors (red, blue, and green) on the color monitor. As the three cameras are silmultaneously scanning the three black-and-white photos, the operator can "level select" for the tone or brightness range on each of the three images for the purpose of highlighting and color coding certain unique spectral densities (shades of gray on the black-and-white photos that are used as input). Ideally, this process results in a unique color coding on the TV monitor for each type of feature that is to be identified. An example of this kind of image enhancement will be found in Fig. 7-10(b).

Means by Which "Synthetic Stereo" Can Be Produced from Landsat Imagery

Fig. 7-11 demonstrates a capability that is of great interest to those who seek to inventory, monitor, and manage the earth's land resources. Such persons usually have become accustomed to using conventional *aerial* photography, rather than *space* photography, for such purposes. Hence, they are likely to find that the greatest limitation to the use of space photography lies in the fact that it ordinarily cannot be viewed three dimensionally. Because of this limitation, one of the greatest clues to the identification of a resource feature, i.e., the topographic site that normally is associated with it, cannot be discerned with sufficient certainty.

The stereogram comprising Fig. 7-11 is a Landsat Multispectral Scanner image (bands 4, 5, and 7) that has been merged with digital terrain data to produce a "synthetic stereo" image at a scale of 1:500000. The area shown is a portion of the Paradox Basin in Utah. The *right* member of this stereo pair has been produced in the conventional manner, merely through the use of the MSS digital data. As a result, features within that image have essentially no displacement due to relief, yet relief displacement is the attribute required in a stereoscopic pair of photographs if a three-dimensional impression of the terrain is to be obtained through a study of them. When the *left* member of this stereo pair was being produced, however, the MSS data were differentially displaced to the left or right, through use of elevational data obtained, pixel by pixel from the digital terrain data scan, by an amount equal to the elevation times 0.6 (the base-to-height ratio). By this means, stereoscopic parallax was synthetically introduced into the scene so that features could be perceived three dimensionally when the scene was viewed through a stereoscope. This stereoscopically enhanced product reveals a variety of lithologic, structural, and botanical phenomena not readily apparent on other enhancement products. Specifically,

1) Mancos shale appears as light blue over a large area in the north and it also outcrops at various places throughout the image;

2) stereo viewing reveals an anticline-syncline-anticline sequence proceeding southward from the LaSal Mountains;

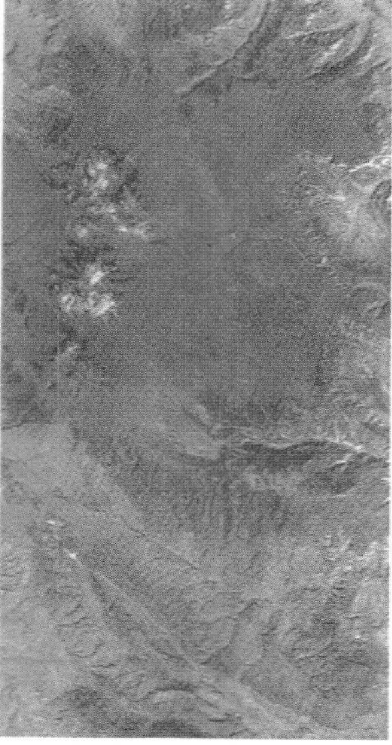

Fig. 7-11: A decorrelated Landsat MSS (4, 5, and 7) image of a portion of the Paradox Basin, Utah, has been merged with Digital Terrain Data to produce this synthetic stereo imagery at a scale of 1:500000. (North is at the top of the photo) The MSS data are differentially displaced east or west in the resampled DTD scan format by an amount equal to the elevation times 0.6 (the base-to-height ratio). As explained in the text, this product reveals a variety of lithologic, structural, and botanical phenomena not readily apparent on other enhancement products. (Prepared by Earth Satellite Corporation for Japan Petroleum Exploration Company, Ltd. and Earth Resources Satellite Data Analysis Center. Used by permission.)

3) the eye-shaped anticlinal structure in the southern third of the image is the Lisbon Valley anticline, notable for the Mancos outcrops along its southern limb and for the bleached formation at its northwestern end, associated with known underlying hydrocarbon reservoirs; and

4) across the LaSal Massif are NE-SW and NW-SE trending fault traces, enhanced by the stereo effect.

This example was prepared by Earth Satellite Corporation for Japan Petroleum Exploration Company, Ltd., and the Earth Resources Satellite Data Analysis Center.

The Construction of Oblique Perspective Views Using a Merger of Landsat MSS and Digital Terrain Tape Data

Just as the above mentioned kinds of data can be merged to produce, in the vertical view, a synthetic stereo image (as described in the preceding section) they also can be used to construct oblique views of the terrain and its resources, as seen from almost any vantage point, and at almost any degree of obliquity. Examples of such oblique perspective views appear in Fig. 7-12(a)–(c).

Description of a Process for Incorporating Many Kinds of Image Enhancement in a Digitally Acquired Space Photograph

The term "Geopic" is applied to a proprietary procedure that has been developed by Earth Satellite Corporation for producing enhanced and highly interpretable color composite imagery from Landsat MSS computer-compatible digital tape data. Many remote sensing scientists consider Geopic-enhanced Landsat imagery to be the most interpretable of all the forms of enhancement that have been attempted of Landsat MSS data to date. In view of the above, we deem it appropriate to provide the following description of the Geopic process, limited though the description must be because of the proprietary nature of the process.

EarthSat's "Geopic" images are produced from Landsat computer compatible tapes using algorithms for geometric and radiometric corrections, edge enhancement, gray scale adjustments, and scan-line suppression (de-striping). These corrections and enhancements are briefly described as follows:

1) *Geometric corrections* are made by using a series of algorithms in order to produce maximum geometric accuracies in the Geopic images. Systematic correction parameters include satellite pitch, roll, yaw, heading, and altitude variations; image skew caused by both earth rotation (by latitude) and finite scan time; spectral band offsets; mirror scan velocity; and panoramic corrections for earth curvature. Scale accuracy is maximized in all directions for easy superimposition of overlay maps, charts, or multidate Landsat scenes of the same area. [An example of the importance of making geometric corrections will later be seen by comparing Fig. 7-13(a) and (b).]

(a)

Fig. 7-12: (a) The merging of Landsat MSS data with digital terrain data, pixel-by-pixel, has permitted this oblique perspective view of the San Francisco Bay area to be produced, the better to portray the topographic site that is occupied by certain land resource features (e.g., the San Andreas Fault, along which highly linear lakes are aligned just above the center of the photograph). Water depths and water turbidity patterns also are accentuated by this color enhancement. Compare with (b). (b) A second color-enhanced oblique view prepared from the same input data, but looking in almost the opposite direction. In this color enhancement, differences in water conditions are largely lost but a very useful color coding to separate grasslands (yellowish-brown), forested areas (red), and urbanized areas (blue) has been employed. (Photos courtesy of Ralph Bernstein, IBM Scientific Center.) (c) This oblique view, made from Landsat MSS data merged with digital terrain data, provides an element of terrain appreciation that is not obtainable from non-stereoscopic vertical view, such as that shown for this same area in Fig. 13(a) and (b). All three figures are readily compared when it is realized that 1) Owens Lake, one of the most conspicuous features in all three views, is in the upper left quadrant of each photo, identifiable in each case by its nearly circular shape and its conspicuous island, and 2) this oblique view looks on the area from south to north with the snow-capped Sierra Nevada and White Mountains occupying the left and right edges of the photo, respectively. (Photo courtesy of the Remote Sensing Laboratory, Stanford University.)

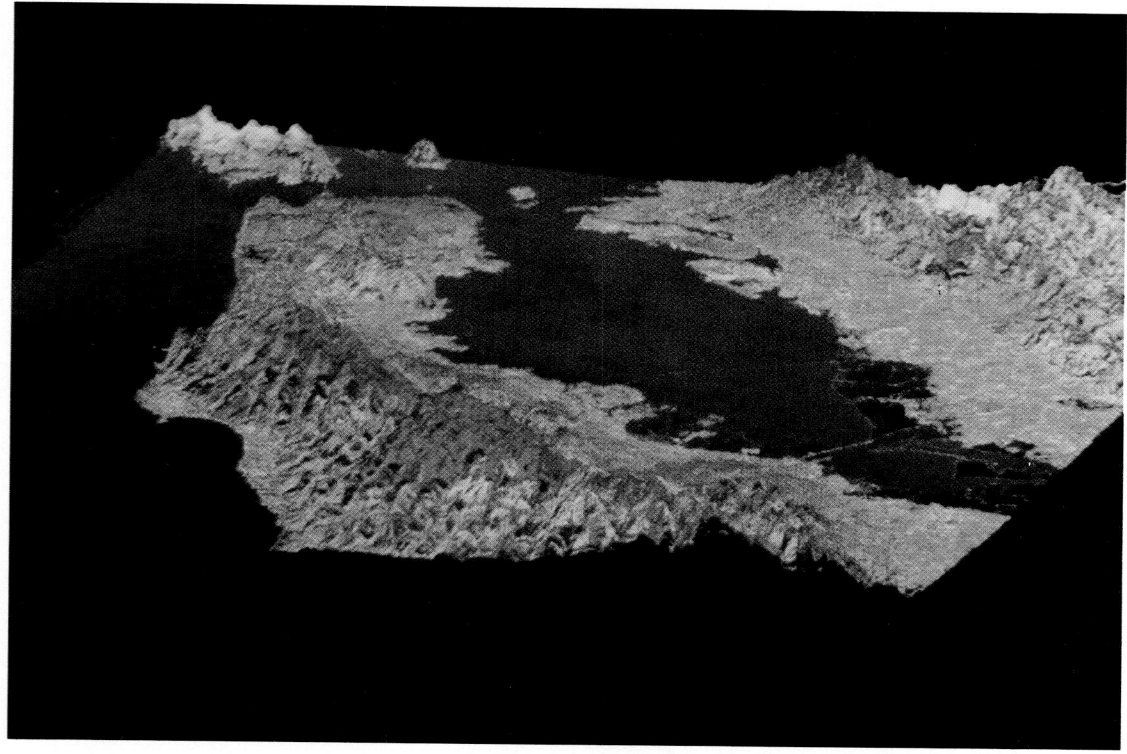

(b)

(c)

Fig. 7-12 (*Continued.*)

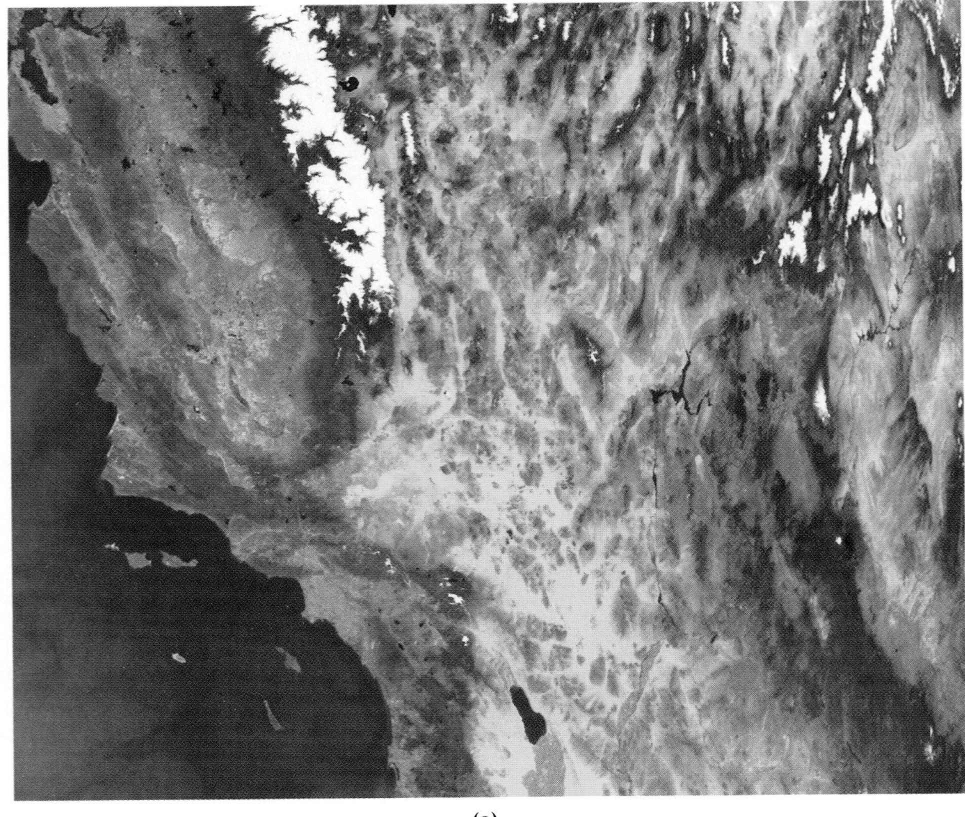

(a)

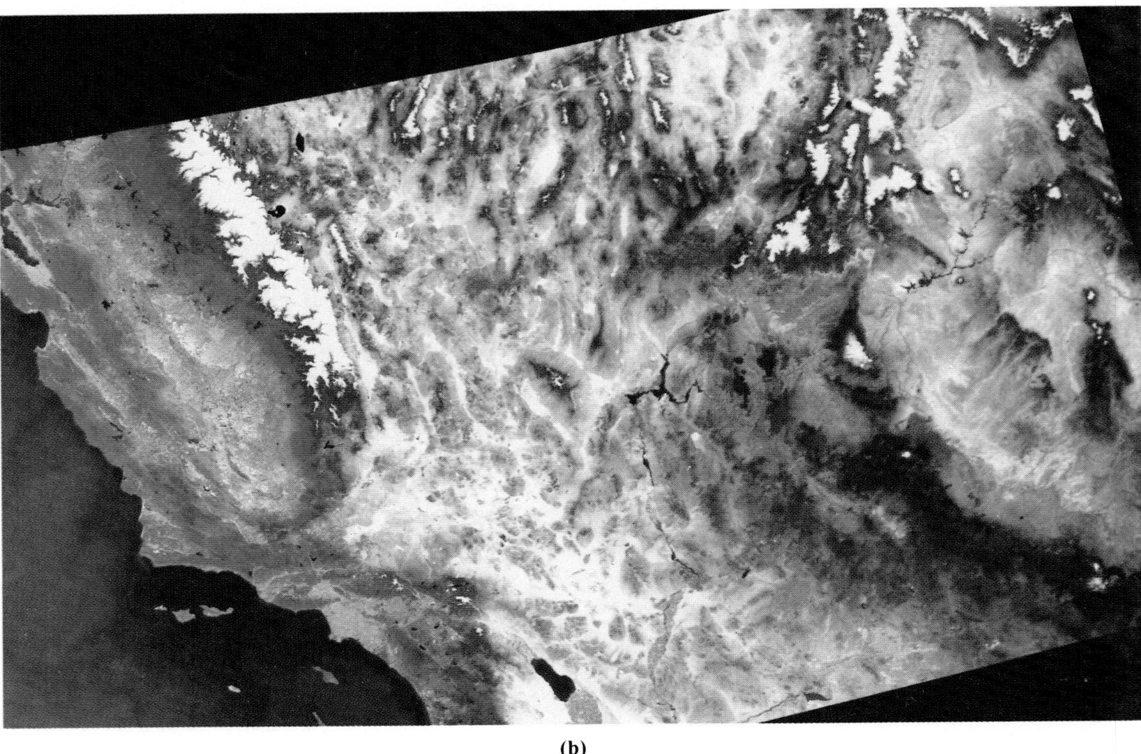

(b)

Fig. 7-13: (a) New and very useful methods are being developed for portraying vast portions of the earth's surface form space, as exemplified by this imagery of the southwestern United States. Near the top left of this scene is San Francisco Bay; near the far right center is the "four corners" area where the boundaries of Utah, Colorado, New Mexico, and Arizona intersect; near the bottom center a part of northern Mexico can be seen. Other prominent features include the snow-covered Sierra Nevada Mountains (top left quadrant); Los Angeles and the Channel Islands (bottom left quadrant); aircraft vapor trails over Nevada's "basin and range" topography (top right quadrant); and the Roosevelt Lake area of southeastern Arizona (bottom right quadrant). Personnel of the Environmental Research Institute of Michigan (ERIM) generated this imagery from scene

2) *Radiometric corrections* are made in order to maximize the uniformity of the tone and/or color with which features of a given category are imaged throughout any given Landsat scene. Such corrections are necessary, partly because of the fact that the six rows of four-band detectors used in the Landsat Multispectral Scanner System (each subject to its own variable gains) are operative with each oscillation of the MSS scanning mirror. To correct for the consequent apparent variations in scene brightness within any given Landsat scene, representative scene brightness histograms are made separately for six adjoining scan lines as registered by the six detectors. A comparative analysis of these histograms permits calibration corrections to be made as necessary to achieve maximum scan-line suppression, thereby improving radiometric fidelity.

3) *Edge enhancement corrections* are made in order to increase contrasts in color and density along feature boundaries within a Landsat image. The objective of this process is to improve the analyst's ability to rapidly identify linear and other border features such as field boundaries, land-use interfaces, marked relief or geologic structural changes, water-body limits, soil-type boundaries, and lithologic changes.

4) *Gray scale adjustments* are made by using a contrast enhancement algorithm which stretches the range of reflectance values to maximize image information content. Gray scale adjustments can also be manipulated to visually highlight image areas of special interest.

5) *Scan-line suppression:* An additional algorithm is employed in the "Geopic" image enhancement process to suppress regularly spaced scan lines (i.e., striping) in an image caused by repetitive malfunctions of the MSS detectors.

6) *Color coding:* Once the above corrections have been applied individually to MSS bands 4, 5, and 7 for a given Landsat scene, various color codings of the composite multiband image can be accomplished. One of these, designed to produce a simulated color infrared composite, is produced of that scene by assigning blue, green, and red dyes to bands 4, 5, and 7, respectively, and optically combining them by an additive color process into a single color composite image. Both the color composite image and the black-and-white single-band images from which it was derived can be produced at scales up to 1:100000, or even larger, and in either opaque print or transparency form. In addition to the color code that is used in making the simulated color infrared composite, other color codes are frequently employed in order to increase the interpretability of specific features that are of primary interest to the user.

AN INTEGRATED SET OF ILLUSTRATIONS

Most of the users or potential users of space photography for the inventory, monitoring, and management of land resources are concerned with making *regional* or *subregional* applications of such photography. In addition, they usually are concerned with the making of these applications in some *political entity*, such as a county or state, because certain resource management laws and regulations that must be adhered to often are different in one political entity than another. Hence, in order to avoid a fragmented presentation within this section, all of the examples of space photography that are contained in it have been taken from a single region—that extending from the southern part of the Rocky Mountains, westward to the Pacific Coast, with the state of California serving as both the *subregion* and the *political entity* dealt with in most of the examples. Generally speaking, the examples are arranged in such a way as to provide a progression from vast area coverage at limited spatial resolution to progressively more limited area coverage, but at progressively higher spatial resolution.

Three Examples of NOAA Satellite Imagery

Fig. 7-13(a) consists of a color coded presentation for almost all of the entire region encompassed by the examples that follow it, i.e., Figs. 7-13 to 7-21. That figure was produced as follows.

Three bands of imagery acquired by the Coastal Zone Color Scanner on board the Nimbus 7 spacecraft were used as input data. A positive image from the blue band (0.43–0.45 μm) was printed in green; a positive image from the scanner's near-infrared band (0.70–0.80 μm) was printed in red; and a negative image from the scanner's thermal infrared band (10.5–12.5 μm) was printed in blue. From its orbital altitude of 955 km, this wide angle scanner covered a swath width of approximately 1800 km. The consequent foreshortening of features at the edge of the swath is particularly noticeable on the right edge, but can be geometrically corrected as in Fig. 7-13(b).

Fig. 7-13(b) shows essentially the same area as that appearing in the preceding figure, but with two differences: 1) the scene has been geometrically corrected in order to provide a nearly uniform scale throughout its entire extent, and 2) the area has been differently color coded. Specifically, in this instance, only two bands have

brightness data acquired simultaneously in three bands by the Coastal Zone Color Scanner on board the Nimbus 7 spacecraft. The bands used and the color assigned to each band were as stated in the accompanying text. (b) Shown here is the result of applying a geometric correction to (a), as necessary in order to portray the entire area at a uniform scale. In the process of making the correction, only two bands were employed, as follows: the blue band (0.43–0.45 μm) was color-coded in cyan and the near-infrared band (0.70–0.80 μm) was color-coded in red. The resulting color scheme, as compared with that used in (a), increases some contrasts (e.g., between urban and non-urban parts of the San Francisco Bay and Los Angeles areas) while reducing others (such as those in the basin-and-range topography near the center of each of the figures). (Photos courtesy of NOAA and the Environmental Research Institute of Michigan.)

been used, i.e., one from the blue part of the electromagnetic spectrum using wavelengths ranging from approximately 0.4 to 0.5 μm (color coded in red); and a near-infrared band with wavelengths ranging from approximately 0.7 to 0.9 μm (color coded in blue).

Fig. 7-14 shows a subset of the area shown in the preceding two figures and employing the same data-acquisition tape in conjunction with another one, taken at a later time of year, in order to color code major changes that occurred between the two dates.

Landsat Multispectral Scanner Image of an Area in Central California

By way of indicating the great potential usefulness of space photography for the inventory, monitoring, and

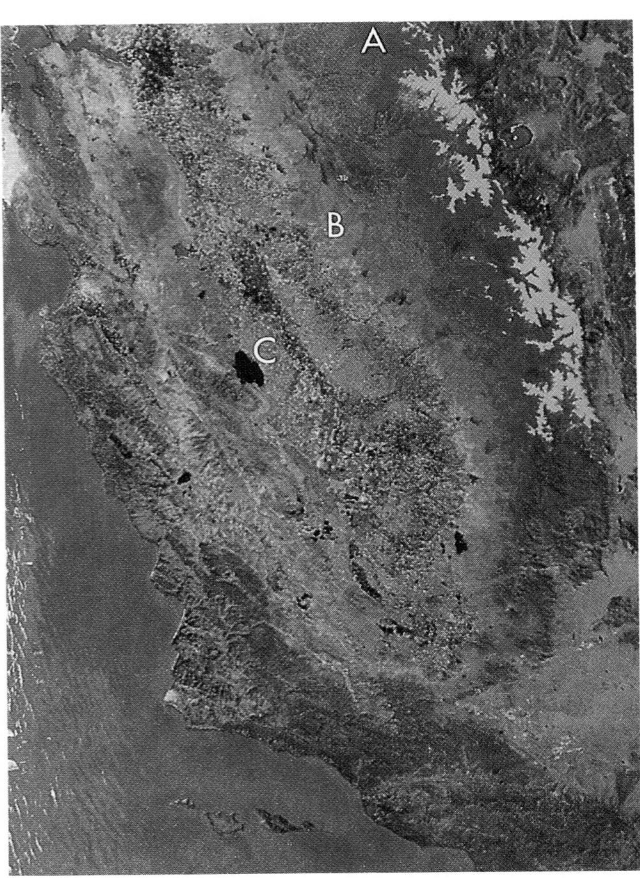

Fig. 7-14: This is a geometrically-corrected change-detection image prepared from two acquisitions of Nimbus 7 Coastal Zone Color Scanner data, covering western California from San Francisco to Los Angeles. It indicates changes that occurred between May 10 and June 13, 1979. It was prepared from a digitally merged two-data file of CZCS data. It illustrates what appears to be a promising way of cost-effectively monitoring rapid changes over large areas. The image was prepared using the following color code for the three separates:

	Color Code
June CH2 (0.51–0.53 μm)—May CH2	Red
May CH6 (10.5–12.5 μm)	Blue
June CH5 (0.70–0.90 μm)	Green

Among the major changes that have taken place are (A) Dark blue—reduction in snow pack; (B) Orange/yellow—senescence of green grassland vegetation; and (C) Black—wildfires. (Image courtesy of John E. Colwell of the Applications Division, Resources and Technology, Environmental Research Institute of Michigan.)

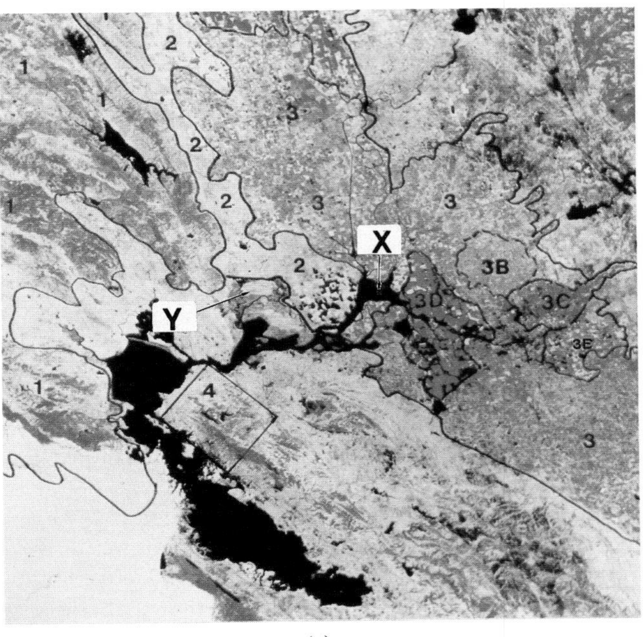

(a)

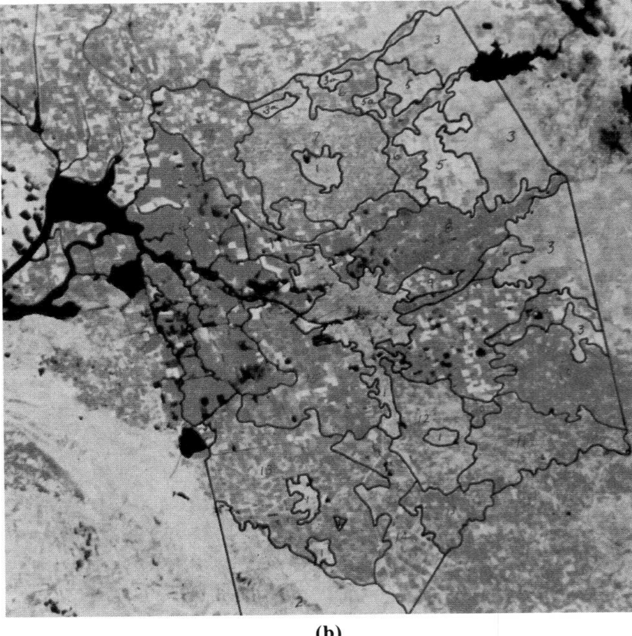

(b)

Fig. 7-15: (a) On this single frame of Landsat multispectral scanner imagery, a major portion of north central California is shown, extending from the San Francisco Bay region in the lower left corner (partly obscured by fog) across the Great Central Valley, and into the Sierra Nevada Mountains in the upper right. For a discussion of the annotated features, see text. (b) Enlargement of the right central portion of (a). (Photos courtesy of NASA.)

management of land resources, let us attempt to make a multidisciplinary analysis of one representative frame of it, i.e., the one comprising Fig. 7-15. This particular frame, as acquired by the Landsat Multispectral Scanner (MSS), covers a vast portion of central California, extending from the fog-enshrouded Pacific Coast and San Francisco Bay (lower left) across the Great Central Valley to the foothills of the Sierra Nevada mountains (upper right). This imagery was obtained on July 26, 1972, less than 72 hours after Landsat 1 had been launched. In terms of image quality, it is highly representative of that still being obtained from Landsat MSS imagery.

Scientists are studying the extent to which various kinds of earth resource information can be extracted from such imagery. Some of the significance of these studies comes from the fact that California serves as a multistage, multidisciplinary test site. During the past several years, tremendous amounts of information have been compiled for the area shown in Fig. 7-15 through the conducting of conventional on-the-ground resource surveys. A comparison of such information with the image appearance of resource features helps the analyst develop the ability to interpret similar features in other parts of the globe that are highly analogous to this area in terms of climate, geology, and topography, but where the earth resources are far less well inventoried and developed.

From a study of the broad area labeled 1 in Fig. 7-15, the experienced photo interpreter could readily discover a series of four or five mountain ranges and consequently conclude that the topography is too rugged and the soil too shallow throughout most of the area to permit the production of agricultural crops. In this area, the bright red coloration of certain features, as seen in this "false color" enhancement, indicates the presence of trees and shrubs that have remained green even during the dry summer period when this photograph was taken. Such color characteristics contrast sharply with the lighter toned, yellowish sections of Area 1, most of which are covered with various annual grasses that have turned brown by this season of the year.

The healthier and more succulent the vegetation, the brighter red it appears on this imagery. This is an excellent example of the usefulness for highlighting certain significant ecological relationships which the broad synoptic view from space provides. Specifically, within Area 1 the succession of mountain ranges generally parallels the Pacific Coast (which is covered by fog in the lower left of the photo). With each range farther inland, the coastal influences that tend to keep the vegetation from drying up during the hot California summer become progressively less pronounced and hence the mountain ranges appear progressively less red and more yellowish.

In Area 2 of Fig. 7-15, the characteristic midsummer signature of dry-land agriculture as practiced on this rolling foothill topography and on the upper slopes of the valley floor is seen. Most of the individual fields are large and exhibit, on this July 26 imagery, the characteristic light tone distinctive of mature crops. Interspersed with them are fields that are dark in tone owing to the recent plowing of stubble from the annual cereal crops (wheat, oats, and barley) that comprise the dominant dry-land agriculture of such areas.

Area 3 shows a portion of the Great Central Valley of California—a valley nearly 400 miles long and 50 miles wide—which has been called "the largest and most fertile flat land area in the world." Near the center of Area 3, an area comprised of the southern limits of the Sacramento Valley and the northern limits of the San Joaquin Valley (which collectively constitute the Great Central Valley) is defined by the confluence of the south-flowing Sacramento River with the north-flowing San Joaquin River, forming a large delta region. Much of the delta was flodded in the spring of 1972 when one of the levees of the San Joaquin River gave way. The portion that still remained flooded on July 26 of that year is marked with an *X*. From sequential space photographs obtained from Landsat 1 every 18 days, the rate at which the flood waters were receding could be determined, as seen in Fig. 7-16. On-the-ground observations combined with the interpretation of these space photos provided a valuable measure of the correlation between 1) the length of time during which an area was inundated at this time of year, 2) reductions in crop yield, and 3) the mortality of orchard trees.

Within Area 3 of Fig. 7-15 several soil types and associated crops can be accurately delineated. Only a few of these sub-areas have been outlined with a dashed line on Fig. 7-15, but all of them are highly significant in relation to the inventory and management of agricultural resources. For example, sub-area 3*A* consists of heavy clay soils devoted mainly to the production of rice. Because this irrigated rice was by far the most succulent crop still present in late July, the fields are readily definable by their bright red coloration. In marked contrast, sub-area 3*B* has sandy soils, and vineyards are the predominant crops grown there.

The soils of sub-area 3*C* are unusually deep and well drained, with the result that fruit and nut orchards predominate, since orchards produce highly valuable crops per unit area and usually require such soils. Sub-area 3*D* is a delta region formed at the confluence of the Sacramento and San Joaquin Rivers. Its "peaty-muck" soils support mainly corn, sugar beets, and asparagus. Sub-area 3*E* contains soils of intermediate fertility, depth, and irrigability, and safflower as well as small grains predominate in this part of the valley. (Fig. 7-17 shows an enlargement of one small part of this area.)

Area 4 of Fig. 7-15 includes two small reservoirs, the necessity of which can be inferred from the broad synoptic view provided by this single Landsat frame of imagery. These two reservoirs, and several similar ones nearby, are used primarily to provide temporary storage of water transported by pipelines from the main freshwater source, the Sierra Nevada Mountains (upper right), for use in the large centers of population that fringe San Francisco Bay. Without such a storage facility, the pipelines would have to be of far greater size and number to obviate water shortages in the Bay Area cities during periods of peak use, such as the occasional three- to five-day hot spells during the summer.

Area *Y* contains one type of feature that is of particular interest to the geologist who is interested in prospect-

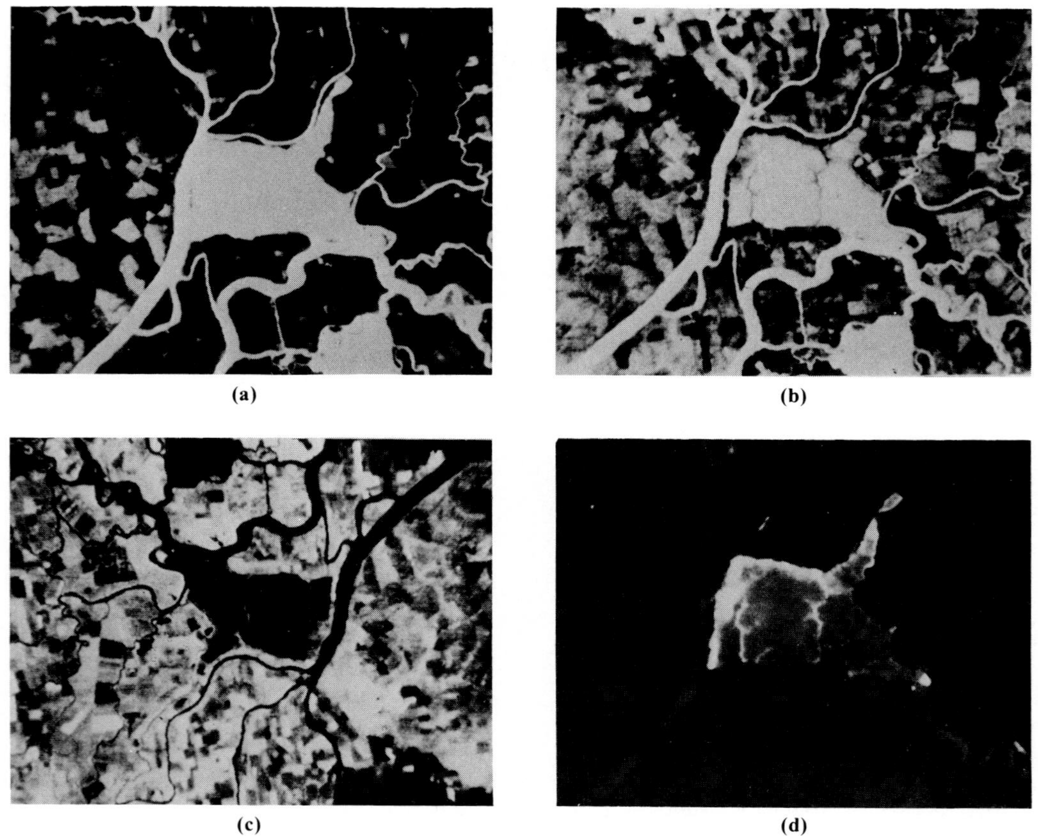

Fig. 7-16: Change detection as accomplished by the comparison of two dates of Landsat MSS imagery. These four Landsat photos pertain to the area in California's Great Central Valley that is labeled *X* on Fig. 15(a), an area within which flood waters, from a levee break along the north bank of the San Joaquin River, were in the process of receding rapidly. All four photos were obtained from Band 7 (a near-infrared band) of the multispectral scanner. (a) Negative rendition of the image acquired on July 26, 1972. (b) The same area as imaged by the Landsat MSS 36 days later. (c) A positive rendition made from the negative image that appears in the top right illustration. This rendition was made so that it could be superimposed over the negative transparency made from the top left illustration, thereby facilitating "change detection." (d) The composite image formed by this superposing of the two images, both in transparency form, over a light table. Note that the light toned areas are the ones from which flood waters have receded during the interim.

ing for certain minerals or for petroleum. It is readily recognized as a "topographic high" because it is almost completely surrounded by marshland, readily identified by the meandering streams and the characteristic reddish brown coloration registered by its sedges, rushes, and other marshland species of vegetation. Furthermore, the main axis of this feature trends E–W, which is about 45° out of phase with most other features in this area, such as the mountains, valleys, and reservoirs. Of further interest to the geologist is the fact that, within this large feature of elliptical shape, he could discern, even on the original space photograph, the smaller concentric ellipses that are indicative of a geologic structure known as an "anticline." When a closer look at this feature (as in high resolution *aerial* photos) shows the geologist that it is, indeed, an anticline, he becomes even more interested because of the increased likelihood that he might find petroleum or natural gas beneath the center of it. This increased likelihood results from the fact that the upwarped rock strata of the anticline would tend to trap these relatively light petroleum products, if any are present in the area, when they floated to the top of the subsurface water table. Consistent with predictions of this type, which a geologist frequently is able to make, this anticline is one of the few oil and gas producers in the entire area shown on this space photograph. Consequently, this example provides yet another illustration of the value of a space photograph in providing the first stage "look" at a vast area, thereby permitting the image analyst to select the few specific portions of it which merit more detailed study.

Enlargement of One Small Part of a Landsat Multispectral Scanner Image, Showing Agricultural Crops in California's Great Central Valley

Each portion of a typical frame of Landsat imagery, such as that appearing in Fig. 7-15, shows far greater detail than might have been suspected until any such portion is greatly enlarged, as in Fig. 7-17. From a study of this particular enlargement, a great many kinds of identifications can be made with an accuracy of nearly 100 percent, once the various kinds of features within the scene have been "calibrated" by means of direct on-site

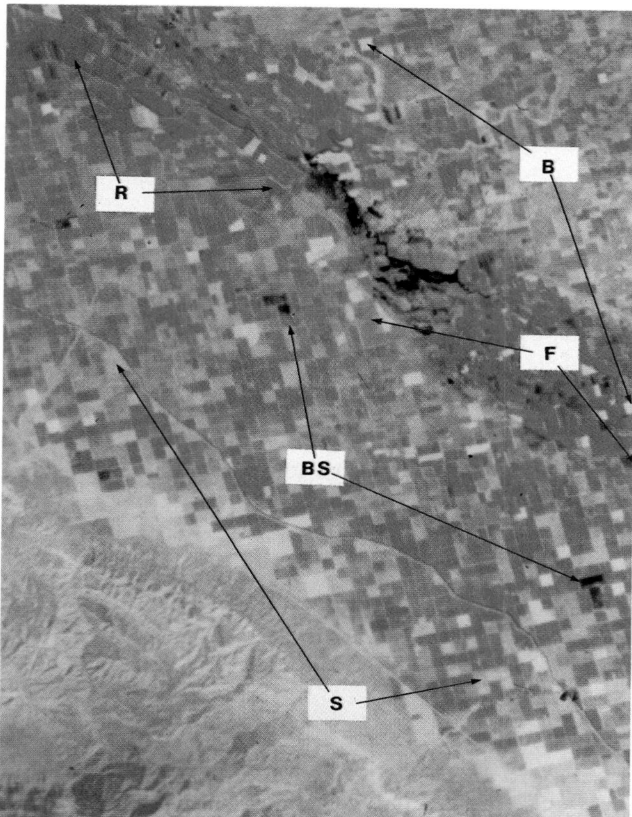

Fig. 7-17: Simulated Infrared Ektachrome imagery, greatly enlarged, of an agricultural area in California's Great Central Valley, taken from an altitude of 570 mi by the Landsat multispectral scanner (MSS). For an explanation of the means by which major crop types can be identified on imagery such as this, see text. Scale approximately 1:100 000. (Photo courtesy of NASA.)

inspection of a few representative examples of each of them. Examples of such identifications are as follows:

1) All fields that exhibit a *light yellow* color in the middle of the growing season, such as the fields labeled 1, are *barley* fields that have matured and are ready for harvest.

2) All fields that exhibit a more *saturated yellow* color, such as those labeled 2, are mature *safflower* fields.

3) All fields that have a *highly saturated red* color, such as those labeled 3, are *rice* fields, resulting from the fact that rice is essentially the only crop that is still in the succulent green stage this far into the growing season.

4) All fields that exhibit a *gray to black* coloration are *recently harvested barley* fields within which the stubble has since been burned, in conformity with standard practice throughout this region.

5) All fields that have a *blue* coloration, such as those labeled 5, are in the *fallow* state, although this category may include some fields that have been planted so late in the growing season that there is not yet sufficient green foliage to register on the image (e.g., cotton which is planted late because it would have been highly susceptible to frost damage had it been planted too early in the spring).

6) Most of the remaining fields are registered here in *various shades of red,* merely indicating that *healthy crops* are growing in them, but the crop in any such field might be any of several different kinds.

Further and sometimes more positive identifications can be made with respect to such fields, however, from a comparative analysis of one Landsat image, such as that shown in Fig. 7-17, with a second image of the same area as recorded by Landsat 18 days later (assuming, as with Landsats 1, 2, and 3 that the repeat cycle is every 18 days) or 16 days later (as with Landsats 4 and 5). Thus, for example:

1) Some of the fields which in Fig. 7-17 were in unidentifiable shades of red, will have turned a unique shade of *yellow* at the later date and virtually every such field, on ground checking, will be found to contain a mature crop of *seed alfalfa.*

2) Some of the fields that were fallow in Fig. 7-17, and therefore had a blue coloration, will exhibit a *pink* coloration on the later data and in most cases will prove to be *cotton.*

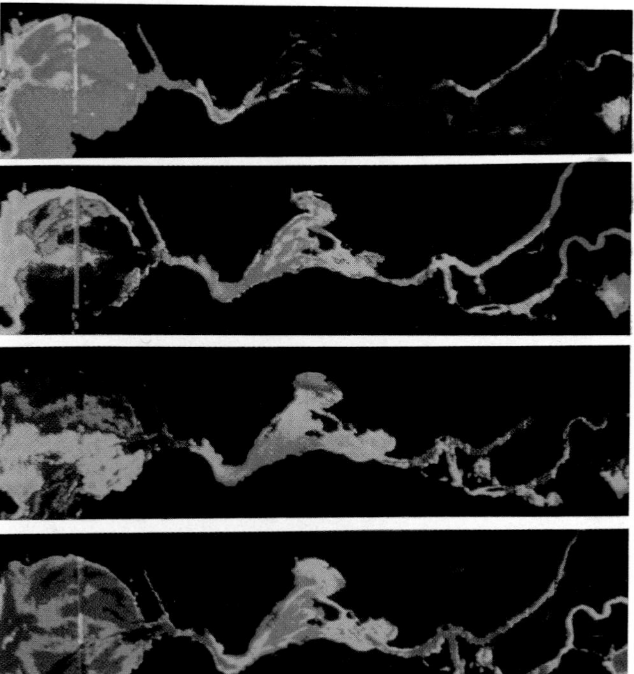

Fig. 7-18: Four Landsat MSS color-coded images of California's Sacramento River and the upper end of the San Francisco Bay into which it flows. From top to bottom these figures have been color-coded in terms of the water's salinity, chlorophyll content, suspended solids, and turbidity, respectively. In each case, "boat truth" was simultaneously acquired with respect to the water-quality parameter of interest. Such data, as collected for many representative portions of the area shown here, permitted regressions to be established that related the concentration of the parameter in question to Landsat MSS spectral reflectance values in bands 4, 5, and 7. The progression, from low to high concentration of the parameter dealt with, is from blue to red in each of these images. (Courtesy of [22] and the *Manual of Remote Sensing,* 2nd ed., 1983. Copyright the American Society for Photogrammetry and Remote Sensing, used by permission.)

3) Any given field that was correctly identified in Fig. 7-17 as a barley field (from its light yellowish appearance) is likely to exhibit, on the later date, a gray to black color associated with such a field in the "burned stubble" state, thereby confirming the accuracy of the earlier identification of it.

The Color Coding of Parameters that Govern Water Quality

As indicated by the captions that accompany Figs. 7-18 to 7-20, it is possible to color-code certain water-quality attributes in bay and delta regions using "water truth," as acquired simultaneously from boats of representative portions of the area at the time of the overflight by a spacecraft. The color-coding techniques that were employed in producing these figures were the same as previously described in the section dealing with electrooptical methods of image enhancement.

Applications of Space Photography in the Inventory, Monitoring, and Management of Rangeland Resources

As indicated by the caption that accompanies Fig. 7-21, it is possible to inventory and monitor forage and related attributes of rangelands with the aid of space photography, augmented as necessary with limited

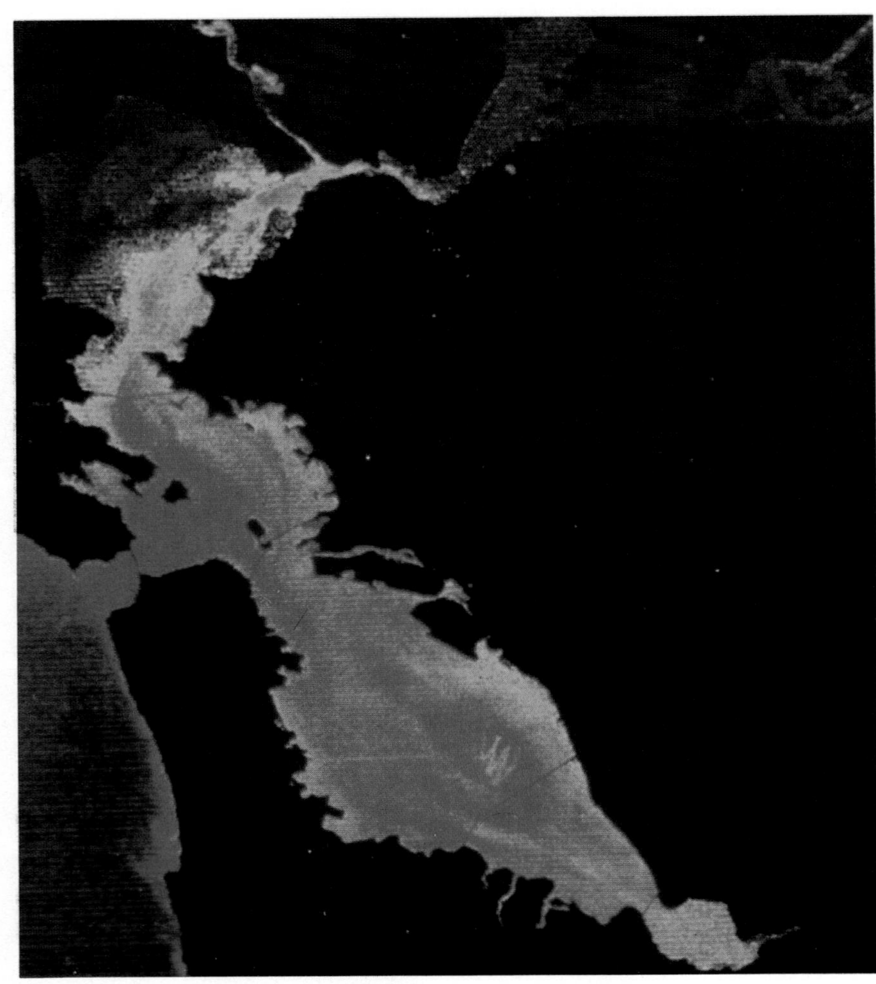

Fig. 7-19: Color-coded Landsat-derived map of the entire San Francisco Bay, together with that of the Pacific Ocean (left) and lower reaches of the Sacramento River (top one-fourth of the photo). Land is registered here in black. Note the San Francisco peninsula in the lower left quadrant and bridges spanning San Francisco Bay. As indicated by the following legend, the color-coding used here provides a progression from the blue to the red end of the visible spectrum that is indicative of increasingly higher concentrations of water salinity. (From the Manual of Remote Sensing, 2nd ed., 1983. Copyright by the American Society for Photogrammetry and Remote Sensing, used with permission.)

Color	Salinity in parts per thousand
Blue	0–5.0
Light blue	5.1–9.8
Light green	9.9–14.8
Green	14.9–20.0
Yellow	20.1–24.9
Red	25.0–29.9
Brown	30.0–35.0

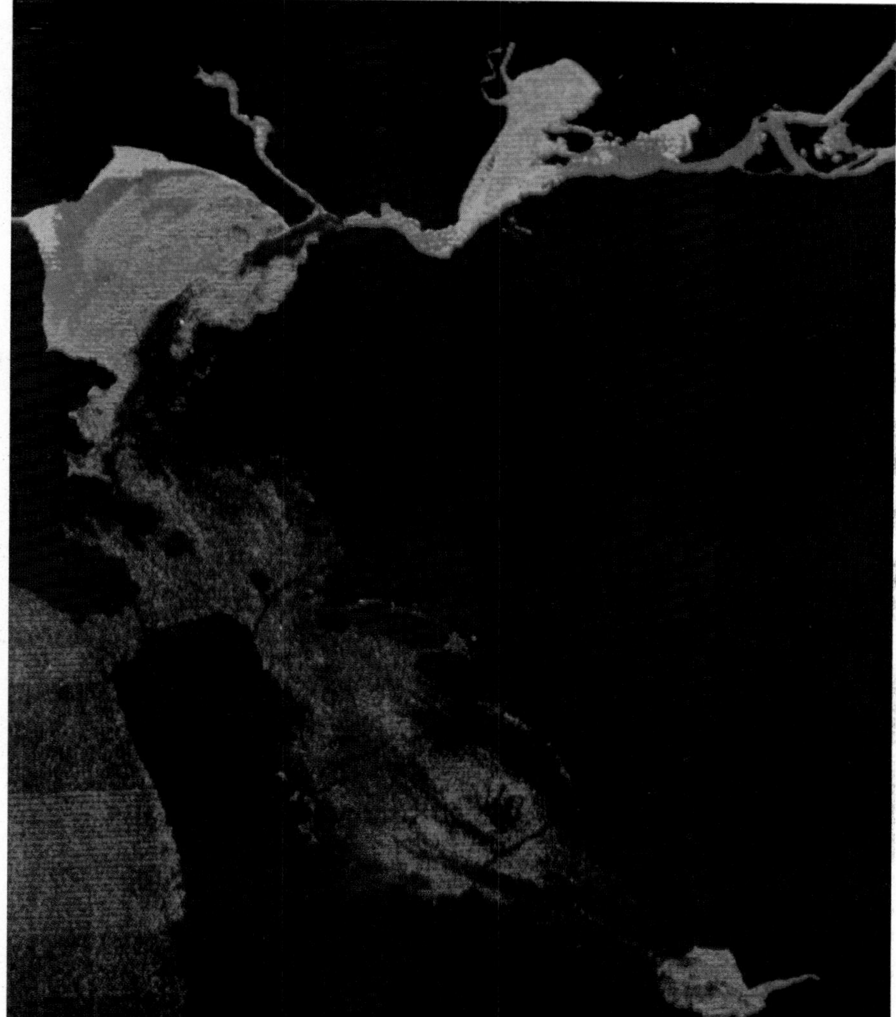

Fig. 7-20: Color-coded Landsat-derived *water turbidity* map of the entire San Francisco Bay and the lower reaches of the Sacramento River (top one-fourth of the photo). In keeping with the color-coding used in Fig. 7-19, the progression of color from the blue to the red end of the spectrum is indicative of the increasingly higher concentrations of water turbidity. (From the Manual of Remote Sensing, 2nd ed., 1983. Copyright by the American Society for Photogrammetry and Remote Sensing.)

Color	Turbidity in NTU
Dark blue	0–1.4
Blue	1.5–3.5
Light blue	3.6–5.6
Light green	5.7–7.6
Green	7.7–10.4
Yellow	10.5–15.9
Orange	16.0–20.7
Red	20.8–25.4
Brown	25.5–73.5

amounts of low-altitude aerial photography and simultaneously acquired ground observations.

In many wildland areas dead vegetation continues to accumulate each year, assuring the potential for high-intensity wildfires that are difficult to control and are highly damaging to resources, lives, and property.

Managing this hazard can be accomplished by fuel modification methods that fall under five broad classifications: biological, chemical, manual, mechanical, and prescribed fire. These methods can be applied in various combinations to any vegetative type to obtain specific desired objectives within existing physical, environmental, ecological, social, and legal constraints. To this end, the managers of wildland areas are developing and implementing intensive plans for fuel management, entailing construction of firebreaks and/or greenbelts to break up large expanses of highly flammable vegetation. The plans also cover safer access for firefighters and their equipment to enable them to control wildfires at small sizes and with less damage to resources.

A key element in developing these plans is the use of remote sensing by which space images are acquired for

Fig. 7-21 (*Caption on following page.*)

(e)

Fig. 7-21: The potential usefulness of Landsat multispectral scanner imagery for the inventory of range grass and the monitoring of "range readiness" is shown on this multidate photography of an area in west central California. For each of the five dates the Landsat image is accompanied by a terrestrial photo taken from within the area on the same date. Note that the period of maximum greenness of the vegetation, as seen on the terrestrial photo, corresponds to the period of maximum redness on the Landsat imagery. By May 10 of this particular year, most of the range grass in the right half of the photo (an area of rain-and/or fog-shadow) has turned brown and has passed the optimum period of readiness for grazing by livestock. (Courtesy of David Carneggie; from the *Manual of Remote Sensing,* 2nd ed., 1983. Copyright by the American Society for Photogrammetry and Remote Sensing, used with permission.)

those areas where severe fuel management problems are known to exist. Remote sensing data, augmented with ancillary data, provide a base for developing plans. Information derived from remote sensing is desirable because it can be gathered quickly and frequently for extensive inaccessible areas, and can be easily manipulated to meet fuel management objectives.

Recently, in California's Mendocino County, the Board of Supervisors, recognizing the need for a comprehensive wildland fuel management plan, established a Fuel Control and Brush Range Management Committee that includes private land owners as well as representatives from county, state, and federal agencies. Its purpose is to develop and implement fuel management plans for areas of high fire hazard throughout the county. Remote sensing experts are assisting the FMC in this effort.

An experimental wildland area of 476000 acres has been established in Mendocino County. Space imagery of this area was taken from an altitude of 570 miles by a Landsat satellite. Detailed classification of vegetation (fuel) types has been made from this imagery [see Fig. 7-22(a)] through the use of computer-assisted techniques at a cost of approximately one cent per acre. An overlay to this map, showing property ownerships, is then superimposed. In addition, tabular information giving the acreage for each fuel class within the study area is provided. As shown in Fig. 7-22(a), it is possible from computer analysis of Landsat data to identify and map, in color-coded form, the seven wildland fuel classes that have been defined by fuel specialists from the California Department of Forestry and the U.S. Forest Service, who have major fire control responsibilities in the area. Of particular interest are the stands of highly volatile chamise (yellow), especially when they occur adjacent to valuable stands of commercial conifer (dark green). In addition, the location of areas classified as barren/grassland (blue) is of interest in relation to fuel management planning because it is within these areas that man-made firebreaks can best be tried.

The classified map has been further simplified, as seen in Fig. 7-22(b), so that only the chamise fuel class (dark gray) is displayed. Such a presentation graphically illustrates where this highly volatile fuel is concentrated within the study area and where fuel management planning should therefore begin. Based on an analysis of this and other remote sensing derived information, augmented by such ancillary data as soil and topographic maps, portions of the heavy stands of chamise are presently being modified through the use of prescribed burning, i.e., the burning of vegetation in a predetermined manner under carefully controlled conditions. Further plans are being developed to modify the remaining chamise stands.

Future work in wildland areas will focus on maximum integration of remote sensing technology with the government-sponsored Coordinated Resource Planning, a program aimed at improved resource management through the cooperative efforts of private land owners and public agencies under the leadership of the U.S. Forest Service.

Achieving the following objectives of Coordinated

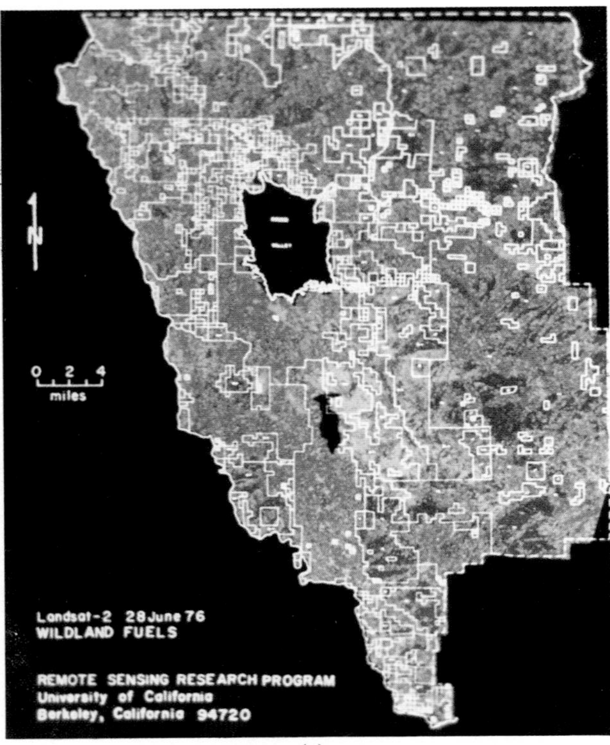

(a)

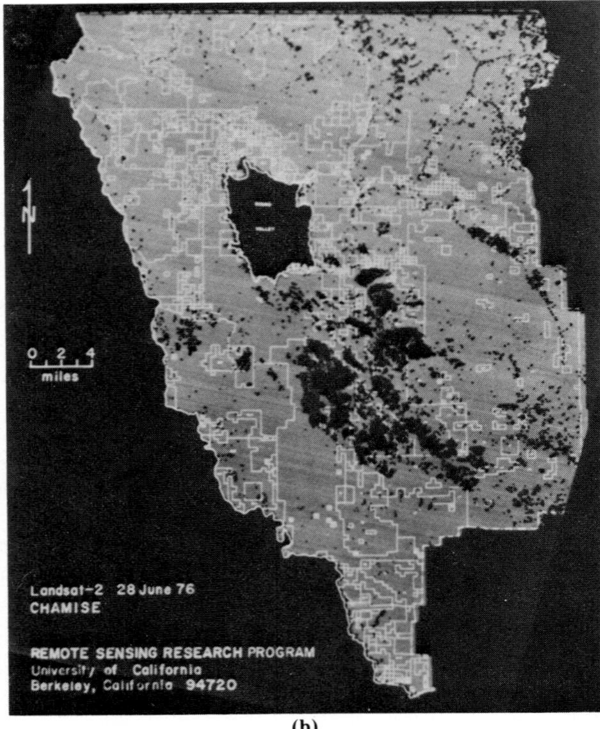

(b)

Fig. 7-22: (a) This wildland fuels map for a 476 000-acre area in the California Coast Range was produced through the use of computer-assisted techniques applied to an image acquired by the Landsat satellite. The areas of high fire hazard (because of their dense stands of volatile chamise brush) are shown here in yellow and are seen, for the most part, to be in close proximity to valuable stands of commercial conifer (shown in green). (b) This simplification of (a) emphasizes by means of a dark gray tone the locations of highly flammable stands of chamise. Such a map indicates where active fuel modifications should begin, and hence is referred to as a "thematic map" for that particular attribute. As compared with the complexly color-coded presentation of (a), this simply-coded map is much more readily interpreted in terms of the single attribute that it seeks to emphasize.

Resource Planning, as applied to wildland areas, will rely heavily on remote sensing as the primary source of valuable information:

1) Improving the quality and quantity of forage and habitat for domestic animals and wildlife.

2) Maintaining and improving the harvest of forest products compatible with other resource values.

3) Managing the watershed so as to
   a) increase the quantity and quality of water; and
   b) prevent or reduce pollution, siltation, and erosion.

4) Providing for maximum public benefit from the land and its resources, including recreational benefits.

5) Improving the economic status of each ranch unit or other land parcel that is involved.

6) Reducing substantially the potential for disastrous fires.

Color Infrared Photograph Taken by Skylab Astronauts of the San Francisco Bay Area and Environs

It is to be emphasized that, unlike Fig. 7-15, which was produced from digital records of scene brightness by the means previously described (as recorded by scanner systems aboard unmanned satellites), a great many excellent photos have been taken in the past by Mercury, Gemini, and Apollo using conventional cameras and photographic films. Then, through a daring maneuver known as "splashdown" these brave astronauts have brought the exposed film back to earth. There it has been processed in the conventional manner to produce excellent photographs such as those shown in Fig. 7-24(a). Although this picture covers much the same area as that shown in Fig. 7-15, it was taken in the month of February, whereas Fig. 7-15 was taken in July. Hence, many features exhibit a much different seasonal appearance on this photo. Furthermore, Fig. 7-23 exhibits much better spatial resolution than does Fig. 7-15, and hence permits the interpretation of finer details. Specifically, the major road system is much more clearly seen in Fig. 7-23, as is the anticlinal feature that was previously referred to as "Area 5" in Fig. 7-15. Water turbidity patterns, resulting from the heavy runoff from recent storms, also can be clearly discerned in Fig. 7-23.

A Matching Natural Color Photo Taken by the Skylab Astronauts of the San Francisco Bay Area and Its Environs

Through the use of a second camera that was loaded with a natural color (Ektachrome) film, the Skylab astronauts took the matching photo shown in Fig. 7-24. The overall bluish cast of this photo provides an accurate record of how the earth appeared to the astronauts as they viewed it directly at the time of photography and is in keeping with the fact, as stated in Rayleigh's Law, that the scattering of light by atmospheric haze particles is inversely proportional to the fourth power of the wavelength of the light. Consequently, there is far more

(a)

(b)

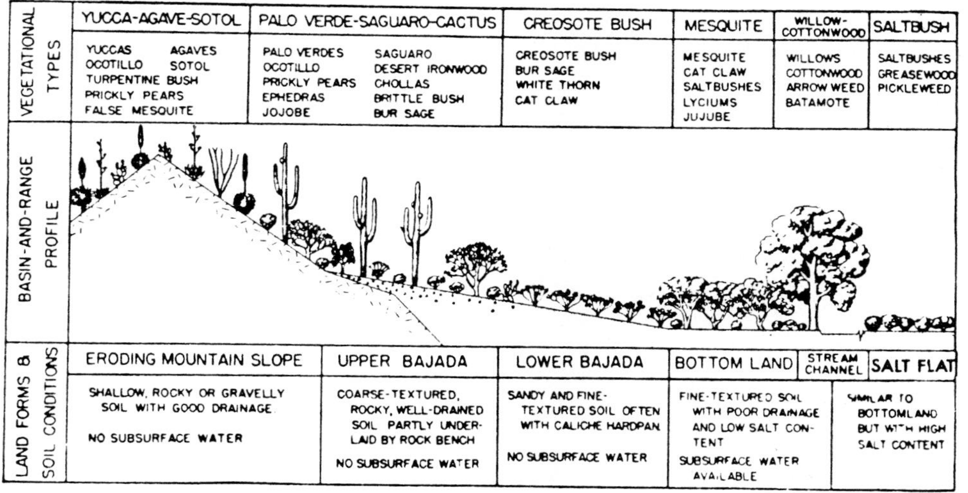

(c)

Fig. 7-23: (a) Color infrared photo taken in March 1969 by the Apollo 9 astronauts from an altitude of 120 miles above Phoenix, AZ. Compare the central portion of this photo with (b) and (c). (Photo courtesy of NASA.) (b) Oblique aerial photo, also taken in March 1969 from an altitude of 1500 ft above the terrain and showing the central portion of (a) as viewed from the southeast. Despite the much higher spatial resolution, this photo is no more useful than the space photo of (a) for identifying vegetation types or inferring soil and hydrologic conditions. The missing link, however, is provided by (c). (c) This diagram prepared in 1954 by Benson and Darrow of the University of Arizona, indicates the close association that exists among the following land resource-related attributes throughout much of the arid western United States, such as the area in (a) and (b): landforms [readily interpreted on a space photograph such as that comprising (a), vegetation types, soil conditions, and hydrologic conditions. As an example, a 1 in long transect (in terms of its photo distance) trending from southwest to northeast through the center of (a) clearly permits one to identify, in the following sequence, each of the landforms listed in the above diagram: eroding mountain slope; upper bajada; lower bajada; bottom land; salt flats (note their light tone); and stream channel. The correctness of these identifications is established by observing the same area on the low altitude oblique photo of (b). Field checking of this area established that for each portion of this transect there were, associated with the landform feature identified there on the space photo, vegetation types, soil conditions, and hydrologic conditions that are indicated for that portion in (c). From this and numerous similar tests made elsewhere throughout Arizona's Sonoran Desert, it is concluded that the mere ability to recognize landforms on space photography of an area such as this can permit the knowledgeable person to correctly infer the identity of associated vegetation types and soil/hydrologic conditions. It is the combination of these attributes together with proximity to urban areas that determines the proper land use for each part of the area. (From [11].)

(a)

(b)

Fig. 7-24: (a) Infrared Ektachrome photo of San Francisco Bay and environs taken from an altitude of 270 mi by the Skylab astronauts. (Scale approximately 1:1 000 000). Compare with (b) and also with Fig. 7-15, which shows somewhat the same area as imaged by the Landsat multispectral scanner. (b) Conventional Ektachrome photo of San Francisco Bay and environs taken simultaneously with (a). (Photos courtesy of NASA.)

scattering of light by such particles within the *blue* part of the visible spectrum than in the remaining (longer wavelength) parts of it.

Simulations of High Resolution Space Photography That Soon Will be Acquired by the French Satellite Known as "SPOT"

Figs. 7-25(a), 7-26(a), and 7-27(a) show areas in northern California as they are expected to appear (with 10 m resolution) when, in 1985, they are imaged by the sensor system that will be on board the French satellite, "SPOT." These simulations were obtained from an altitude of 40000 ft with a line scanner similar to the one that will be used on SPOT and with the same size of pixel ("picture element") in terms of ground area covered per pixel. As a result, these three figures show essentially the same spatial resolution as anticipated from SPOT when its sensor is operating in the 10 m (narrow swath width) mode, rather than in the 20 m (wide swath width) mode. For direct comparison purposes, Figs. 7-25(b), 7-26(b), 7-27(b), respectively, show the same areas as imaged, also in the summer of 1983, with the Landsat Thematic Mapper system. From this comparison it becomes apparent that, for the making of many kinds of land-resource inventories, such an improvement in spatial resolution can be of very substantial benefit, whether offered in the future by the SPOT system or by any of several other systems currently being contemplated. Hence, this section will be extensively illustrated and rather fully discussed. The emphasis in this section on SPOT imagery results merely from the fact that 1) the SPOT system is likely to be the first of these 10-m resolution systems to be launched (SPOT 1 is currently scheduled for a fall 1985 launch) and 2) in consequence, a substantial amount of research has just been completed, based on SPOT simulation imagery, which is considered to be highly indicative of what can be done with high resolution space acquired imagery in the near future.

Monitoring Urban/Suburban Features in the Davis, California, Area

Fig. 7-25(a) covers an area centered around the town of Davis, in California's Great Central Valley. On that imagery, the following kinds of features can be seen clearly enough to be identified and mapped. Certain of these features will later be seen even more clearly on the enlargements of this area appearing in Fig. 7-25(b).

1) Major freeways such as Interstate 80 (labeled 1 in Fig. 7-25(a).

2) Major railroads such as the Southern Pacific Route (labeled 2).

3) The primary commercial or business district of Davis at 3, identifiable by its characteristic blue coloration.

4) The oldest part of the residential district of Davis (labeled 4 and identifiable by its coarse texture and red coloration, caused by the fact that the area is largely over-topped by large, mature shade trees).

5) Progressively younger (more recently constructed) urban areas labeled in sequence as 5*A*, 5*B*, and 5*C*, identifiable because of their absence of mature trees and the more modern, arcuate layout of streets, complete with courts and blind lanes.

6) Individual buildings such as the Davis High School at 6*A*, and buildings devoted to classroom instruction

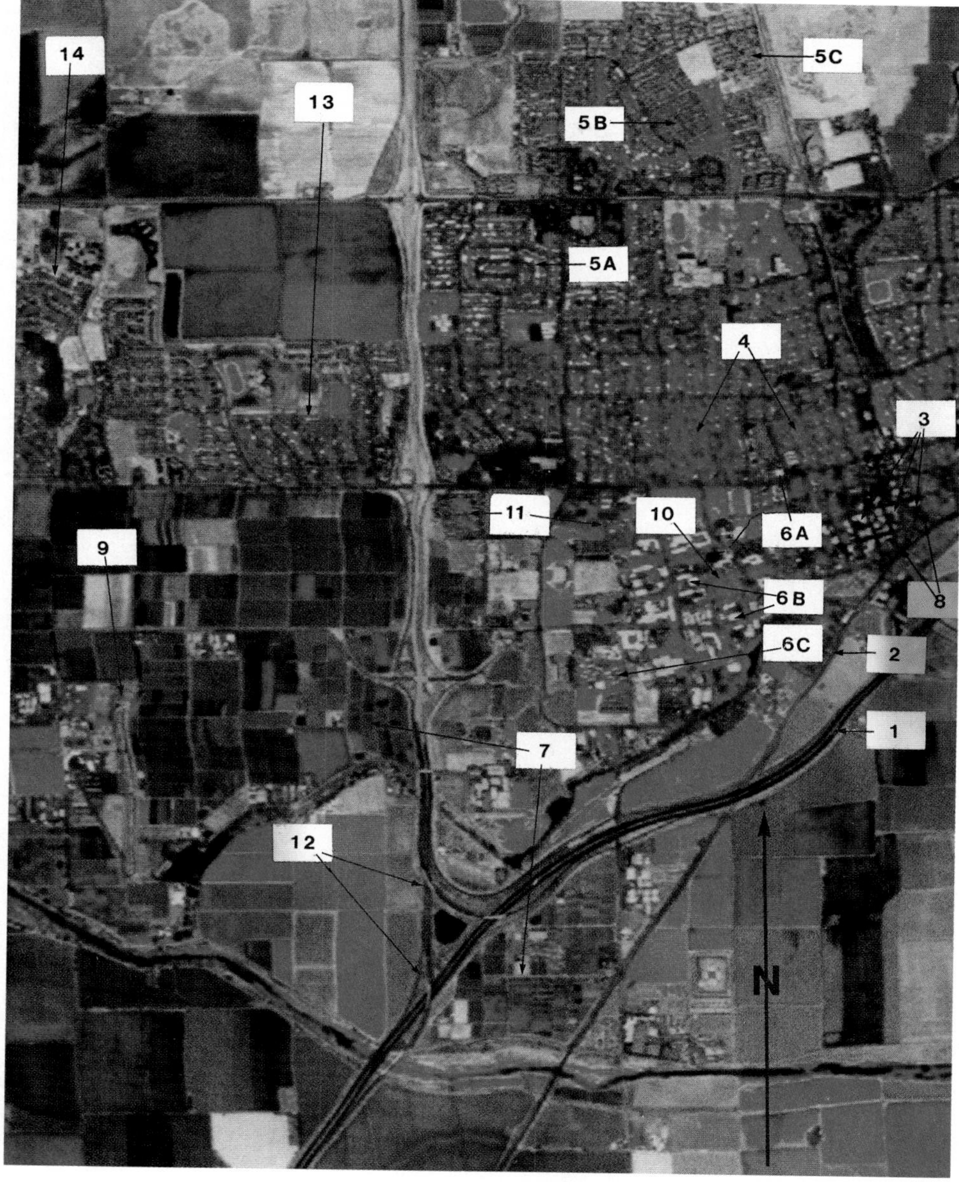

(a)

Fig. 7-25: (a) Simulated SPOT satellite color infrared imagery, scale 1:24 000, acquired on June 24, 1983, of the NASA *agricultural* test site near Davis in California's Great Central Valley. Note that even small fields and their boundaries can be readily delineated on this simulated space photograph—a fact that is all the more apparent in (b), which is an enlargement of the central portion of the area shown here. As indicated by the annotations in (b), each of the six crop categories used in the initial classification of such an area can be identified, on occasion, although not consistently so: O = orchards; V = vine and bush crops; C = continuous cover crops; R = row crops; I = irrigated pasture crops; F = fallow. As previously indicatd by Figs. 7-9 and 7-16, together with the text supporting those figures, it seems probable that, through the use of *multidate* SPOT imagery and properly constructed "crop calendars," not only these six major crop categories but also many of the individual crop types within each category, could be identified to a high order of accuracy. (Photo courtesy of the SPOT Image Corporation and the University of California, Berkeley. © SPOT Image Corporation.) (b) Enlargement of the central part of the simulated SPOT satellite imagery shown in (a). Many important details are far more clearly seen here than in (a). (c) Landsat 4 thematic mapper imagery acquired on August 12, 1983, of the same general area as is imaged in (b). Top: Simulated color infrared imagery. Middle: A second type of false-color rendition made by using the thematic mapper's band 4 to govern *intensity*, band 3 to to govern *hue*, and band 1 to govern *saturation*. Bottom: A third type of false-color rendition made by using the thematic mapper's bands 5, 3, and 2 to govern the red, green, and blue intensities, respectively. (d) Landsat 4 thematic mapper imagery acquired on August 12, 1983, of a large portion of California's Great Central Valley, including the area shown in (a)–(c), which is at the left center, above. Top left: the customary color infrared simulation. The other three photos in this figure show other false-color renditions, each being optimum for some feature identifications but almost uninterpretable for the identification of certain other features, thereby illustrating the validity of the "multi-enhancement" concept to which reference is made in the text. (Enhancements courtesy of Ralph Bernstein, IBM Scientific Center.)

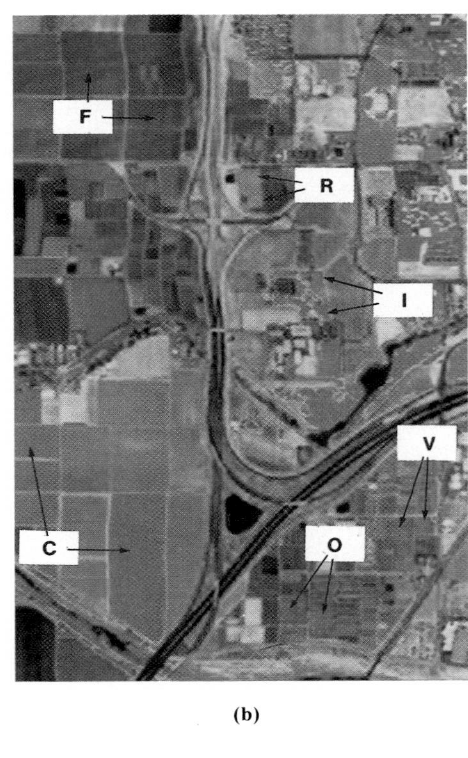

(b)

(c)

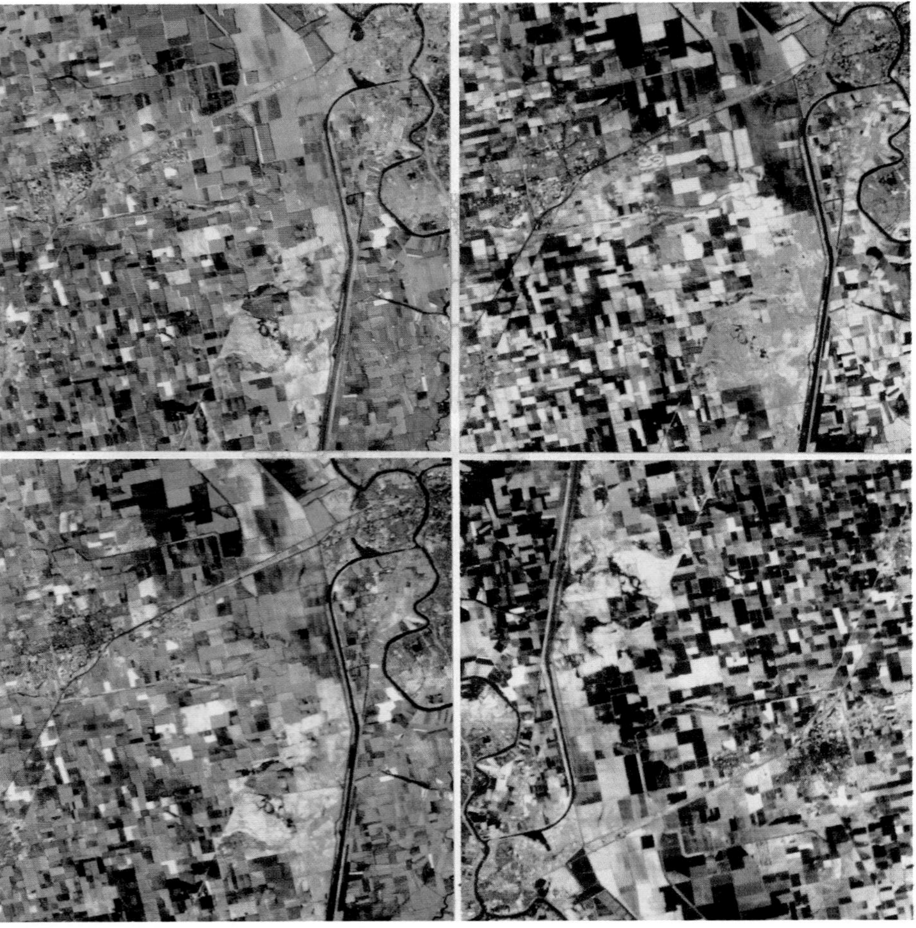

(d)

Fig. 7-25 (*Continued.*)

Fig. 7-26: (a) Simulated SPOT satellite color infrared imagery, scale 1:24 000 acquired on June 24, 1983 of the NASA wildland test site near Woodside on California's San Francisco Peninsula. Note that firebreaks (top, left), secondary roads and individual trees are clearly discernable on this simulated space photography—a fact that is all the more apparent in (b), which shows an enlargement of the top third of this photo. (b) Enlargement of the northern third of the simulated SPOT imagery shown in (a). Many important details are far more clearly seen here than in (a). Each of the six categories used in the classification of vegetation in such an area is consistently identifiable: R = redwood timber; D = Douglas fir timber; M = mixed conifer; H = hardwoods; S = shrubs; and G = grass. The area shown here is a representative portion of the California Coast Range. (Photos courtesy of the SPOT Image Corporation and the University of California, Berkeley. © SPOT Image Corporation.) (c) Landsat 4 thematic mapper imagery acquired on August 12, 1983, showing the same area that appears in (a), using the same simulated color infrared enhancement scheme. Boundaries between grasslands (blue) and woody vegetation (red) appear to be as interpretable in the lower half of this figure as in the corresponding part of (a). Details in the urban-suburban area in the top right quadrant are less interpretable here, however, than on the simulated SPOT imagery of (a), which has higher spatial resolution. (Photo courtesy of NASA.) (d) One of the many color-codings that have been applied to the Landsat 4 thematic mapper data of the area shown in (c). Note that this particular enhancement accentuates the cement freeway in red but is less interpretable than (c) (despite the larger scale here) for differentiating among vegetation tapes. (Enhancement courtesy of Ralph Bernstein, IBM Scientific Center.)

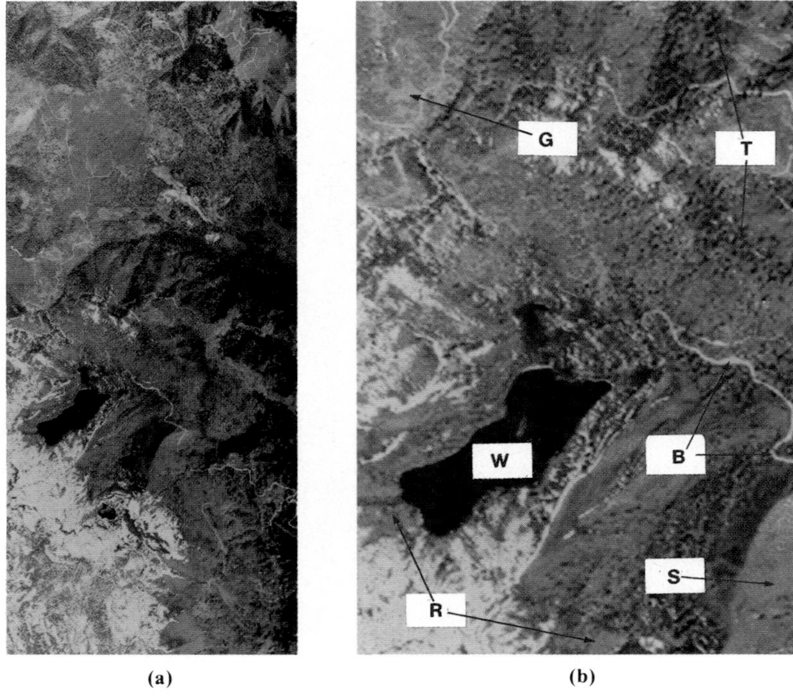

(a) (b)

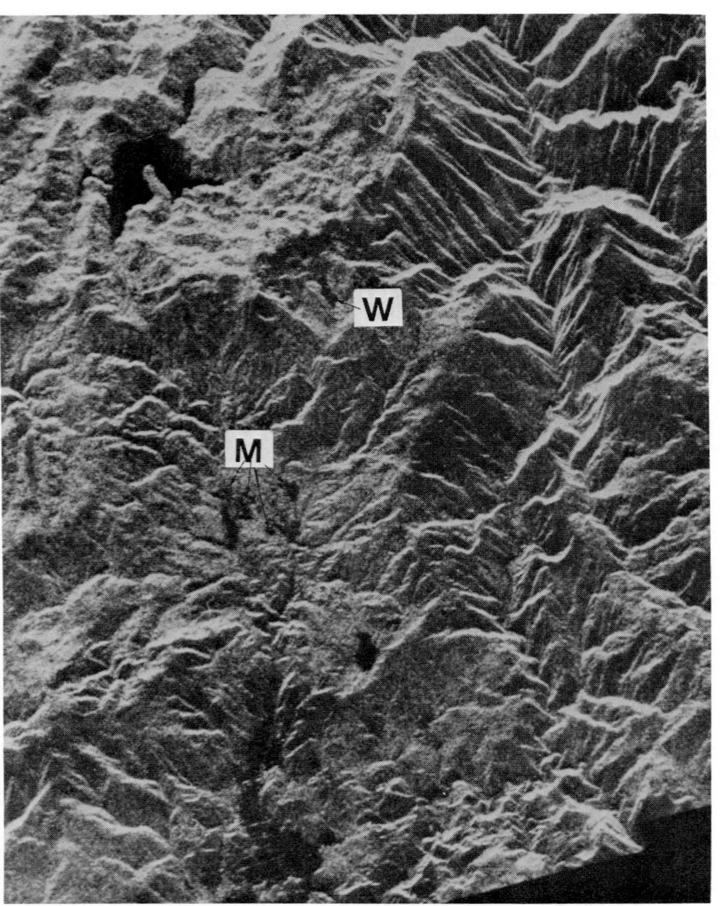

(c)

Fig. 7-27: (a) Simulated SPOT satellite color infrared imagery, scale 1:24 000, acquired on June 24, 1983, of the NASA wildland test site near Quincy in California's Sierra Nevada mountains. Note that the one-lane logging roads and even individual trees of merchantable size are clearly discernable. At the time when this imagery was acquired, snow still covered most of the area at elevations higher than 6000 ft, as shown here in the lower left portion of the photo. As indicated by the annotations in (b), each of the six major categories used in the initial classification of wildland areas in the Sierra are consistently identifiable on this imagery. (Photo courtesy of the SPOT Image Corporation and the University of California, Berkeley. © SPOT Image Corporation.) (b) Enlarged SPOT imagery for the central part of (a). As indicated by the above annotations, each of the six major categories used in the initial classification of wildland areas in the Sierra are consistently identifiable on this imagery, i.e., T = timber; S = shrubs; G = grass; R = rock; S = bare soil; and W = water. The whitest areas are snow covered. (c) Seasat SAR (synthetic aperture radar) imagery of a large part of the NASA wildland test site near Quincy, including the area shown in (a). (Silver Lake, the largest lake seen in (a), is clearly seen in the top center of this radar imagery.) This radar imagery was acquired from an altitude of approximately 570 mi. It not only accentuates topographic features but also exhibits high spatial resolution for many features, including lakes and moist areas. If this figure is rotated 180° radar shadows are made to fall away from the observer and the terrain appears to be reversed in that high features appear to be low and low features appear to be high. (Photo courtesy of NASA and NDAA.)

and to administration on the Davis campus of the University of California at 6*B*, as well as individual dormitories for students at 6*C*.

7) Portions of the area known as "University Farm" (labeled 7) in which experimental crops are being grown in much smaller fields than those comprising the nearby commercial farmland area.

8) A golf course at 8.

One primary purpose of including Figs. 7-25(a) and 7-26(a) here is to evaluate the usefulness of SPOT simulation imagery for urban/industrial monitoring. Data from Flightline 3, "San Mateo County at Woodside," and from Flightline 76, "Yolo County at Davis" were used in the comparisons. The data were acquired on June 24, 1983, along with 9″ × 9″ color infrared aerial photography at an approximate scale of 1:42 000. The visually interpretable SPOT Simulation Data were produced in 10 m resolution at an average scale of 1:24 000. Both of these are test sites formerly used in various NASA and other remote sensing applications programs so that an abundance of background information was already available. In addition, low-elevation aerial oblique photography was taken on the same day and selected ground photographs were taken in color and color infrared either on the same day or within one week of June 24th.

For various reasons, it was not possible to do computer analyses of the digital data. It is felt, however, that this is not a serious fault in an initial evaluation for the reason that the analysis of urban data requires a large amount of information input other than the spectral signature (image geometry, image elements, and texture plus associated and convergent evidence) in making the interpretation decisions. There are in fact some interesting experiments that could be done in digital analysis of the SPOT Simulation Data given the financial resources to see it through.

Monitoring and Planning the Urban–Forest/Range Interface in the Woodside, California, Area

The monitoring of urban–forest/range interface relationships was studied on data from Flightstrip 3, "San Mateo County at Woodside" and comparisons were made with TM and MSS data of the same area. (Compare Fig. 7-26(a) with (b) and (c).) [9].

This test site is in the foothills of the Santa Cruz Mountains which form the backbone of the San Francisco Peninsula and which are bounded on the east by the San Andreas Fault. The area overlooks the south and east bay from redwood-Douglas fir forests, oak woodlands, or various shrublands and grassland areas depending on soil type, direction of slope, and elevation. Thus, the area is in high demand for secluded suburban living with fantastic views and for larger acreage suburban living in the grasslands and oak woodlands of the lower foothills created by San Andreas Fault activity. The San Francisco and Silicon Valley bedroom communities are of high density residential development and tend to be situated around the various light industries and business districts. Because of the high-intensity use of this part of the San Francisco Peninsula, there is pressure to expand high density residential areas over the open foothills in the trough of the San Andreas Fault and up the slopes into the oak woodlands, the redwood-Douglas fir forests, and the high mountain grassland parks of the Santa Cruz Mountains. This diversity of environments and the development that has already taken place provide an ideal area in which to determine the usefulness of SPOT simulation imagery for monitoring these urbanization processes, not to mention the assessment of forest, woodland, shrub, and grassland environments out of which the urban expansion must be carved.

The evaluation was approached by first making a complete delineation of all vegetation and urban types on the SPOT image false color 1:24 000-scale print [Fig. 7-26(a)] and comparing this to the types discernable on the simultaneous CIR aerial photography taken at a scale of 1:40 000. For the scale and intensity of mapping that is appropriate at a scale of 1:24 000, both my colleague, Dr. Charles Poulton, and I found the SPOT image to be equal to the aerial photography in terms of delineations that could be made and the features differentiated. Obviously, there were advantages in the stereo evaluation of the aerial photography. There were features, such as individual small and medium size houses as well as large buildings and parking lots, where the stereo capability often provided the only reliable means of discriminating even on the aerial photography. But, the key point is that for those features that were feasible to map at a scale of 1:24 000, the SPOT image seemed fully adequate for identifying and delineating most of the vegetation types and broader urban/industrial categories. There were, of course, many small features which could be mapped only as point locations which required stereo examination, and some ground examination for identification. We found, for example, in contrast with both MSS and TM, that the resolution provided by the SPOT simulation imagery permitted sufficiently good detection of individual trees and small clumps of trees or bushes that we could develop an index of proportional conifer and hardwood cover in the forested and wooded areas (Table 7-8).

This led us to compare capability to discriminate individual and small clumps of trees in grassland areas (highly contrasting in spectral signature) among the SPOT, TM, and MSS image types (Table 7-9). We found, similarly, that thin hedgerows much less than 10 m wide were detectable when they occurred on spectrally contrasting backgrounds; and practically all of the secondary, two-lane roads and a few single-track roads were detectable where they were not overhung by trees.

In terms of mapping vegetation types and inferring landform and soils for purposes of land-use planning and possible urban expansion, we found that all of the conditions in this test area that we considered important (except for specific soil features, slopes, and soil stability

TABLE 7-8
PERCENT HARDWOOD IN MIXED CONIFER-HARDWOOD STANDS[a]

Item	Stand 1	Stand 2	Stand 3	Stand 4	Stand 5
	Dots on Hardwood in Random Set of Ten Dots				
Number of Observations	6	7	14	16	16
Sum	24	37	20	40	69
Mean	4.00	5.29	1.43	2.50	4.31
S.E. mean	1.53	1.48	1.18	1.32	2.64
Percent Hardwood	40	53	14	25	43

[a]Based on the interpretation of simulated SPOT imagery and found to be in close agreement with the interpretation of simultaneously acquired CIR aerial photography, scale 1:42 000.

TABLE 7-9
OAK TREES DISCERNIBLE IF ON A DRY GRASS BACKGROUND

Number of Oak Trees Counted by Interpreter					
Air Photography		SPOT Imagery		TM	MSS
Positive	Likely	Positive	Likely	Likely	Likely
20	1	19	3[c]	0	0
8	0	8	1[c]	0	0
23	0	14	6	0	0
7	2	12[a]	0	0	0
10	0	7[b]	0	0	0
8	2	15[c]	4[c]	0	0
17	0	17	2[c]	0	0
93	5	92	16[c]	0	0

[a]Includes shrub patches that look like trees.
[b]Border trees blend with hedgerow.
[c]Difference probably due to shrub patches that look like trees on the SPOT image.

characteristics not inferable from vegetation) were detectable by and mappable on the SPOT simulation imagery.

One of the important land-use changes in this area is development of very expensive homes on large lots within the redwood-Douglas fir type. This is important to monitor and control because of the especially high demands that this type of development places on various public services including those associated with fire control, police protection, and the installation and maintenance of public utilities. We selected three developing areas and counted habitations on the ground on existing maps, on aerial photography, and on SPOT, TM, and MSS imagery. The results of this comparison are shown in Table 7-10. These data clearly show the superiority of SPOT simulation imagery over TM imagery and, of course, over Landsat MSS imagery.

In these kinds of areas, watershed runoff can be a serious problem, primarily because of the increase in impervious surface that results from residential development. Table 7-11 shows the results of a test where a random dot grid of 100 dots per square inch was used to record hits on housing units in the largest of the three test areas. This could not have been done with TM or MSS data with any acceptable degree of reliability. A comparison was not done on the aerial photography because the scale of 1:40 000 was too small and the results would have been somewhat scale dependent. Given data sets like these, one could develop an index of increased runoff by field sampling to determine the average square footage of roofs, impervious driveways, and sidewalks. SPOT data produced at a scale of 1:24 000 or analyzed for percentages of these openings in the timber from the digital data would provide a useful index for planning runoff-water control from developed areas within critical watersheds.

While flightline 3 did include an extensive industrial and commercial area in the north, simulated SPOT imagery for that part of the area was not included in the false color imagery prepared for that flightline. The false color data provided only the options of looking at various residential classes and institutional buildings such as schools and churches. We did make comparisons of counts on different kinds of these institutional buildings and made some comparative observations on the black-and-white strip printout of band 3 in comparison with the SPOT-provided aerial photography. This informa-

TABLE 7-10
INCREASE IN HOUSING IN TIMBERED AREAS, 1976–83

Area	Number of Houses in 1976	Number of Houses in 1983[a]			
		Air Photography	SPOT	TM	MSS
A	11	16	11	0	0
B	39	56	56	3	1
C	13	18	14	7	5
Total	63	90	81	10	6

[a]Count includes "Positive" plus "Likely" interpretations, but excludes the lowest confidence class.

TABLE 7-11
PERCENTAGE OF AREA IN HOME CLEARINGS, AREA A (FROM 5 REPETITIONS OF 100 RANDOM DOT COUNTS)

Repetition Number	Percent in Home Clearings
1	15
2	13
3	15
4	16
5	11
Average	14

tion is summarized in Table 7-12. The results confirm the general useability of SPOT simulation imagery for acquiring building-by-building data on institutional, commercial, and industrial areas. The comparison was between SPOT imagery and the aerial photography because neither TM nor MSS provided images useful for this purpose. Generally speaking, the differentiation of certain classes of residential, commercial, and industrial areas can be made with the TM and MSS data, but accuracy would surely drop for the commercial-industrial discrimination, especially with MSS.

Finally, a few general observational notes are summarized in Table 7-13 that illustrate the capabilities of using SPOT Simulation imagery for visual interpretation of urban/industrial features in a region where the urban–forestry/range interface is important.

In summary, the potential of the SPOT simulation imagery for urban-industrial monitoring and planning is obviously very great. One can do just about as much with this data as with small scale CIR aerial photography. Furthermore, had we been able to analyze the tapes for some special applications, significant refinements might well have been developed. In this setting, we see computer analysis for these purposes essentially as a backup for visual interpretation which probably should drive the analysis procedures. This is true because of the inordinate amount of dependence on geometric and spatial relationships and on detailed shapes of features, as well as on convergent and associated evidence in the identification of features in the urban/industrial setting. It would also be important to use simultaneous aerial photography on a sample basis to support the SPOT image interpretations—to say nothing of the ever-present requirement for a high level of ground knowledge and understanding of what to expect in any project area.

Under the leadership of my colleague, S. D. DeGloria, two other studies were conducted to determine the interpretability of simulated SPOT imagery as applied to agricultural and forest resources, respectively [12], [13]. In both studies the improved spatial resolution of SPOT imagery greatly improved the resource inventories. Furthermore, a recent detailed study of urban/suburban resources based on simulated SPOT imagery [10] has led to a similar finding.

TABLE 7-12
INSTITUTIONAL BUILDING COUNT

	Air Photography		SPOT Imagery		
Type Building	Positive	Likely	Positive	Likely	Comment
Schools	6	0	6	2	1
Churches	1	0	4	0	2
Business 1	0	0	7	0	3
Business 2	9	0	5	0	4
Golf Club	1	0	2	0	5

Comments:
1) Confused parking lot for building on SPOT.
2) Confused parking lot for building on SPOT.
3) Orchard and roadways looked like business area on SPOT.
4) Confused parking lot for building on SPOT.
5) Building appeared same dark blue as parking lot on SPOT.

TABLE 7-13
NOTES ON MISCELLANEOUS FEATURES INTERPRETATION FROM SPOT

Item	Note
1	Median in freeway clearly discernible.
2	Two- and four-lane sections of freeway distinguishable.
3	Differentiate black-top from concrete road surfaces (also on TM and MSS if the road/street is detected).
4	Differentiate concrete overpasses on black-top sections of freeway.
5	Most urban streets visible when not tree lined.
6	Two-lane roads visible through woods and grasslands if not obscured by overhanging trees.
7	SPOT is almost as good as 1:42 000 aerial photography for the interpretation of urban/suburban features that are feasible to map at a scale of 1:24 000.

But one must recognize that there are still needs, especially in the developing nations of the world, for small scale, highly generalized resource analyses and maps—appropriate to scales of 1:1 000 000 in some instances and 1:250 000 and 1:200 000 in others. In these cases, the 80 m resolution of the Landsat MSS imagery is still to be preferred because it generalizes to a desirable degree by washing out the noise that is present in simulated SPOT imagery because of its higher resolution. One can, in fact, overkill a developing nation with information—even with information visually interpretable from Landsat MSS data, and all the more so with that interpreted on SPOT imagery.

SPOT Image, Inc. should therefore plan to put out not only high resolution data and imagery but also generalized editions having lower spatial resolution in which pixel clusters would be averaged to create a data set with a coarser resolution on the order of 50–80 m. In our judgement, an averaging approach would be preferred to a data sampling approach that skipped rows and columns to degrade the imagery. One of the obvious advantages of SPOT is going to lie in its stereo imaging capability for selected areas and in the fact that, with one data set, the user has the flexibility of setting his resolution requirements (from 10 m up) to match the level of generalization at which he is working in the interpretation and analysis phases of resource assessment and monitoring projects.

Monitoring and Planning the Urban–Agriculture Interface in the Davis, California, Area

One of the most critical land-use problems in North America and in many other heavily populated regions of the world is urban-industrial sprawl into the best food-producing land. One reason why this is difficult to control is the lack of sufficiently strong-impacting pictures of the consequences—i.e. pictures which would facilitate passage and defense of effective land-use laws and ordinances. Original human settlement tended always to be on the very best of the agricultural cropland, the best timberland or the best rangeland. These settlements, initially established as an operating headquarters or service center for supporting business, invariably grew with population expansion, and uncontrolled sprawl. This process, determined almost alone by "land developer" interests, has been the order of the day in guiding such expansion. The only effective solution is to counter with sound, visually impacting information and feasible alternatives within the framework of rational enabling legislation. Because of its increased spatial resolution (compared to that of Landsat MSS and TM imagery), SPOT imagery can have a role to play in the important world problem of guiding urban-industrial expansion.

For a final example of the potential applicability of high resolution space photography, let us return to the SPOT simulation test site near Davis, California, designated as "Flight Line 76, Yolo County at Davis." The site was imaged with 10 m data. An interpretation of these data provides excellent confirmation of historical facts in the urban growth of Davis, a university-farming community, and illustrates how such interpretations could be used to document and reinforce urban–agricultural planning needs.

Merely from an examination of the simulated space image comprising Fig. 7-25(a), much can be inferred as to the nature and appearance of this area in times past and some of the factors that have influenced the patterns of expansion. For a period of 50 years or more, centered around the beginning of the present century, when railroads were the primary mode of long distance transportation, Davis was primarily a "railroad" town. In fact, the railroad *Y* at point 1 of Fig. 7-25(a) is still an important attribute of Davis and explains why the original business district of Davis, at 2, was located nearby as was the original residential district at 3, evident by the prominence of large trees. The original Davis High School building at 4 and the associated track and playing fields at 5 were no doubt built a short distance out of the original residential area to allow for growth. Even in those earlier days, part of the area, 6, was reserved as a park for recreational use and that area remains preserved somewhat in its original tree-covered, park-like state today.

In 1909, the site immediately southwest of the original Davis urban area was established as the primary agricultural component of the statewide University of California. In keeping with the Federal Land Grant educational system, a 2600-acre agricultural experiment station was established there. The fact that such an experimental area is still very much in existence today can be readily inferred by the smaller-than-normal agricultural fields and the intensity of its secondary road system as compared to surrounding areas. Note that this road system is clearly evident at the 10 m SPOT resolution. Partly because of the importance of agricultural aviation, we are not surprised to find that the University established its own airport at 7 in Fig. 7-25(a) within the confines of that portion of the campus known as "University Farm."

By 1940, approximately 1100 students were enrolled at U.C., Davis, and the University's complex of buildings devoted to classroom instruction and to University administration was developing around the "Quad" at 8—all close and handy to the business district a few blocks to the east. Again, it is significant that these institutional buildings are clearly evident at 10 m resolution, as are the large buildings in the business district, thereby allowing accurate interpretation of these features.

In the late 1940's, with the tremendous influx of students whose education had been interrupted by World War II, a very sizable expansion of the Davis campus took place and the construction of a large number of modern student dormitories began at areas labeled 9 and 10 in Fig. 7-25(a). By this time, a decline in the importance of railroading, combined with burgeoning of student enrollment, caused University-related activities to replace railroad-related activity as the primary industry associated with the town of Davis. Consistent with the replacement of railroad transport by highway transport, the original Davis *Y* at point 1, marking the junction of two major railroad routes, became secondary to the new Davis *Y* at point 11, marking the junction of two major highway routes (Interstate 80 east and west and State Highway 113, an important bypass interconnection with Interstate 5 north of Davis). This caused a new focal point for expanding urbanization near those parts of Highway 113 that would be close enough to Davis to offer realistic commute possibilities. But normally expected commercial development at the *Y* was restricted by the University Farm, and both the Freeway and the watercourse (Putah Creek) restricted easy residential expansion to the south and southeast. Thus, further residential expansion was destined to take place north of "Old Town" and the University Farm.

Davis and its environs are notoriously flat. Flatlands are ideal for bicycling; and bicycles are an extremely useful means of overcoming congestion in such populous areas. The heavy dependence of students on the bicycle, and the related simplification of parking space needs, probably explains why most of the post-war housing developments surrounding Davis—including those in Areas 10, 11, and 12 of Fig. 7-25(a)—are so closely clustered to the primary industry of Davis, the University. That these housing developments are of recent vintage is readily inferred from their relative absence of large trees, compared to "Old Town" Davis at 2. Naturally, the first of these developments were adjacent to Old Town and in close proximity to the already existing high school, a strong drawing card. Note Areas 10 and 11. It was also most logical for the next area—more recent than either Areas 10 and 11 as evidenced in the smaller trees thanks again to SPOT resolution—to be Area 12 immediately north of the high school. Note that the attractiveness of this area was enhanced by a large greenway associated with the golf course that winds through the development with a parking lot and clubhouse plus addition of a shopping area in the southeast corner of the development along the railroad and another on the southwest.

Apparently somewhat contemporary with development of Areas 10 and 11, as seen in Fig. 7-25(a), was the eastern half of Area 13 adjacent to both the University Farm and Highway 113. This placed the development in the next closest proximity to the central activity of the Davis community as well as on a major street into the main business district and to the University. The increased population resulting from development of Areas 10 and 13 resulted in a new school on the northwest corner of the first development in Area 13—thus a strong attractor for more expansion of Area 13 to the west.

This expansion has become a fact as one of the most elegant of the suburban housing areas in Davis. Note that it is complete with artificial lakefront property, a park and activity center, and an additional shopping area. From what one can see in this image, however, one might wonder whether residents of this particular area will be distressed by increasing airplane traffic in and out of the Davis airport occasioned by the fact that 1) such traffic appears to be increasing and 2) Area 13 is in direct line with the airport's only runway and only a scant mile away. Whether this potential problem could have been avoided by more forward-looking planning is academic at this point.

What might the future hold? Factors already mentioned will probably continue to restrict southward expansion of Davis across Interstate 80. This likelihood is further reinforced by positive attractors to the north, including the following:

1) Given land availability at a favorable price, the next most logical expansion area is northeast of Area 13 out of the airplane overflight path. An important consideration should be, however, the productivity and uniformity of the soil in this area for crop production.

2) The SPOT image pattern one and one-half to two miles north of town along Highway 113 suggests greater soil variability and thus the possibility of lower priority for agricultural use. If true, this may be the most judicious area for future urban expansion.

3) A golf course has already been developed 2 mi north of town along Highway 113. This will be a drawing force tending to move residential expansion northward along 113 as land—hopefully the poorer agricultural soil areas—becomes available for urban development.

4) One small, approximately 30 acre development is already in place 1 mi north of town. This will function as a drawing force and a focal point for expansion of urban residential and even small commercial development to service the area as it grows, eventually even motivating an additional school and churches.

5) The existence of the north–south railroad will limit expansion of Area 12 to the east, although additional urbanization could spread north from existing developments east of the railroad track if it were not for the fact that this area lacks a major highway to facilitate north–south traffic as 113 does to the west.

6) To some degree, further expansion of Area 12 northward along the railroad is minimized by the presence of a sewage treatment plant and by the convergence of the railroad and Highway 113 as one moves northward.

Landsat TM Comparison, Davis Site

As shown in Fig. 7-25(c) and (d), several different color enhancements were made of the TM data to compare interpretability. The false color infrared rendition is shown in Fig. 7-25(c). Examination of this figure in comparison with Fig. 7-25(b) will reveal the following:

1) Color renditions with TM are comparable to SPOT and are adequate for differentiating features where the spectral signature is the primary criterion.

2) Where spatial discrimination and ground resolution are important, TM is leagues behind the SPOT simulation. Specifically:

a) The secondary road system in the University Farm is not discernable.
b) Streets within the urban area are mostly evident because of the pattern in tree distribution and none of the smaller houses can be seen as with the SPOT simulation.
c) Large buildings are indistinct although evident. Building shapes are not as clear and, therefore, interpretability of institutional and commercial features is limited with TM.
d) The playing field at the high school could be interpreted as being either a lawn area surrounded by buildings or a combination of buildings and parking. In the SPOT image, its identification is positive because the surrounding track is clearly evident.
e) Highway interchange detail is fuzzy and the median strip in the Freeway is not discernable on TM.

The experienced urban interpreter will appreciate that the above disadvantages severely limit the usefulness of TM data where so much of the interpretation decision depends on resolution detail which preserves the integrity of building shape, discriminates the various features of ground cover, allows detection of major and minor streets and secondary roads, and allows one to draw effectively on spatially related, convergent, and associated evidence for the correct identification decision. It is the judgement of these writers that SPOT data of the quality of this simulation experiment can come close to taking the place of 1:40 000 to 1:50 000 scale aerial color infrared photography, especially when one considers the added advantages of being able to digitally analyze the data for special problems and applications, including 1) the quantifying of certain components of an urban/industrial area, 2) determining the area occupied by each such component, and 3) numerical discriminations to support the visual interpreter.

The Inventory of Wildland Resources in the Bucks Lake-Quincy Area of California

Simulated SPOT Imagery: Fig. 7-27(a) covers an area in California's Sierra Nevada near the town of Quincy. On this SPOT simulation of the area, six major wildland categories are frequently identifiable on SPOT imagery. These are Rock (*R*), Water (*W*), Bare Soil (*S*), Trees (*T*), Shrubs or Brush (*S*), and Grass or other herbaceous vegetation (*G*). In addition, the main road net is clearly discernable as exemplified by points labeled *R*, even though these are dirt roads which, for the most part, will only accommodate one lane of traffic in each direction.

When the same points are checked on the matching frame of Landsat TM imagery [Fig. 7-27(b)], it is apparent that there would be a much lower degree of certainty with which most of the above-mentioned features could be identified and mapped.

Space-Acquired Radar Imagery: Fig. 7-27(c) is radar imagery acquired from space by one of the space shuttle vehicles of a portion of California's Sierra Nevada Mountains, including the area shown in Fig. 7-26(a) and (b). [Note, for example, that the same water body, Silver Lake, has been labelled "W" here as on Fig. 7-27(a) and (b).] While the radar imagery is excellent for portraying topographic relief and for highlighting both water bodies and moist meadows (such as the ones labelled *M*), it is almost totally unsuitable for the mapping of rock, bare soils, trees, shrubs, or herbaceous vegetation.

Color-Coded Landsat Imagery: Fig. 7-28 shows color-coded Landsat imagery of various parts of the Bucks Lake-Quincy area. The model that was used in producing this figure incorporated various kinds of remote sensing and ground data as independent variables and was used to predict thematic dependent variables such as:

1) timber production capability maps
2) daily potential evapotranspiration
3) average daily radiation.

From these data, thematic maps were produced illustrating each of the dependent variables. These maps appear in full color in the following figures of *The Manual of Remote Sensing* [4]: Figs. 32-38, 34-61, 34-62, and 34-63.

The Multiple-Use Concept in Relation to Space Photography of Forested Areas

Foresters have decided that they need essentially six categories of information, the same as the agriculturists require, as seen previously in Table 7-2. Then, using a somewhat different approach, they decided that remote sensing might be especially helpful to them by providing information on which to base "multiple-use" decisions relative to each part of the forest. The multiple-use concept is a complex one, and it is far more applicable in forestry than in agriculture. Some users of the forest would like to have all parts of the area managed primarily with a view to maximizing timber production. At the other extreme are those who want to preserve our forests as primeval museums to be enjoyed by posterity. In between are those who would condone each of these uses for specific parts of the forest so long as it did not interfere with their objectives of using the forest primarily as a source of water for domestic use; or of minerals for industrial use; or of fish, game, boating, hiking, skiing, etc., for recreational use. The true complexity of the forester's decision-making problems as he seeks to satisfy these many user requirements is suggested merely from a reading of the forester's official definition of the term "multiple use" as set forth in the Multiple Use Act of 1960.

> The management of all the various renewable resources of the national forests so that they are utilized in the combination that will best meet the

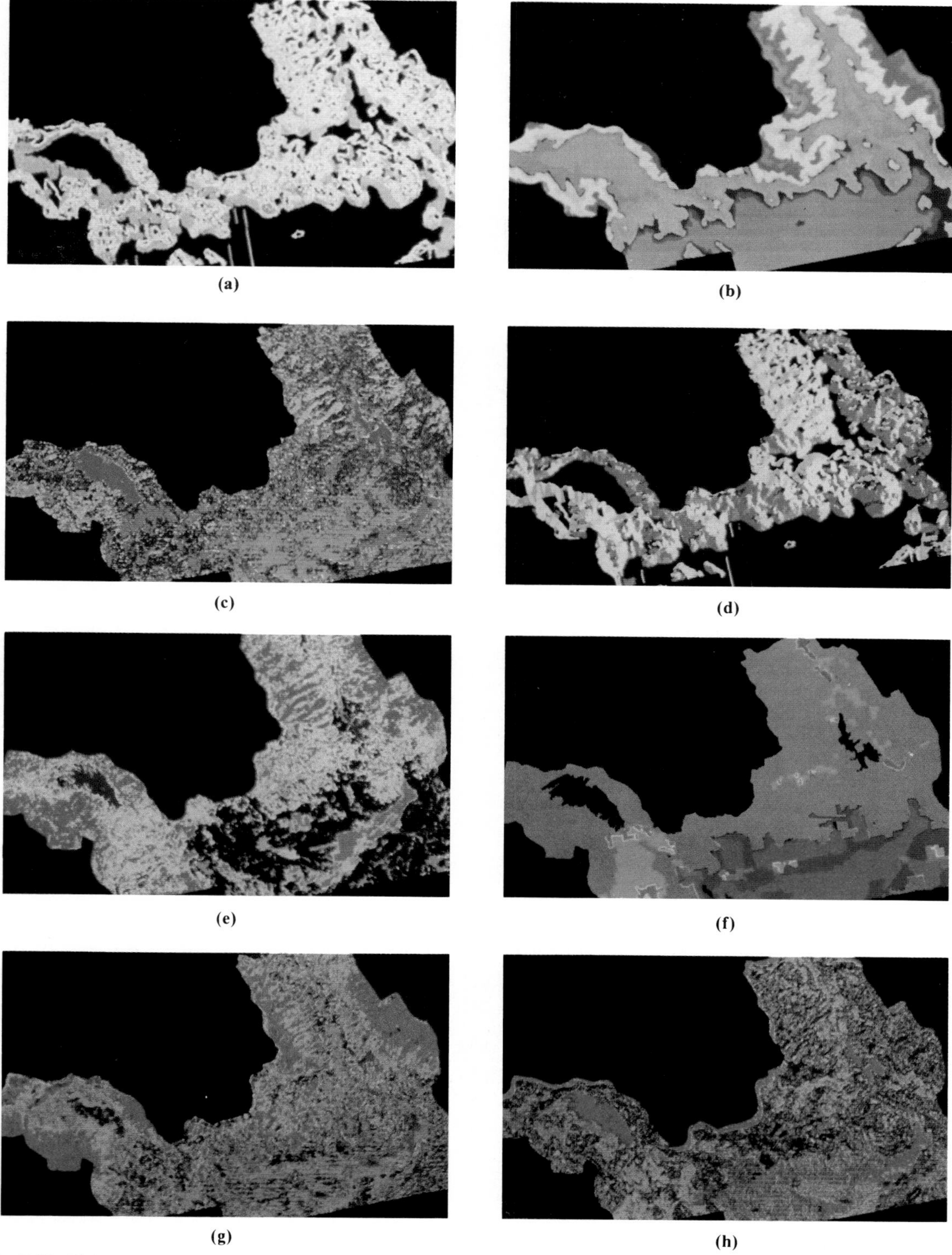

Fig. 7-28: Eight thematic maps showing a part of the NASA Bucks Lane Forestry Test Site in California's Sierra Nevada mountains. In the making of maps (a)–(c), use was made of USGS-supplied digital terrain data to provide a color-coded presentation of elevation, slope, and aspect, respectively. Map (d) is of the type that often is most useful to the land resource manager in that, by a merging of the elevational data with a Landsat-derived vegetation classification, it provides an accurate vegetation/terrain classification, pixel-by-pixel, for the entire area. Map (e) shows the ouput resulting from a pixel-by-pixel calculation of the Landsat MSS "band 7/band 5 ratio" for the area. Since healthy green vegetation has a uniquely high band 7/band 5 ratio, the color coding on this map is quite similar to that of map (d). The remaining three all of which are important factors for the land resource manager to consider, for example, in selecting the areas in which reforestation measures are most likely to succeed. Obviously, the manager of resources in this area finds many other important applications for these thematic maps when using them either individually or in appropriate combinations. (Courtesy of the Remote Sensing Research Program, University of California, Berkeley.)

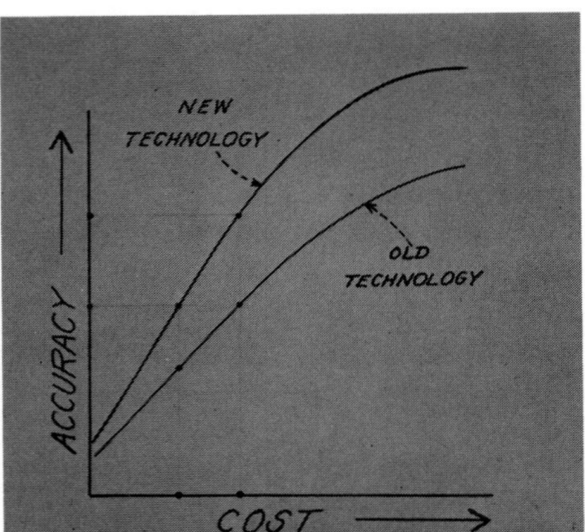

Fig. 7-29: A major factor in gaining the resource manager's acceptance of modern, remote sensing based capabilities is to be found by presenting to him reliable "accuracy/cost" curves with which he can compare the "old" technology that he presently is using for the inventory and monitoring of land resources with a technology that is based on modern remote sensing capabilities. By comparing such curves he can readily see possibilities for 1) achieving higher inventory accuracy at a given cost; 2) the same accuracy as at present, but at reduced cost; or 3) some combination which will provide somewhat great accuracy at somewhat less cost. There are many other important factors that can govern his acceptance of modern remote sensing technology, however, as indicated in Table 7-8.

needs of the American people; making the most judicious use of the land for some or all of these resources or related services over areas large enough to provide sufficient latitude for periodic adjustments in use to conform to changing needs and conditions; that some land will be used for less than all the resources, and harmonious and coordinated management of the various resources, each with the other, without impairment of the productivity of the land, with consideration being given to the relative values of the various resources, and not necessarily the combination of uses that will give the greatest dollar return or the greatest unit output.

This definition is so complex that the concept of multiple use may easily escape the average reader. However, Draeger [14] has rendered a service by providing the following more comprehensible statement of this multiple use concept:

> The allocation of lands to various uses such that the combination of uses best meets the needs of those for whom the land is being managed, provided that in the process of developing this allocation, values and costs of each of the many possible patterns of exclusive uses and compatible co-uses are at least taken into consideration. In some instances, depending on the circumstances, several uses may be derived simultaneously from the same parcel of land, while in other cases a particular area may be allocated to only a single use but managed as a portion of a greater whole composed of many such exclusive-use parcels.

A Look to the Future with Emphasis on "Technology Acceptance"

If we were to reflect for a moment on the information that has been provided thus far in this chapter by means of the various remote sensing figures and the supporting text, we might prematurely conclude that we have provided an adequate indication of the present state of the art. After all, the term "adequate" is a highly subjective one as evidenced by reactions already received by the present writer from those who have examined a considerably expanded version of what is presented here. I refer to the *Manual of Remote Sensing* (2nd ed.) that was recently published by the American Society of Photogrammetry [4]. Reactions to that document already have ranged from "Why did you omit the following 5 important topics" to "Although I have not attempted to digest this formidable two-volume manual, I can't help but have the feeling that it contains, in its 2700 pages, considerably more about the subject of remote sensing than is actually known."

Earlier in this chapter, we cited the highly perceptive comment of Abraham Lincoln as he recognized that it is important for us to know "where we are and whither we are trending" if, as we look to the future, we are to better know "what to do and how to do it." We emphasized that such an observation can be fully as germane today with respect to the remote sensing of land resources from space as it was in Lincoln's time with respect to the opportunities and problems that confronted his generation.

Be that as it may, we certainly would be remiss if we were to conclude this chapter without expanding a bit on the subject of "technology acceptance" as applied to the probable future use that is likely to be made of remote sensing from space as an aid to the inventory, monitoring, and management of the earth's land resources.

Table 7-14 constitutes a brief summary of most of the considerations that are involved in gaining the general acceptance and beneficial use of modern remote sensing technology. Emphasis throughout that table is on the use of such technology for the inventory, monitoring, and management of the earth's land resources.

As explained by the caption to Fig. 7-29, cost-effectiveness curves of the type shown there, if they are both reliable and believable, can do much to provide the rationale needed by a land resource manager in deciding to switch to the new technology. As indicated by Table 7-14, however, many other factors can govern his degree of acceptance or rejection of modern remote sensing based technology.

If this chapter has accomplished its primary purpose, it has indeed given an adequate indication of "where we are" with respect to the potential use of modern remote sensing technology for the inventory, monitoring, and management of land resources. In any event, the greatest need now, as we consider "whither we are trending," would seem to lie not in the development of still greater technology, even though such continuing progress could

TABLE 7-14
SOME CONSIDERATIONS INVOLVED IN GAINING THE GENERAL ACCEPTANCE AND BENEFICIAL USE OF MODERN REMOTE SENSING TECHNOLOGY

1) *The Concept of "Appreciation Versus Adoption":* In any given organization, modern remote sensing technology has not been adopted until the following condition exists: the top-level decision makers are sufficiently convinced of the potential usefulness of that technology for solving certain resource-related problems that a) they *routinely* consider using it, as each resource inventory/monitoring/management problem arises, and b) they then forthrightly direct that it be used in each appropriate instance.
2) *The Concept of "Requirements for Adoption":* If a particular aspect of modern remote sensing technology is to be adopted in any given instance, the decision maker needs to be convinced of the *simplicity, direct applicability, continuing availability,* and *economy.* In many instances, the *timeliness* with which the desired information can be obtained through its use is also a very important consideration.
3) *The Concept of "Progression Toward Adoption":* Remote sensing technology will not transfer itself. If the top-level decision maker is to become convinced that he should actually *adopt* modern remote sensing technology, he ordinarily must progress toward that viewpoint through the following succession of steps: *awareness, interest, evaluation, trial,* and *adoption.*
4) *The Concept of "Deterrents to Adoption":* The primary deterrents to the adoption of modern remote sensing technology by an organization's top-level decision makers are a) *Oversell* (resulting in the raising of user expectations to levels beyond what current capabilities can deliver); b) *Overkill* (as when the user is urged to use elaborate techniques of computer-assisted analysis when the desired information could have been derived quite adequately through the use of simple, inexpensive, manual interpretation techniques); c) *Undertraining* (most commonly exemplified when a novice who has just completed an "appreciation" course in remote sensing is required to plunge directly into the demanding tasks that are involved in making operational use of modern remote sensing technology; d) *Underinvolvement* (as when the user agency, plagued by a lack of qualified and/or motivated personnel, turns over the bulk of the work to consultants or others who lack familiarity with the user agency's resource problems, information needs, and perhaps even the resource itself); e) *Spurious Evaluation* (as when the user agency, forced by higher authority or others into a "rush to judgement" produces premature, incomplete, incestuously-validated, and overly-optimistic appraisals); f) *Misapplication* (resulting in part from the sheer glamour of the shiny new tool known as remote sensing, and perhaps best metaphorized by the saying "give a small boy a hammer, and he soon discovers that everything needs pounding"); and g) the *"Not-Invented-Here" Syndrome* (exemplified by the reluctance of a top-level executive to admit that some other person or organization came up with a great idea, before he or his organization did).
5) *The "Maintenance of Momemtum" Concept:* Too often a resource manager or his boss is given only a superficial exposure to the marvels of modern remote sensing technology through an "appreciation" type of course from which he graduates with great enthusiasm. Almost always there is no follow-up to ensure that his organization will begin to put this technology to practical use. Consequently, only a short time later he likely will have lost all of the momentum that has been instilled in him during the course. The key to solving this problem often is as follows: By prior arrangement, each attendee brings with him to the course a specific unsolved resource-inventory/management problem of importance to his organization. Also by prior communication with the instructional staff for the course, he brings with him (or the staff procures for him) the Landsat MSS or TM data tapes and/or aerial photography covering his problem area. During the concluding phases of the course, under supervision of the staff, he applies to this specific problem the data-enhancement and data-analysis techniques that have been dealt with earlier in the course. In this way he leaves the course having "solved" the problem. Almost invariably he can scarcely wait to field check his analyses, refine and correct them as necessary and then implement the management measures that are dictated by such analyses. From this success, he can then progress to additional remote sensing-related problems. Thus, his momentum is maintained. If *procedural manuals* and *technical assistance* also are made available to the recent attendee of the course, these can be factors which will aid him still further in maintaining his momentum, once he encounters difficult remote sensing-related projects while on the job.

be welcomed. Rather, the primary need is for gaining a higher degree of acceptance than we currently enjoy, with respect to the recently developed technology, by those who can benefit most greatly from its intelligent use, namely the land resource managers themselves.

REFERENCES

[1] *Manual of Photogrammetry,* 2nd ed. Amer. Soc. Photogrammetry, ch. 12, 1952.
[2] *Manual of Photographic Interpretation.* Amer. Soc. Photogrammetry, 1960.
[3] "Renewable resource management—applications of remote sensing," *Proc. Renewable Natural Resources (RNRF) Symp. on Applications of Remote Sensing to Resource Management,* Amer. Soc. Photogrammetry, Seattle, WA, 1983.
[4] *Manual of Remote Sensing,* 2nd ed. Amer. Soc. Photogrammetry, 1983.
[5] R. G. Best, "Remote sensing applications for wildlife management," in *Proc. RNRF Symp. on Applications of Remote Sensing to Resource Management,* Amer. Soc. Photogrammetry, 1983, pp. 55–96.
[6] R. N. Colwell, Report of Commission VII (Photographic Interpretation) to the International Society of Photogrammetry, pt. I, *General Photogrammetric Engineering,* vol. 18, no. 3, pp. 375–400, 1952.
[7] ——, "Determining the prevalence of certain cereal crop diseases by means of aerial photography," Univ. Calif., Hilgardia Bulletin 26, pp. 223–286, 1956.
[8] R. N. Colwell, "Remote sensing of natural resources," *Sci. Amer.,* pp. 54–69, Jan. 1968.
[9] R. N. Colwell and C. E. Poulton, "SPOT simulation imagery for

urban monitoring—A comparison with Landsat TM and MSS imagery and with high altitude color infrared photography," *SPOT Symp. Proc.*, Scottsdale, AZ, May 1984.

[10] R. N. Colwell, "The potential usefulness of various SPOT image enhancements for monitoring the urban and peri-urban environment," in *Proc. SPOT 3 Simulation Conf.*, Reno, NV, Mar. 1985.

[11] R. N. Colwell and D. M. Carneggie, "Applications of remote sensing to arid land problems," in *Food, Fiber and the Arid Lands.* Tucson, AZ: Univ. Arizona Press, 1971.

[12] S. D. DeGloria, "Evaluation of SPOT imagery for the interpretation of agricultural resources in California," in *SPOT Symp. Proc.*, Scottsdale, AZ, May 1984.

[13] ——, "The interpretation of forest resources in California on simulated SPOT imagery," in *SPOT Symp. Proc.*, Scottsdale, AZ, May 1984.

[14] W. C. Draeger. "Applications of remote sensing in multiple use wildland management," Ph.D. dissertation, School of Forestry, Univ. Calif., Berkeley, 1970.

[15] I. R. Hoos, "Technology transfer as social process—A sociological perspective," in *Proc. 16th Space Congress*, Cocoa Beach, FL, 1979.

[16] ——, "Dynamics of forestry management, the view from space," in *Proc. Int. Short Course/Conf. on Remote Sensing*, Humboldt State Univ., Aug. 1979.

[17] S. Khorram, H. G. Smith, E. F. Katibah, R. W. Thomas, J. M. Sharp, and A. Kaugars, "Remote sensing as an aid in watershed-wide estimation of solar radiation, water loss to the atmosphere, areal extent of snow, and water content of snow," in *An Integrated Study of Earth Resources in the State of California Using Remote Sensing Techniques*, Space Sci. Lab., Univ. Calif., Berkeley, series 17, issue 53, 1978.

[18] S. Khorram and H. G. Smith, "Use of digital terrain data in topographic analysis," in *Case Studies of Applied Advanced Data Collection in Management.* New York: Amer. Soc. Civil Engineers, 1980, pp. 368–383.

[19] S. Khorram, "A remote sensing-aided procedure for site-specific estimation of net radiation over large areas," *J. Appl. Photogr. Eng.*, vol. 8, no. 1, pp. 31–35, 1982.

[20] S. Khorram and E. F. Katibah, "Use of Landsat multispectral scanner data in vegetation mapping of a forested area," in *Proc. 1981 Annu. Conv., Amer. Soc. Photogrammetry*, Washington, DC.

[21] S. Khorram and H. G. Smith, "Use of Landsat and environmental satellite data in evapotranspiration estimation from a wildland area," in *Proc. 13th Int. Symp. Remote Sensing of Environment*, Ann Arbor, MI, 1979, pp. 1445–1454.

[22] S. Khorram, "Water quality mapping from Landsat digital data," *Int. J. Remote Sensing*, vol. 2, no. 2, pp. 145–153, 1981.

[23] S. Khorram, "A remote sensing-aided procedure for site-specific estimation of net radiation over large areas," *J. Appl. Photogr. Eng.*, vol. 8, no. 1, pp. 31–35, 1982.

[24] P. G. Langley, R. C. Aldrich, and R. C. Heller, "Multistage sampling of forest resources by using space photography" in *Proc. 2nd Annu. Earth Resources Aircraft Program Status Review*, NASA-JSC, Houston, TX, 1969, pp. 19-1–19-21.

[25] C. E. Poulton and R. I. Welch, "Where remote sensing fits in resource planning, development and management: What the resource administrator needs to know or appreciate," NASA Ames Research Center, Moffett Field, CA, 1980.

[26] C. E. Poulton and R. I. Welch, "Work flow in remote sensing applications—A conceptualization leading to the design of an integrated system to acquire information," NASA Ames Research Center, Moffett Field, CA, 1980.

[27] P. A. Murtha, "Users' manual of remote sensing for vegetation damage detection and assessment—A preliminary report," in *Proc. Renewable Natural Resources (RNRF) Symp. on Applications of Remote Sensing to Resource Management*, Seattle, WA, 1983, pp. 368–468.

[28] National Aeronautics and Space Administration, *Proc. Conf. of Remote Sensing Educators (CORSE '78)*, Stanford Univ., 1978.

[29] National Aeronautics and Space Administration, "Monitoring earth resources from aircraft and spacecraft," NASA SP-275, 1971.

[30] N. M. Short, *The Landsat Tutorial Workbook*, NASA Ref. Publ. 1078, 1982.

[31] H. G. Smith and S. Khorram, "Remote sensing aided procedure for conifer growth modeling in northeast Sierra Nevada," in *Proc. Fall Technical Meeting, Amer. Soc. Photogrammetry*, 1980, pp. RS-2F-1–14.

[32] W. G. Spann, N. J. Hooper, and D. J. Cotter, "An analysis of user requirenments for operational land satellite data," in *Proc. 15th Int. Symp. on Remote Sensing of Environment.*, Univ. Michigan, 1981.

[33] P. T. Tueller, "Rangeland remote sensing, an approach," in *Proc. RNRF Symp. on Applications of Remote Sensing to Resource Management*, Amer. Soc. Photogrammetry, 1983, pp. 42–54.

8

OCEAN REMOTE SENSING

Robert L. Bernstein and Payson R. Stevens

Traditional ocean measurements, from ships and moored buoys, are generally expensive to collect. Vast areas of the ocean are hardly ever observed this way. Yet the need to make global and continuous ocean measurements exists, and grows more critical with time. Remote sensing from space has been recognized for many years as the principal means by which this need may be ultimately met. During the past 15 years, NASA has developed sensor technology, and then built and launched several spacecraft dedicated in whole or in part to ocean measurements. The Seasat satellite, devoted entirely to oceanography, carried four new microwave-based sensors, during its brief three month operating life in 1978. Other satellites, with primarily geodetic and meteorological missions, have also contributed greatly to advancing the state of the art in ocean remote sensing. In this chapter, we review the areas where remote sensing has demonstrated its potential, and describe how the entire field is now poised to move from the initial demonstration phase, which is now ending, to the more mature phase of routine utilization, such as now characterizes other fields.

APPLICATION AREAS FOR OCEAN REMOTE SENSING

The needs for ocean remote sensing data are very broad. From a scientific point, the ocean interacts with the atmosphere, with the land, and with the ice-covered regions. These interactions are complex, often global in scale, and constantly changing, and require spacecraft for adequate measurement coverage. Increasing polar region mineral exploration and transportation operations require remote sensing for both safety and economy. Fisheries applications of remote sensing include the careful management of what is ultimately a limited but renewable resource. Safe and efficient ocean transportation requires accurate forecasts of wind, wave, and strong current. Naturally occurring and man-made substances introduced into populated coastal regions can be observed in detail from space, and need to be monitored more carefully. It may even prove possible to assist in marine geophysical exploration for offshore oil, gas, and minerals through some remote sensing techniques. Each of these application areas will be described in more detail, with emphasis on their remote sensing aspects.

The Ocean's Role in Weather and Climate

The ocean and atmosphere interact with each other in many different ways. The ocean responds to the force imposed by the atmosphere, with waves and currents arising from the action of the wind, and with sea surface temperatures decreasing under conditions of cold dry air blowing over warmer water. But in addition, the atmosphere itself responds to changes in ocean conditions. The ocean is able to store far more heat than the atmosphere, and retain it for a much longer time, as a result of the great differences in heat capacity and density between air and water. The upper few meters of water in the ocean can hold and release as much heat as the entire atmosphere above it. Ocean currents can then transport heat absorbed in the tropics to the higher latitudes where it may be released. In this way the gradient of temperature between the equator and the poles is moderated considerably. Fluctuations in ocean currents and temperatures, which are linked to fluctuations in atmospheric conditions, thus play an important role in producing variations in the earth's climate.

A long-term goal of oceanographers and meteorologists is to construct computer numerical models of the coupled ocean–atmosphere system, and supply these models with observations of both fluids on a global basis to make more accurate forecasts of weather and ocean conditions. Satellite remote sensing offers the promise of providing the needed observations on a frequent yet global scale.

Since a later chapter covers the remote sensing of the atmosphere and its weather, we will focus here on those aspects of remote sensing which are most important to oceanography, and the impact of the ocean on the atmosphere. For oceanography, surface wind is probably the single most important measurement which can now be made remotely from space. In 1978 the Seasat satellite carried a radar device called a scatterometer, designed to transmit microwave signals and measure the amount scattered back from small wavelets (a few centimeters in wavelength) and other roughness features of those dimensions on the ocean surface which are strongly correlated with wind. The radar backscatter measurements, which were made looking in two directions at each point on the ocean, are convertible to wind speed and direction, and agree with nearby individual ship and buoy reports. Unlike ship reports, which are concentrated on the major shipping lanes, virtually absent in the tropics and southern hemisphere, and biased away from the worst weather, scatterometry provides uniform repeatable coverage. With such an improvement in coverage, models of ocean waves and ocean currents, which are driven by surface wind, ought to provide better and more detailed results.

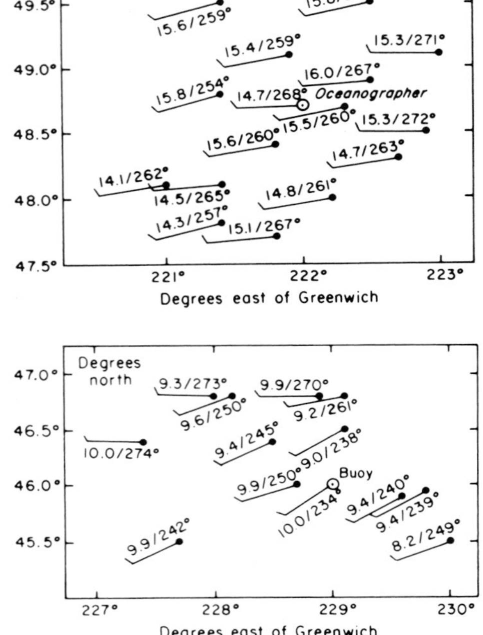

Fig. 8-1: Seasat scatterometer wind vectors (filled circles) show good agreement with North Pacific ship and buoy measurements (open circles) [7].

Solar energy absorbed in the upper ocean may be transported great distances by currents before being released back to the atmosphere. Traditional surface current measurements, from ships and buoys, are difficult and expensive to collect. Radar altimetry has attracted the attention of oceanographers for the last 15 years as a means of making this measurement from space, on a continuous global basis. The measurement principle is straightforward, but requires altimeter data of extreme precision and accuracy on the distance between an orbiting satellite and the sea surface. In the absence of any currents, tides, or other disturbances, the ocean surface would be a level, or equipotential surface of the earth's gravitational field. When a surface current is flowing, coriolis acceleration induced by the earth's rotation causes the ocean surface to tilt very slightly at right angles to the direction of flow. Thus, for example, a 1 m height difference exists across the Gulf Stream. To be useful on a global basis, altimeter measurements must be accurate to a few centimeters. Remarkably enough, it has been possible to build and demonstrate such accuracies. The Seasat radar altimeter was designed to give 10 cm accuracy, and in fact yielded data a factor of two better than that.

In order to obtain surface current information from radar altimetry, two additional pieces of data are required. First, the orbit of the satellite must be determined to comparable accuracy, and second, the earth's equipotential surface, usually referred to as the geoid, must also be known to the same accuracy. While it is not possible yet to determine orbits and the geoid to the requisite accuracies, significant progress is being made. For example, modest improvements to the tracking system used on Seasat would today provide 14 cm accuracy in the radial component of the orbit for a single overpass, and would give an even lower value through time-averaging of multiple passes. The next generation of NASA radar altimeter missions will use the global positioning system satellites, which are expected to provide orbit accuracies of a few centimeters. For the geoid determination problem, NASA has proposed GRM, the Geophysical Research Mission, which should give a dramatic improvement in knowledge of the geoid.

Yet even before such improvements in orbit and geoid are in hand, altimetry is still capable of providing valuable information on the time-fluctuating part of currents. The best data set to demonstrate this capability of altimetry was obtained during the Seasat mission. NASA's next generation altimeter mission, called TOPEX for ocean topography experiment, is expected

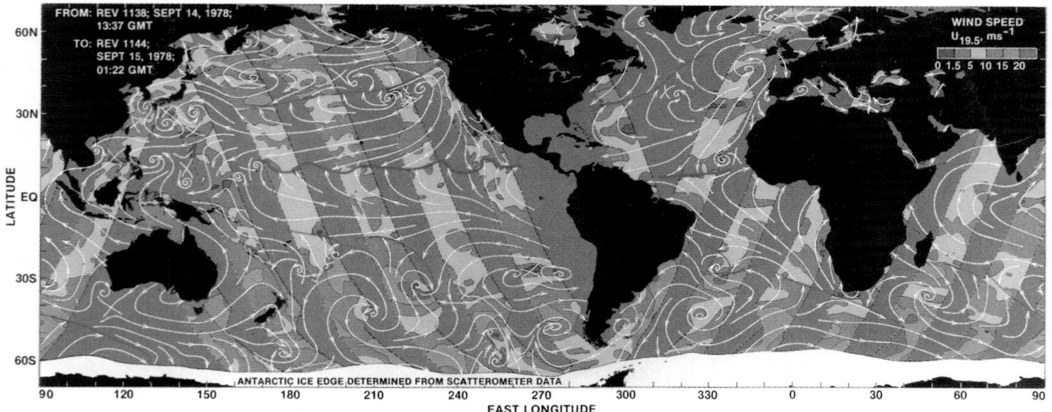

Fig. 8-2: Map of the surface wind field constructed from 12 hours of Seasat scatterometer data. Measurements of wind speed and direction are restricted to the colored swaths, where dark green denotes speeds less than 1.5 m/s, and red areas are in excess of 20 m/s; prepared by P. Woiceshyn (JPL), M. Wurtele (UCLA), S. Peteherych (AES-Canada), D.H. Boggs (dB Enterprises), and R. Atlas (NASA/GSFC).

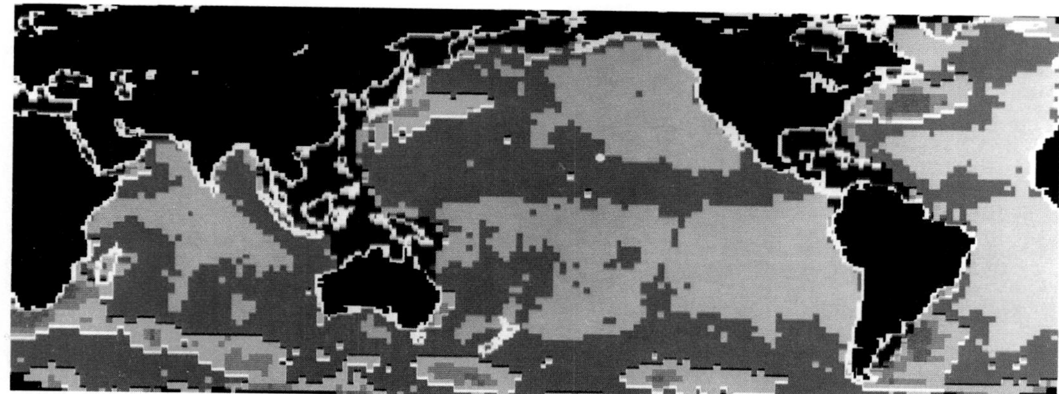

Fig. 8-3: Global map of surface height variability, showing local maxima in the Gulf Stream, Kuroshio, and Antarctic Circumpolar currents [2].

to fly in the late 1980's, and further the state of the art of radar altimetry, demonstrate that surface currents can be mapped globally on a regular basis, and provide three or more years of this data.

In addition to surface wind and surface current, the total amount of heat stored in the upper ocean is of great importance to climatological studies. This requires measurements of ocean temperature extending from the surface down to a depth of several hundred meters or more. While altimetry may possibly be used to give the vertical integral of heat storage, the vertical distribution of the storage is most important, and no remote sensing technique can provide that information. Yet the surface temperature itself can now be determined to useful accuracy, and may, in concert with altimetry and scatterometry, provide sufficient constraints on models of upper ocean processes to adequately estimate the vertical distribution of stored heat.

Sea Ice

At any one time, sea ice covers approximately 13 percent of the world's oceans, and its thickness, extent, and composition affect global climate and weather. Remote sensing permits measurements to be made in regions that may be inaccessible by any other means. Extreme cold, cloud cover, winds, dark polar winters, and constantly moving ice floes make the polar seas the most hazardous on earth. Prior to satellites, little was known about seasonal or year-to-year variations in sea ice cover. Satellite surveys also provide the practical advantage of finding shorter and safer sea routes for military and mineral exploitation operations in polar regions.

Visual and Infrared Techniques

The Landsat multispectral scanning system (MSS) and the NOAA weather satellite advanced very high resolution radiometer (AVHRR) have both been used to collect imagery in the visual and infrared portions of the spectrum. Landsat offers a higher resolution (30–80 m) than the AVHRR (1–4 km), and has proved useful for studies of ice conditions in local areas. AVHRR images have been used to monitor large-scale pack ice behavior.

Visual and thermal infrared systems rely on sensing reflected sunlight or the thermal emission from ice or snow surfaces, but cannot readily distinguish first-year from multi-year ice. Persistent cloud cover over much of the polar oceans limits coverage substantially. Instead, such imagery is primarily used for large-scale observations of sea ice movement and extent.

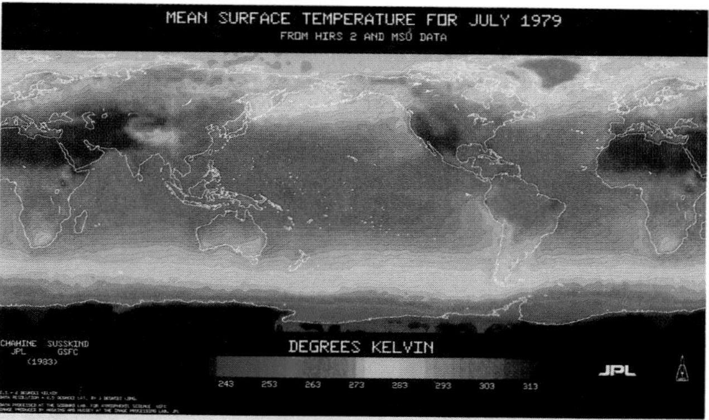

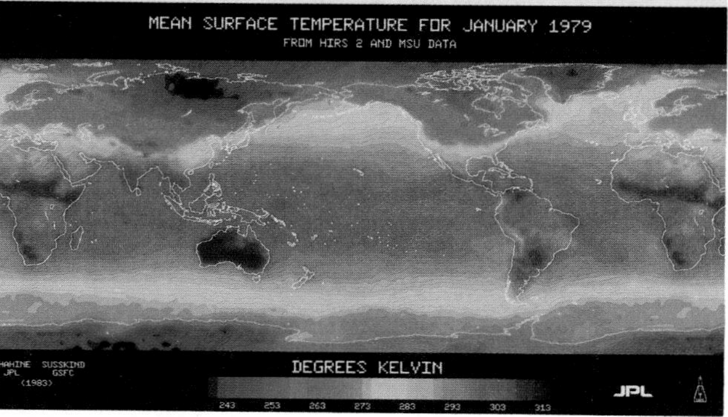

Fig. 8-4: Global surface temperature maps for July (top) and January (bottom) 1979, constructed from NOAA weather satellite infrared and microwave radiometer data; provided by J. Susskind and D. Reuter (Goddard Space Flight Center), M. Chahine and K. Hussey (JPL). For further details see [8].

Microwave Sensing Techniques

Both visible and infrared systems are severely limited by darkness and cloud cover, with consequently infrequent opportunities for imaging. Satellite microwave techniques fortunately do offer all-weather, day and night observations of the polar regions. Microwave observing systems employ either active or passive techniques. Active systems (radars) observe the energy scattered back from snow, ice, and open water surfaces. Passive systems (microwave radiometers) measure the electromagnetic energy emitted by matter. Emission properties at various microwave wavelengths differ markedly from open water, ice of various ages, and land surfaces with or without snow cover.

Passive Microwave Measurements

Two main passive microwave sensors have been used for observing sea ice. In 1972, NASA launched the electrically scanning microwave radiometer (ESMR), the first satellite sensor to provide global data on sea ice extent. ESMR provided data until 1983, covering 1400 km wide swaths with a resolution of about 25 km. It has provided large-scale mapping in the polar regions showing the percentage of ice cover from month to month and from one year to the next. Images constructed from ESMR data have provided the first clear picture of sea ice changes around Antarctica and in the Arctic Ocean.

The scanning multifrequency microwave radiometer (SMMR), another passive sensor, was launched in 1978 aboard the Nimbus 7 satellite. By observing at five different microwave frequencies, it eliminated certain ambiguities associated with the ESMR's single frequency, and has provided excellent sea ice data. SMMR has a 780 km swath with a spatial resolution between 30 and 150 km, depending on the frequency employed. SMMR has successfully provided images of sea ice concentration and the multiyear ice fraction in the Arctic Ocean. The SMMR ice data are now routinely used to guide civilian and military polar operations. In 1986 the U.S. Department of Defense polar orbiting meteorological satellites will begin carrying a microwave radiometer (termed SSM/I, for special sensor microwave/imager), which will provide ice data of even higher quality. Between ESMR, SMMR, and SSM/I, it will be possible to examine sea ice cover and its variability over decade time scales, with important implications for climate studies, and the long-term global warming hypothesized to result from the steady increase in atmospheric carbon dioxide. Modeling results predict that this warming will be most pronounced in the polar regions. Thus, the total amount of sea ice, as observed by these sensors from 1972 to 1986 and beyond, may be the first clear indication of such warming.

Active Microwave Measurements

Though passive microwave sensors provide synoptic data, their low resolution (tens of kilometers) restricts their utility to large-scale applications only. In contrast, synthetic aperture radar (SAR) is an active microwave sensor with very high resolution. An *L*-band wavelength (25 cm) SAR was flown on Seasat in 1978, and provided imagery at 25 m resolution. During its 100-day life the Seasat SAR provided detailed ice data as it covered its 100 km swaths. During the last 30 of those 100 days, exactly repeating coverage was possible every three days. Radar returns give information, independent of any cloud cover, on the roughness of the surface, thus identifying distinctive features such as floes, open-water leads,

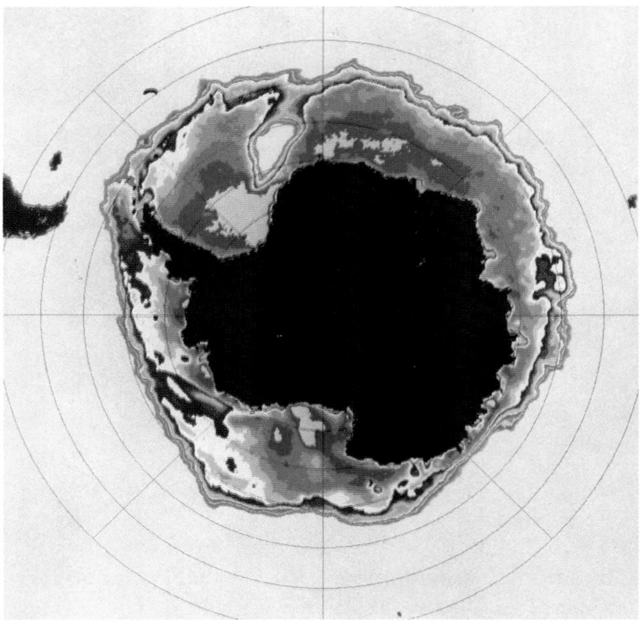

(a)

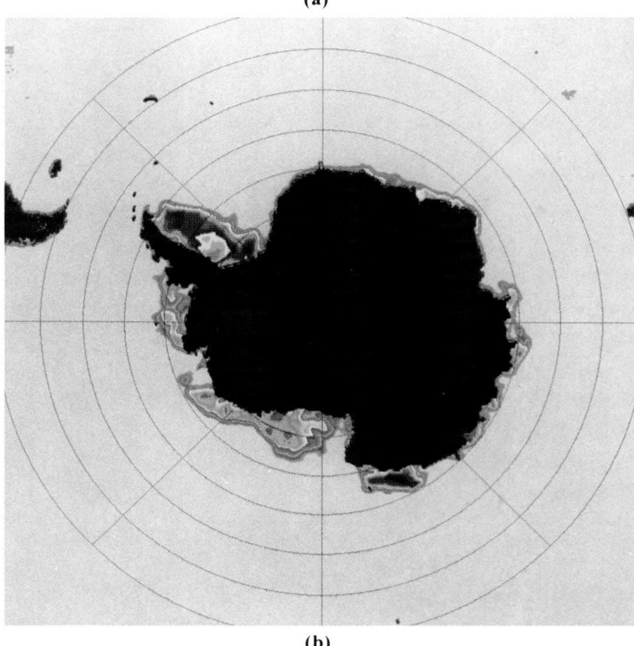

(b)

Fig. 8-5: Seasonal variation in Antarctic sea ice cover, from the NASA Nimbus 5 ESMR sensor, (a) between the winter maximum in August 1976, and (b) the summer minimum in February 1976. Antarctica is overlaid in black; ice concentration decreases from high near the coast (reds) to low near the ice margins (blues) [9].

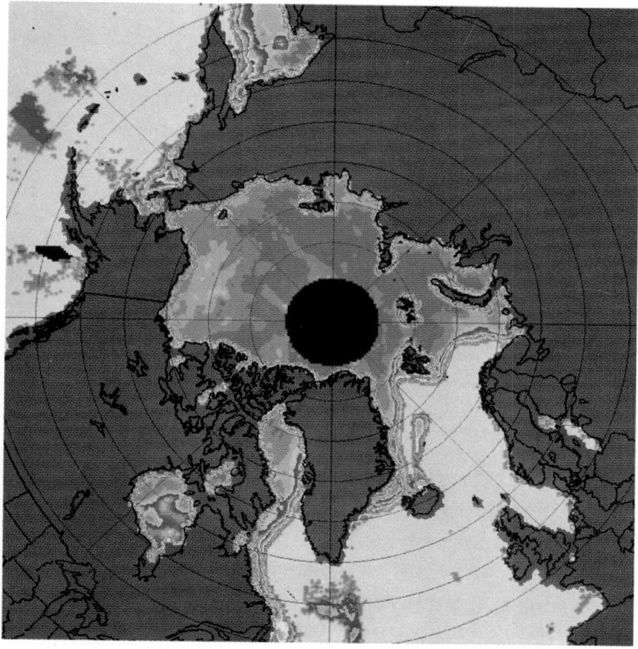

(a)

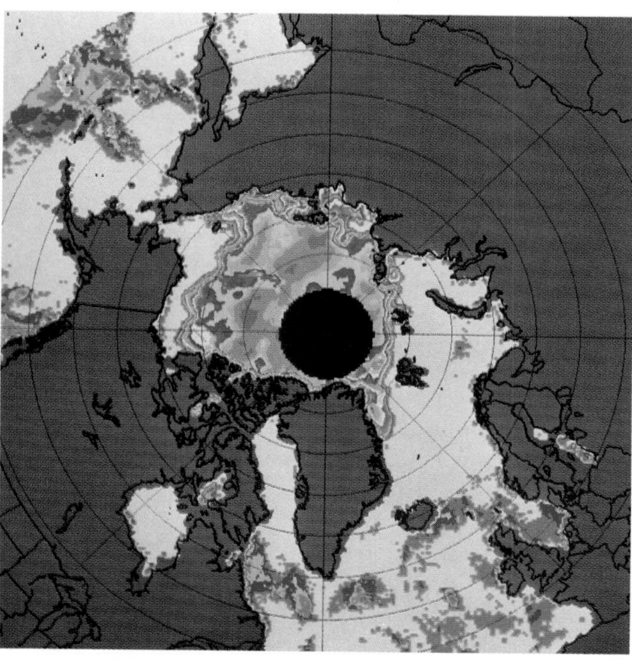

(b)

Fig. 8-6: Seasonal variation in the arctic ice cover, from the NASA Nimbus 7 SMMR, between the winter maximum in April(a), and the summer minimum in September 1979(b). Color coding shows maximum ice concentrations in dark purple thinning to red and yellow. Light blue areas are open ocean. Dark blue patches over ocean indicate heavy clouds or rain. The dark spot over the North Pole lacks data as Nimbus 7 does not traverse this area [1].

and ridges. Smooth new ice cannot be distinguished from open water by SAR, but such conditions do not constitute a navigation hazard. Sequential SAR data have identified changes in ice pack movement over a period of days.

Radar Altimetry

The altimeter and scatterometer offer the other two active microwave techniques for observing sea ice. Radar altimeters were aboard NASA's GEOS-3 (April 1975 to December 1978) and Seasat (July to October 1978) satellites, and had a footprint of about 12 km radius. On land the altimeter measured over 600 000 elevations on the Greenland and Antarctic ice sheets, revealing the surface topography of previously unexplored regions. Altimetry data over sea ice provided accurate estimates of the boundary between ice-covered and open water, along track lines directly below the spacecraft. Other altimetry missions include the U.S. Navy's Geosat (1985), the European Space Agency's ERS-1 (1988/89), and the U.S. Navy/NASA/NOAA joint mission NROSS (1989). The combined passive and active microwave measurements of sea ice provide both large, regional, and local views. These offer powerful tools which ultimately will give new insights into the impact of sea ice on global climate, as well as aid exploration activities in polar regions.

Living and Non-Living Marine Resources

Fisheries

The biological marine environment is characterized by high variability in time and space. As with most remote sensing applications, satellites have a major advantage for understanding marine ecological resources through their ability to observe synoptically large areas of the ocean at repeated intervals. In fisheries applications, the key questions for effective management are what are the distributions and abundance of the fish, their prey,

Fig. 8-7: Arctic pack ice in the Beaufort Sea, from the Seasat Synthetic Aperture Radar. Most pack ice is less than a few meters thick. The brighter object near the center of this 100 km area image is the ice island T-3, a 30 m thick tabular iceberg. Such images demonstrate the great potential for monitoring sea ice formation and movement [4].

and their predators? At the moment, satellites cannot answer most of these questions—for they represent basic marine ecological interactions, about which still little is known. But satellites do offer an important option to one problem biological oceanographers have always faced: the severe and costly limitations of ships as platforms for sampling the ocean.

The fluctuations in fish stocks are largely due to changing ocean conditions. Part of the goal of fishery management is to understand these variations, rather than average conditions, so that they can be modeled and related to environmental variables affecting fish populations.

Satellites enable scientists to view the entire oceans in a single day, and large regional areas in a few minutes. Fisheries applications use data from active and passive sensors aboard different satellites. Primary observations related to fisheries management have been surface color, temperature, and current, the latter based on the measurement of marine winds. From a research perspective the goal is to identify oceanic conditions which affect fisheries recruitment, distribution, abundance, and harvest. From the fisherman's practical perspective, there are a number of satellite remote sensing aids that are being developed which will help maximize their yields and minimize costly search time. And of course, fishermen need forecasts of general wind, wave, ice, and other environmental conditions, along with navigation, communication, and search and rescue systems, all of which depend to an increasing degree on spacecraft operations.

Ocean Color

Most marine organisms—from plankton to whales—tend to have patchy and irregular distributions. This feature further complicates identifying their distribution and abundance since ship searches are limited. At sea, *in situ* observations depend on nets and sonar. Though satellites are limited to observing only the sea surface, the launching of the coastal zone color scanner (CZCS) aboard Nimbus 7 in October 1978 offered biological oceanographers for the first time a powerful satellite tool in understanding ocean biological processes. The CZCS, however, does not directly measure fishery abundance and distribution. Instead it offers a more basic biological approach by measuring the intensity of sunlight backscattered from the upper meters of the ocean in four spectrally narrow color bands, in the blue, green, yellow, and red portions of the spectrum. Ocean color is primarily determined by the photosynthetic pigment, or chlorophyll concentrations, of microscopic drifting marine plants. These algae or phytoplankton are the basis of the entire marine food chain, and their pigments tend to absorb blue and red light. Therefore, the greener the water the greater the phytoplankton abundance. The CZCS scans and measures variations in ocean color from space at a resolution of 800 m, and computer processing of the resulting digital image data yields quantitative pictures of phytoplankton pigment concentrations.

The CZCS can measure pigment with an accuracy of about 35 percent, for water relatively free of suspended sediments. Since chlorophyll and pigment concentrations vary over three orders of magnitude, from hundredths of a milligram per cubic meter in most open ocean regions, to values exceeding ten in biologically active coastal areas, 35 percent accuracy measurements are exceedingly useful. Thus, the CZCS offers the first large-scale synoptic views of phytoplankton abundance and distribution. When combined with ship data, and data from other satellite sensors, these regional views will help us to understand processes affecting fish populations and to assess coastal water productivity.

Applications

Fisherman have trade secrets which help them locate prey. One useful indicator has been the abrupt ocean frontal boundaries which can occur between regions of different ocean color; these are zones where certain fish may congregate. Often these transition zones separate coastal from offshore regions. Coastal waters usually have higher phytoplankton abundance and productivity due to land runoff and coastal upwelling nutrient enrichment. The greater phytoplankton abundance results in more organisms higher up the food chain that feed on them, including commercially valuable fish. CZCS images have already been successfully used by commercial fisherman to help them increase their yields and decrease their search time by locating ocean fronts. Fisheries such as albacore tuna on the West Coast, the blue fin fishery on the East Coast, and the shrimp fishery on the Gulf Coast have all demonstrated the efficiency of directing operations using remote sensing.

Remote Sensing of Suspended Sediments and Pollutants

Surface pollutants, such as oil or other organic slicks, can be detected under certain conditions by remote sensors operating at electromagnetic wavelengths. At ultraviolet and visible wavelengths, oil slicks tend to have a higher index of refraction than background water. Depending on sun and viewing geometry, the slick can then reflect either more or less light than the neighboring areas back to the sensor. In the thermal infrared, the emissivity of oil differs from that of water. Oil slicks dampen small waves, causing microwaves to backscatter more vigorously from open water than from oil slick covered areas. To detect oil particles emulsified in the upper few meters of the water column, airborne laser fluorosensors are used. Satellite and airborne remote sensors are also being used to verify oil drift and dispersion models and study the capture of oil slicks by oceanographic fronts.

Substances further than a few meters down into the water column are more difficult to detect since water strongly absorbs most wavelengths outside the visible region, and scatters even visible light. Some pollutants are not detectable, but can be deduced only by associa-

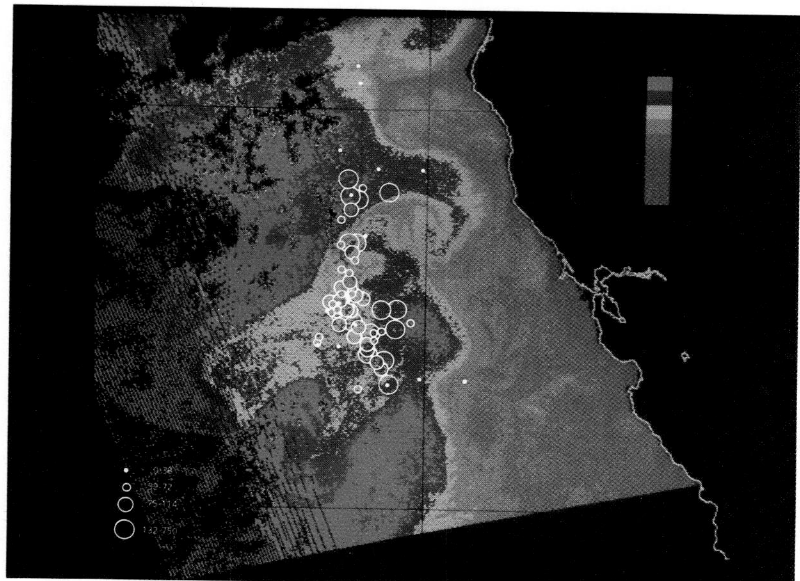

Fig. 8-8: Nimbus 7 CZCS water color image off the California coast on September 21, 1981, with greener near shore waters color-coded in red. Albacore tuna catches between September 19 and 24 noted by circles. The tuna are visual feeders, preferring clearer (blue) water areas immediately adjacent to more biologically active regions [6].

tion with other materials. Only a few substances can be detected and unambiguously identified by remote sensors alone. Generally, dissolved substances absorb certain spectral bands and thus change the color of the backscattered light. Suspended particulates, especially inorganic sediments, tend to reflect more energy, giving turbid waters a brighter appearance. Water bodies having different compositions can be identified by their colors and turbidities using simple color film in aerial cameras. Suspended sediment can also be used as a natural tracer to map current circulation patterns in turbid estuaries and coastal waters. However, to obtain a quantitative measure of pollutant concentrations, multispectral scanners must be used together with some carefully analyzed water samples obtained from boats.

Ocean-dumping wastes, such as industrial acid and municipal sewage sludge, have also been observed by Landsat and aircraft sensors. Principal component analysis of Landsat imagery obtained over New York Bight and off the coast of Delaware has enabled investigators to differentiate the types of ocean wastes dumped, and to determine their drift and dispersion. It is reasonable to conclude that aircraft and satellite sensors, when properly used in conjunction with boat measurements, enhance the ability to monitor certain pollutants and natural substances in water.

Marine Geophysics

Landsat and aircraft multispectral scanners have already shown their value in exploiting mineral resources on land. Remote ocean sensing for non-living marine resources, however, is still in a developmental stage. The two areas of exploration that are of greatest commercial interest are mineral and oil development. At present, the ocean remote sensing applications for offshore oil exploration and the submarine mining of minerals are rather limited. As we will show below, the limiting factor is the scale of resolution available in the currently available radar altimeter data; higher resolution will be required to identify details of important features such as rift zones and ridges showing submarine tectonic activity.

From its brief three-month life in 1978, the Seasat radar altimeter data has been used to generate global maps of the earth's gravity field. The radar altimeter generates pulses which reflect off the sea surface. From careful timing of their arrival back at the altimeter, the distance between the satellite and sea surface may be determined to a precision of about 6 cm. The actual relief of the sea surface is largely a measure of fluctuations in the earth's gravity field, since seamounts, or denser rocks beneath an otherwise flat sea bottom will cause the ocean surface to bulge upward by a measurable amount. Seasat data have been used to produce maps of the average sea surface topography, revealing valuable information about the composition and bathymetry of the ocean floor.

The structural trends which appear in these gravity maps may help direct oil exploration efforts. Their primary benefit is for a general reconnaissance prior to undertaking costly shipboard gravity and seismic surveys. Oil companies may wish to extend their knowledge of certain geological structures beyond previously explored areas. For example, if a fault has been mapped which seems promising for drilling, altimeter data may indicate how far that structure extends horizontally.

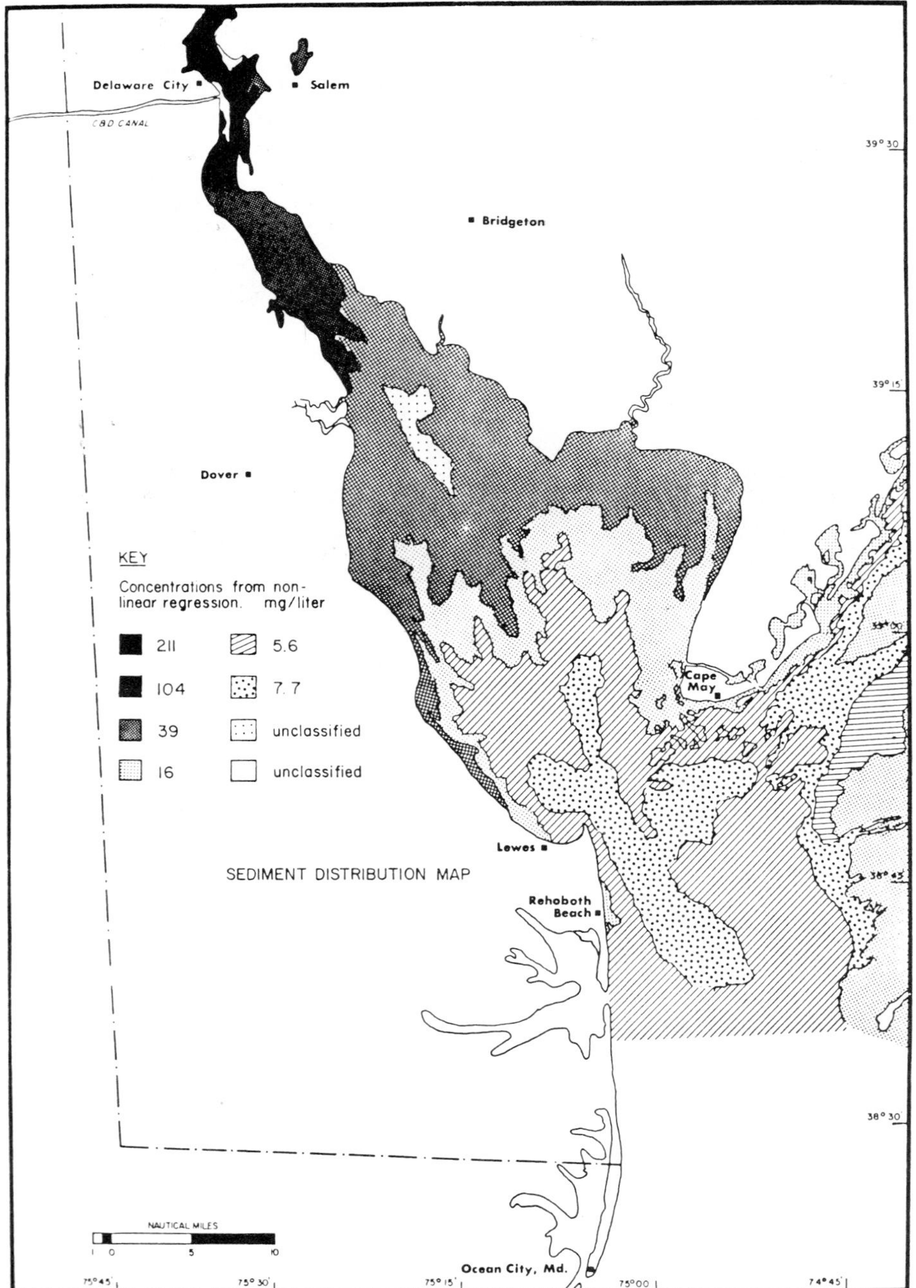

Fig. 8-9: A map of suspended sediment concentration for Delaware Bay, obtained by digital analysis of Landsat multi-spectral scanner imagery and correlation with 16 water samples obtained from boats and helicopters [5].

This would enable surveys to concentrate and maximize their search efforts. Identifying local gravity lows is also important, for they generally indicate thick sedimentary columns which may have oil-bearing potential.

From the Seasat mission, the presently available altimeter data coverage of the globe is limited to about 150 km resolution. (Resolution along any one trackline is only a few kilometers, but the separation between tracklines is 150 km). A U.S. Navy altimeter mission, Geosat, launched in 1985, will have an orbit designed to yield 18 km resolution, but the data will be classified and thus not available for a wide variety of potential applications. The European Space Agency is now considering a plan to operate its ERS-1 mission (1989 launch) in an orbit which would give 50 km resolution coverage. France and Japan also plan missions incorporating radar altimeters during the early 1990's. It thus appears that sometime within the next ten years, sufficiently dense coverage of the oceans by radar altimeters will have accumulated so that offshore oil and mining applications can be pursued.

Important Discoveries and Results

Only some of the many applications of ocean remote sensing, those that we believe are the most important or

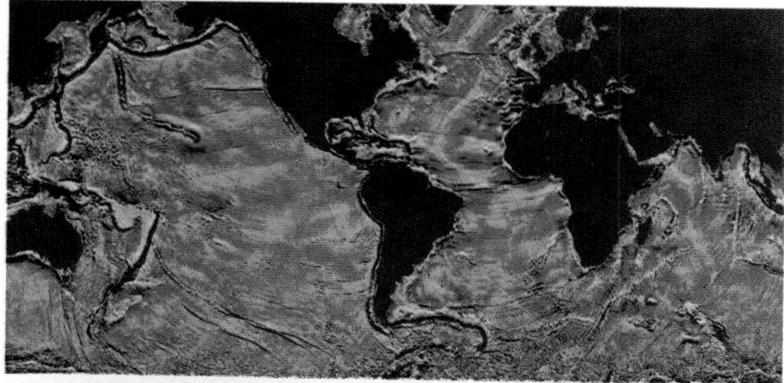

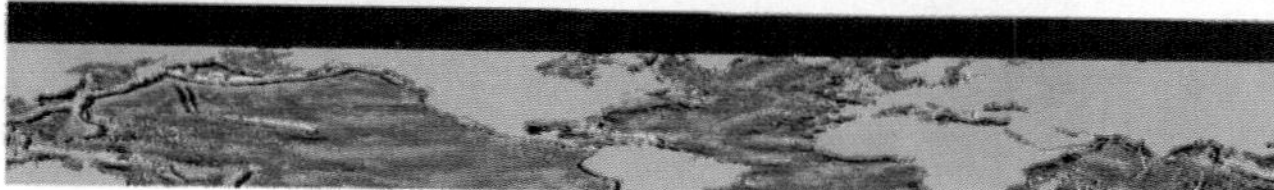

Fig. 8-10: Sea level variations, measured by the Seasat radar altimeter, are well-correlated with bathymetric relief and the earth's gravity field. Here the altimeter sea level data show the various midocean ridges, fracture zones, trenches, and chains of seamounts of the world ocean. Image produced by W. F. Haxby, Lamont-Doherty Geological Observatory.

illustrative, were selected for discussion in this chapter. The major point we wished to make was that the entire area of ocean remote sensing has made substantial progress along a very broad front of sensor technology, with applications extending to many aspects of ocean science and applications. No single discovery stands out; rather, a collection of techniques and capabilities from NASA research spacecraft has been demonstrated. The promises of the late 1960's and early 1970's have for the most part been tested and validated during the late 1970's and early 1980's. The result has been to lay a foundation for the next crucial step in making this technology available. This step will be putting in place systems for delivering sensor data on a long term, consistent, and easily accessible basis. Both ocean science and applications require the uninterrupted flow of data, the former to construct long-term series of global properties, and the latter to always have the data necessary to make tomorrow's forecast of marine conditions. NASA is now entering upon the next phase, of working in collaboration with other U.S. agencies, as well as with other nations, to design, build, launch, and operate the necessary spacecraft and sensors.

REFERENCES

[1] D. J. Cavalieri, P. Gloersen, and W. J. Campbell, "Determination of sea ice parameters with the Nimbus-7 SMMR," *J. Geophys. Res.*, vol. 89, pp. 5355–5369, 1984.

[2] R. E. Cheney, J. G. Marsh, and B. D. Beckley, "Global mesoscale variability from collinear tracks of Seasat altimeter data," *J. Geophys. Res.*, vol. 88, pp. 4343–4354, 1983.

[3] T. H. Dixon, M. Naraghi, M. K. McNutt, and S. M. Smith, "Bathymetric prediction from Seasat altimeter data," *J. Geophys. Res.*, vol. 88, pp. 1563–1571, 1983.

[4] L. Fu and B. Holt, "Seasat views oceans and sea ice with synthetic-aperture radar," Jet Propulsion Laboratory, Pasadena, CA, JPL Publication 81–120, 1982.

[5] V. Klemas, *Proc. 9th Int. Symp. Remote Sensing of Environment*, Ann Arbor, MI, 1974, pp. 1289–1317.

[6] R. M. Laurs, P. C. Fiedler, and D. R. Montgomery, "Albacore tuna catch distributions relative to environmental features observed from satellites," *Deep-Sea Res.*, vol. 31, pp. 1085–1099, 1984.

[7] W. J. Pierson, Jr., "The measurement of the synoptic scale wind over the ocean," *J. Geophys. Res.*, vol. 88, pp. 1683–1708, 1983.

[8] J. Susskind, J. Rosenfield, D. Reuter, and M. T. Chahine, "Remote sensing of weather and climate parameters from HIRS2/MSU on TIROS-N," *J. Geophys. Res.*, pp. 4677–4697, 1984.

[9] H. J. Zwally, J. C. Comiso, C. L. Parkinson, W. I. Campbell, F. D. Carsey, and P. Gloersen, "Passive microwave observations," in *Antarctic Sea Ice, 1973–1976*, NASA Scientific and Technical Information Branch, Washington, DC, NASA SP-459, 1983.

9

GEOPHYSICAL REMOTE SENSING

Robert L. McPherron

PRELIMINARY REMARKS

Since the space age began in 1957, there has been a continually increasing use of spacecraft for remote sensing of the earth. Early efforts were devoted to obtaining images of the surface in visible or near visible wavelengths. As time has progressed, other wavelengths in the vacuum ultraviolet, thermal infrared, and microwave regions have been used as well. Today, parameters other than reflected electromagnetic radiation are being used to produce earth images. Such parameters include the earth's static magnetic field, its gravity field, and self-luminosity produced by the loss of particles into the atmosphere.

In this brief review of geophysical remote sensing, we provide examples of the various types of images which have been used to obtain information relevant to geophysical studies. We begin with an image illustrating how information about large-scale geological structure is obtained from shuttle radar images. Next, we show one example of a specially processed image from Landsat and Seasat that reveals information about crustal composition. Following this we present a very recent map (in contour form) from the Magsat project that shows magnetic perturbations caused by crustal magnetization. Next, we discuss an image of the topography of the sea surface derived from Seasat altimetry, which reflects the nature of the earth's gravity field. Finally, we discuss a Dynamics Explorer image of the aurora produced from vacuum ultraviolet emissions of particles bombarding the atmosphere. In each section, our presentation is in three parts—a brief introduction, a summary of a recent spacecraft mission, and a brief discussion of an image produced by that mission.

REMOTE SENSING OF GEOLOGICAL STRUCTURE

Introduction

The regional coverage, minimal geometric distortion, low angles of solar illumination, and repetitive coverage have helped make satellite images from the Landsat spacecraft particularly useful in geologic mapping. One of the most obvious and most useful features of these images is "lineaments," long linear trends in the earth's surface. Most lineaments represent zones of weakness. In many cases, they exhibit structural offset showing that faulting has occurred. Another frequently observed feature of Landsat imagery is circular trends, many of which are associated with centers of igneous activity. Additional features include color and texture from which information concerning rock types can be derived. The relationship between these features in the images makes it possible to generate maps of large and inaccessible regions.

An example illustrating how such images have been applied to geological mapping of plate tectonics effects is discussed by Sabins [28]. The Afar Triangle region of Ethiopia is a complex region resulting from the convergence of three spreading axes at a triple junction. In the Landsat image described by Sabins, structural trends are highlighted by low-angle illumination of topographic relief created by normal faults more or less parallel to the trends of the three spreading axes.

There are many advantages to the use of the Landsat images in geologic mapping, particularly in desert regions where there is relatively little cloud cover. In tropical regions, however, the persistent cloud cover and heavy vegetation make Landsat images much less useful. An alternative method of imagery applied in such situations is synthetic aperture radar. In this technique microwave radiation transmitted from an antenna on an airplane or spacecraft is reflected from surface features and the returned signal is recorded on film or magnetic tape. In subsequent processing the signal is converted to an image corresponding to that which would be produced by a transverse beam width.

As discussed below, one of the first payloads carried into orbit by the space shuttle was a synthetic aperture radar.

The Shuttle Imaging Radar Mission

The second flight of the space shuttle Columbia in November 1981 carried a scientific payload consisting of a number of remote sensing experiments [30]. The orbit of the spacecraft was inclined at 38° to the earth's equator and flown with the shuttle bay pointing earthward so that the instrument package could view the earth. Although a number of technical problems were encountered, several of the experiments returned high-quality data. The Shuttle Imaging Radar, a side-looking synthetic aperture radar, operated extremely well and its data have led to several new discoveries [7]. This instrument utilized a microwave signal with 23 cm wavelength, and a beamwidth such that from the shuttle altitude of 262 km it had a ground resolution of 40 m. The median incidence angle of the radar beam from the vertical was 50°. A single scan of the antenna produced a ground swath width of 50 km.

Results from the Shuttle Imaging Radar

A major advantage of the synthetic aperture radar is its ability to image through heavy cloud cover. These conditions are a persistent problem in tropical areas, where it has not been possible to obtain an adequate set of visible images with the Landsat multispectral scanner. The radar is also useful in heavily vegetated areas, where geological structure is revealed by subtle changes in the height of the forest canopy.

To illustrate the use of synthetic aperture radar under these conditions, Sabins [27] has geologically mapped several areas in Indonesia. Fig. 9-1, taken from this article, was prepared as part of this study. The radar image and the corresponding interpretation map and cross section show the northwest Vogelkop region of Irian Jaya. As apparent from the image, the radar easily penetrates the cloud cover to reflect from the vegetation. Since the vegetation surface conforms to the underlying topography, the image reveals geologic structure through the presence of lineaments, attitudes of strata (dip and strike), circular features, and texture, as in the case of Landsat imagery. For example, two major faults are present in the left half of the image. Between the two

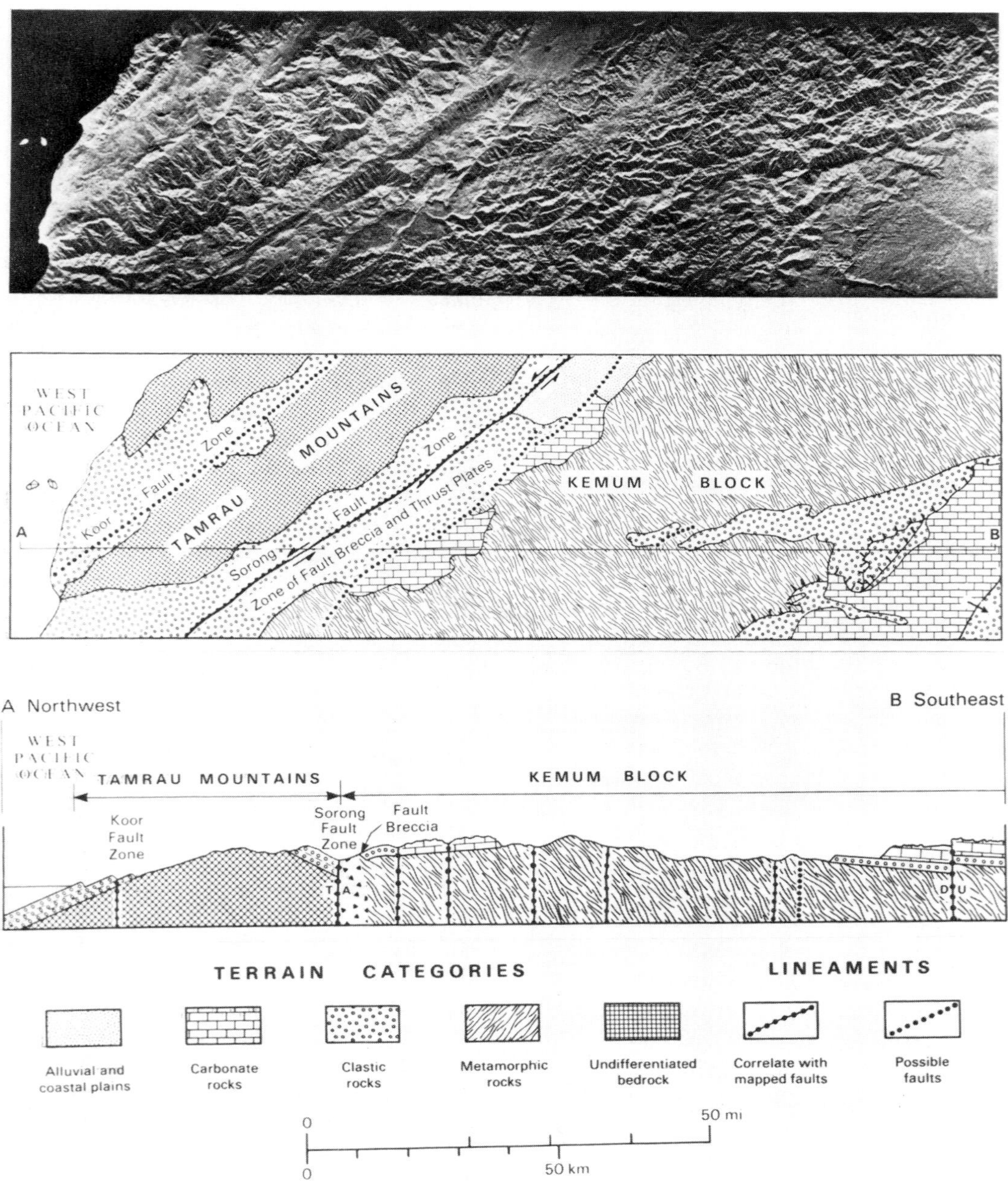

Fig. 9-1: Vogelkop Region of Irian Jaya as seen by the Shuttle Imaging Radar. Two major fault zones are evident in the left side of the image (best seen in proper relief by inverting the image) [27].

lineaments defining these faults there is a major mountain range. In this image five different rock types can be identified; for example, the pitted terrain at the lower right corner is a karst region produced by the weathering of carbonate rocks.

Using such images, Sabins [27] has created geologic maps of a small portion of this part of the western Pacific. By applying image-processing methods to digital data derived from the images, e.g., spatial filtering, it is possible to create color composite images which distinguish between rock types. For example, the pitted karst terrain with its characteristic higher spatial frequency was distinguished from clastic terrain with its lower spatial frequency (see front cover of the journal containing Sabins' article).

Based on his mapping of the few available images, Sabins concludes that radar imaging is extremely valuable in penetrating the perpetual cloud cover and enhancing the expression of geologic structure and rock types. As an aid to development of the hydrocarbon resources in this area, he recommends a complete mapping of Indonesia by future shuttle missions.

REMOTE SENSING OF CRUSTAL COMPOSITION

Introduction

Remote sensing methods provide a tool for mineral exploration that is becoming increasingly valuable as improvements are made in sensing technology. These methods provide data for geologic mapping of unexplored regions, mapping of fracture systems that control mineral deposition, and direct detection of alteration of host rocks associated with ore deposits.

Mapping from satellite imagery is an obvious extension of previous work with aerial photography. The broad coverage of satellite images, however, facilitates studies of regional scale. The digital nature of the data, together with improved methods of image processing, makes it possible to emphasize features of interest such as lineaments, surface roughness, etc. Direct detection of surface alteration is more complicated and depends on improvements in the spectral and spatial resolution of satellite sensors.

Early examples of the use of satellite imagery in resource exploration have been reviewed by Sabins [28]. As illustrated in his review, Landsat images of Nevada have been used to map lineaments which, in turn, have been shown to correlate well with the location and value of known mining districts. The same technique applied to central Colorado reveals a number of possible areas of interest, some of which are known to be mineralized and some which remain to be evaluated.

Landsat images have also been used to locate zones of hydrothermal alteration. In a study of the Goldfield, Nevada, mining district, four bands from the Landsat multispectral scanner were ratioed and then superimposed, with each ratio represented by a different color. In the color composite image (see Plate 6 of [28]), altered areas appear with unique colors. Although there was no one-to-one correspondence between known mining areas and color anomalies, the results were sufficiently good to justify further work.

As discussed by Sabins, the detection of alteration depends on the reflectance spectra of the products of alteration. Unfortunately, the Landsat MSS is far from optimum for such work, and numerous attempts have been made to develop improved scanners. One improvement of the thematic mapper system (described in the following section) was the addition of bands that record reflectance at the 1.6 and 2.2 μ atmospheric windows; these bands are especially important for mapping hydrothermally altered rocks. Another improvement was reported by Kahle and Goetz [14] who describe a six-channel, thermal infrared scanner that can record variations in silica content and distinguish between carbonate and clay-bearing units. In a test flight over Death Valley, California, the new scanner was quite successful in mapping siliceous rocks. In a second flight over the Cuprite mining district in Nevada, the scanner correctly identified known areas of alteration.

To illustrate the nature of current work with existing spacecraft data we discuss an image produced from a combination of Landsat and Seasat data.

The Landsat and Seasat Missions

The unmanned Landsat mission consists of a series of five spacecraft, the first launched in July 1972. The spacecraft have been placed in near-circular, sun-synchronous orbits at 918 km altitude and 80° inclination. The phase of the orbit was chosen such that southward passes cross the equator at approximately 10:00 A.M. local time. This orbit provides optimum angles of solar illumination to delineate subtle features in topography.

All five Landsat satellites have carried a multispectral scanner (MSS); in addition, Landsats 4 and 5 carry a thematic mapper (TM). Both imaging systems employ an oscillating scan mirror that sweeps a small field of view across the terrain at a right angle to the satellite orbit path. Energy of visible, reflected infrared, and thermal infrared wavelengths is separated by a spectrometer into discrete wavelength intervals (spectral bands) which are digitally recorded on separate images. Significant characteristics of MSS and TM systems are listed in Table 9-1.

The Seasat mission is described briefly in a later section with reference to the radar altimeter experiment. Here we note only that it also carried a synthetic aperture radar which differed from that in the shuttle by illuminating the ground with a median incidence angle 20° away from vertical.

Results from the Landsat–Seasat Missions

The use of imagery to define crustal composition is illustrated in Fig. 9-2. This image, taken from recent

TABLE 9-1
LANDSAT IMAGING SYSTEMS

	Multispectral Scanner	Thematic Mapper
Missions	Landsat 1-5	Landsat 4, 5
Ground Coverage	185 by 185 km	185 by 185 km
Spatial Resolution	80 m	30 m
Spectral Bands	4	7
Spectral Range	–	–
Visible and reflected IR	0.5–1.1 μm	0.45–2.35 μm
Thermal IR	None	10.4–12.5 μm

work of Merifield *et al.* [23], demonstrates how digitally registered combinations of Landsat and Seasat images can emphasize small differences in surface properties. The image shows a mountainous area in the eastern Mojave Desert near Vidal Junction, California. The Colorado River appears in the lower right. In Fig. 9-2, the bluish rectangular insert was prepared in the following manner: the Seasat image was projected in blue; MSS band 4 (visible green) was projected in red; MSS band 7 (reflected IR) was projected in green.

The two Landsat bands discriminate geologic units on the basis of reflectance, while the Seasat radar data provide information about surface roughness. As an example, debris flows made up of relatively large fragments of rock are evident from the blue coloration near the tops of alluvial fans (center of image). Such discrimination cannot be made from Landsat data alone, as desert varnish produces nearly uniform coloration (see, for example, fans along top and right margins where no Seasat data are available).

Using this composite image, the authors were able to map over 30 surficial units, as compared to about 8 units visible in the Landsat MSS data alone. A comparison with field mapping revealed several units in the composite image that could not be distinguished in the field, and only one unit was discovered in the field which was apparent in the image. The authors conclude that composite images including surface roughness data derived from radar are capable of discriminating subtle differences between surficial deposits that may go undetected by a geologist in the field.

SATELLITE MAPPING OF THE EARTH'S MAGNETIC FIELD

Introduction

Accurate models of the earth's magnetic field are needed for a variety of applications, including studies of

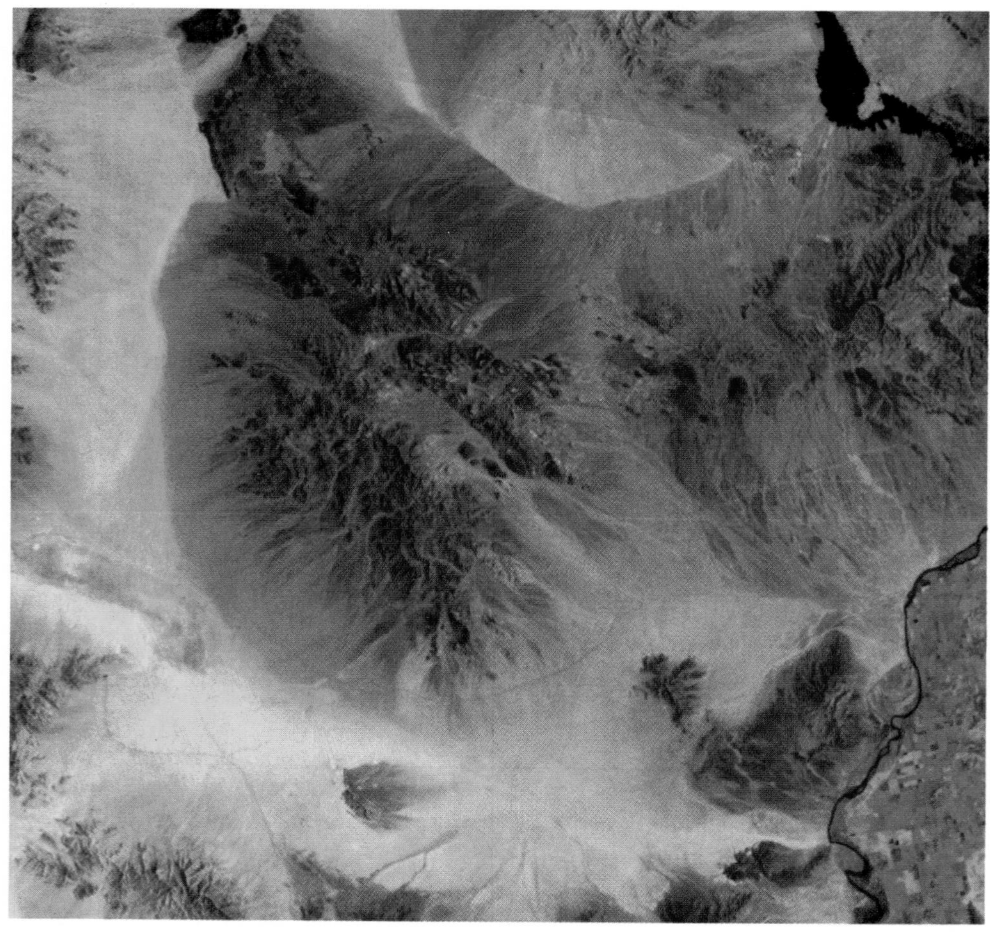

Fig. 9-2: Digitally coregistered Landsat and Seasat images of Vidal Junction, California, in the Eastern Mojave Desert. Landsat 3 band 4 is red, band 7 is green, and the Seasat data are blue. The combination facilitates geological mapping of surface materials in the desert terrain.

the origin of the main field and the causes of crustal magnetization. They also include more practical goals such as better maps for navigation, surveying, or describing the motion of energetic particles in the Van Allen radiation belts.

Until recently, magnetic maps of continental and global scale were produced from a world-wide distribution of permanent magnetic observatories supplemented by a number of repeat stations temporarily occupied in critical areas. On a more local scale, aeromagnetic surveys, tied to the observatory data, were used to provide additional detail.

In the 1960's, shortly after the first spacecraft were launched, the POGO spacecraft [21] carried total field magnetometers measuring the earth's field in the altitude range 400–1500 km [4]. Magnetic surveying with satellites has an obvious advantage over older methods. By placing the satellite in a polar orbit, the entire field of the earth can be sampled in a very short time.

Models of the main field produced from POGO data were not entirely consistent with those constructed from ground data, and it was suggested [33] that the lack of agreement might be a consequence of using only scalar data in their construction. In spite of such problems, the models were used to calculate residual fields along spacecraft trajectories. After removing effects of external sources, these residuals were found to occur systematically in geographic coordinates. Global maps of these magnetic anomalies were found to be correlated with tectonic features of the earth [26]. This success eventually led to approval of the Magsat mission described below.

The Magsat Mission

The Magsat (magnetometer satellite) mission officially began in 1977 with two primary objectives: improving models of the earth's main field, and constructing better maps of continental-scale crustal magnetization [18]. The entire spacecraft was devoted to these ends and was constructed so that precise measurements of the vector magnetic field were possible [24]. The spacecraft consisted of two modules—an instrument module and a base module. This allowed parallel development and, after launch, helped to isolate the instrument from effects of spacecraft operations.

The instrument module included the spacecraft attitude determination system, an attitude transfer system, a stable boom, and vector and scalar magnetometers. The base module contained the usual spacecraft services as well as tape recorders and a magnetic attitude-control system. Novel aspects of the Magsat mission include the use of star cameras to determine precise attitude, a precision sun sensor and pitch gyros for redundant attitude information, an optical system relating sensor orientation to spacecraft attitude, and mechanical and thermal isolation of the optical system. This system provided an overall accuracy of 20 arcsec in the orientation of magnetometer sensors.

The scalar magnetometer was a cesium vapor type designed to measure the field magnitude with an absolute accuracy of 1.5 nT. The vector magnetometer was a ringcore fluxgate which, after pre-flight and in-flight calibrations, measured each component of the field with an absolute accuracy of about 3 nT. To achieve this accuracy, the sensor assembly was constructed from special ceramics and was temperature stabilized.

The Magsat spacecraft was launched on October 30, 1979, into a sun-synchronous orbit with 97° inclination, 561 km apogee, and 352 km perigee. It survived for 7½ months, terminating on June 11, 1980. Initial results of this mission have been reported in a special issue of *Geophysical Research Letters* (vol. 19, no. 4, 1982). Several highlights of the mission are mentioned in the following section.

Results from the Magsat Mission

As discussed by Langel [18], the first result of the Magsat mission was a new model of the earth's magnetic field [17] based on a global distribution of scalar and vector field measurements on a quiet day. This model, MGST(6/80), was a 13th degree and order spherical harmonic fit to the observations with approximately an 8 nT rms residual in all components. The model demonstrated that the dipole component of the earth's field is continuing to decrease by roughly 26 nT per year. The rate is so great that, if it continues, the dipole component of the field will reverse sign in 1200 years.

Another important result is that at satellite altitudes, the core field dominates the observations for all spherical harmonic terms of degree/order less than 14 and the crust dominates at higher degree/order. The model also shows that the core/mantle boundary, defined as the spherical surface at which the rate of change of the surface integral of the radial component of the field goes to zero, agrees with seismic determinations within 2 percent.

The crustal field can be obtained from Magsat observations by subtracting a model of the core field, a model of external effects, and linear trends along the orbit. This has been done for the scalar field [19] and for the vector components [20]. Fig. 9-3, taken from the latter reference, illustrates the results for the south component. Using a combination of contour lines and colors, the map shows crustal anomalies observed at satellite height (~350 km). The anomalies typically have amplitudes of order ±12 nT or less, and scales of lengths of more than 250 km. One of the largest anomalies is in central Africa (the Bangui anomaly). Comparison of this map with corresponding maps of other components and the scalar field leads to the conclusion that most of the sources of crustal magnetization are "induced" by the earth's core field and not caused by remanent magnetization.

Although not a primary goal of the Magsat mission, its data have contributed substantially to studies of electrical currents linking the ionosphere to the magnetosphere. As discussed by Zanetti *et al.* [34], the accuracy of the magnetometer and the main field model have made it

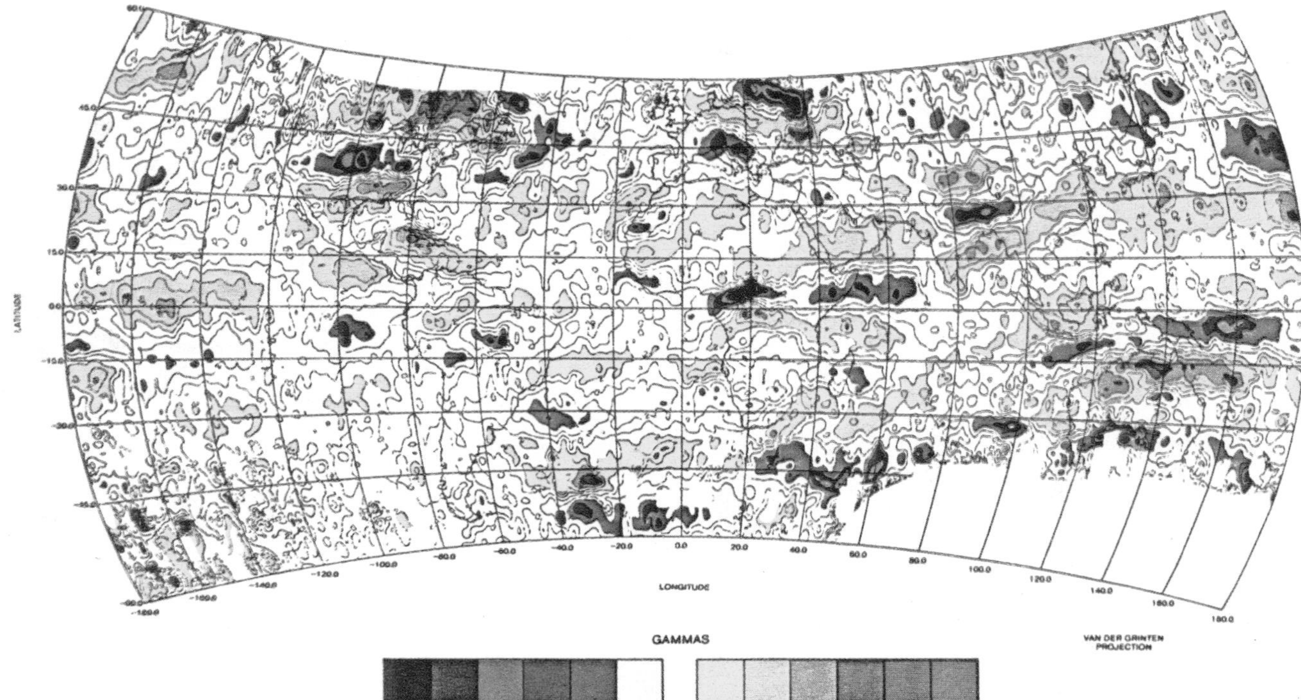

Fig. 9-3: A world map of crustal magnetization as measured in the south component by the Magsat spacecraft. A large negative anomaly in central Africa is quite evident.

possible to remove the main field and quantitatively study perturbations caused by ionospheric and field-aligned currents. For example, Zanetti *et al.* [35] have demonstrated that under certain conditions in the solar wind (see discussion of theta aurora in a subsequent section), a two-celled ionospheric current system driven by field-aligned currents flows entirely within the polar cap. At other times, the two-celled system closes at lower latitudes and is closely associated with auroral and magnetic disturbances. Iijima *et al.* [13] have used multiple orbits of Magsat to produce "images" of the field-aligned and ionospheric currents projected onto the polar cap. These images are considerably blurred by registration problems caused by variability in the location and strength of the currents, but the general pattern is clear.

SATELLITE MAPPING OF THE EARTH'S GRAVITY FIELD

Introduction

The gravitational field of the earth, like its magnetic field, is far from simple. For an observer at the earth's surface, or a spacecraft in low orbit, the field varies with location as a consequence of lack of spherical symmetry, differing densities of materials, varying elevation of the surface, etc. A spacecraft responds to these variations by subtle changes in its orbit [16]. Classical satellite geodesy determines the earth's gravitational field by tracking a spacecraft from fixed ground stations. Since orbits depart at most from true Keplerian orbits (those around a point mass) by only a few kilometers, tracking must be done very accurately.

Initially, camera images of the spacecraft against a field of stars were used to track spacecraft. Today, position is derived primarily from satellite-to-ground-station range rate, which is measured from Doppler changes in radio frequencies transmitted from the satellite to ground stations. Positions are also determined from ground-station-to-satellite range by reflecting a laser signal off the spacecraft.

These techniques provide the primary source of data for determining the gravity field over land areas. Unfortunately, the uneven distribution of tracking stations, and their virtual absence from the oceans, creates serious problems. In an attempt to solve these problems, many satellites and hundreds of ground stations are used in computer programs which determine simultaneously the locations of both the spacecraft and stations. A major problem with this approach is the lack of sensitivity of spacecraft orbits to detail in earth structure.

A new technique of gravimetry that has just begun to be used is radar altimetry. In this technique, a high-frequency radar emits a short-duration pulse whose vertical travel time to the earth's surface and back is measured. Combined with satellite position information, this gives the elevation of the earth's surface relative to a reference surface. Over land this information is of little use for gravimetry, but over sea it provides a description of the earth's geoid. By definition, the geoid is an equipotential surface in the earth's gravity field on which a point mass experiences no gravitational force parallel to the surface. The normal to the geoid at any point is the local vertical. It may depart significantly from the normal to the reference surface used to characterize the earth. The reason that radar altimetry is valuable over the oceans is that, apart from minor or predictable effects due to storms, tides, currents, etc., the mean sea surface corresponds to the geoid. In contrast to spacecraft orbits, spacecraft altitude above the sea surface is more sensitive to small-scale features.

In recent years, two spacecraft, the Geodynamics Experimental Ocean Satellite (GEOS 3), and Sea Satellite (Seasat), have carried radar altimeters. These two missions are described briefly in the following section.

The GEOS 3 and Seasat Missions

The GEOS 3 spacecraft was launched into a polar orbit on April 9, 1975 [32]. It was placed in a nominally circular orbit at 843 km altitude and 115° inclination. During its 3½ year lifetime, it returned a large quantity of useful altimetry data. The spacecraft was gravity-gradient stabilized with a long boom along the local vertical by an internal-momentum wheel orthogonal to the plane of the orbit. The primary instrument to which the spacecraft was dedicated was the radar altimeter. This instrument measured the spacecraft-to-sea surface separation with an accuracy of 40 cm. The spacecraft also carried a variety of instruments to aid in determining the spacecraft location, including transponders, laser reflectors, Doppler systems, and satellite-to-satellite communication links. The altimeter was found to be useful for measurements other than altitude, including wave height, wind velocity, and location of sea ice.

Seasat was launched into a polar orbit on June 28, 1978 [3]. Its orbit was near circular at an altitude of 800 km, with inclination 108°. The spacecraft functioned properly for three months, failing from a short circuit in the electrical system on October 10, 1978. During this time its radar altimeter returned approximately 70 days of data, which covered roughly 95 percent of the world's oceans every 36 hours. Its instrument complement included, in addition to the radar altimeter, a scatterometer, a synthetic aperture radar, a visible and infrared radiometer, and a scanning multichannel microwave radiometer. In addition, it carried a variety of devices for tracking and locating the spacecraft.

Results from the GEOS 3 and Seasat Missions

Scientific results from the GEOS 3 and Seasat missions have been published in several special issues of the *Journal of Geophysical Research* (vol. 84, no. B8, 1979; vol. 87, no. C5, 1982; vol. 88, no. C3, 1983). The results are too numerous and too varied to be summarized here. Instead, we illustrate the nature of the work concerned with the earth's gravity field by a brief description of an image of the South Pacific seafloor topography only recently published [29].

As mentioned above, radar altimetry is particularly useful in delineating small-scale undulations in the earth's geoid over the oceans. These undulations, in turn, are related to the topography of the seafloor. As discussed by Sandwell [29], this relationship depends on the precise mechanism responsible for isostatic compensation (adjustments required to float the crust on the mantle), but is nearly linear and depends on the wavelength of the topography.

Using general characteristics of the various linear relations, Sandwell has designed filters that emphasize the topographic geoid signal. Fig. 9-4 is an image which illustrates the results of his analysis and reveals previously unknown characteristics of the South Pacific.

To produce the image, both GEOS 3 and Seasat altimeter data were used. Radial errors in spacecraft position were suppressed by taking along-track derivatives of ascending and descending profiles. At profile crossovers, the along-track derivatives were used to calculate the direction of the geoid gradient (local vertical). Increased accuracy was obtained by combining data from two spacecraft using an interpolation function based on the geoid–topography relationship. The orientation of the local vertical was displayed by the "hill shading" technique. In this technique, the angle between an assumed direction to the sun and the local vertical is used to assign an intensity to the light reflected to an observer. Although the geoid undulations are small, their effects can be amplified by placing the sun direction close to the horizon plane. The two images in Fig. 9-4 utilize illuminations from NNE in the upper panel, and SSW in the lower panel.

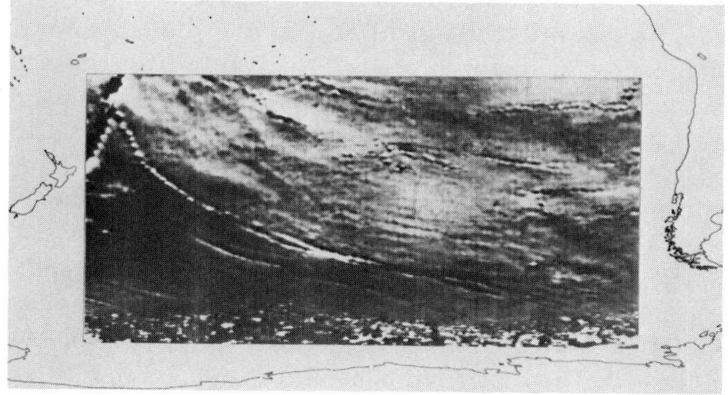

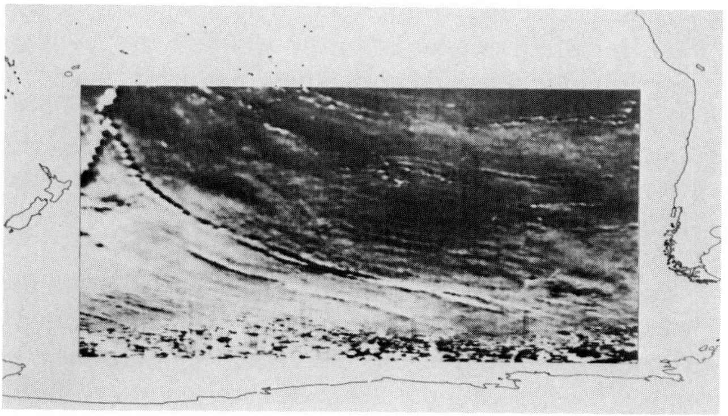

Fig. 9-4: An image of the mean sea surface in the South Pacific produced from GEOS 3 and Seasat altimeter data. The illumination is nearly horizontal from the NNE in the top frame and from the SSW in the bottom. The long curving feature starting in the upper left is caused by the Louisville ridge which joins to the Eltanin fracture zone. Many seamounts are also evident.

A number of features are apparent in the two panels and can be identified by comparison with a bathymetric chart of the South Pacific. The most obvious is the Louisville Ridge and Eltanin Fracture Zone which begin in the upper left corner and curve downward across the images. The ridge can be distinguished by its uniformly bright SW face and dark NE face in the bottom panel. The fracture zone, on the other hand, appears as a pair of light and dark bands that change polarity at the ridge spreading axis. The juncture of the ridge and fracture system was previously suspected, but was definitely established by this image.

In addition to such macroscopic features, a number of seamounts can be discerned in the images by their isolated signatures, with a bright spot always facing the sun's direction. A number of them have never been charted.

SATELLITE IMAGING OF THE POLAR AURORA

Introduction

Throughout history man has been awed and perplexed by the polar aurora. Seen from the earth's surface, the aurora often fills the sky with rapidly moving draperies of light. So extensive and so complex are these auroral forms that it was not until the International Geophysical Year (1957) that a comprehensive description of a typical auroral disturbance was developed. This phenomenological model was called the auroral substorm [1] because the disturbances are most frequent and intense when a magnetic storm is in progress. A major advance which made it possible to develop the auroral substorm model was the deployment during the IGY of All Sky Cameras at many locations in the polar regions. By piecing together images derived from many different events, a statistical picture of an auroral disturbance was created.

As soon as the first spacecraft was launched, it became evident that the aurora would be best viewed from space, where a view of the entire polar region could be obtained. The first attempt to do this was on the ISIS 2 spacecraft launched into a low orbit (~1400 km). This instrument [2] consisted of a photometer and filters sensitive to certain emissions in the visible band. A two-dimensional image was created once per orbit by a combination of spacecraft rotation and orbital motion. A major discovery made by this imaging system was the existence of a belt of diffuse aurora equatorward of the brighter and more active forms characterized in the substorm model [22].

A second imaging system was developed as part of the Defense Meteorological Satellite Program (DMSP). Originally built to image cloud cover at night, this system was carried on polar orbiting spacecraft at altitudes of ~850 km. It produced broad-band visible images by a combination of orbital motion and a scanning mirror. This system greatly stimulated studies of auroral morphology and led to a number of new discoveries [25], [31].

Both of the foregoing systems depended on satellite motion to produce an image. The first attempt to produce images rapidly from a fixed point in space was made on the Japanese satellite Kyokko [15]. This instrument used an image-memory tube sensitive to emissions in the vacuum ultraviolet. It was designed to produce images every two minutes from elevations between 650 and 4000 km.

The first truly successful global auroral imaging instrument producing images at a rapid rate was flown on the Dynamics Explorer Mission. This mission and an illustrative result are discussed in subsequent sections.

The Dynamics Explorer Mission

The Dynamics Explorer Mission consists of two spacecraft devoted to studies of the coupling between the ionosphere and the magnetosphere [11]. The two spacecraft DE 1 and DE 2 were launched simultaneously on August 3, 1981, into the same orbital plane [12]. The first spacecraft, DE 1, was placed in a higher orbit with apogee at 4.9 Re and perigee at 1300 km. The second was placed in low orbit (305–675 km). Their orbital motion was controlled so that the low-altitude spacecraft was near the foot of magnetic field lines passing through the higher spacecraft.

Both spacecraft carried a wide variety of particle and field instrumentation to study phenomena related to auroral disturbances. DE 1 included a spin scan auroral imager to produce global images of the aurora. This instrument [8] consisted of three separate photometers, two in the visible and one in the vacuum ultraviolet. The instruments use scanning mirrors and spacecraft rotation to produce two-dimensional images. Since the spacecraft is often sunlit while imaging the aurora, one of the primary design problems was elimination of scattered light from the daylight ionosphere, which in the visible spectrum can be a million times as intense as the auroral emissions. Depending on the mode of satellite telemetry, successive images are produced by all three photometers every 3–12 min. Each photometer also has a command selectable filter which allows the experimenter to generate simultaneous images in different wavelength bands. This allows deduction of such information as the energy of the particles exciting the auroral emissions, or the electrical conductivity of the ionosphere produced by their impact.

Results from the DE Mission

Initial results of the Dynamics Explorer Mission have been published in a recent issue of *Geophysical Research Letters* (vol. 9, no. 9, 1982). An exciting new discovery is the existence of a unique configuration of aurora named the "Theta aurora." Fig. 9-5 produced by the imaging instrument illustrates this phenomenon [9]. At the time this image was taken, the DE 1 spacecraft was ~3.6 Re above the north polar cap descending at about 19 hours local time. The dusk terminator passes through central Europe.

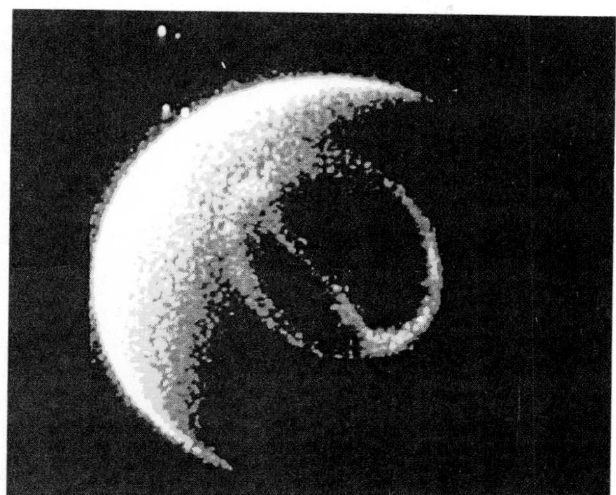

Fig. 9-5: An image of the Theta aurora produced by the spin scan auroral imager on Dynamics Explorer 2. Dynamics Explorer 1 was approximately 3.6 earth radii above the north polar cap. The circle is the auroral oval and the bar is a particular phenomenon which occurs only when the solar wind magnetic field is northward.

The auroral configuration shown in Fig. 9-5 includes the well known ring of aurora, the "auroral oval" which surrounds the magnetic pole. In addition, however, it includes a bar of auroral emission across the polar cap which, together with the oval, creates the "Greek letter, theta." Although elements of this picture have been previously seen from the ground and from other imaging spacecraft, this is the first time it has been seen in its entirety.

The explanation for this phenomenon is presently controversial. However, the fact that it occurs only when the magnetic field of the solar wind is strongly northward suggests an explanation. In this model first proposed by Dungey [5], the solar wind magnetic field is pressed against the earth's field at high latitudes (both north and south) and the two fields merge. The bright spot in the auroral oval near noon would be at the feet of field lines involved in this process. Subsequently, the solar wind drags the feet of these newly merged field lines around the earth to the night side. Particles lost to the ionosphere along these field lines produce the aurora in the dawn and dusk sectors of the oval. Later still, the field lines on opposite sides of the earth move towards each other in the nightside equatorial plane, with particle loss creating the night sector portion of the oval. Finally, the field lines meet near midnight and flow sunward across the center of the polar cup. Again, the earth's field lines merge with the solar wind field, releasing the line originally merged and connecting with a new one.

An interesting by-product of the foregoing process is that everywhere in the two dark cells outlined by the parts of the theta aurora, electrical currents flow along magnetic field lines into, or out of, the polar cap. In Fig. 9-5 the currents are in on the left side (dusk) and have been directly measured by the Magsat magnetometer described in a previous section [13] and have a total strength of the order 1 million A.

CONCLUDING REMARKS

Satellite imagery is becoming progressively more important in studies of the earth. This applies to geophysical studies of the earth's gravity and magnetic fields and to studies of atmospheric luminosity, as well as to studies of crustal geology and composition where images have been used for a longer time. The images are important because they bring together, in one display, vast amounts of information. Without the images it would be extremely difficult to assimilate the information and perceive the relationships that exist within them. Because most images today are in digital form, it is possible to manipulate them so as to emphasize features of interest. Scales may be changed, viewing and illuminating angles altered, contrast enhanced, other information superimposed, etc. These techniques of image processing enhance the value of images, converting subtle differences to forms more readily perceived by the human visual system.

There are many applications of space-borne imaging systems in geophysical studies, both fundamental and applied. As discussed above, new information concerning the motion of crustal plates has been derived from visible and "altimetric" images of the earth's surface. Similarly, pictures of the aurora have shown the existence of phenomena only dimly perceived in previous studies and have unambiguously shown the existence of a strong interaction of the solar wind with the earth during magnetically quiet times. The images have also been used for rather practical purposes such as geological mapping for hydrocarbons in tropical jungles beneath heavy cloud cover, or for classifying different types of surface units in mineralized regions.

As time progresses, it can be expected that increasingly detailed information about the earth will be derived from satellite imagery. This information will serve ever more practical needs. At the same time, these techniques will be applied to data acquired by spacecraft orbiting other planetary bodies. The insight into planetary processes derived from such images will provide greater understanding of how our own planet came to be as it is today.

ACKNOWLEDGMENT

The author would like to thank his colleagues at UCLA who, through their own expertise, have made it possible to summarize work in fields other than his own. Special thanks are due to W. G. Ernst, W. M. Kaula, F. F. Sabins, and P. M. Merifield in this regard. The author also thanks the original authors, acknowledged in the text and figure captions, for providing copies of images and comments on the summary, and thanks the various journals that have granted permission to use the published images. The author's time and efforts have been supported by a grant from the National Aeronautics and Space Administration (NGL-05-007-004).

REFERENCES

[1] S.-I. Akasofu, "The development of the auroral substorm," *Planet. Space Sci.*, vol. 12, pp. 237–282, 1964.

[2] C. D. Anger, T. Fancott, J. McNally, and H. S. Kerr, "The ISIS 2 scanning auroral photometer," *Appl. Opt.*, vol. 12, pp. 1753–1766, 1973.

[3] G. H. Born, J. A. Dunne, and D. B. Lame, "SEASAT mission overview," *Science*, vol. 204, no. 4400, pp. 1405–1406, 1979.

[4] J. C. Cain and R. A. Langel, "Geomagnetic survey by the polar orbiting geophysical observatories, World Magnetic Survey, 1957–1969," Int. Assoc. Geomagn. Aeron., Paris, Bull. 28, pp. 65–74, 1971.

[5] J. W. Dungey, "Interplanetary magnetic fields and auroral zones," *Phys. Rev. Lett.*, vol. 6, p. 47, 1961.

[6] C. Elachi, "Spaceborne imaging radar: Geologic and oceanographic applications," *Science*, vol. 209, no. 4461, pp 1073–1082, 1980.

[7] C. Elachi, W. E. Brown, J. B. Cimino, T. Dixon, D. L. Evans, J. P. Ford, R. S. Saunders, C. Breed, H. Masursky, J. F. McCauley, G. Schaber, L. Dellwig, A. England, H. MacDonald, P. Martin-Kaye, and F. Sabins, "Shuttle imaging radar experiment," *Science*, vol. 218, no. 4576, pp. 996–1003, 1982.

[8] L. A. Frank, J. D. Craven, K. L. Ackerson, M. R. English, R. H. Eather, and R. L. Carovillano, "Global auroral imaging instrumentation for the Dynamics Explorer Mission," *Space Sci. Instrum.*, vol. 5, pp. 369–393, 1981.

[9] L. A. Frank, J. D. Craven, J. L. Burch, and J. D. Winningham, "Polar views of the earth's aurora with Dynamics Explorer," *Geophys. Res. Lett.*, vol. 9, no. 9, pp. 1001–1004, 1982.

[10] E. A. Guinness, R. E. Arvidson, J. W. Strebeck, K. J. Schulz, G. F. Davies, and C. E. Leff, "Identification of a Precambrian rift through Missouri by digital image processing of geophysical and geological data," *J. Geophys. Res.*, vol. 87, no. B10, pp. 8529–8545, 1982.

[11] R. A. Hoffman and E. R. Schmerling, "Dynamics Explorer program: An overview," *Space Sci. Instrum.*, vol. 5, pp. 345–348, 1981.

[12] R. A. Hoffman, G. D. Hogan, and R. C. Maehl, "Dynamics Explorer spacecraft and ground operations systems," *Space Sci. Instrum*, vol. 5, pp. 349–367, 1981.

[13] T. Iijima, T. A. Potemra, L. J. Zanetti, and P. F. Bythram, "Large-scale Birkeland currents in the dayside polar region during strongly northward IMF—A new Birkeland current system." *J. Geophys. Res.*, vol. 89, no. A9, pp. 7441–7452, 1984.

[14] A. B. Kahle and A. F. H. Goetz, "Mineralogic information from a new airborne thermal infrared multispectral scanner," *Science*, vol. 222, no. 4619, pp. 24–27, 1983.

[15] E. Kaneda, *Proc. Int. Workshop Selected Topics Magnetospher. Phys.*, Japanese IMS Committee, Tokyo, p. 15, 1979.

[16] W. M. Kaula, "Method of experimental physics: Geophysics, Part B: Field measurements," unpublished report, 1984.

[17] R. A. Langel, R. H. Estes, G. D. Mead, E. B. Fabiano, and E. R. Lancaster, "Initial geomagnetic field model from Magsat vector data," *Geophys. Res. Lett.*, vol. 7, no. 10, pp. 793–796, 1980.

[18] R. A. Langel, G. Ousley, and J. Berbert, "The MAGSAT mission," *Geophys. Res. Lett.*, vol. 9, no. 4, pp. 243–245, 1982.

[19] R. A. Langel, J. D. Phillips, and R. J. Horner, "Initial scalar magnetic anomaly map from MAGSAT," *Geophys. Res. Lett.*, vol. 9, no. 4, pp. 269–272, 1982.

[20] R. A. Langel, C. C. Schnetzler, J. D. Phillips, and R. J. Horner, "Initial vector magnetic anomaly map from MAGSAT," *Geophys. Res. Lett.*, vol. 9, no. 4, pp. 273–276, 1982.

[21] G. H. Ludwig, "The orbiting geophysical observatories," *Space Sci. Rev.*, vol. 2, p. 175, 1963.

[22] A. T. Y. Lui and C. D. Anger, "A uniform belt of diffuse auroral emission seen by the ISIS-2 scanning photometer," *Planet. Space Sci.*, vol. 21, pp. 799–809, 1973.

[23] P. M. Merifield, R. S. Hazen, and D. L. Evans, "Discrimination of desert surficial materials for engineering/environmental geology applications utilizing coregistered LANDSAT MSS and SEASAT SAR data," in *Proc. Int. Symp. Remote Sensing Environ.* (Third Thematic Conference, Remote Sensing in Exploration Geology), Colorado Springs, CO, Apr. 16–19, 1984.

[24] F. F. Mobley, L. D. Eckard, G. H. Fontain, and G. W. Ousley, "MAGSAT—A new satellite to survey the earth's magnetic field," *IEEE Trans. Magn.*, vol. 16, pp. 758–760, 1980.

[25] C. P. Pike, and J. A. Whalen, "Satellite observations of auroral substorms," *J. Geophys. Res.*, vol. 79, no. 7, p. 985, 1974.

[26] R. D. Regan, J. C. Cain, and W. M. Davis, "A global magnetic anomaly map," *J. Geophys. Res.*, vol. 80, no. 5, p. 794, 1975.

[27] F. F. Sabins, Jr., "Geologic interpretations of space shuttle radar images of Indonesia," *Amer. Assoc. Pet. Geol. Bull.*, vol. 67, no. 11, pp. 2076–2099, 1983.

[28] ——, *Remote Sensing—Principles and Interpretation.* San Francisco: Freeman, 1978.

[29] D. T. Sandwell, "A detailed view of the South Pacific geoid from satellite altimetry," *J. Geophys. Res.*, vol. 89, no. B2, pp. 1089–1104, 1984.

[30] M. Settle and J. V. Taranik, "Use of the Space Shuttle for remote sensing research: Recent results and future prospects," *Science*, vol. 218, no. 4576, pp. 993–995, 1982.

[31] A. L. Snyder, S.-I. Akasofu, and T. N. Davis, "Auroral substorms observed from above the north polar region by a satellite," *J. Geophys. Res.*, vol. 79, no. 10, pp. 1393–1402, 1974.

[32] H. R. Stanley, "The Geos 3 project," *J. Geophys. Res.*, vol. 84, no. B8, pp. 3779–3783, 1979.

[33] D. P. Stern and J. H. Bredekamp, "Error enhancement in geomagnetic models derived from scalar data," *J. Geophys. Res.*, vol. 80, pp. 1776–1782, 1975.

[34] L. J. Zanetti, T. J. Potemra, and M. Sugiura, "Evaluation of high latitude disturbances with MAGSAT (The importance of the MAGSAT geomagnetic field model)," *Geophys. Res. Lett.*, vol. 9, no. 4, pp. 365–368, 1982.

[35] L. J. Zanetti, W. Baumjohann, T. A. Potemra, and P. F. Bythrow, "Three-dimensional Birkeland-ionospheric current system, determined from MAGSAT," Amer. Geophysical Union, Washington, DC, Magnetospheric Currents, Geophysical Monograph 28, 1984.

10

WEATHER AND ATMOSPHERE REMOTE SENSING

S. Ichtiaque Rasool

WEATHER FORECASTING

May 20, 1960, will remain an important date in the history of the science of weather forecasting. On that date Tiros 1, the very first weather satellite, took an image of the earth centered over the eastern Pacific which showed a cloud system organized in a unique pattern extending over 6000 km, from the north of Hawaii to the Great Lakes [Fig. 10-1(a)]. When this cloud image was overlaid over a map of the area it was clear that we had just acquired a technique to map the weather systems from space, over land and oceans alike [Fig. 10-1(b)].

Until that date the major problem in weather forecasting had been our almost complete ignorance of weather patterns over the large expanses of the ocean, obviously because of the lack of observing stations there. It is, however, the same weather systems which evolve and, in a few days, move over to the land areas. At the latitudes of U.S. and Europe, these systems usually travel from west to east at a speed of ~500 km a day. Tracking these storms and fronts continuously from space therefore not only gives us a means of forecasting their arrival time over a given region but, more importantly, provides a data set to study their evolution in time. Today, almost 25 years later, satellites of several nations image the entire globe every half-hour throughout the day and night (Fig. 10-2). Information derived from these images on the location of storms, fronts, and the wind systems has now become the principal real-time input to most of the short term weather forecasts now issued around the world.

As for the longer term weather forecasts, the problem is more complex. As shown in Fig. 10-3, a meteorologist can usually proceed in two different but rather complementary ways to formulate a weather forecast. One way is by simple projection into the future of what is happening now. Knowing the correct state of the atmosphere at that time and knowing exactly at what speed and in which direction the fronts and storms are moving, one can extrapolate the state of the atmosphere, often very correctly, over the next several hours. Because the atmosphere has a certain amount of "persistence" in its behavior, forecasts produced by this method could remain valid for a few hours. The satellite imagery is of considerable help here to define the initial state of the atmosphere, to determine the speed and direction of motion of the weather systems, and to update the forecast every few hours because the satellite measurements are so frequent. This method has been usefully employed for tracking severe weather, such as hurricanes over the oceans, and correctly predicting when and where they will hit the land. For longer term weather forecasts, however, the problem becomes different. Here one has to predict not just the movement of weather systems, but the evolution of fronts and storms. This can only be done if one understands the physics of atmospheric dynamics and can simulate the behavior of the atmosphere correctly in a computer. In this case one follows the other route shown in Fig. 10-3: namely first put in as correct a description of the global atmosphere as possible in the computer, write down the fundamental equations of motion, gas dynamics, energy, mass and momentum conservation; and given the initial pressure, temperature, and wind, let the computers calculate what the conditions will be a few hours or even a few days later.

Since a model can never be a perfect simulator of the real atmosphere, the physics we put in cannot be complete, and the description of the initial state of the atmosphere cannot be perfect at micro- and meso-scales; the errors in forecast grow as a function of the increasing

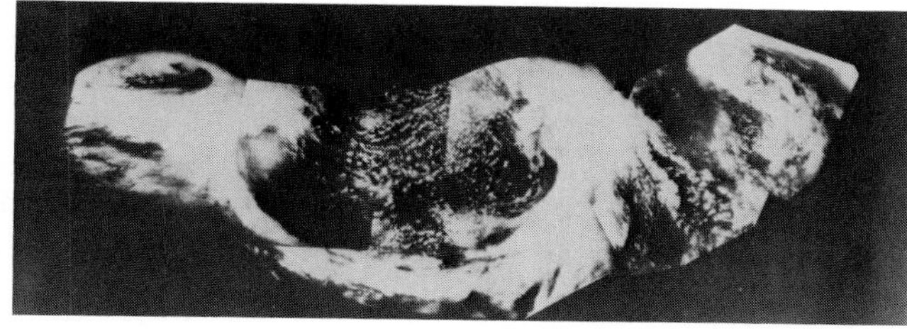

(a)

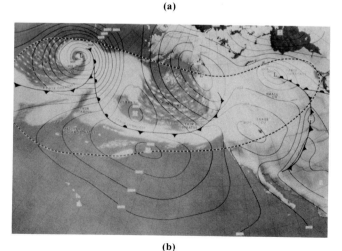

(b)

Fig. 10-1: (a) Experimental cloud depiction chart prepared from Tiros pictures superimposed on NAWAC 00002 map analysis of May 20, 1960. (b) Actual Tiros photographs taken on May 20, 1960.

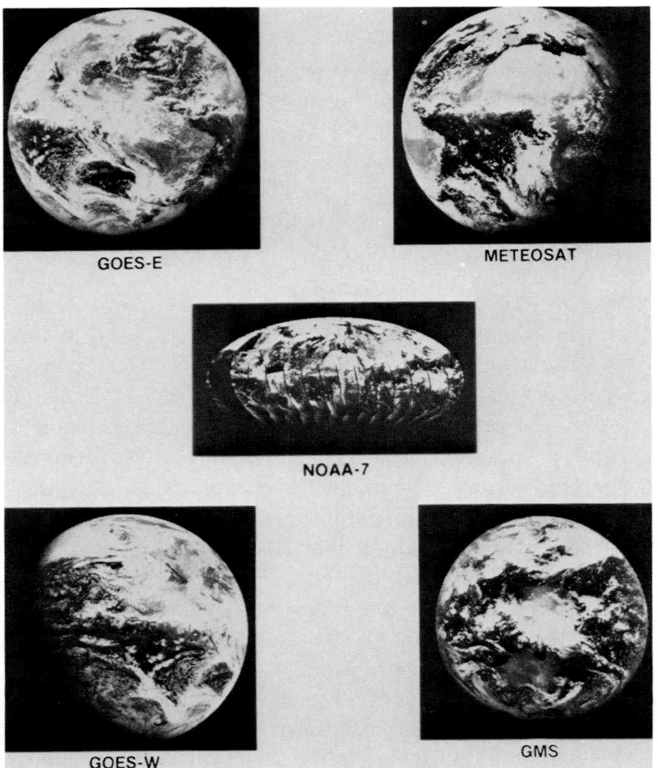

Fig. 10-2: Earth as a whole photographed on July 1, 1983, by a family of five satellites, GEOS E and W (U.S.), Meteosat (European), NOAA 7 (U.S.), and GMS (Japanese). Such images are now routinely used in tracking fronts and storms throughout the world (NASA).

length of the forecast with a doubling time of two to three days. But there are ways to minimize the errors which depend on 1) the capacity or the power of the computer to handle more than 80 million bits of weather data received every day, 2) speed of the computer to perform the manipulation fast enough (~50 million operations per second so that, for example, one day's global weather can be simulated in less than an hour), 3) the accuracy of the initial description of the atmosphere around the globe, and 4) the way the various variables in the *real* atmosphere are parameterized in the model with which the computer is supposed to work.

Simultaneous progress in both satellite and computer technology has slowly improved the accuracy of five-day forecasts (Fig. 10-4). According to statistics prepared by the European Center for Medium Range Forecasts, in three years, between 1980 and 1983, the range of useful forecast increased from five days to six days. This became possible by a combination of circumstances. Satellite observations were able to fill the gaps in data which existed because of the absence of radio sondes over the oceans, especially for the southern hemisphere; more powerful and faster computers enabled more realistic representation of the atmosphere in the computer and some improvements in the models themselves were realized during the last few years. However, we are still far from the theoretical limit of weather predictability which is supposed to be about 14 days. This is approximately the time span in which the very small scale (~100 m) disturbances (eddies) develop into large scale weather patterns. Because of our inevitable inability to correctly describe the very small scale structures of the atmosphere-ocean system, a two week forecast can be safely assumed to be the limit of the numerical prediction technique.

The question remains, however, of how to improve the accuracy of the current five-day forecast and extend the time of useful forecasts to a week or ten days. Here, again, one needs parallel progress in modeling, computer assimilation of the data, accuracy of satellite observation of the relevant parameters, and a deeper insight into the physics of the ocean–air interactions. In the case of satellites, the improvement has to come in the accuracy of *temperature sounding* and in *wind speed determinations* in the lower atmosphere. These parameters appear to be crucial in describing the initial state of the atmosphere;

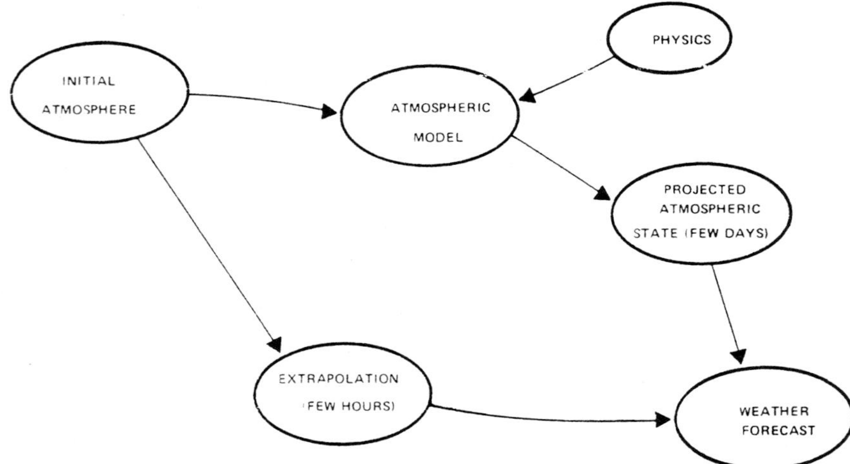

Fig. 10-3: Elements of weather forecasting. Starting from a description of the atmosphere, two ways to forecast the weather (short range, medium range) [7].

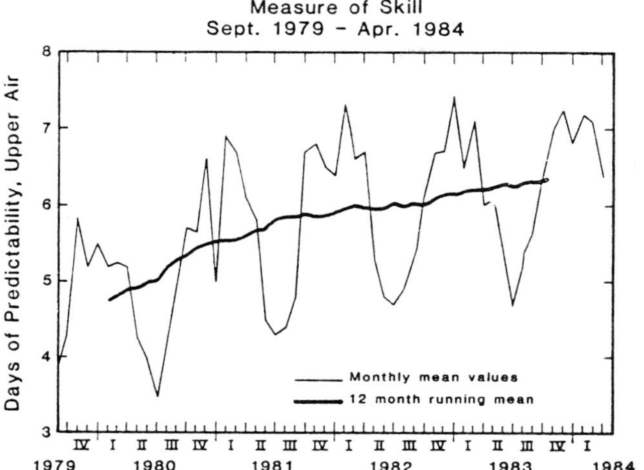

Fig. 10-4: A measure of the skill of European Center for medium-range weather forecasting (ECMWF) northern hemisphere forecasts for September 1979 to April 1984 [2].

the more accurate the input, the better the forecast. A test for the potential of satellite observations of these parameters and of their impact on weather forecasting was performed as part of Global Atmospheric Research Program in 1979, when seven satellites imaged and probed the atmosphere simultaneously with weather buoys in the oceans and drifting balloons in the atmosphere measuring the same parameters locally. A careful analysis of data from satellites and on-the-spot measurements revealed that the accuracy of temperature profiles in the atmosphere retrieved from the satellite measurements is much lower than expected and that the heights assigned to winds derived by the satellite observations of cloud motion may also be in substantial error (see Table 10-1). Improvements in both these domains are crucial before full potential of the satellite data can be realized in medium- and long-range forecasting systems. Attempts in this direction have already been initiated by a number of groups and agencies around the world. New instrumentation will be tried on the space shuttle, more sophisticated algorithms for temperature retrieval are being developed, and new innovative techniques to accurately measure winds in the lower and middle atmosphere are also being discussed.

In addition, techniques to monitor rainfall and soil moisture from space continue to be developed. In any weather forecast, precipitation is of course the most difficult parameter to predict because its occurrence depends on so many other complex interactions including the moisture present in the soil and in the vegetation. An accurate monitoring of both the precipitation and the soil humidity will certainly not only make the input to the models more realistic, but also help check on the predictions so that the physics of the models can be improved. This could be a major technological challenge for the next decade.

CLIMATE MONITORING AND PREDICTION

Climate is a "synthesis" of weather situations over a time period long enough (months, years, decades) so that its statistical properties can be established. Occasional departures from the statistical mean are what we call the climate anomalies and these are what we would like to be able to predict. As in the case of weather forecasting, again, there are two ways to proceed. First, if we have accurate and long enough data sets on climatic statistics and if, from them, we are able to delineate trends and/or periodicities, one can try to predict the future climatic situations just by extrapolations along the trend and periodicity lines. The gross seasonal changes in climate

TABLE 10-1
OBSERVATIONAL ERRORS FOR DIFFERENT OBSERVING SYSTEMS

	Temperature (°C)				Wind (ms^{-1})		
	TIROS-N				Clouddrift wind		
Level (mb)	Radiosonde	Clear/ partly cloudy	Microwave	Radiosonde pilot ASDAR AIDS	NESS WISCONSIN	ESA LMD	HIMAWARI
10	4.5	2.8	2.8	6	8	8	13
20	3.8	2.6	27	6	8	8	13
30	3.2	2.5	2.6	6	8	8	13
50	2.7	2.4	2.5	6	8	8	13
70	2.3	2.2	1.4	6	8	8	13
100	2.1	2.0	1.6	6	8	8	13
150	2.1	2.0	1.7	6	8	8	13
200	2.0	1.9	1.8	6	8	8	13
250	1.8	1.9	1.9	6	8	8	13
300	1.6	1.8	20	6	8	8	13
400	1.5	1.8	2.2	5	7	8	10
500	1.2	1.7	2.2	4	7	8	10
700	1.1	1.8	2.5	3	5	8	6
850	1.1	2.0	3.9	2	4	7	6
1000	—	—	—	2	4	7	6

Sea surface pressure SYNOP/SHIP 1.0 mb; buoy 2.0 mb; COLBA/DROPWINDSONDE/TWOS-NAVAID observation errors are calculated from the level II-b quality information. Temperature given as layer means.

and the onset of summer or winter monsoons every year are good examples in this regard. However, to be able to predict the year-to-year fluctuations in the intensity of the winters or the summers or to be able to forecast, for example, the exact arrival date of the Indian monsoon in advance, is of course the more difficult problem. Here one not only needs the statistics, but also better insight into the mechanism of climate change. However, in order to study this mechanism, the model should not only incorporate the atmospheric dynamics, but its interactions with the oceans, with ice and snow masses, with land surface, with solar radiation, and with atmospheric gases and constituents such as CO_2, dust, and O_3 which can actively modulate the energy input to the atmosphere. All these parameters vary with time and place at different rates. It will be close to impossible to account for the variability of each one of these parameters before issuing a climate forecast. However, with the limited insight that we do have in the theory of climate dynamics we can begin to make a few simplifying assumptions, depending on the time and space scale for which climate prediction is being sought.

When climatic changes over a few weeks to months are concerned, one can safely ignore, for example, the deep oceans, the polar ice, and variations in CO_2 and O_3. For this time frame the more important parameters, apart from the solar inclination, seem to be upper ocean temperatures, atmospheric dust and volcanic aerosols, extent of snow in the mid-latitudes, and soil moisture or evapotranspiration, especially over the semi-arid regions. Today, climate modelers are experimenting with these parameters by inserting hypothetically large anomalies in the values of each of these variables separately in their models to study the impact on predicted climate. It is becoming very clear that a precise and consistent monitoring of these four parameters on a global scale and over several years would not only make our models more realistic, but even improve our ability to make climatic prediction on seasonal time scale.

Let us now examine the state-of-the-art in monitoring these parameters from satellites.

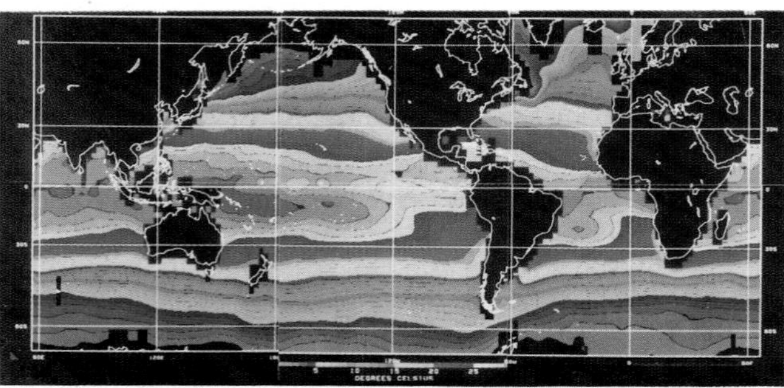

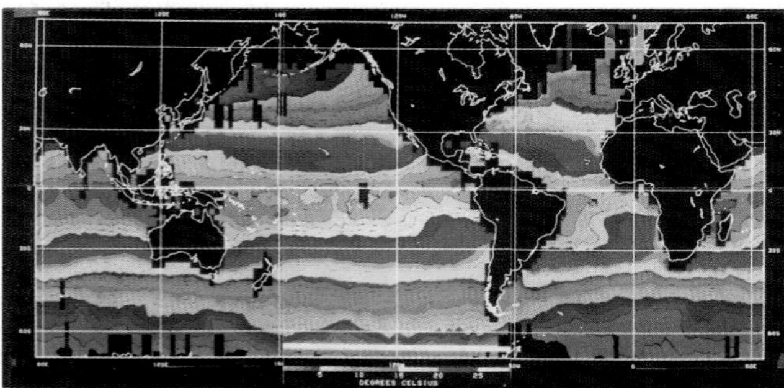

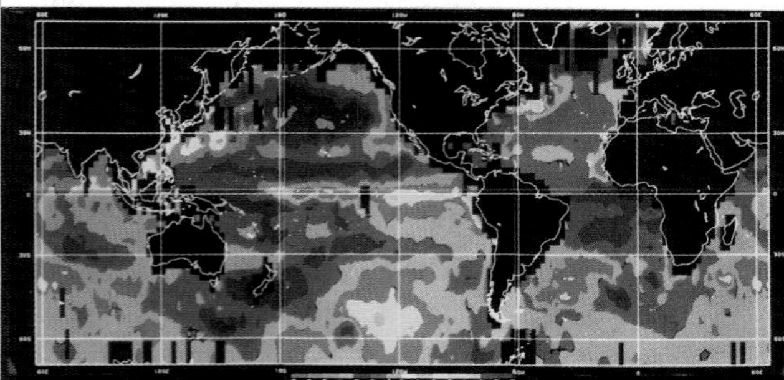

Fig. 10-5: (Top) Robinson–Bauer climatology of sea surface temperature for March. (Middle) Sea surface temperature for March 1983 measured by NOAA/AVHRR multichannel derivation procedure. (Bottom) The anomalies [8].

SEA SURFACE TEMPERATURE

Because of the very high heat capacity of the oceans and the fact that the latent heat of water is as high as 560 cal/g, relatively small changes in ocean temperature (1–2° C) and in evaporation rates (a few millimeters per day) produce major energy fluctuations in the atmosphere. These inputs get translated into large changes in circulation and rainfall patterns around the world. Accurate monitoring of the sea surface temperature with a precision of ~1 °C therefore has been the principal objective of the space program. Although the radiometers on the spacecraft have become quite sensitive, the main problem has been to correct for the absorption and emission produced in the intervening atmosphere, particularly by clouds and dust. Slowly the algorithm to retrieve the sea surface temperature from the satellite measured radiances is getting more realistic, and recently a review of this topic revealed that, although some problems remain, we may be close to achieving 1 °C accuracy in sea surface temperature measurements from satellites (Fig. 10-5). This advance in space measurements is quite significant because of what we have learned from the recent event of El Niño (1982–1983) which impacted the global climate in a significant and unpredictable manner for more than a year (Fig. 10-6). It will be sometime before we really understand the exact mechanism of the El Niño phenomenon and especially its impact on global climate. But satellite monitoring of such events can certainly begin to give us an advance notice of climatic fluctuations which may be imminent.

ATMOSPHERIC DUST AND VOLCANIC AEROSOLS

On an average, the earth's atmosphere is quite turbid. The "normal" amount of dust in the lower atmosphere is ~50 particles/cm^3 of diameter = 1 μm. However, the concentration increases inversely with size as d^{-4} which

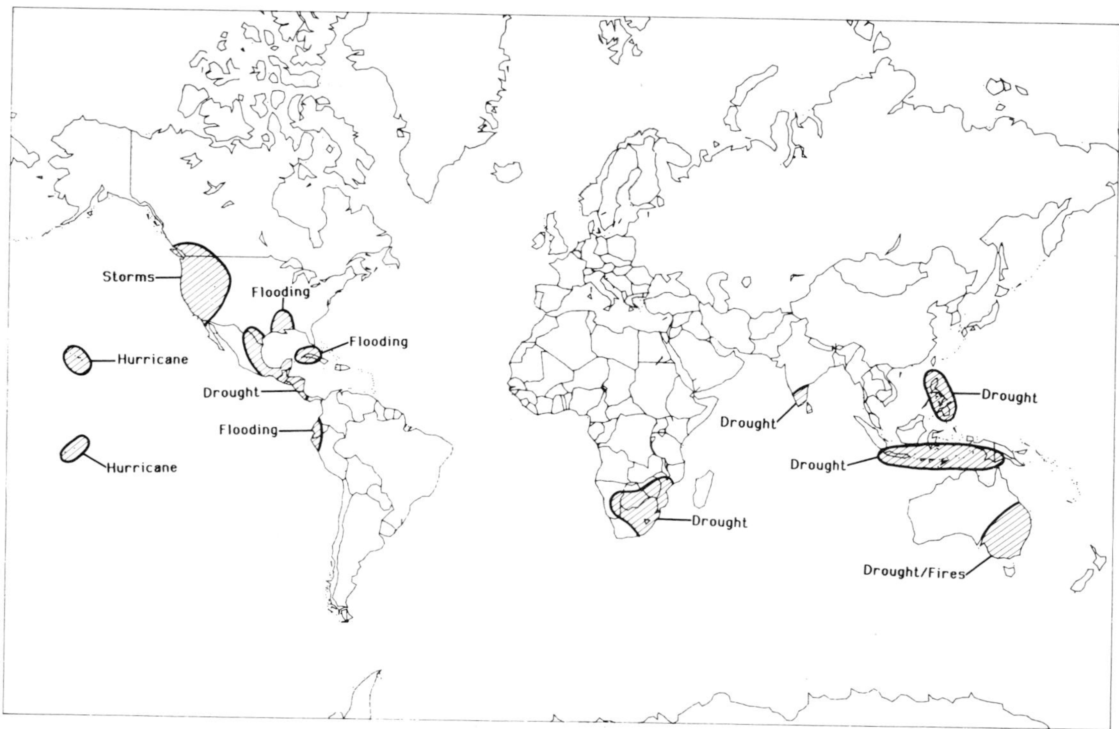

Fig. 10-6: In early 1983, the Equatorial Pacific warmed up by about 4° C. This impacted the energy balance of the earth and consequently the global rainfall pattern was modified.

means that there are close to 10^6 particles/cm^3 of size $d = 0.1$ μm. Depending on the optical properties of these particles and the magnitude of changes in their concentration, the impact on climate could be anywhere from negligible to very large. One model calculation suggests that the average effect on these aerosols is to cool the earth's surface globally by 1.3° C, implying that if all dust were to be removed from the atmosphere, the earth would warm up by more than 1° C [4]. This is a relatively large impact when one considers that the average temperature of the earth as a whole has not varied by more than 1° C in the last century.

The impact on climate is different for different kinds of aerosols depending whether they are continental, maritime, volcanic, or urban in origin. Also, if they are located in the troposphere their lifetime is not much longer than about 10 days and their impact is regional, while if they reach the stratosphere they can circle the earth for several years and the effect is global. Two examples are in order. In the troposphere the most abundant kind of aerosols are those of maritime origin and those which are raised by wind over land. Very often the dust raised over Sahara during large storms is transported all the way to Florida and to the coast of South America. A layer of Saharan dust lingers over the coastal Atlantic for weeks at a time and must change the radiation balance of the atmosphere considerably. Satellite monitoring of this dust is now possible and researchers are beginning to assess the climatic impact of large scale changes in tropospheric aerosols on regional climate. The impact of volcanoes on global climate is much more profound. This is because often the volcanoes exhale SO_2 which, as gas, is easily transported to the stratosphere. Once it reaches those levels it combines with H_2O to form H_2SO_4 which, because of low temperatures, immediately condenses into small shiny droplets. These droplets can circle the earth for months to years and because of the fact that their albedo is very high, they reflect a large fraction of sunlight back into the space and thereby have a cooling effect on the earth's surface. It is now believed that the earth's climate in the last two decades has been largely modulated by two big volcanic events: Mt. Agung in 1963 and El Chichon in 1982. Both explosions contained considerable amounts of sulphur and analysis of past data indicates that the Northern hemisphere temperature did go through a minimum in 1963–1964 (Fig. 10-7). The interesting question is whether the cold winter of 1983–1984 in the U.S. was the result of El Chichon. No judgment can be made as yet because the effect of El Chichon should be global rather than regional, and those analyses are still to be completed. In any case, satellites have proven to be ideal in measuring the spreading of volcanic aerosols around the globe (Fig. 10-8), and at the same time to assess its impact on the surface by measuring the surface temperatures over the oceans and land alike. Although the problem in the case of El Chichon was complicated by the occurrence of El Niño during the same time frame, this provides us with a challenge to deconvolve and sort out the cause and effect relationship of two major perturbers of the global climate system: volcanoes and ocean temperature anomalies. This is also an exceptional

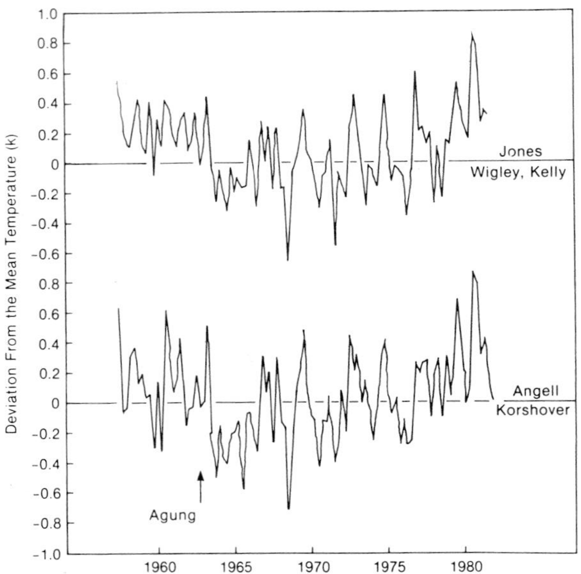

Fig. 10-7: Northern hemisphere annual mean surface temperature anomalies, departures from 1946–1960 mean [10].

opportunity because both these events took place when the global network of observations was at its densest, satellite coverage of the atmosphere and surface of the earth were close to optimum and computing capacity and speed were more than sufficient to process large quantities of data rapidly.

SNOW AND ICE

The seasonal variations in the extent of snow and ice cover, especially in the Northern Hemisphere, are probably the largest changes that occur on earth's surface over a year. Impact of these changes on surface radiation balance is large because the albedo of fresh snow is as high as 0.80 while that of vegetated land is only 0.20, involving a change of more than a factor of two in solar input when land becomes snow covered. These changes have implications on the atmospheric temperature, weather system, and climate of the following season. For example, it has been shown that the extent of snow cover over the Himalayas during the winter is correlated with the intensity of the Indian monsoon the following summer (Fig. 10-9). Interestingly, this idea was put forth 100 years ago by Blanford and later by Walker but could not be substantiated until satellite remote sensing data became available in the 1970's [5]. On longer time scales, extent of the polar ice (sea ice) will become very relevant especially when we discuss the climatic impact of increasing CO_2 in the atmosphere later in this paper.

EVAPOTRANSPIRATION AND SOIL MOISTURE

Because of the relatively high latent heat of water, the atmosphere receives more energy by condensation of clouds than it does directly from the sun. The major source of evaporation is, of course, the oceans, but the variability, both in space and in time, of the rate of evaporation from land could be a major source of perturbation to climate. Numerical experiments have been performed to demonstrate the importance of land evaporation on global distribution of rainfall (Fig. 10-10). One of the challenges to the current space technology is to be able to measure the evaporation rate or the soil moisture accurately from space. Despite a number of recent attempts using radars, passive microwave, and thermal infrared, there is currently no one technique which can be adopted for operational use by satellites. An indirect way of estimating the soil "wetness index" from data already available routinely from satellites is to plot the diurnal excursion of land surface after correcting the values for solar angle and other atmospheric effects such as clouds. Fig. 10-11 shows one such attempt using data from the NOAA satellite. It is difficult to visualize a ground validation technique at this scale of resolution. Only when data, such as these, are analyzed for several years and correlated with precipitation history, will one be able to translate this "wetness index" into a physical parameter.

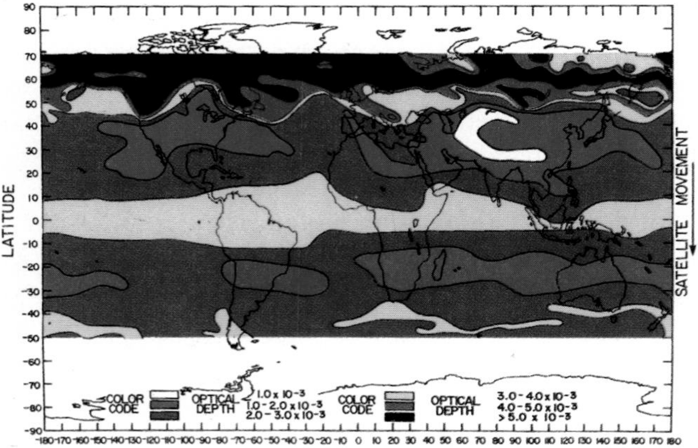

Fig. 10-8: Stratospheric dust concentrations as measured by a NASA satellite (SAGE) three months after the eruption of Mt. St. Helens in 1980 [11].

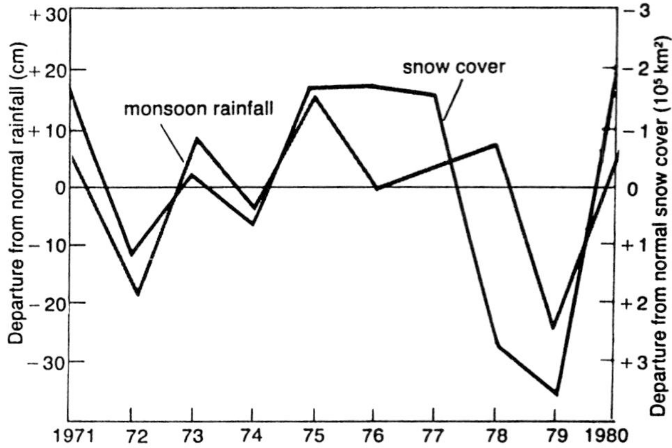

Fig. 10-9: Positive correlation between the extent of Himalayan snow cover and the intensity of the monsoon over India in the following summer [5].

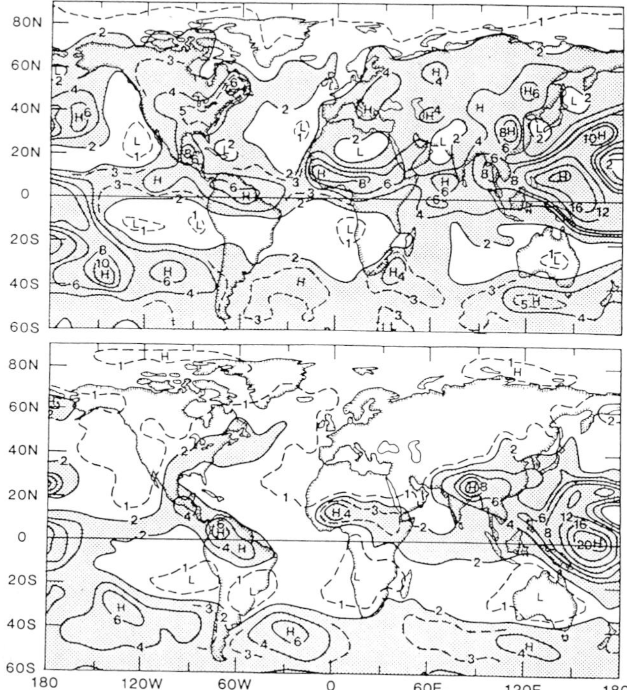

Fig. 10-10: A general circulation modeling experiment demonstrating the effect of wet soil (top) versus dry soil (bottom) on global distribution of rainfall [12].

MAN'S IMPACT ON CLIMATE

Human activity may be affecting the future climate in two different ways. First, the burning of fossil fuels is adding large quantities of CO_2 to the atmosphere, so much so, that in the last 35 years the amount of CO_2 in the atmosphere has increased by about 10 percent. Second, large scale deforestation in the tropics may have an implication on the climate of the future.

For the CO_2 problem, the increase in the atmosphere represents roughly half of what the fossil fuels have put in, the other half either being taken up by the oceans or by the biomass. With the current rate of increase, it is possible that atmosphere CO_2 will someday increase to twice its current value, provided of course the rate of removal from the atmosphere does not increase also. The time span in which this could happen is very uncertain and estimates range from 50 to 200 years. The climatic impact of the increase has been studied exhaustively in the last decade and the consensus seems to be that if one considers the effect of CO_2 alone, which is of course unrealistic because other climatic variables are changing as well, the earth should warm up by 1–2 °C during the next century. This predicted warming is, interestingly, of the same magnitude as the predicted coolings due to large volcanic events, although it is true that the volcanic activity is transitory while CO_2 buildup is a problem of longer term. While it is difficult to visualize a method to reduce the buildup of CO_2 in the atmosphere, it is quite important that in the next few years we are able to test climatic impact theories by global experimentations. It is presumed on the basis of theoretical calculations that the warming of the earth because of the increase of CO_2 will be accentuated at the polar latitudes. Also it is predicted that because of increasing CO_2 in the stratosphere, the temperature at 30–40 km altitudes should *decrease* by as much as 10 °C. Both of these predictions of the model can be tested by today's space technology. First attempts in this direction have already been made (Fig. 10-12). The direction of change observed by satellites in the extent of the Antarctic Sea ice is still inconclusive, if not opposite of what was predicted. As for the stratosphere, a *warming* has been observed mainly because of the El Chichon dust in 1983 [6].

In deforestation our knowledge of the magnitude of current change and of its potential impact on climate is very meager. Clearing of forests changes the albedo of the surface, and therefore the solar energy input to the lower atmosphere. It also changes the evaporation rate of moisture from the surface. Both imply a change in climate and in rainfall patterns over large areas. However, actual quantitative data on both the cause and effect are lacking. Again, satellites offer the opportunity to measure these changes quantitatively in time. In the case of tropical deforestation the problem has been that those regions are quasi-permanently cloud covered and therefore "invisible" from the satellites. However, with the advent of space-born radar technology it is possible that we will begin to monitor long-term changes in the forest cover of the earth. Simultaneous measurements of rainfall patterns and surface albedo will finally provide

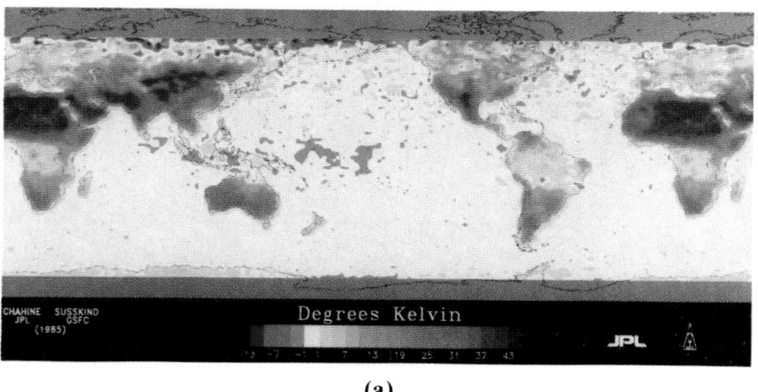

(a)

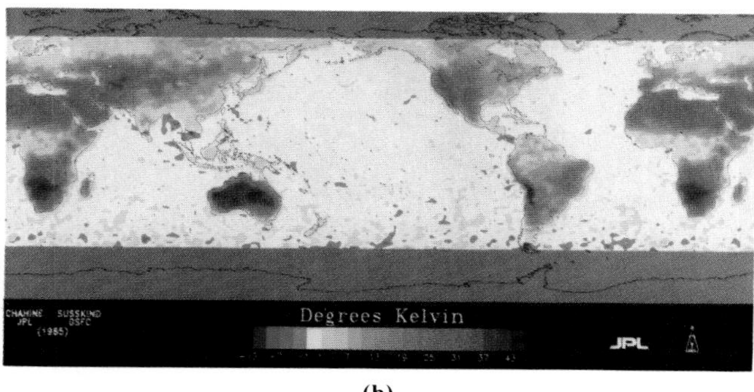

(b)

Fig. 10-11: A "wetness index" map for the world derived from diurnal temperature excursions as measured by NOAA and corrected for solar elevation and cloud cover. (a) January 1979. (b) July 1979. (From data processed by Chahine, JPL; and Susskind, ASFC.)

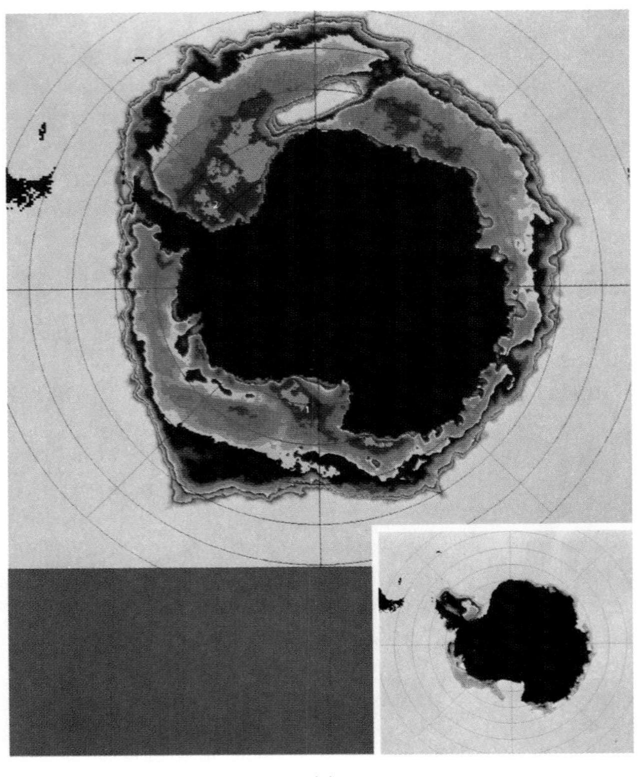

(a)

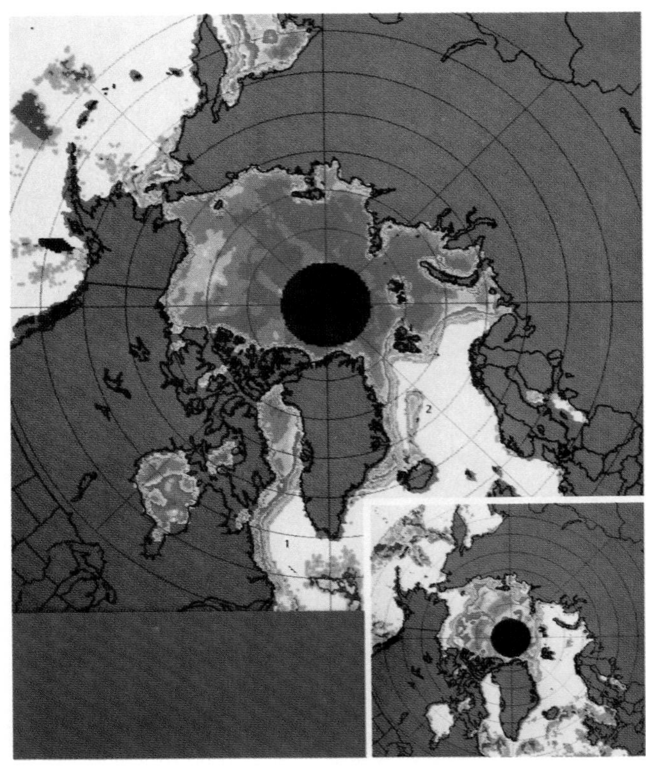

(b)

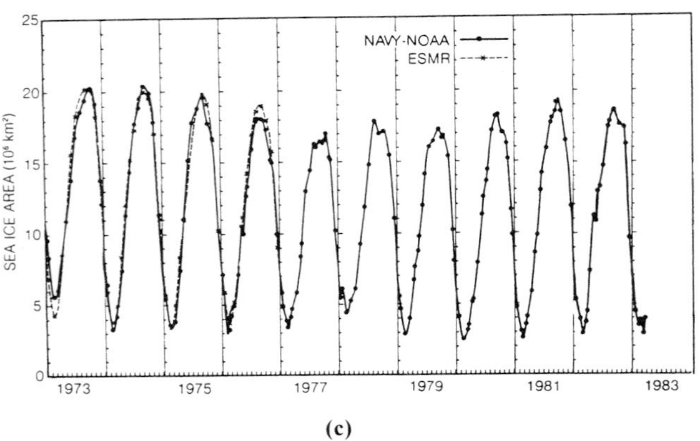

(c)

Fig. 10-12: (a) Mean monthly sea ice concentration around Antarctica during 1976. (b) Mean monthly sea ice concentration around the Arctic during 1976. (c) A ten-year time series of Antarctic sea ice extent [13].

some quantitative data to help resolve, at least partly, the long-standing controversy on what really is man's impact on climate on a global scale.

REFERENCES

[1] L. Bengtsson, *Bull. Amer. Meteorol. Soc.*, vol. 63, p. 277, 1982.
[2] ——, GARP Special Report, no. 43, WMO, Geneva.
[3] E. Lorenz, *Tellus*, vol. 34, p. 505, 1982.
[4] J. Hansen, NY Acad. Sci., 1980.
[5] J. Walsh, *Amer. Sci.*, 1983.
[6] K. Labitzke, *Geophys. Res. Let.*, 1984.
[7] S. I. Rasool, "Weather forecasting, past, present and future," Congressional Hearing, Comm. on Science and Technology, no. 96, 1978.
[8] A. Strong, *Ocean Air Int.*, 1985.
[9] Jones *et al.*, *Monthly Weather Rev.*, 1982.
[10] J. K. Angell and G. V. Gruza, "Climate variability as estimated from atmospheric observations," in *The Global Climate*, 1985, p. 25.
[11] P. McCormick, NASA SP 458, Washington, DC, 1982.
[12] J. Shukla and Y. Mintz, *Science*, vol. 215, p. 1498, 1982.
[13] H. J. Zwally, *Annals Glaciol.*, vol. 5, p. 191, 1983.

PART III:
MATERIALS PROCESSING IN SPACE

11
MATERIALS PROCESSING IN SPACE: AN OVERVIEW

Louis R. Testardi

The NASA Materials Processing in Space (MPS) Program has roots which go back to early observations of fluid materials behavior in low gravity which were made during the Apollo flights of the late sixties. A few years later, the MPS program became established formally in organization and budget. Its goals today are to improve our fundamental understanding of the role of gravity in the processing of materials, to provide facilities for low gravity experimentation with, and low gravity processing of, candidate materials, and to assist American industry in the early commercialization of space processing. A continuing element of this program is thorough ground-based research to yield new knowledge, to establish a need for flight experimentation, and to identify the conditions, controls, and data for maximum use of this presently limited resource. To this program element, which must continue, we have now joined the period of space experimentation with the successful inauguration of the space shuttle flights.

Gravity is responsible for convection, sedimentation, and buoyancy, all of which affect the processing of materials in the fluid state. It may, in special cases, also cause material deformation, and almost always leads to the requirement for a processing container and the concomitant risk of chemical reaction and contamination. Most of the research presently carried out in the MPS program relates to one of these phenomena. The scientific bases, the prior findings, and the present activities and plans are the subject of the chapters which follow.

In Chapter 12, "Materials Science and Engineering in Space," John R. Carruthers, the former Director of the NASA MPS Program, describes the ubiquitous role of gravity in a wide variety of fluid phase processing including vapor deposition, casting, solidification, and crystal growth. And with its presence woven in the complex interrelation of the processing and performance of materials, Dr. Carruthers describes how the science of processing, both terrestrial and low gravity, can benefit from the findings of this program.

In Chapter 13, Robert J. Naumann, Chief of NASA's Space Processing Division, Marshall Space Flight Center, describes the low gravity experiments conducted on Skylab, Apollo Soyuz, and the Space Processing Applications Rocket (SPAR) program. These were man's first opportunities for extended-time low-gravity experimentation and the results have been important both in improving our knowledge of the role of gravity in processing, and in the design of new or more detailed experiments.

New experimental opportunities for space processing aboard the space shuttle, and the current MPS research program (now named Microgravity Science and Applications) are described by William A. Oran, Manager of Science Projects at NASA Headquarters, in Chapter 14 "Current Program to Investigate Phenomena in a Microgravity Environment." Limited duration low *g* facilities which include drop tubes/towers, KC-135 aircraft in parabolic trajectories, and, again, SPAR capabilities, are described along with some recent data relevant to low gravity processing. Both the ground-based research and the shuttle flight programs are described.

Finally, Charles F. Yost, Manager of Commercial Applications at NASA Headquarters, describes the NASA program to assist industrialization in Chapter 15, "Commercialization of Materials Processing in Space." The assist is provided through technical outreach programs, and in three formal levels of cooperative involvement for private enterprise. While the coexistence of the basic research and the commercialization goals within the same program is sometimes questioned, no other arrangement could lead as effectively to the increased benefits and reduced risks and costs for both.

The commercialization of MPS is in its infancy. Its future is as much in the hands of entrepreneurs as it is in the still to be understood laws of nature. NASA's overall program, however, scientifically pursued with good talent, will yield a return in the form of improved understanding of the processing of materials, whether processed in space or on earth.

12

MATERIALS SCIENCE AND ENGINEERING IN SPACE

John R. Carruthers

INTRODUCTION

The processing of materials into useful forms usually involves the liquid or gas phases at some point. Processes such as chemical vapor deposition, casting, solidification, and crystal growth are basic to the separation and synthesis of a wide class of materials including electronic, metallurgical, glass/ceramic, and biological. Although major efforts have been devoted to understanding the solid state of these materials, relatively little attention has been paid to the fluid state and its influence on producing desired properties in the solid state. An example of these complex interactions in electronic materials includes control of dopant or compositional distributions through segregation and convection processes at the growth interface. It has been apparent that, as increasingly stringent requirements are being placed on materials by new or more complex applications, these processes must be controlled down to the atomic level and over larger macroscopic dimensions simultaneously.

The ubiquitous role of gravitational acceleration body forces on the fluid state is of major interest if these processes are to be understood and controlled. In actuality, the flows of both heat and mass are monitored and changed to control phase transformations. However, such flows produce changes in the density of the fluid media which, in turn, will move or convect in a gravitational environment. The consequences of these uncontrolled motions are many and varied. However, the basic result is to make the required spatial and temporal control of the phase transformations very difficult. Further effects of gravity are associated with the need for containers which often result in uncontrolled contamination as, for example, occurs when oxygen is introduced into silicon from the dissolution of the quartz crucibles into silicon melts.

The capability to study these gravitational effects on complex materials processes through routine access to the low-gravity environment of space has been of interest to the materials science community for many years. However, the scientific basis for understanding these effects has been very sparse. Consequently, much work has been undertaken to provide this basis for fluid behavior and shape change effects responsible for compositional, morphological, and structural imperfections in materials that may influence properties and affect performance. Some of the general progress in this work is described next. More detailed accounts are available elsewhere [1],[2].

LOW-GRAVITY EXPERIMENTS

There are four primary areas in which gravity influences fluid and liquid shape behavior in materials processing:

Fluid Behavior

1) density-gradient driven convection such as thermal and/or solutal convection.

2) sedimentation and buoyancy effects in multiphase fluids.

Liquid Shape Behavior

1) hydrostatic pressure effects that cause deformation of liquid and some solid shapes.

2) containerless technologies in which molten, reactive materials can be processed or measured with container contamination.

In addition to the influence of these factors on material processing, they also are important in a wide variety of other scientific endeavors, such as geophysical fluid dynamics, combustion science, cloud physics, and critical point phenomena. In combination with these and other basic science areas, the materials science program is leading to the establishment of a national microgravity facility to study the behavior of fluids and liquid shapes as well as to prepare limited quantities of precursor materials for study on earth. The experiments in materials processing are paradigms on which much of the future activities will be based.

The approach required to make effective use of low-gravity experiments is similar to that of many other facilities in which the traditional limits on materials processing phenomena are extended. Extensive use must be made of modeling and simulation experiments as well as physical scaling principles. These experiments, in turn, provide important insights into the boundary conditions and measurement techniques necessary in actual materials processing experiments in both earth-gravity and low-gravity environments. Access to short-duration low-gravity experiments is also an important precursor to more extensive space experiments of longer duration. The use of drop towers and tubes as well as ballistic trajectory aircraft and sounding rockets have provided invaluable early experience.

At this rather early stage in the use of microgravity to study materials processing, it should be remembered that the U.S. program has accumulated fewer than 200 h of total space experiment time since the early 1970's. Furthermore, many of the early experiments were poorly conceived, inadequately implemented, and improperly analyzed [3]. Therefore, much recent emphasis has been focused on acquiring the necessary fundamental understanding. Previous space experiments in materials processing have been reviewed elsewhere [4].

Fluid Behavior

Processing areas in which fluid behavior have been studied include:

1) crystal growth from melts, fluxes, solutions, and the vapor phase.
2) dendritic solidification and casting.
3) unidirectional solidification of multiphase alloys.
4) electrophoretic separation of charged particles and biological molecules.

In crystal growth, thermal and solutal convection exists due to temperature and composition gradients that are not aligned in the direction of gravity. There are many schemes that have been used or proposed to control such convection on earth [5], but clearly the most straightforward approach is to reduce the *g* level. Most other schemes introduce other perturbations into the system or are not applicable to all materials. For example, static magnetic fields, which damp thermal convection in semiconductor or metal melts, are ineffective for non-conducting melts or for inhibiting solutal convection. Furthermore, by damping thermal convection, temperature profile nonuniformities occur which interfere with the growth-rate behavior in many systems. Interactions between thermal and solutal density gradients have also been shown to cause unstable fluid flow which in fact is coupled to morphological (shape) instabilities of the growth interface [5]. Only by recourse to low *g* levels can such phenomena be studied, understood, and possibly eliminated.

An example of sedimentation effects in crystal growth occurs for the vertical Bridgman growth of the solid solution system, $Hg_xCd_{1-x}Te$, where the diffusion boundary layer becomes enriched in mercury, which in turn flows along the interface and settles in the position of lowest gravitational potential. This position is usually located at the container wall because of radial heat flow. However, this locally enriched volume also possesses a lower liquidus temperature so that the interface forward motion with the moving isotherms is retarded. The end result of this cumulative redistribution of mercury in the melt is a phenomenon known as "interface runaway" where the actual growth interface becomes highly nonisothermal and nonplanar. Furthermore, constitutional supercooling is unavoidable and dendritic solidification results. Equivalent behavior is observed for flux and solution growth when the liquid phase is confined.

In crystal growth from the vapor, the influence of vapor phase thermal convection on transport rates and crystal morphology has been studied in closed tubes by orienting them at different angles with respect to gravity [7]. However, this range of fluid transport still cannot account for the observed rates in space experiments where GeI_4 crystals were grown. Furthermore, the ampoule orientation exerts a strong influence on the morphology of the growing crystal with more stable flows resulting in larger, more faceted crystals.

In dendritic solidification and casting, convection and sedimentation play important roles in determining the structure of metal alloys. Quantitative measurements of thermal and solutal convection effects on dendritic growth velocity have been measured in transparent systems which model metals [8]. Much has been learned by performing experiments by varying the dendrite growth direction with respect to gravity. For low thermal undercoolings, dendritic growth velocities are strongly increased by thermal convection for downward growth. It is now possible to test various kinetic models of dendritic solidification by understanding and reducing such convection effects. Additional features of dendritic growth behavior have been seen in casting structures frozen under low gravity; reduction of dendrite fragmentation and sedimentation produces very large and highly aligned columnar dendrites in castings [9].

Planar front solidification, used to prepare aligned structures of multiphase metal alloys, is also an area where convection and sedimentation are important. The important feature of these systems is the requirement for diffusion coupling between the various phases at the growth interface. In order to provide this coupling, planar interfaces must be produced by high temperature gradients in the melt which, in turn, cause excessive thermal convection and a resultant loss of the desired diffusion coupling. In other types of systems, liquid phase separation occurs as part of the solidification process, and the action of sedimentation and buoyancy results in massive separation of the melt into separate phases, resulting in destruction of the alloy structure.

Another important process in which gravitational limitations exist is electrophoresis, the transport of electrically charged particles or molecules in a dc electrical field. The process is used to separate the charged constituents. However, the joule heating in the process causes hydrodynamic instabilities and disruption of the flow, which interferes with the separation [10]. Furthermore, at high throughputs, sedimentation of the separated constituents occurs. Various schemes to reduce these adverse effects have been proposed over the years. However, the routine availability of low gravity has resulted in studies of these complex phenomena and even to the development of a commercial process to separate and purify biological materials in space.

In all these areas, new phenomena and difficulties became important in reduced gravity. For example, thermocapillary convection at free liquid surfaces becomes significant [1]. Also, additional segregation phenomena have been revealed at curved crystal growth interfaces [1]. As more experience is acquired in low-gravity experiments, this list of new phenomena will undoubtedly expand. Furthermore, there are other scientific disciplines where gravitational acceleration effects are important such as cloud physics and combustion sciences. These and other areas will be the subject of extensive study in low-gravity facilities.

Liquid Shape Behavior

In materials processing and property measurement, the ability to manipulate molten reactive materials without the need for containers is an important advantage of a reduced gravity environment. This capability eliminates container contamination and reaction effects, as well as enables the preparation and study of spherical liquid drops, shells, and other multiphase configurations.

Although liquids do not require containers at low gravity, it is necessary to constrain their position and shape. Consequently, containerless positioning technologies have been under intensive development and the responses of liquid flows and shapes to the positioning fields have been studied [11]. Four different types of positioning fields can be used singly or in combination: acoustic, aerodynamic, electromagnetic, and electrostatic.

Acoustic fields of high intensity and frequencies from 1 to 100 kHz can be used in either resonant or interference modes to levitate or position any solid or liquid objects in a gas. Various configurations have been developed to conform to different applications. In resonant chambers, the object is located at the minimum pressure mode and can be rotated, oscillated, or moved by changing the resonance conditions.

Aerodynamic levitation and positioning schemes have been used with both subsonic and supersonic flows with gas nozzles which are designed to provide stable positioning from the spreading action of the jet around the object. However, this method has severe limitations of cooling, vaporization, and instabilities which have restricted its use for materials processing.

Electromagnetic containerless limitation has been used on earth for many years to melt conducting metal samples. However, the necessity to provide levitation at G_e results in large radio-frequency power requirements which cause uncontrolled specimen heating. A wider variety of conducting materials may be melted at low gravity.

Electrostatic levitation and manipulation is a relatively new method developed for processing glass shells at G_e inertial confinement fusion targets. The method may be used for nonconductors and also works in a vacuum or gaseous environment. In the method, an electrically charged object is subjected to alternating electric fields between (at least) two suitably positioned and shaped electrodes. Static stability cannot be achieved by this method and the alternating field is used to produce dynamic positioning through position monitoring and feedback control.

Important applications of containerless positioning technology in low gravity have been developed over recent years. Glass and amorphous metal systems where purity and/or nucleation avoidance are desired are excellent candidates for study and preparation by containerless methods [12]. Liquid drops, shells, and compound drops have been studied extensively by both containerless positioning and neutrally buoyant systems. The properties of molten, reactive materials may be studied at high temperatures by containerless methods: surface tension, viscosity, density, thermal expansion, heat capacity, and thermal conductivity.

Other processes are also sensitive to hydrostatic pressure effects. Liquid floating zones, used widely in crystal growth at G_e, are limited in length by the ability of surface tension to counteract the hydrostatic pressure. At reduced gravity, no such constraint exists and cylindrical zones are stable up to lengths equal to the circumference, regardless of the surface tension [13]. Under these conditions, many of the limitations of the floating-zone crystal growth at G_e, including non-uniform compositional segregation and point defect clusters in silicon crystals, may be reduced or removed at reduced gravity. Another influence of hydrostatic pressure is the self-deformation of solids at high temperature. One such material, mercury iodide, possesses an extremely low critical resolved shear stress for plastic deformation at the growth temperature. The resultant generation of dislocations degrade the performance of this material in applications as nuclear, X-ray, and gamma ray detectors.

An additional feature of low gravity is the ability to prepare or deploy drops and drop arrays with more control over spatial position for long periods of time. One example of such a capability is the preparation of monodisperse latex spheres from polymer solutions. Such spheres are prepared at small sizes (less than 5 μm) at G_e and are used as size standards because of the very narrow particle size distribution. At larger sizes, sedimentation causes collision and coalescence. Preliminary quantities of large-size monodisperse latex spheres have been prepared in space and appear to possess very high uniformity [14].

FUTURE DIRECTIONS

The environment of space provides opportunities other than reduced gravity for studying materials processes and preparing materials. The unlimited availability of solar energy and the possibility of a high pumping speed vacuum may also enable new and innovative approaches to processing materials. Direct access to solar energy requires high inclination and/or high altitude earth orbits to provide direct line-of-sight for extended periods of time. In addition, stringent pointing requirements, combined with the need to maintain low gravity and provide protective atmospheres around the material at high temperature make this opportunity somewhat complex, and space systems designs and operation to satisfy requirements for materials science and engineering would be both complex and expensive. The achievement of a high pumping speed vacuum behind an orbiting molecular wake shield is a concept that has also been considered. In this concept, reduced gas molecular densities are achieved because the orbiting shield travels faster than the mean gas molecular speed and thus sweeps out a wake region. In near-earth orbit, the pressure level is comparable to an equilibrium level of 10^{-12}–10^{-13} torr. However, the pumping speed is extremely high and the molecular distribution is highly non-equilibrium, thus providing opportunities for molecular beam experiments. As opposed to earth vacuum systems, this type of vacuum becomes less expensive as the vacuum volume increases. Furthermore, the density drops to far lower equivalent pressure levels if the shield travels in deep space as opposed to earth-orbit space. Nevertheless, the complexity and expense of studying such a capability must await sufficient scientific interest and justification.

The ability to process materials in a space environment will also possess interest for the future construction and operation of space systems. For example, simple metal fabrication and joining operations may be understood and adapted to space. On a longer term basis, raw materials for use as propellants may be obtained and processed to enable sample return missions in deep space exploration. Interest in the raw material available in near-earth crossing asteroids is also increasing. The current planetary exploration program of NASA includes missions to perform astronomical charting and geological assessments of these space materials resources.

In the near term, interest is focused on the influence of gravitational acceleration on materials processes and fluid phenomena of scientific and technological interest. Thus, the components of a national facility to perform experiments under a range of reduced gravity times and levels are being developed, tested, and demonstrated for these applications. As with other national facilities, increasing use will occur as access to the facility becomes more routine and an experience base develops. It appears that with routine operation of the NASA Space Transportation System now demonstrated and the increasing level of scientific sophistication regarding gravity as a variable now underway, this evolution of interest is occurring.

REFERENCES

[1] J. R. Carruthers and L. R. Testardi, *Ann. Rev. Mater. Sci.*, vol. 13, p. 247, 1983.

[2] G. E. Rindone, Ed., *Materials Processing in the Reduced Gravity Environment of Space*, Materials Research Society, vol. 9. New York: Elsevier.

[3] W. P. Slichter *et al.*, "Materials processing in space" (Report of the Committee on the Scientific and Technological Aspects of Materials Processing in Space, Space Application Board), Nat. Acad. Sci., Washington, DC, 1978.

[4] R. J. Naumann and H. W. Herring, "Materials processing in space: Early experiments," NASA SP-443, 1980.

[5] J. R. Carruthers, in *Preparation and Properties of Solid State Materials*, vol. 34, W. R. Wilcox and R. A. Lefever, Eds. New York: Marcel Dekker, 1975.

[6] S. R. Coriell, M. R. Cordes, W. J. Boettinger, and R. F. Sekerka, *Adv. Space Res.*, vol. 1, p. 5, 1981; also *J. Crys. Growth*, vol. 49, p. 13, 1980.

[7] H. Wiedemeier, in *Materials Processing in the Reduced Gravity Environment of Space*, G. E. Rindone, Ed., Materials Research Society, vol. 9. New York: Elsevier, p. 433.

[8] M. E. Glicksman, N. B. Singh, and M. Chopra, in *Materials Processing in the Reduced Gravity Environment of Space*, G. E. Rindone, Ed., vol. 9. New York: Elsevier, p. 461.

[9] M. H. Johnston and R. A. Parr, *Metall. Trans.*, vol. 13B, p. 85, 1982.

[10] D. A. Saville, in *Physicochemical Hydrodynamics*, D. B. Spalding, Ed., vol. 2, p. 893, 1978.

[11] M. Barmatz, in *Materials Processing in the Reduced Gravity of Space*, G. E. Rindone, Ed., vol. 9. New York: Elsevier, p. 25.

[12] D. R. Uhlmann, in *Materials Processing in the Reduced Gravity of Space*, G. E. Rindone, Ed., vol. 9. New York: Elsevier, p. 269.

[13] J. R. Carruthers, E. G. Gibson, M. G. Klett, and B. R. Facemire, *Progr. Astronaut. Aeronaut.*, vol. 52, p. 207, 1977.

[14] R. J. Naumann, Marshall Space Flight Center, NASA, private communication.

13

MATERIALS PROCESSING IN SPACE: REVIEW OF THE EARLY EXPERIMENTS

Robert J. Naumann

INTRODUCTION

In the conceptual years of spaceflight, it was recognized that the orbital environment was in many ways different from that of earth and that the design and construction of spacecraft to operate in this environment would be a formidable challenge. The primary environmental factors that had to be considered were the vacuum of space, energetic radiation from the sun and other sources, and the near absence of apparent gravitational effects resulting from the free-fall condition of orbital flight. Of these factors, the long exposure to a virtually zero-gravity environment is a truly unique situation that cannot be duplicated or even approximated for any length of time on earth.

Microgravity Environment

A zero-gravity environment is an ideal situation that, in practice, can never be completely realized in an orbiting spacecraft. There are a number of kinetic effects associated with an actual spacecraft that produce artificial gravity-like forces. Any unconstrained object in a spacecraft is actually in its own orbit around the earth. Only if this object is located at the center of mass of the spacecraft will it have exactly the same orbit. The accelerations associated with these effects are on the order of $GM_e r/a^3$ where G is the gravitational constant, M_e is the mass of the earth, a is the semimajor axis of the spacecraft orbit, and r is the distance of the object in question from the center of mass [1]. These accelerations have a magnitude of approximately 10^{-7} g_0 (earth's gravity) per meter displacement from the spacecraft center of mass, and vary periodically at twice the orbital frequency.

In addition to these inertial accelerations, the spacecraft will experience accelerations that arise from atmosphere drag, crew motion, and attitude control thruster firings. The drag produces more or less continuous low-level accelerations typically in the range of 10^{-7}–10^{-5} g_0, depending on the orbit, the day–night cycle, the solar activity, and the configuration and the attitude of the spacecraft. Crew motion and thruster firings produce more or less random impulses that are on the order of 10^{-7}–10^{-5} g_0 s.

These low-level accelerations must be considered in the design and analysis of experiments, but since they are typically six orders of magnitude lower than earth's gravity, dramatic changes in fluid behavior can be expected in an orbiting spacecraft. Also the term "microgravity" is a more accurate description of the acceleration environment than the more commonly used term "zero gravity."

Motivation for Microgravity Experiments

Interest in space processing began to evolve in the late 1950's from several different disciplines. The behavior of fluids in spacecraft was the object of a number of research efforts in order to design propellant management systems and other fluid systems required by the then emerging space technology. The development of spacecraft thermal control systems that utilize change of phase of materials for heat storage prompted questions concerning solidification phenomena in zero gravity. The possibility of erection and repair of large structures in space by brazing and welding raised issues concerning the flow of liquid metals when dominated by capillary forces.

In considering the likely behavior of liquids and solidification processes in the low-gravity environment of an orbital spacecraft, it was recognized that this environment might be useful for a variety of unique processes. Obviously, buoyancy-driven convective flows would be eliminated or drastically reduced. This would allow many processes such as crystal growth, alloy solidification, combustion, and other chemical reactions to be studied under diffusion-controlled conditions which are quiescent, easier to control, and vastly simpler to analyze. Also, many secondary phenomena such as thermocapillary flows, electroosmotic flows, Stefan flows, Soret diffusion, etc., could be studied without the complicating and sometimes overwhelming effects of buoyancy-driven convection.

Sedimentation or Stokes settling is also virtually eliminated in the microgravity environment. This allows fluids with multiphase components to be held in suspension more or less indefinitely. The most obvious application of this aspect of the microgravity environment to materials processing would be the solidification of systems with a liquid-phase miscibility gap or with a dispersed second phase. This may also be useful for investigating processes such as nucleation and growth phenomena, studies of systems near their critical points, performing experiments to test various ripening theories, and investigating the stability of foams, flocs, and other dispersions. Of course, the lack of sedimentation also poses additional problems that must be dealt with, such as the removal of bubbles from a system.

The absence of hydrostatic pressure permits the shapes of liquids to be dominated by surface tension. This should allow liquid bridges or molten floating zones to be extended to the Rayleigh limit (length of the zone equal to the circumference). One obvious application is the extension of the float zone crystal growth process to materials that have too low a surface tension to support a reasonable size zone in earth's gravity. The extended zone length would also allow more area for heat input which could result in better control of the temperature field in the vicinity of the growth interface. It should be possible to obtain flatter isotherms, which should result in less radial segregation, and steeper axial gradients, which would provide more stability against interfacial breakdown and allow higher growth rates and/or dopant concentration.

In the virtual absence of gravity, it is also possible to suspend droplets or high temperature melts without physical contact. Some small controlling force such as electrostatic, electromagnetic, or acoustic radiation pressure is required to overcome the residual spacecraft accelerations and maintain the position of the droplet or melt. This offers unique opportunities to process materials at high temperatures without a crucible. Since many materials become highly corrosive above their melting points, this technique may be useful for measuring thermophysical properties or for producing ultrapure materials. Also, the elimination of heterogeneous nucleation sites should allow melts to be deeply undercooled before solidification. Such undercooling may have dramatic effects on the final microstructure of the solid. Metastable phases, inaccessible to equilibrium solidification, may be obtained in bulk samples. Amorphous phases may be obtained in systems that do not ordinarily form glasses. It may also be possible to test various theories of homogeneous nucleation by using containerless techniques to eliminate heterogeneous nucleation sites.

Pre-Skylab Experiments

The drop towers at the Marshall Space Flight Center (MSFC) and at the Lewis Research Center were used extensively to predict the behavior of propellants in tanks during low-gravity coasting periods. Equilibrium configurations and oscillations of liquids in partially filled tanks were studied. Also, a number of experimental concepts such as various levitation and position control techniques for later space processing experiments were developed and tested in these facilities. A small sample of uniformly dispersed monotectic alloy (Ga-50At%Bi) was prepared in a drop tower experiment at the Marshall Space Flight Center [2]. Fig. 13-1 illustrates the resulting microstructure, and Fig. 13-2 shows the resistive properties of the composite material.

The first opportunities to perform experiments in a long-duration, low-gravity environment came during the last several Apollo flights. Several simple "suitcase" experiments were added to the flight manifest to be performed by the crew during the trans-earth coast period.

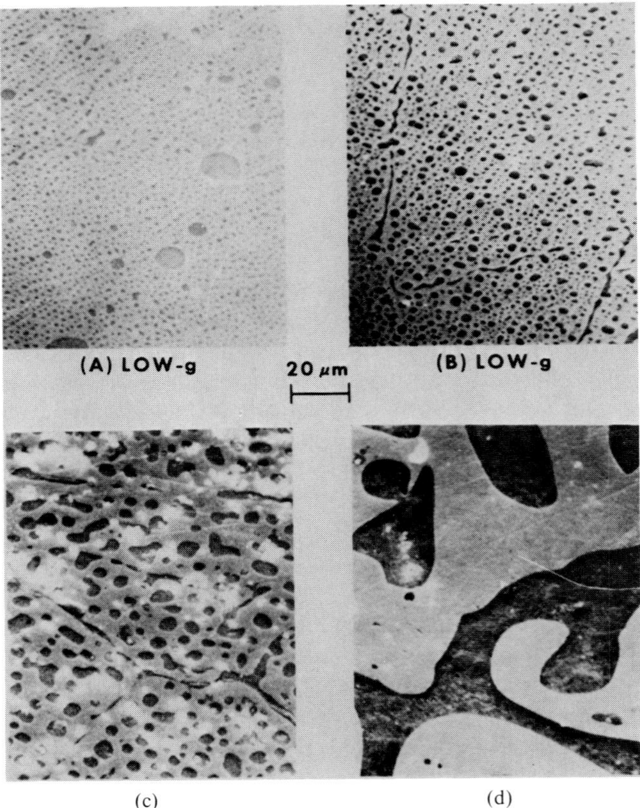

Fig. 13-1: Ga–Bi solidified at various quench rates during free fall compound with sample solidified in unit gravity. (a)–(c) low *g*. (d) 1 *g*.

One series of experiments was carried out in a small low-temperature casting furnace. Samples consisted of composites of fibers, particles, and gases in a matrix of In-Bi, and a model monotectic system consisting of sodium acetate and paraffin. On the ground, these samples exhibited almost complete density segregation. In the space processed samples, the second phase material was more or less dispersed throughout the sample, but the dispersions were by no means uniform. Considerable clumping and aggregation were observed, which indicated that simply removing gravity did not eliminate all of the problems in producing uniformly dispersed composites [3].

Another group of experiments was conducted on Apollo 14 and 17 to investigate heat flow and convection in the low-gravity environment. The apparatus is shown in Fig. 13-3. A thin layer of Krytox 143 A2 (a perfluoroalkyl polyether) was heated from below. Vivid Bernard cells were observed, driven by the instability of resulting from the warmer fluid with lower surface tension replacing the cooler fluid at the free surface. The onset of this instability was compared with the ground control experiment in which buoyancy also contributes to the instability [4].

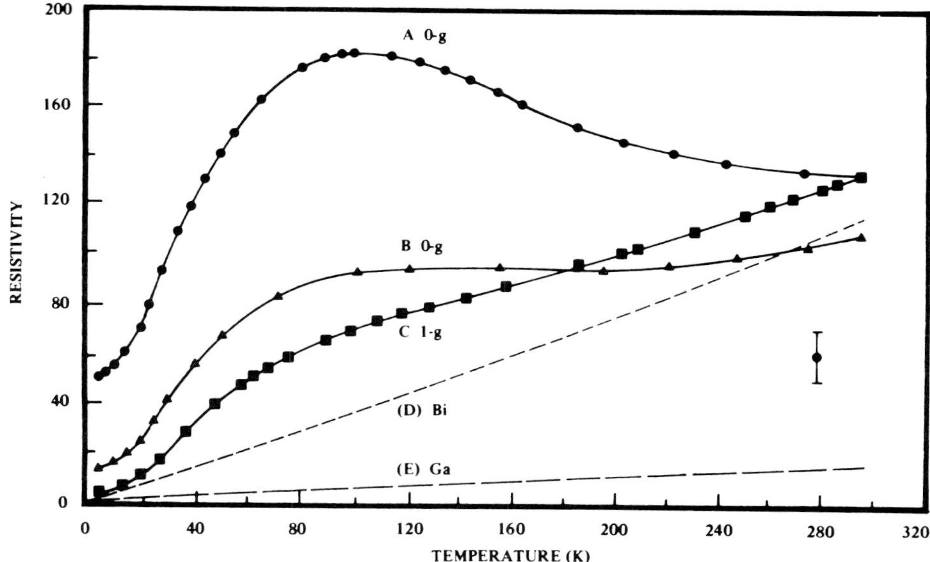

Fig. 13-2: Electrical resistivity of the immiscible alloys Ga–Bi processed in the drop tower. Curves *A* and *B* correspond to zero-*g* processed samples with different microstructures obtained by varying the cooling rate. *C* corresponds to the 1-*g* control sample, and *D* and *E* are the resistivities of the pure elements.

In the conduction part of the experiment, heat transfer was measured in gases and liquids. On Apollo 14, the temperature distribution in the test cells indicated 10–30 percent more heat transport than could be accounted for by pure conduction. The experiment was repeated on Apollo 17 during a period of reduced crew activity. In this case, the temperature distribution indicated that the heat transfer was totally conductive.

The third type of demonstration experiment conducted on Apollo was an attempt to perform an electrophoretic separation in a liquid column. Convective flow resulting from joule heating severely distorts such separations on earth unless a gel or other supporting technique is used to suppress the liquid motion. The use of a supporting medium restricts types of separations that can be performed and the amount of material that can be processed. For example, particulate material such as cells or cell components generally cannot pass through the gel.

The Apollo experiments consisted of electrophoretic columns containing a borate buffer and a variety of samples including a mixture of red and blue dyes, human hemoglobin, deoxyribonucleic acid (DNA) extracted from salmon sperm, and different sizes of submicron latex particles [5]. The apparatus is shown in Fig. 13-4. The objective of the experiment was to observe the shape and distortions of the sample band as it moved under the influence of the applied field. It was possible to determine that the samples were transported along the tube, but the shapes of the sample bands were severely distorted by the electroosmotic flow along the column walls. This flow results from the fact that the walls of the tube adsorb ions which attract a layer of oppositely charged ions in the buffer. In an electric field this layer of charge in the fluid moves, which causes the buffer to flow along the walls. This flow, combined with the return flow in the center of the tube, produces a bullet-shaped distortion. It was later found that this flow could be eliminated by a suitable coating on the walls that effectively has a zero surface charge. Again, this serves to illustrate the care that must be taken to eliminate

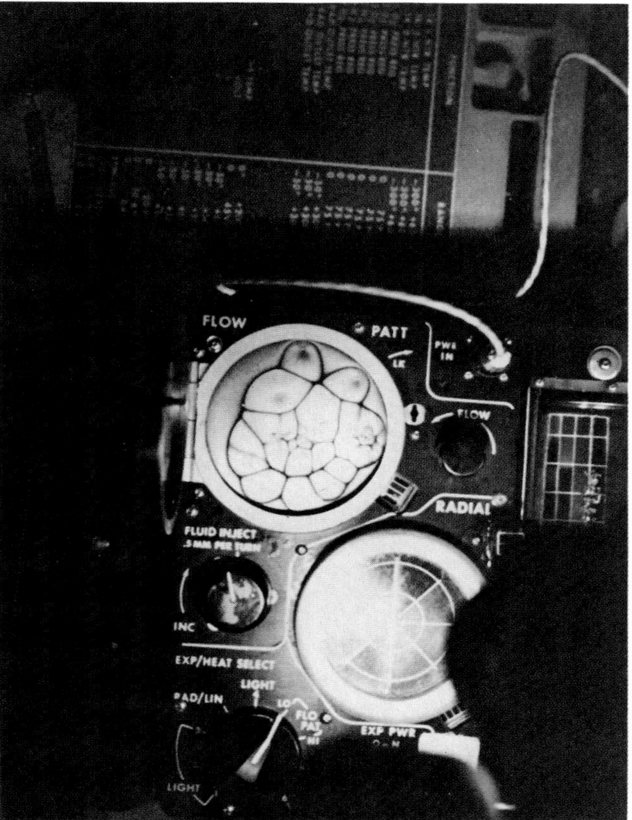

Fig. 13-3: Benard cells observed in the heat flow and convection experiment flown on Apollo 17.

Fig. 13-4: Static column electrophoresis apparatus flown on Apollo 16.

extraneous flows in designing low-gravity experiments to take advantage of the lack of buoyancy-driven convection.

SKYLAB EXPERIMENTS

Skylab offered the first opportunity to perform space processing experiments in a dedicated facility, although considerable constraints on apparatus and experiment development were imposed by the available development time and on-board power. Also, a series of demonstration experiments was conducted during the last portions of the Skylab mission. These "ersatz" experiments were conceived during the mission and were performed using available on-board equipment. In some cases, special equipment was sent up in the Apollo capsules transporting the crew.

Welding and Brazing Experiments

The first materials processing experiments approved for Skylab were the welding and brazing demonstrations. These were originally motivated by the desire to fabricate and repair large structures in space.

The brazing experiment developed by J. R. Williams [6] (MSFC) is noteworthy in that it used an exothermic reaction to produce the heat necessary to melt the braze alloy. This is an excellent means for obtaining high temperatures quickly without a large source of electrical power. Since the quality of a brazed joint depends on capillary action to cause the liquid braze alloy to completely fill the braze joint, careful control of the gap to be filled must be exercised in earth's gravity. In space, the gap width should be less critical since there is no

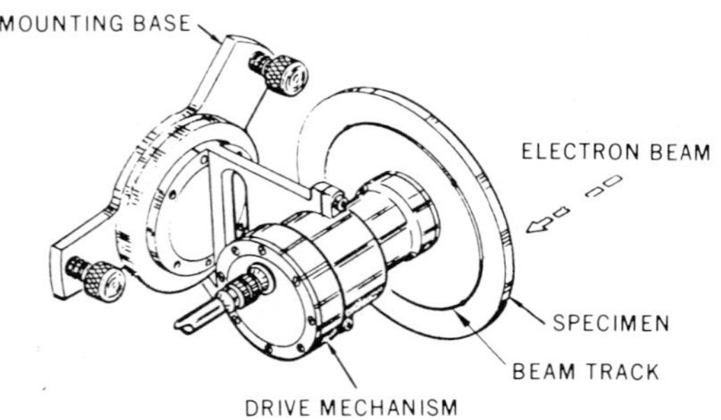

Fig. 13-5: Schematic of Skylab welding experiment.

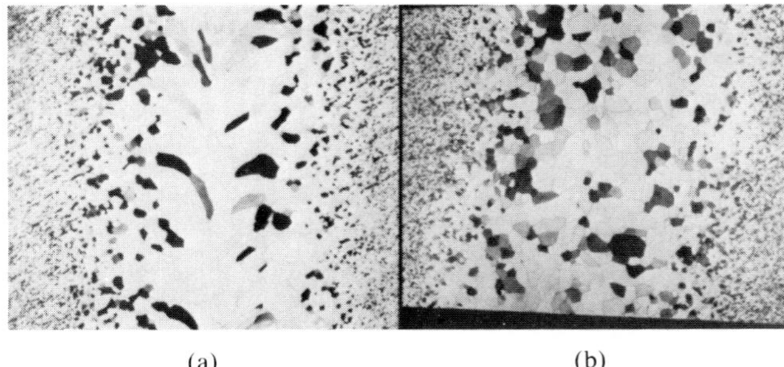

Fig. 13-6: Comparison of the microstructure of welds performed in space with those on earth. Note (a) the columnar micro structure in the sample processed on earth and (b) the equiaxed structure in the flight sample.

gravitational flow to oppose the capillary flows. Indeed, it was demonstrated that successful braze joints were formed on nickel and stainless steel tubing with gaps between the sleeves as large as 0.5 mm, and it appears that even larger gaps could be tolerated. This is substantially larger than the tolerance on braze joints in earth's gravity, indicating that brazing may be a most useful technique for joining tubing or other structures in space.

The welding experiment, developed by E. C. McKannan and R. M. Poorman [7] (MSFC), was performed with an apparatus shown in Fig. 13-5. The test disks were fabricated from 304 stainless, 2219-T87 aluminum, and pure (99.5 percent) tantalum. A sharply focused electron beam impinged on the disk, which was slowly rotated. The thickness of the disk was graduated to vary the penetration of the beam. At the end, the beam was defocused and the disk halted to form a large melt pool which was allowed to solidify. Little difference was observed in the external appearance of the welds made in space as compared to those on the earth. The microstructure, however, revealed some striking differences. Large columnar grains oriented perpendicular to the interface characterized the ground control welds, whereas the space welds had an equiaxed structure (Fig. 13-6). The reason for this difference is not clear. One speculation is that partially melted grains form the nucleation sites. These would fall to the bottom of the weld seam in earth's gravity where they nucleate columnar structures, whereas they would remain suspended in space and nucleate the equiaxed structure.

Crystal Growth Experiment

Some potential advantages of growing crystals in a microgravity environment were demonstrated by A. F. Witt *et al.* [8] (M.I.T.). Samples of InSb doped with Te (10^{18} atoms/cm^3) were grown on earth by the conventional Czochralski technique. These were melted back over a portion of their length in space using a gradient furnace and allowed to regrow by lowering the furnace temperature. After a brief growth transient required to achieve steady-state growth conditions, very uniform

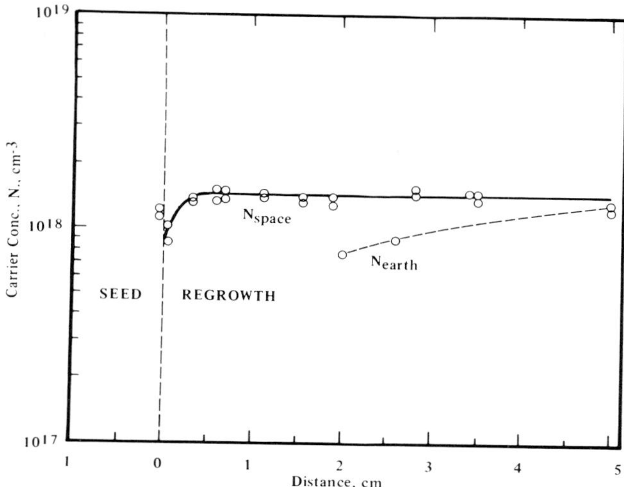

Fig. 13-7: Dopant profiles of Te in InSb obtained by Witt and Gatos. After the initial growth transient, the space processed samples have uniform dopant profiles characteristic of diffusion-controlled transport, whereas the earth processed samples have profiles characteristic of convective mixing.

dopant concentration was obtained along the axial direction. This indicates that diffusion-controlled growth conditions were indeed achieved, as may be seen in Fig. 13-7. Also, the space-growth sample was completely free of the dopant striations usually seen in earth-grown samples (Fig. 13-8). These striations originate from growth rate fluctuations which are generally believed to be associated with convective flows in the melt.

A novel crystal growth experiment, developed by H. U. Walter [9] (University of Alabama, Huntsville), involved a quasi-containerless growth of InSb. A portion of an earth-grown single crystal was melted back as was done by Witt and Gatos, except the crucible was larger than the rod diameter so that the melt could form a spherical shape as illustrated in Fig. 13-9. As the melt was solidified in the gradient freeze mode, an unanticipated combination of effects from volume change on freezing and contact angle between the melt and solid formed from a tear-drop shape which ultimately encountered the end of the crucible. This caused the

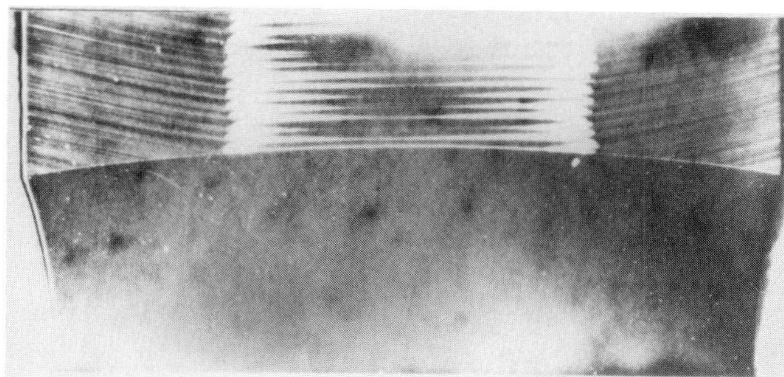

Fig. 13-8: Region of melt-back and regrowth of Witt and Gatos In–Sb sample. Note the rotational striations characteristic of Czochralski growth material and the complete absence of growth rate striations in the portion of the sample regrown in space. Also note the reduced diameter of the space-grown portion as the sample pulled away from the crucible.

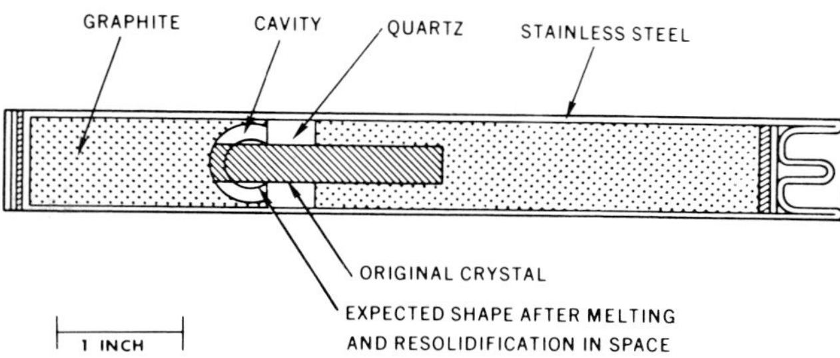

Fig. 13-9: Configuration of the InSB sample grown in a quasi-containerless fashion by Walter.

peculiar bent tip shown in Fig. 13-10. More important are the very smooth crystallographic planes on the unconfined growth portion of the crystal. X-ray topographs indicate that this portion of the crystal is virtually free of defects. There are dopant striations; however, this is puzzling in light of the results obtained by Witt and Gatos. One possible explanation is that an unconfined melt may be subject to Marangoni (surface tension-driven) convection which gives rise to growth rate fluctuations similar to those produced by gravity-driven convection on earth. Also, a melt confined only by its contact with a solid, which in turn is fixed to the spacecraft, will be affected more by small high-frequency accelerations (*g* jitter) associated with crew motion than a melt constrained by a container. Oscillations produced by such residual accelerations could conceivably result in striations. Clearly, more research is needed to resolve such questions.

J. T. Yue and F. W. Voltmer [10] (Texas Instruments) grew germanium crystals with various dopants by the same technique used by Witt and Gatos. The objective of this experiment was to investigate the distribution of

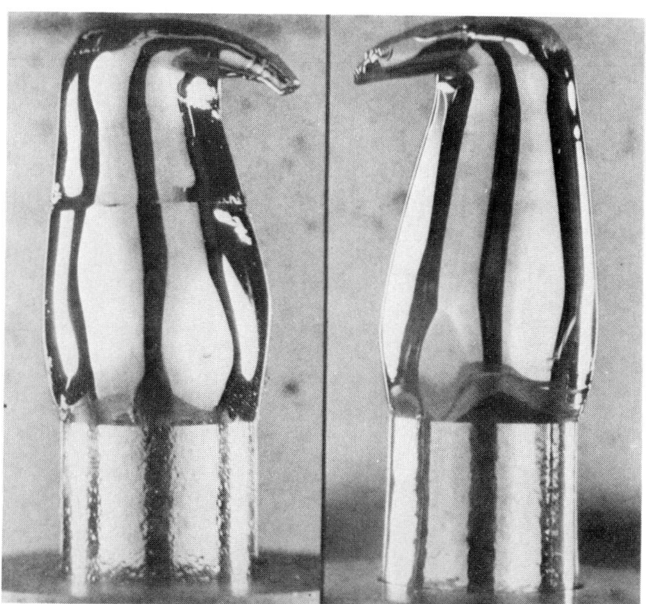

Fig. 13-10: Resulting crystal grown by Walter. Note the facets on the portion of unconfined growth. Also note the bent tip as the crystal grew into the end of the growth cavity.

dopants in the space-grown samples. The results were somewhat inconclusive because crucible contaminants produced some unwanted doping. Also, the amount of crystal grown in space was not sufficient to achieve steady-state growth conditions. There appeared to be more uniform radial distribution of dopants in the space-grown sample shown in Fig. 13-11. The earth-grown control samples seem more subject to random compositional fluctuations which are probably caused by unsteady growth conditions associated with gravity-driven convection. The space-grown samples had a more uniform composition in the interior, but did show a marked increase in dopant concentration near the surface. The growing crystal in this experiment pulled away from the crucible wall (as did several other experiments) which indicates that the melt may also have had a free surface. This would permit Marangoni flow that could account for the observed radial dopant profile. However, a curved solidification interface imposed by the thermal conditions could also account for the observed effect.

InSb–GaSb solid solution ingots were melted back and directionally resolidified in space in a crystal growth experiment developed by W. R. Wilcox (Clarkson College of Technology). Twins from the original cast material that was not melted in space propagated into the directionally solidified material. In space, very few new twins formed during solidification, while many formed in the ground control experiment [11]. This indicates that uncontrolled convection is at least partially responsible for twin formation. Later experiments conducted on earth in a magnetic field tended to confirm this finding [12].

Prof. H. Wiedemeier (Rensselaer Polytechnic Institute) flew a closed-ampoule vapor crystal experiment in which iodine was used as a chemical transport agent to grow GeSe and GeTe [13]. The crystals grown in space were generally larger with a more compact growth habit and smoother facets than the control samples grown on earth. They also exhibited a much greater perfection as determined by etch pit analysis. A major surprise in the experiment was the fact that the total material transported from source to seed was substantially greater in

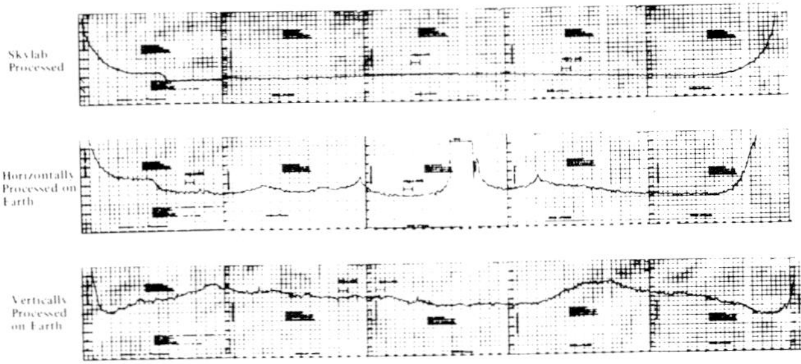

Fig. 13-11: Radial dopant profiles obtained by Ge by Yue and Voltmer. All samples have increased dopant concentrations near the perifery; however, the space-grown samples appear to be free of the variations in their interior.

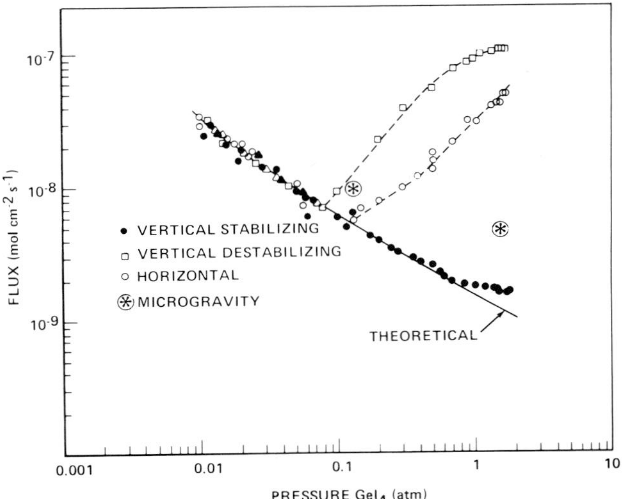

Fig. 13-12: Transport rates observed by Weidemeier as a function of pressure for various orientations relative to the gravity vector. Also the computed transport rate for diffusion-controlled conditions is seen to compare favorably with the rates observed for vertical stabilizing geometry. Note the significant departure of the observed transport rates in space from what was thought to be diffusion-controlled conditions.

space than on earth when the growth ampoule was in the vertical stabilizing configuration (hot over cold) [14]. This is shown in Fig. 13-12. The reasons for this anomalous transport rate are not clear, although we now realize that considerable convection takes place even in a thermally stable system because of radial thermal and solutal gradients caused by wall effects. It is conceivable that convective cells driven by these effects actually inhibit diffusive transport in a thermally stable configuration in earth's gravity.

Metallurgical Experiments

The metallurgical experiments on Skylab generally produced results that were not significantly different from the ground control experiments. Reger (TRW) solidified in microstructure several immiscible systems (Au–Ge, Pb–Zn–Sb, Pb–Sn–Sb) [15]. Little if any difference could be discerned between the flight and ground control samples. An anomalous superconductive transition was reported in the preliminary analysis of the Au–Ge flight sample, but this was not confirmed by later measurements. We have recently learned from splat cooling experiments that there does exist an unstable non-equilibrium superconducting phase in this system.

Takahashi (Japanese National Research Institute for Metals) solidified SiC fibers in an Ag matrix [16]. The space-solidified samples exhibited somewhat better homogeneity than the ground control samples, and did have some improvement in mechanical properties.

Several eutectics were directionally solidified in the gradient-freeze furnace. A. Yue (U.C.L.A.) found improved alignment and continuity in NaF fibers imbedded in an NaCl matrix [17]. Al–Cu solidified in the experiment of E. Hasemeyer showed little or no difference between the flight and ground control samples [18].

Demonstration Experiments

An excellent example of one of the demonstration experiments was the Liquid Floating Zone Experiment conceived by John Carruthers (Bell Laboratories) during the Skylab mission [19]. With the help of Astronaut Gibson, sufficient equipment was found on board to perform a remarkably sophisticated study of the stability of liquid floating zones. Fig. 13-13 illustrates the experimental setup. Several interesting results were obtained. It was demonstrated that under static conditions, the zone could actually be extended 5 percent beyond the theoretical Rayleigh limit before it separated into two spherical droplets as shown in Fig. 13-14. Also, an unexpected non-axisymmetric "jump-rope" instability, shown in Fig. 13-15, was discovered when the zone was rotated. Earlier experiments in a Plateau tank[1] had found only axisymmetric "hour-glass" rotational instabilities. This illustrates that classical simulation techniques cannot be relied upon to predict what will actually happen under microgravity conditions.

[1] *Plateau investigated various "weightless" phenomena more than 100 years ago by using various immiscible fluids with equal densities.*

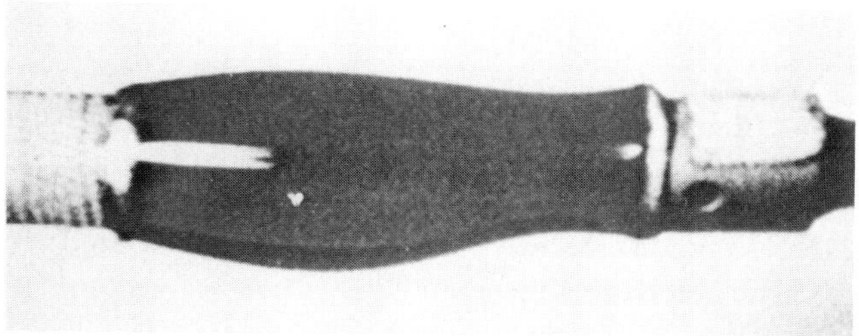

Fig. 13-14: Extended zone of colored water near the Rayleigh limit.

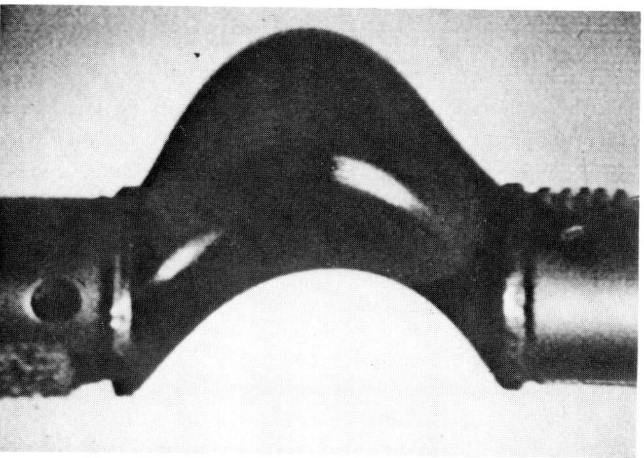

Fig. 13-15: Jump-rope or *C*-mode instability observed on Skylab when zone is rotated above a certain critical velocity.

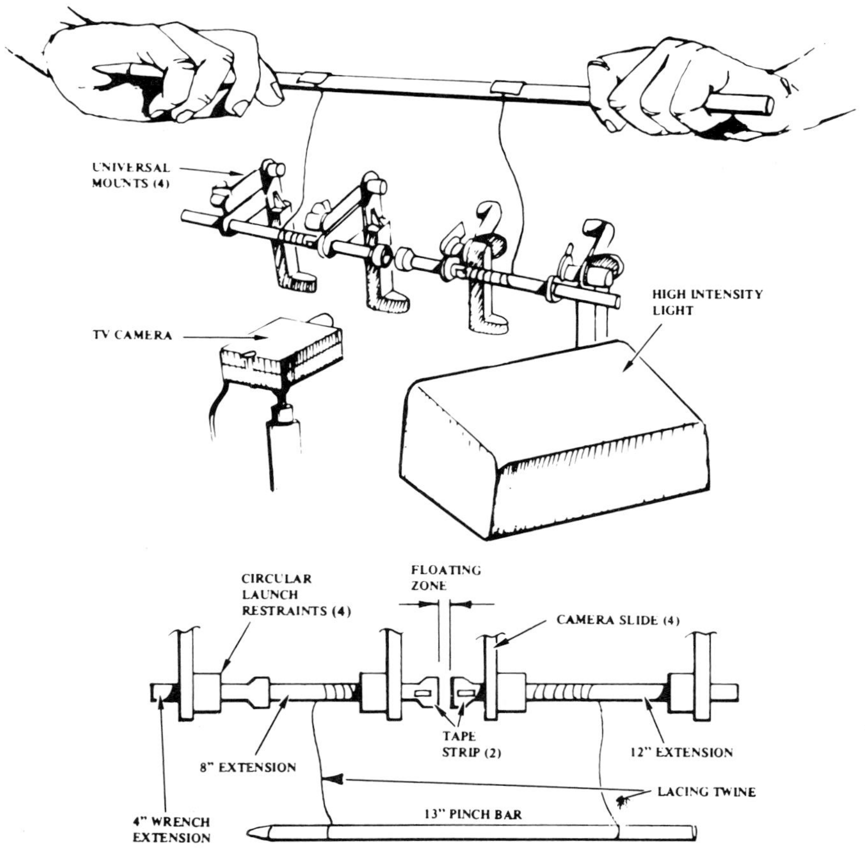

Fig. 13-13: Apparatus used to investigate stability of liquid zones using available equipment on Skylab such as socket wrench extensions, camera mounts, tape, and string.

APOLLO SOYUZ EXPERIMENTS

The Apollo Soyuz Test Project (ASTP) offered an opportunity to repeat some of the more interesting Skylab experiments and to accommodate several new investigations. Unfortunately, the time to prepare these experiments was short and the resources offered by ASTP in terms of power, weight, volume, and on-orbit time were much less than on Skylab. Therefore, the experiments on ASTP in many cases were performed under more primitive conditions than on Skylab.

Crystal Growth Experiments

Witt and Gatos repeated their Skylab experiments with Te-doped germanium in a new furnace that was equipped with Peltier pulsing for interface demarcation. This allows the interface shape and position to be correlated with time and gives a direct measure of growth rate. Contrary to the Skylab experiment, uniform dopant distribution was not achieved and there were indications that asymmetries in the furnace module produced unequal radial heat flow [20]. Similar effects may also have occurred in their Skylab experiment, but were not as apparent because of the difference in segregation coefficients and the sensitivity of the technique used on the ASTP experiment.

Wiedemeier repeated his Skylab experiment using different materials and transport agents. $GeSe_{0.99}$ $Te_{0.01}$ was grown using GeI_4 as the transport agent. Other samples included $GeS_{0.98}$ $S_{0.02}$ with $GeCl_4$, and GeS with $GeCl_4$ and Ar. The Ar was added to vary the total pressure independently of the partial pressure of the transport agent. The results were similar to those obtained in Skylab; i.e., significantly enhanced growth rates as compared to what was then believed to be diffusion-controlled growth on the ground, and improvements in growth habit and perfection suggestive of quiescent, diffusion-controlled growth conditions in the space-grown crystals [21].

Several co-precipitation crystal growth experiments were conducted by M. D. Lind [22] (Rockwell Science Center). In these experiments, two reactive solutions were allowed to interdiffuse. The reaction produced an insoluble precipitate which, in the microgravity environment, remains suspended. The larger crystallites grew at the expense of the smaller particles by Ostwald ripening in order to lower the total surface energy of the system. Crystals of calcium tartrate, calcium carbonate, and lead sulfide were grown by this technique and compared with similar crystals grown by the gel technique on earth. The calcium tartrate crystals grown in space exhibited both prismatic and platelet growth habits, whereas those grown on earth are predominantly prismatic. Most of the platelets were approximately 5 mm in diameter and appeared to be flawless. The calcium carbonate crystals were rhombohedral in shape and approximately 0.5 mm in size. For these crystals, little difference between space-grown and earth-grown in gel was observed. The lead sulphide crystals in each case were only 0.1 mm in size and accompanied by a fine precipitate which indicated insufficient time was available for the growth process.

Metallurgical Experiments

The halide eutectic growth experiment of Yue was repeated using LiF fibers in an NaCl matrix [23]. Again, some improvement in optical properties of the matrix was found, which could be attributed to better alignment and continuity of the structure.

Ang (Aerospace Corporation) and Lacy (MSFC) attempted to prevent density-driven phase separation in the Pb–Zn and Al–Sb systems. Pb–Zn is a monotectic system which has a clearly defined region of liquid-phase immiscibility. Cooling through this region results in the nucleation of droplets of the minority liquid second phase in the primary liquid host phase. Such droplets grow rapidly by diffusion and quickly separate by Stokes settling (or rise) in earth's gravity. The system was melted and soaked above the consolute temperature in space in order to homogenize the two liquid phases into a single liquid phase. After cooling and solidification, Lacy and Ang were surprised to find almost complete separation of the two phases [24] (see Fig. 13-16). We

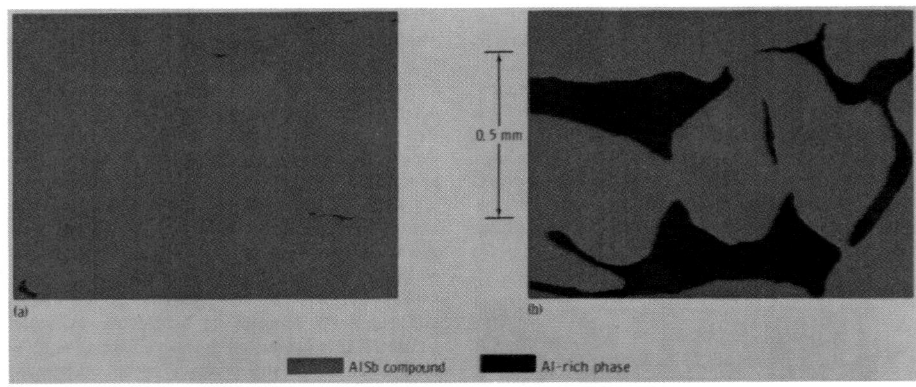

Fig. 13-16: Comparison of the amount of Al–Sb compound obtained in the space sample (a) and in the ground control experiment (b).

now know that there are several mechanisms, such as critical wetting and droplet migration caused by surface tension-driven flows, that can also produce phase separation. These mechanisms are usually masked on earth by density-driven separation.

The difficulty in obtaining the compound AlSb by direct solidification of a stoichiometric mixture of the two components suggested that a small region of liquid phase immiscibility might exist near the melting point of the compound. Ang and Lacy [25] carefully prepared a stoichiometric mixture and compared the material solidified in space with that solidified on the ground. A considerable amount of off-compound phases was found in the ground-control sample, whereas the flight sample formed almost pure compound throughout as shown in Fig. 13-17. More recent phase diagrams indicate that AlSb is in fact congruently melting; therefore, the formation of off-compound phases in unit gravity may be caused by compositional inhomogenetics and concomitant density separation during the solidification process.

Larson (Grumman Aerospace Corporation) directionally solidified MnBi/Bi eutectic in the gradient-freeze furnace and was surprised to find considerable enhancement in the coercive strength of the flight sample [26]. Several factors apparently contributed to this fortuitous result. MnBi/Bi is a low-volume fraction eutectic in which faceted MnBi rods grow in a nonfaceted Bi matrix. Post-flight refinements of the phase diagram revealed that the flight samples had in fact contained less Mn than the eutectic composition. This would lead to a changing volume fraction in the ground control sample because of convective mixing, whereas the flight sample can adjust to steady-state growth conditions by building a diffusion layer ahead of the growth front. Also, for reasons that are not completely clear, solidification of MnBi/Bi eutectic in an environment with less convection produces a finer, more closely spaced rod structure. This was demonstrated in later SPAR rocket flights and by solidification experiments in strong magnetic fields. All of these factors conspired to produce a finer, more regular microstructure which was apparently near the optimum size for elongated single magnetic domains.

In several of the Skylab experiments involving directional solidification, the portion of the sample that solidified in the microgravity environment was smaller in diameter than the containing ampoule. This suggested that melts that do not wet the container may pull away from the container walls in the absence of hydrostatic pressure. Reed (Oak Ridge National Laboratory) conducted an experiment to establish whether or not surface tension-driven convection would be important under these circumstances [27]. An Au–Pb slug was pressure bonded to one end of a Pb sample which was placed in a cylindrical graphite container. These samples were melted and soaked at different temperatures for several hours, and solidified in the isothermal portion of the furnace. The effect of a convectively stable system versus that of an unstable system can be readily seen in Fig. 13-18 by comparing the sample in which the heavier Au was on the bottom with the sample in which it was on the top. However, even in the solutally stable configuration, considerable mixing took place as is evident from the presence of some gold throughout the lead sample. This was apparently driven by unavoidable radial thermal gradients in the cooling process. Compare these results with the samples melted and solidified in space (Fig. 13-19). Here, the distribution of gold appears to be dominated by diffusion, although some curvature of the isoconcentration profiles is evident. The experimenters suggest that this may be due to solutally driven Marangoni flow which occurs because the non-wetting fluid pulls away from the container walls in low gravity, thus exposing a free liquid surface to which no-slip boundary conditions no longer apply. This explanation appears questionable, however, since there is no evidence

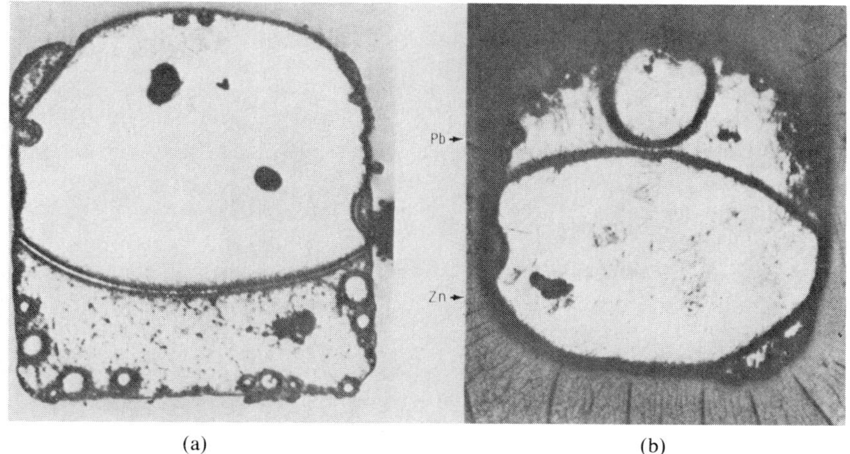

(a) (b)

Fig. 13-17: Massive separation of Pb and Zn obtained when solidifying the monotectic system. In the ground control sample (a), the denser Pb-rich phase settles to the bottom of the crucible. In space (b), the Zn-rich phase apparently wets the crucible, and the surface tension of the Pb-rich phase is apparently lower than the interfacial energy between the Pb and Zn-rich phases.

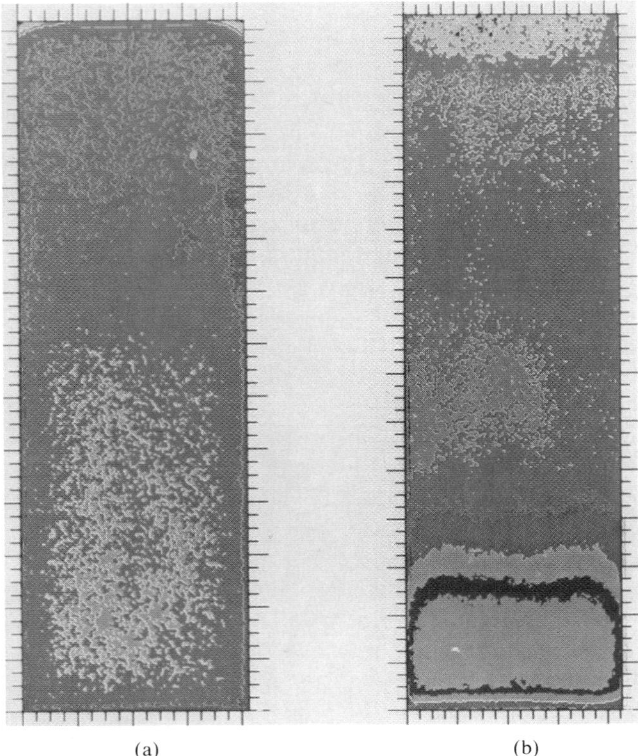

Fig. 13-18: Distribution of Au in Pb during solidification in unit gravity. Lighter colors indicate Au-rich, whereas black denotes pure Pb. Mixing is almost complete for the solutally unstable case (Au over Pb) (a). Considerable mixing also occurs for the solutally stable case (b).

that a free liquid surface actually existed in this particular experiment and various alternative mechanisms for these curved isoconcentration profiles, e.g., segregation effects, thermal expansion effects, etc., have not been satisfactorily ruled out. What is perhaps more significant is that diffusive transport dominated in the space processed samples despite residual accelerations and rotational effects from the spacecraft motion.

Electrophoresis Experiments

Two electrophoresis experiments were also conducted on the ASTP. Hannig (Max Planck Institute for Biochemistry, Munich) developed a continuous flow electrophoretic separator, whereas Allen *et al.* (MSFC) developed a static column device. The Hannig device attempted to separate rat bone cells, mixtures of human and rabbit red blood cells, rat spleen cells, and mixtures of rat and human lymphocytes. No collection was attempted; instead, the separation was monitored by an optical array to detect the UV absorption of the biological material. The experiment was compromised by the fact that the quartz halogen lamp burned brighter than expected in the low-gravity environment and saturated the detector [28]. This was apparently caused by the absence of convection in the gas inside the lamp.

The static electrophoresis device processed eight samples consisting of mixtures of different red blood cells (rabbit, horse, and human), human lymphocytes, and kidney cells. A special coating was developed for the columns to suppress the electroosmotic flows that had plagued earlier static electrophoresis experiments. The individual columns were electrophoresed for a predetermined length of time and frozen in place using a thermoelectric cooler. The frozen columns were then placed in an LN_2 dewar for return to earth and analysis. Photographic analysis of the columns containing the red blood cells showed that the sample bands remained nearly planar, indicating that electroosmosis had indeed been controlled. Some enhancement in production of Urokinase and other cell products was observed from portions of the column containing the kidney cells, indicating that separation of cells according to their function may have been achieved [29].

SPACE PROCESSING APPLICATIONS ROCKET (SPAR) EXPERIMENTS

The SPAR rocket is a Nike-boosted, Black Brant vehicle equipped with a recovery system and a stabilized attitude control system to prevent rotational-induced accelerations. Nominal low-gravity times of 5 min are

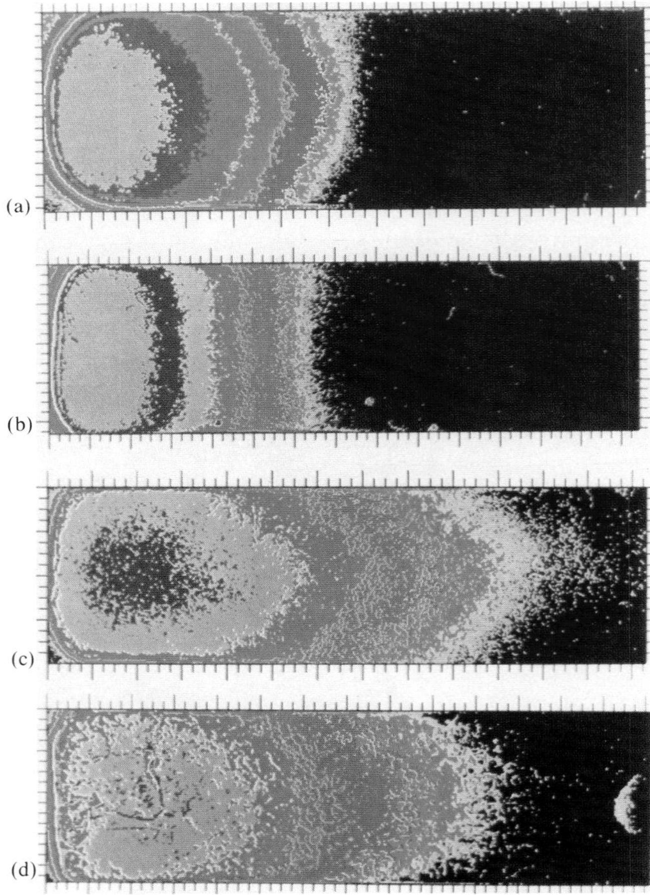

Fig. 13-19: Distribution of Au in Pb in flight samples. Transport seems to be dominated by diffusion. Some distortion near the walls is evident, especially in the samples processed at higher temperatures [(c) and (d)].

obtained with this system for payload weights up to 300 kg. The short times available are not conducive to crystal growth, biological separations, or other experiments that require carefully controlled conditions over a long time. Certain fluid and solidification experiments, however, can be conducted reasonably well in such a facility. The following paragraphs will describe some of the more significant results obtained from the SPAR project.

Liquid Mixing Experiments

One of the first experiments conducted in the SPAR program was designed by C. F. Schafer (MSFC) to characterize the low-gravity environment and its effect on fluid mixing. Mixtures of In and Pb–In in Al cylinders were melted and resolidified during the five-minute low-gravity portion of the flight. The samples were located near the outer perimeter of the rocket. One sample was oriented with its axis perpendicular to the longitudinal axis of the rocket with the Pb-rich end on the inside. It experienced a mixing flow over the length of the sample. The sample whose axis was aligned with the rocket axis experienced only a slight deformation of the interface. The flows can be explained by a slow residual roll about the longitudinal axis producing accelerations on the order of $10^{-6}g$ [30].

Solidification Experiments

A series of experiments was carried out by M. H. Johnston (MSFC) to understand the low-gravity solidification process in castings. The first experiments used $NH_4Cl + H_2O$ as a transparent analog for metallic systems. In the flight experiment, the entire solidification was initiated from only a few nucleation sites. There was no evidence of dendritic breakage and multiplication of nucleation sites. Growth was completely columnar with no transition to an equiaxed zone [31].

Johnston continued this investigation with various metallic systems including Sn-15 wt % Pb, and Sn-15 wt % Bi. A special centrifuge furnace was used to investigate the microstructure of castings at elevated g levels also. In general, the grain structure becomes finer as the g level increases, and macrosegregation and freckling become more predominant. At low gravity, the solid consists of relatively few large grains. Macrosegregation and freckling is virtually eliminated. One unexpected result was that the dendrite arm spacing was found to be more uniform and greater in the space processed samples. This is attributed to the steeper and more steady temperature and concentration gradients that result from the reduction of convective flows [32].

Several attempts to produce fine *in situ* dispersions in various compositions of the monotectic Al-In system were made by H. Alborn (University of Hamburg) and by S. H. Gelles (Gelles Associates). For cases in which Al was the host phase, the results were similar to those obtained by Lacy and Ang in their ASTP experiment. Almost complete phase separation was observed. The Al-rich material was surrounded by the In-rich component which apparently preferentially wets the graphite crucible. Also, In-rich regions were found in the center of the ingot, apparently resulting from droplet migration in the thermal gradient [33]. Meanwhile, Cahn (NBS) showed that one of the components of a monotectic

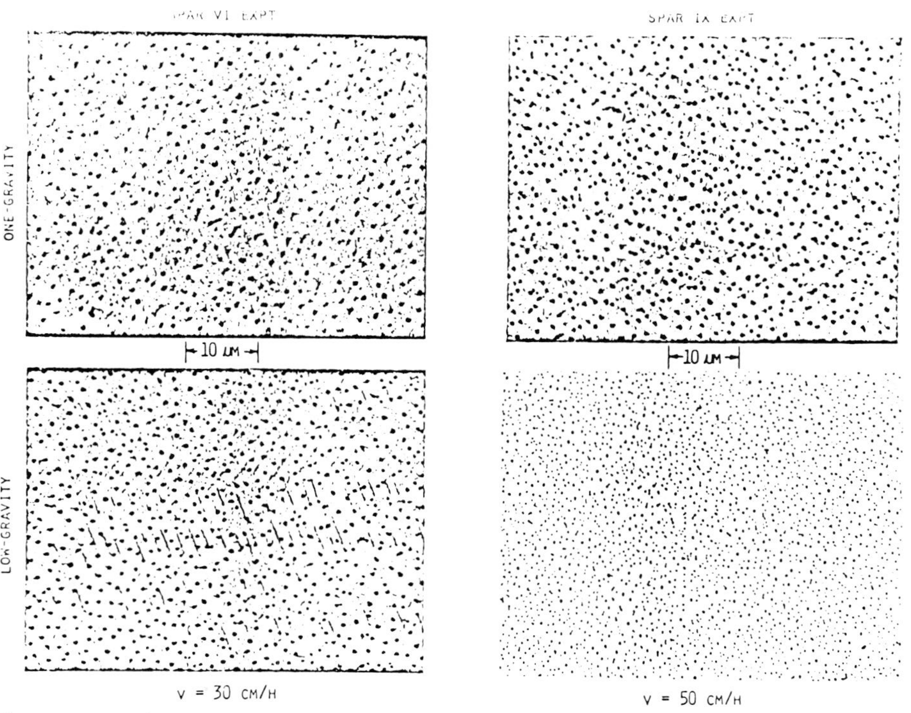

Fig. 13-20: Microstructure of MnBi/Bi eutectic solidified at two different velocities in space compared with the ground control samples. The smaller rod spacing and diameters are especially noticeable in the $V = 50$ cm/h sample.

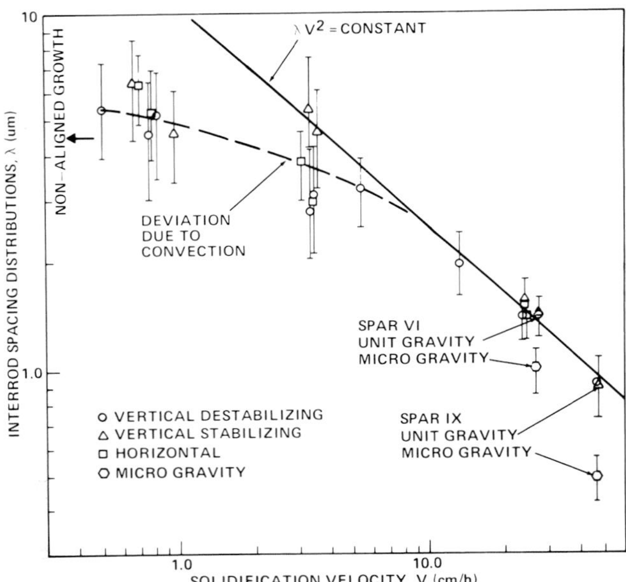

Fig. 13-21: Plot of rod diameter versus growth rate for MnBi eutectic. Note that the samples processed in unit gravity follow λV^2 = constant for growth rates greater than 10 cm/h regardless of orientation. However, the space processed samples depart significantly from this law.

system becomes perfectly wetting at some temperature below the critical consolute temperature [34], and if this component were the minority phase, the system would be unstable; i.e., the minority phase will spread between the majority phase and the container walls [35]. C. Potard (Center for Nuclear Studies, Grenoble, France) repeated the experiment in a SiC crucible which was preferentially wet by the Al-rich host phase. This eliminated the critical wetting instability predicted by Cahn, and produced a dispersed In-rich phase [36]. The dispersion was not completely uniform, however, indicating that agglomeration and other effects must still be controlled.

The intriguing results obtained by Larson were further investigated by several SPAR experiments [37]. Comparison of flight and ground control samples revealed a significant decrease in Mn–Bi rod diameter and spacing in the flight samples, especially at the higher solidification rates, as shown in Figs. 13-20 and 13-21. This was quite unexpected since it is generally assumed that eutectic solidification is controlled by diffusion [38]. Recent experiments in magnetic fields confirmed that suppression of convective effects is indeed responsible for the difference in microstructure [39].

Containerless Processing Experiments

The SPAR project was also useful for developing various containerless processing techniques. Wang (JPL) developed a three-axis acoustic levitator. Liquid spheres were positioned, rotated, and undulated by modulating and phasing the acoustic drivers. The dynamic behavior of droplets was investigated along with other phenomena such as mixing and bubble centering mechanisms [40].

High-temperature containerless melting and solidification experiments were conducted using a single-axis acoustic interference-type levitator furnace developed by Whymark [41] (Intersonics, Inc.). Melting was accomplished, but the acoustic field was not sufficient to retain the molten sample [42]. Additional development is required to perfect this technique.

The electromagnetic levitator, developed by R. T. Frost [43] (General Electric) melted and solidified a sphere of beryllium containing a dispersed phase of beryllia. This experiment, developed by G. Wouch *et al.* (General Electric) investigated the utilization of a dispersed oxide, BeO, as a grain-refining agent to improve the microstructure of cast beryllium. The containerless process prevented any wall-induced nucleation, thus assuring that the dispersed phase would act as the nucleating agent. The sample was successfully positioned by the electromagnetic field during the flight. Melting and solidification were accomplished without any physical contact with the coils. A fairly uniform dispersion of the BeO was obtained in the flight sample. However, the BeO dispersoid did not produce the anticipated fine grain structure [44].

A more complete description of all of the SPAR experiments may be obtained from the SPAR Mission Reports [45]–[51].

REFERENCES

[1] C. A. Lundquist, "Attitude control for experiments in microgravity," in *Proc. AIAA Guidance Contr. Conf.*, Gatlinburg, TN, Aug. 15–17, 1983.

[2] L. L. Lacy and G. H. Otto, "Electrical resistivity of gallium-bismuth solidified in free fall," *AIAA J.*, vol. 13, p. 219, 1975.

[3] I. C. Yates, "Apollo 14 composite casting demonstration," Final Rep., NASA TM X-64641, Oct. 1981.

[4] P. G. Grodzka and T. C. Bannister, "Heat flow and convection experiments aboard Apollo 17," *Science*, vol. 187, p. 165, Jan. 1975.

[5] R. S. Snyder *et al.*, "Free flow particle electrophoresis on Apollo 16," *Sep. Purif. Methods*, vol. 2, p. 259, Sept. 1973.

[6] J. R. Williams, "M552 exothermic brazing," in *Proc. 3rd Space Process. Symp., Skylab Results*, vol. I, M-74-5, June 1974, p. 33.

[7] E. C. McKannan, "M551 metals melting," in *Proc. 3rd Space Process. Symp., Skylab Results*, vol. I, M-74-5, June 1974, p. 85.

[8] A. F. Witt *et al.*, "Crystal growth and steady state segregation under zero gravity: InSb," *J. Electrochem. Soc.* vol. 122, p. 276, 1975.

[9] H. U. Walter, "M560 growth of spherical crystals," in *Proc. 3rd Space Process. Symp., Skylab Results*, vol. I, M-74-5, June 1974, p. 235.

[10] J. T. Yue and F. W. Voltmer, "Influence of gravity-free solidification on solute microsegregation," *J. Cryst. Growth*, vol. 29, p. 329, 1975.

[11] J. F. Yee, M. C. Lin, K. Sarma, and W. R. Wilcox, "The influence of gravity on crystal defect formation in InSb-GaSb alloys," *J. Cryst. Growth*, vol. 30, p. 185, 1975.

[12] S. Sen, R. A. Lefever, and W. R. Wilcox, "Influence of magnetic field on vertical Budgman-Stockbarger growth of $In_x Ga_{1-x} Sb$," *J. Cryst. Growth*, vol. 43, p. 526.

[13] H. Wiedemeier, F. C. Klaessig, E. A. Irene, and S. J. Wey, "Crystal growth and transport rates of GeSe and GeTe in microgravity environment," *J. Cryst. Growth*, vol. 31, p. 36, 1975.

[14] D. Chandra and H. Wiedemeier, "Chemical vapor transport and thermal behavior of the GeSe-GeI_4 system for different inclinations with respect to the gravity vector, comparison with theoretical and microgravity data," *J. Cryst. Growth*, vol. 57, p. 159, 1982.

[15] J. L. Reger, "M557 immiscible alloy compositions," in *Proc. 3rd Space Process. Symp., Skylab Results*, vol. I, M-74-5, June 1974, p. 133.

[16] S. Takahashi, "Preparation of silicon carbide whisker reinforced silver composite material in a weightless environment," *15th AIAA Aerospace Sciences Meeting*, Los Angeles, CA, Jan. 24–26, 1977, AIAA Paper 77-195.

[17] A. S. Yue and J. G. Yu, "Solidification of NaCl-NaF eutectic in space," in *Proc. AIAA/ASME Thermophysics Heat Transfer Conf.*, Boston, MA, July 15–17, 1974, AIAA Paper 74-646.

[18] E. A. Hasemeyer, "M566 copper-aluminum eutectic," in *Proc. 3rd Space Processing Symp., Skylab Results*, vol. I, M-74-5, June 1974, p. 457.

[19] J. R. Carruthers, "Studies of liquid floating zones," in *Proc. 3rd Space Process. Symp., Skylab Results*, vol. II, M-74-5, June 1974, p. 843.

[20] A. F. Witt, H. C. Gatos, M. Lichtensteiger, and C. J. Herman, "Crystal growth and segregation under zero gravity: Ge," *J. Electrochem. Soc.*, vol. 125, p. 1832, Nov. 1978.

[21] H. Wiedemeier *et al.*, "Morphology and transport rates of mixed IV-VI compounds in microgravity," *J. Electrochem. Soc.*, vol. 124, p. 1095, July 1977.

[22] M. D. Lind, "Crystal growth from solutions in low gravity," *AIAA J.*, vol. 16, p. 458, 1978.

[23] A. S. Yue, in C. W. Yeh, and B. K. Yue, "Halide eutectic growth, (MA-131)," in *"Apollo-Soyuz Test Project Summary Science Report,"* vol. I, NASA SP-412, 1977, p. 491.

[24] L. L. Lacy and C. Y. Ang, "Monotectic and symtectic alloys (MA-044)," in *"Apollo-Soyuz Test Project Summary Science Report,"* vol. I, NASA SP-412, 1977, p. 403.

[25] C. Y. Ang and L. L. Lacy, "Gravitational influences on the liquid-state homogenization and solidification of aluminum antimonide," *Met. Trans.*, vol. 10A, p. 519, 1980.

[26] D. J. Larson, "Zero-G processing of magnets (MA-070)," in *"Apollo-Soyuz Test Project Summary Science Report,"* vol. I, NASA SP-412, 1977, p. 449.

[27] R. E. Reed, W. Uelhoff, and H. L. Adair, "Surface-tension induced convection (MA-040)," in *"Apollo-Soyuz Test Project Summary Science Report,"* vol. I, NASA SP-412, 1977, p. 367.

[28] K. Hannig, H. Wirth, and E. Schoen, "Electrophoresis (MA-014)," in *"Apollo-Soyuz Test Project Summary Science Report,"* vol. I, NASA SP-412, 1977, p. 335.

[29] R. E. Allen *et al.*, "Column electrophoresis on the Apollo-Soyuz test project," *Sep. Purif. Methods*, vol. 6, p. 1-59, 1977.

[30] C. F. Schafer and G. H. Fichtl, "SPAR I liquid mixing experiment," *AIAA J.*, vol. 16, p. 425, May 1978.

[31] M. H. Johnston, C. S. Griner, R. A. Parr, and S. J. Robertson, "The direct observation of unidirectional solidification as a function of gravity level," *J. Cryst. Growth*, vol. 50, p. 831, 1980.

[32] M. H. Johnston and R. A. Parr, "The influence of acceleration forces on dendritic growth and grain structure," *Met. Trans.*, vol. 13B, p. 85, Mar. 1982.

[33] S. H. Gelles and A. J. Markworth, "Microgravity studies in the liquid-phase immiscible system: Al-In," *AIAA J.*, vol. 16, p. 431, May 1978.

[34] J. W. Cahn, "Critical point wetting," *J. Chem. Phys.*, vol. 66, p. 3667, 1977.

[35] ——, "Monotectic composite growth," *Met. Trans.*, vol. 10A, p. 119, Jan. 1979.

[36] C. Potard, "Solidification of hypermonotectic Al-In alloys under microgravity conditions," in *Materials Res. Soc. Symp. Proc., Materials Process. Reduced Gravity Environment Space*, vol. 9, G. E. Rindone, Ed. Amsterdam: North Holland, 1982, p. 543.

[37] R. G. Pirich, "Studies of directionally solidified eutectic Bi/MnBi at low growth velocities," *Met. Trans. A*, 1983.

[38] K. H. Jackson and J. D. Hunt, "Lamellar and rod eutectic growth," *Trans. AIME*, vol. 236, p. 1129, Aug. 1966.

[39] D. J. Larson, R. G. Pirich, R. Silbertstein, J. de Carlo, and C. Buscemi, "Effect of applied magnetic field on directional solidification of eutectic MnBi/Bi," in *Proc. of ASM/AIME Fall Meeting*, Philadelphia, PA, Sept. 1983.

[40] T. G. Wang, W. M. Saffren, and D. D. Elleman, "Drop dynamics in space," in *Materials Sciences in Space with Application to Space Processing*. New York: AIAA, 1977, pp. 151–172.

[41] R. R. Whymark, "Acoustic field positioning for containerless processing," *Ultrasonics*, p. 251, Nov. 1975.

[42] R. J. Naumann, W. A. Oran, R. R. Whymark, and C. Rey, "Postflight analysis of the single-axis acoustic system on SPAR VI and recommendations for future flights," NASA TM-82396, Jan. 1981.

[43] R. T. Frost and C. W. Chang, "Theory amd applications of electromagnetic levitation," in *Materials Res. Soc. Symp. Proc., Materials Process. Reduced Gravity Environment Space*, vol. 9, G. E. Rindone, Ed. Amsterdam: North Holland, 1982, p. 71.

[44] G. Wouch *et al.*, "Uniform distribution of BeO particles in Be casting produced in rocket free fall," *Nature*, vol. 274, p. 235, July 20, 1978.

[45] "SPAR I Final Report," NASA TM-X-3458, Dec. 1976.

[46] "SPAR II Final Report," NASA TM-78125, Nov. 1977.

[47] "SPAR III Final Report," NASA TM-78137, Jan. 1978.

[48] "SPAR IV Final Report," NASA TM-78235, Jan. 1980.

[49] "SPAR V Final Report," NASA TM-78275, Aug. 1980.

[50] "SPAR VI Final Report," NASA TM-82433, Oct. 1981.

[51] "SPAR VII Final Report," NASA TM-82535, Oct. 1983.

14

CURRENT PROGRAM TO INVESTIGATE PHENOMENA IN A MICROGRAVITY ENVIRONMENT

William A. Oran

CURRENT RESEARCH PROGRAM

Much of the research sponsored by the Microgravity Science and Applications (MSA) Division of NASA is an outgrowth of the investigations conducted on the Apollo and Skylab vehicles. These activities, which are documented in [1], are briefly discussed in Chapter 13 on early experiments in space. The currently sponsored research can be roughly grouped by processes in a similar manner to the early studies as follows: 1) crystal growth, from solution, from vapor, from melt via directional solidification and float zone; 2) metallic alloy solidification, from melt via directional and isothermal solidification; 3) bioseparation processes; 4) blood (and other non-Newtonian fluid) rheology; 5) containerless processing (i.e., melting and solidifying without use of a container) science, measurement of high temperature thermo-physical properties of materials, and production of new, metastable phases or glass composition; 6) combustion processes; 7) chemical and transport phenomena; 8) cloud microphysics; and 9) fluid behavior in low gravity and free surface phenomena.

Investigations are selected for funding on the basis of peer review of proposals. In addition to requiring a high quality of science in the proposed work, the research must have a relevance to experimentation in a low-gravity environment. Since the area of microgravity science experimentation is relatively new, most new efforts are placed in a two to four year ground-based research program to help quantitatively define the relation of microgravity experiments to the research. At the end of this period, the studies must be reproposed with relevance to low-gravity experiments quantitatively defined as part of the continued effort.

Within each of the major research disciplines, there can be several investigations with their own thrust depending on their ultimate objective. Again, the thrusts can be grouped into major categories of 1) increasing the understanding of processes to improve ground processing systems (e.g., combustion); 2) improving processing techniques to manufacture materials of technological interest in space (e.g., semiconductor crystals); 3) increasing the knowledge/technical base to help understand basic physical phenomena (e.g., critical point measurements); and 4) studies to better understand phenomena influencing low-gravity processing investigations (e.g., free surface Marangoni flow).

The MSA program publishes a description of all research tasks and the reader is referred to [2] and [3] for detailed description of activities. The following is a brief discussion of results of some of these investigations. Information on the results of all investigations can be found in the program bibliographies, [4] and [5].

RESULTS FOR THE GROUND-BASED PROGRAM

Interesting data have been obtained with the 400 foot drop tube at NASA's George C. Marshall Space Flight Center (MSFC). These tubes allow for the containerless solidification of drops of molten material. Small 2–3 mm dia. drops of niobium germanium have been undercooled 500 K (for compositions up to 30 at wt % Ge) before solidification. The results show a change in morphology with undercooling; the greater the undercooling, the finer the microstructure coupled with an increase of the Ge solubility of the beta phase. The experimenters are attempting to produce the metastable A-15 phase of Nb_3Ge in the bulk by solidifying the materials in the hypercooled regime [6]. At this time, a major limitation is instrumental; even after a 4 s fall time, the drops are hitting the bottom of the tube molten or semimolten. The next step may be for these experiments to be conducted in the free-fall environment of space which will allow for a virtually unlimited time for containerless solidification.

Additional drop tubes are sponsored by NASA and include an aerodynamic, zero drag system currently in operation at the Jet Propulsion Laboratory (JPL). This system has been used to containerlessly solidify very thin wall shells. The experiments have led to an understanding of the formation of highly concentric, highly uniform spherical skills (of interest for the fabrication of fusion targets), as well as creating metallic shells having a very fine texture [7].

Depending on the time scale of phenomena, other facilities are available either to do definitive experiments or to conduct precursor studies in preparation for shuttle experiments. These facilities include drop tubes and drop towers, which can give 5 s of low gravity (g), or airplanes flying Keplerian trajectories which can give 30 s of low g and sounding rockets which can give up to 300 s of low g. For an example, Fig. 14-1 shows the activity

Fig. 14-1: Testing of an acoustic positioning device in the cabin of a KC-135 during the low-gravity portion of its flight path.

in the cabin of a KC-135 aircraft in the low-gravity portion of its flight. The program is further augmented by earth-based laboratory instrumentation and apparatus such as aerodynamic containerless processing devices, electromagnetic processing coils, and high-intensity static magnetic fields used to impede the convective fluid motion during crystal growth. Some results obtained with these facilities are presented in [8].

Space Processing Applications Rockets (SPAR) have been used in the study of the directional solidification of MnBi near the eutectic point $Mn_{2.7}Bi_{97.3}$. Fig. 14-2 shows a comparison of microstructure of samples directionally solidified on ground in a free fall. The difference between the low *g* and ground processing is substantial, with the convection occurring during ground processing apparently coarsening the structure. In addition to understanding the influence of convection/non-convection on the morphology and composition, the investigators are trying to improve the magnetic properties by controlling and increasing the fraction of eutectic rod material in the matrix. This may have applications in the design of induction motors. While MnBi is not a preferred magnetic material, the same possibility may hold true for CoSm which does have a wide application in motors. Additional investigation studying the off eutectic growth of MnBi are planned for the shuttle along with studies of the directional solidification of Co_2Sm_{17} and $CoSm_5$ [9].

SHUTTLE FLIGHT PROGRAM

A major facet of the microgravity experiment program was initiated with the successful flight of the space shuttle in April 1981. The first flight experiment sponsored by the MSA Division was the monodispersed latex particle investigation of Professor John Vanderhoff of Lehigh University. This experiment studied the influence of processing parameters (i.e., low gravity) on the quality of latex beads produced in chemical reactors. Previous experiments by Vanderhoff and others indicated that beads with dispersions (standard deviation/diameter) ~1 percent could be obtained in ground processing experiments for bead diameters up to 5 μ. The studies showed that from 5 μ to about 100 μ, severe difficulties in ground processing occurred because of density variations between the beads and the suspension medium (water). This degraded the quality of the beads, significantly increasing the dispersion. This experiment was initially flown on shuttle flight 3 in March 1982 and then subsequently on shuttle flights 4, 6, and 7. The experi-

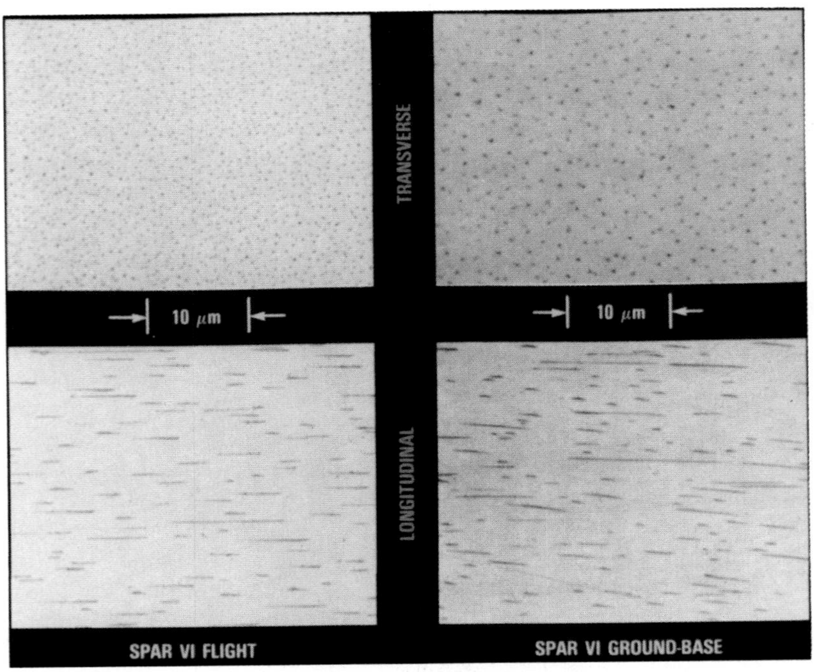

Fig. 14-2: Microstructure comparisons of directionally solidified Bi/MnBi eutectic. Reprinted with permission of authors of [9].

ments to date have shown that by processing in space, a 1 percent dispersion can be obtained for bead diameters up to 30 μ. Samples of the flight processed beads have been analyzed by the Micrometrology Group of the National Bureau of Standards. They have found them to be of high quality, significantly better than corresponding beads processed on earth. They have requested samples of flight processed materials to be sold as primary reference material as part of their Standard Reference Materials Program managed in cooperation with the American Society for Testing Materials (ASTM) [10].

The monodisperse latex reactor was an automated experiment placed in the shuttle cabin. This area can accommodate only relatively simple experiments requiring low power. More complex, higher power experiments require the capabilities of the shuttle cargo bay. These activities were initiated with the first flight of the Materials Experiments Assembly (MEA) on shuttle flight 7 in June 1983. The MEA is designed to carry several rocket class apparatus in the shuttle cargo bay. The first MEA carried two general purpose rocket furnaces (GPRF's), and a single axis acoustic levitator (SAAL), a levitation furnace. The GPRF's were used to study vapor transport/crystal growth using GeSe in an Xe atmosphere, by Prof. H. Wiedemeier of Rensselaer Polytechnic Institute (RPI) and, solidification of immiscible alloys (InAl and TeTh) by Dr. S. Gelles of S. Gelles Associates. The flight of the SAAL was primarily to obtain data on the characteristics (i.e., stability and strength) of the levitation/processing at high temperatures. However, a secondary objective was to process oxide samples provided by Professor D. Day from the University of Missouri (Rolla). At the time this chapter was written, the flight samples were being analyzed and results were not available; however, it appears that instrument malfunctions precluded the obtaining of useful data on the operation of the SAAL.

Another experimental effort involves the analysis of a biomaterial separation technique in microgravity using a continuous flow electrophoresis system (CFES). The CFES was designed and built by McDonnell Douglas Aerospace Corporation (MDAC) with company funds and then flown on the shuttle under a joint endeavor agreement (JEA) to evaluate the commercial viability of separating pharmaceutical materials in space. The history of the JEA's, their concept, and the details of the NASA-MDAC JEA are discussed in another chapter. However, as part of this JEA, NASA has conducted separations on the CFES to help characterize the capabilities of generic continuous flow separation systems in space. Fig. 14-3 presents data showing the distortions in the flow pattern (and by implication, the degradation of the quality of separation) that occurs when the separation electric fields are established across the electrolytic buffer carrying the biomaterials. Additional details on the operation of an electrophoretic system can be found in [11]. The NASA sponsored investigators have separated materials such as hemoglobulin, polysaccarides, kidney cells, and pituitary cells using the CFES. The results are being analyzed to determine not only characteristics of the separations in low gravity, but to help quantitatively measure the benefits of separating pharmaceutical materials in space. Details on MDAC's efforts have been publicized, and some information can be found in [12].

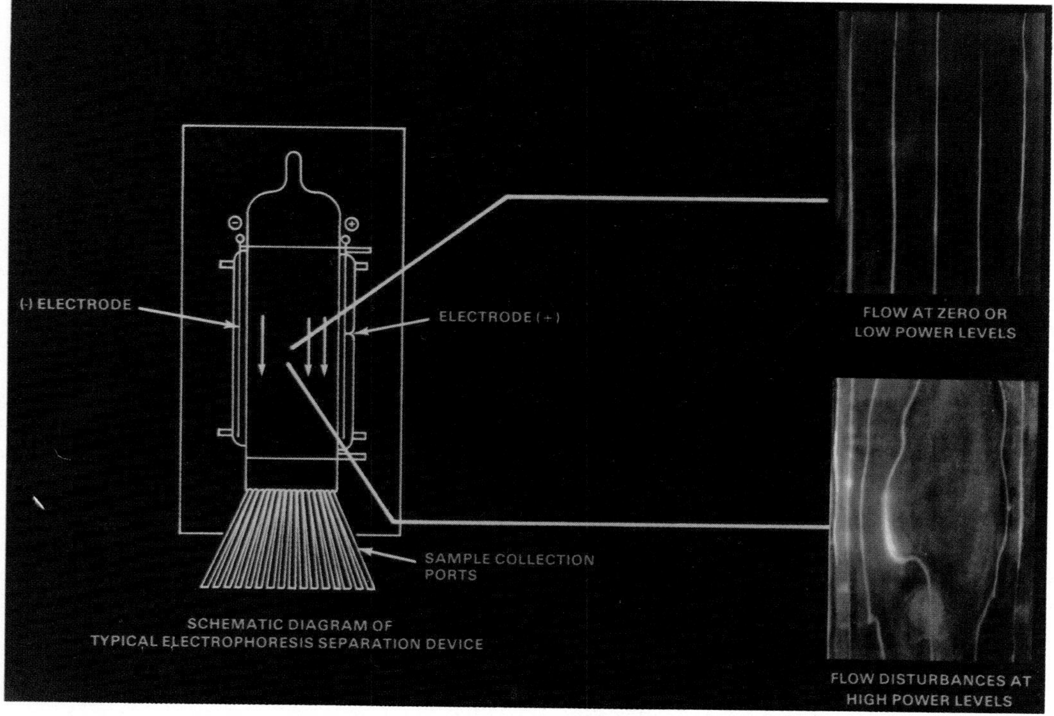

Fig. 14-3: Streamlines of flow patterns at different conditions in a flowing electrophoresis separation system.

The microgravity research program has, because of the shuttle, the ability to conduct a series of iterative experiments to understand and exploit various microgravity phenomena. Every shuttle launch allows for the possibility of flying apparatus to conduct updated experiments for seven days in a microgravity environment. The recent advances of the program must be due, in large part, to the increase in capabilities associated with the shuttle. The MSA program is steadily expanding in use of shuttle-borne facilities. Four MSA sponsored and JEA related experiments were performed in 1982. Eight experiments are scheduled for 1983, 14 are scheduled for 1984, and further increases in the flight experiment activity are projected for succeeding years.

There is a continuing and expanding interest in microgravity studies. During the Apollo and Skylab era, the United States was the only country promoting a microgravity experimentation program. Today, numerous other countries are actively encouraging microgravity experimentation. Based on budgetary comparisons, the total European effort (European Space Agency and member nations) is at least comparable to the American program. Little is known of the Russian activities except that they conducted a series of experiments on their space station (Salyut 6) and are beginning to conduct additional experiments on a new station (Salyut 7). The Japanese are increasing their efforts and (apparently based on the results of a sounding rocket program) have put earnest money on a spacelab/shuttle flight to conduct a series of microgravity experiments.

The ability to perform iterative microgravity experiments coupled with the strong ground-based program (that has been instituted over the past few years) will result in a quantitative understanding of the benefit of microgravity to basic studies, applied science, and ultimately materials production. However, even with the currently projected flight manifest, experimenters will have access to orbital low gravity with the shuttle only about seven days per month. To enhance the transfer/impact of microgravity on earth-based technology, NASA is projecting a Microgravity Sciences Laboratory as part of the space station, which will allow for continuous experimentation similar to that occurring in ground-based laboratories. With the station, the ability to modify equipment hardware in space will allow for the rapid *ad hoc* experimentation that characterizes the materials (and other high technology) industry on earth. A timely dissemination of the low gravity results to emerging technology/problems being studied on earth will build a wide realization of the benefits of microgravity and accelerate the use of space for studies and applications. Very futuristic studies discuss the possibility of fleets of free-flying space platforms and stations producing high-value high-technology items for use on earth.

REFERENCES

[1] R. J. Naumann and H. W. Herring, "Materials processing in space: Early experiments," NASA SP-443, 1980.

[2] E. Pentecost, "Materials processing in space, program tasks," NASA TM-82525, Apr. 1983.

[3] ——, "Materials processing in space, program tasks," NASA TM-82496, Sept. 1982.

[4] ——, "Materials processing in space bibliography 1983 revision," NASA TM-82507, Jan. 1983.

[5] ——, "Materials processing in space bibliography," NASA TM-82466, Mar. 1982.

[6] R. Bayuzick, Vanderbuilt University, private communication.

[7] M. C. Lee, J. M. Kendall, and W. L. Johnson, "Spheres of metallic glass $Au_{55}Pb_{22.5}Sb_{22.5}$ and their surface characteristics," *Appl. Phys. Lett.*, vol. 40, p. 383, Mar. 1982.

[8] G. E. Rindone, Ed., *Symp. Proc. 1982 Materials Res. Soc. Meeting, Materials Proc. Reduced Gravity Environment Space*, vol. 9, Boston, MA.

[9] See article by D. S. Larson and R. G. Pirich, "Influence of gravity driven convection on the directional solidification of Bi/MnBi eutectic composites," on p. 532 of [8].

[10] Letter sent by Dr. D. Johnson, Director, National Measurements Laboratory, National Bureau of Standards, DOC to Dr. B. I. Edelson, Associate Administrator for Space Science and Applications, NASA Headquarters.

[11] M. Bier, O. A. Palusinski, R. A. Mosher, and D. A. Saville, "Electrophoresis: Mathematical modeling and computer simila-tion," *Science*, vol. 219, Mar. 1983, p. 1281.

[12] See article on p. 27, *Aviation Week and Space Technology*, July 1982.

15

COMMERCIALIZATION OF MATERIALS PROCESSING IN SPACE

Charles F. Yost

INTRODUCTION

The Space Act directs NASA to conduct its activities to preserve U.S. leadership in aeronautical and space sciences, and in technology and its applications. One of the unique and new areas of research and technology which has emerged from the space effort is the processing of materials in an environment where the effective gravitational acceleration is very low (10^{-6}–10^{-2} g). Because of its origin in and close ties to the space program, this emerging field is commonly referred to as materials processing in space (MPS).

For short-duration investigations of material properties and process mechanisms (a few seconds to a few minutes), a low-*g* or microgravity environment can be achieved by allowing specimens to free fall in a drop tube or a drop tower, or by flying aircraft and rockets through carefully prescribed trajectories. A spacecraft in orbit about the earth may be viewed as being in a state of continual free-fall which provides a long-term low-*g* environment. Thus, with the advent of the space shuttle in 1981, a low-*g* environment can be sustained for several days at a time. By the end of the decade, proposed free-flying spacecraft, serviced by the shuttle, could be routinely available. These "free flyers" will provide a microgravity environment for long periods of time (months to years). Eventually, manned space stations may be designed for large-scale MPS operations with manned supervision for indefinite periods.

INFLUENCE OF GRAVITY ON PROCESS MECHANISMS

Research to date is providing new insights into the pervasive influence of gravity on materials properties and processes and is suggesting ways in which new materials or processes having possible commercial application can be produced. At least four different mechanisms affect materials properties and processes.

Convection

The elimination of gravity-driven convection in molten materials can preclude the sometimes undesirable stirring and mixing encountered during the growth of crystals, the casting or solidification of alloys and composites, chemical reactions, or the separation of biological materials.

Sedimentation and Buoyancy

The elimination of gravity-induced sedimentation and buoyancy can broaden the spectrum of alloys and composites that may be formed by permitting particles of vastly different density to remain in suspension until solidification occurs. Also, the elimination of sedimentation and buoyancy eliminates the need for mechanical stirring. This is important in instances where the stirring may be detrimental to the materials involved.

Gravity-Induced Deformations

Where hydrostatic pressure controls or limits a process, or the force of gravity (weight) causes deformation or fracture of a material, the elimination of gravity can provide opportunities for investigations of unique materials and manufacturing techniques.

Containerless Processing

The elimination of the necessity to confine liquids and molten materials within a container can open interesting possibilities. Materials, depending upon their electromagnetic characteristics and the influences of processing in a gaseous environment, may be melted, mixed, manipulated, shaped, and solidified in a low-gravity free suspension by use of acoustic, electromagnetic, or electrostatic fields. Surface tension will hold the materials together in a mass. Another form of containerless processing which can be enhanced by microgravity is the float zone process; since the molten zone will be confined by surface tension, much more latitude may be available for crystal production in space.

LAYING A SOUND TECHNOLOGICAL FOUNDATION

The scientific and technological benefits which can be derived from the exploitation of materials processing in low *g* are fundamental in nature and may result in significant improvements in materials utilization and producibility; the potential economic benefits may be both substantive and viable. However, utilization of this technology must be approached with deliberation and realism.

In early work, there was considerable emphasis on investigating materials processes in suborbital and orbital experiments that could rapidly lead to the production of commercially viable products in space. While a number of interesting results were obtained, it became clear that much more sophistication was required in process control and diagnostics, particularly with regard to the control and measurement of thermal gradients and quenching rates used in many of the processes. Sample preparation was found to be especially critical when it was necessary to control oxide formation or to completely homogenize a specimen. In containerless processing, much has been learned about the precise positioning and rotational control needed to prevent the sample from contacting the levitating device, as well as disruptive accelerations on and unwanted stirrings within the sample. Better methods for obtaining flow and temperature fields were found to be necessary in order to observe what is happening during a process. Present scientific and technological work sponsored by NASA is concentrating on identification of basic process mechanisms and gravitational influences on these mechanisms.

It is imperative to understand that MPS is an infant science and that a space-based industry deriving from MPS is non-existent. Prior to the present shuttle flights, the entire experimental base of MPS rested on the use of ground-based facilities such as drop tubes/drop tower, KC-135 and F104 aircraft, and SPAR rockets. During this time, approximately 85 experiments were conducted having an accumulative low-*g* time using these facilities of about 100 h. Nevertheless, the results of these experiments are significant and indicate a very promising future for commercial uses of space-based MPS technology.

GOALS OF MATERIALS PROCESSING IN SPACE

Materials processing in space has two goals: the opportunity now available for longer duration flights on the orbiter to further develop and expand the fundamental science of materials processing in the space environment is the first goal.

The second goal of MPS is a commercial one and is directly and intimately connected to the first goal. The commercial goal of NASA's MPS program is the encouragement of private sector establishment and operation of self-supporting space processing industries. The strategies for achieving this are to:

1) Create awareness of possible commercial opportunities to industry through publications, seminars, and visits to NASA Centers.

2) Encourage private sector participation in space processing by

- utilizing joint endeavor agreements, technical exchange agreements, and industrial gust investigator agreements;
- identifying barriers and disincentives to industrial exploitation of space and taking actions necessary to overcome them.

Experimental processes on which MPS science is based and from which commercial opportunities will be derived are shown in Table 15-1.

TABLE 15-1
MPS PROCESSES

	Furnace Processing	Containerless Processing	Fluids/Chemical Processing	Biological Processing
Research Processes	Isothermal gradient freeze Float zone Directional solidification Vapor crystal growth Electroepitaxy	Acoustic Electromagnetic Electrostatic		Electrophoresis Isoelectric focusing Isotachophoresis
Typical Experiments	Crystal growth from melt Crystal growth from solution Crystal growth from vapor Semiconductor crystals Immiscible alloys IR detectors Eutectics—monotectics Composite materials	Ultra-high temperature materials Glass processing Immiscible glasses Fusion targets	Large-scale chemical processing—polymeric processing Low-temperature crystal growth Measurement of thermodynamic phenomena High-volume biological purification and separation Measurement of transport phenomena Chemical deposition Catalysis Fluid dynamic studies	Separation of pharmaceutical materials Blood properties

INCENTIVES FOR COMMERCIALIZATION

To accelerate technological innovation based on materials processing in low gravity and to provide incentives for commercialization, NASA has initiated a program wherein it will share in the costs and risks of early investigations by commercial industry. Joint endeavor types of arrangements have evolved from NASA policy statements on the "Early Usage of Space for Industrial Purposes"[1] and "Guidelines Regarding Joint Endeavors with U.S. Domestic Concerns in Materials Processing in Space" which were published in the summer of 1979 (see Appendixes).

In these joint activities, NASA and interested, qualified commercial organizations enter into "constructive partnerships" as equals who have sufficient motivation toward common objectives to make independent commitments and to share in the risks and benefits. Activities are selected across the spectrum of materials processes to investigate the low-*g* environment as a valuable tool for isolating and characterizing gravitational effects on ground-based materials processes, or for actually producing unique materials in space for commercial application.

Since market incentives are presently inadequate to bring about technological innovation based on low-*g* technology, under its joint arrangements programs, NASA can provide certain tangible incentives, such as 1) providing flight time on the Space Transportation System (space shuttle) on appropriate terms and conditions; 2) providing technical advice, consultation, data, and use of facilities and equipment; and 3) entering into joint research and development programs where each party funds its own participation. In turn, NASA has the opportunity to have research done by the industry at no cost, gets research samples from the company's apparatus, and acquires data on the capabilities of the apparatus. Joint activities may range from exchanges of technical information and collaboration on low-*g* ground-based and flight experiments to joint projects (joint endeavors) to develop a marketable product.

Three kinds of joint arrangements have been developed: the technical exchange agreement, the industrial guest investigator agreement, and the joint endeavor agreement. Salient features of each follow.

Technical Exchange Agreements

- The Technical Exchange Agreement fosters exploration of low-*g* materials processes or properties for short periods of time on KC-135 or F104 flights and in drop towers or drop tubes. It is an agreement in which the proprietary rights of the company are protected.
- There is no exchange of funds between the participants.

[1] *It should be pointed out that the "Early Usage of Space" policy does not limit these arrangements to MPS.*

Industrial Guest Investigator Agreements

- The industrial guest investigator agreement is one in which a company scientist or engineer can work with a NASA scientist or engineer on a project of mutual interest to explore the effect of low *g* on materials properties or processes.
- There is no exchange of funds between the participants. Each of these agreements enables a company to explore the properties of a particular material or to investigate a process with a relatively small commitment of resources by the company or NASA. If, as a result of these investigations, a company decides that it needs longer duration low-*g* time than is available on the ground-based facilities, it can proceed to a joint endeavor agreement.

Joint Endeavor Agreements

- The Joint Endeavor Agreement is an agreement between equal participants—between NASA and an industry participant in a venture of mutual interest.
- All agreements are characterized by no-exchange-of-funds and a cooperative approach.
- The industrial participant will provide the hardware. If the hardware development is successful, then NASA will provide space on the STS system on a negotiated basis.
- At no point in the venture does the government allocate resources without the private sector participant doing likewise. Spending by the industrial participant equals or exceeds, and precedes, government spending in all arrangements discussed or finalized to date.
- Agreements are negotiated on a case-by-case basis and are tailored to the scientific needs of the venture and to the resources of the participants.
- All agreements exist for a specified length of time.
- If, during the period of the agreement, it is indicated that a commercial product or process will result from the venture, any succeeding flights on the STS system are fully paid for by the industry participants.
- The industrial partner has greater freedom of action than he would under a procurement contract.
- Risk-reward incentives are an inducement to front-end high risk investment by the industrial partner for the following reasons:

  a) promise of unique technological accomplishment with consequent opportunity to penetrate and develop a market;
  b) exclusive position (contractually defined and limited in "process and product") during term of JEA;
  c) proprietary rights of industrial participants are protected.

Commercial-Scale MPS Facilities

It is expected that MPS experiments on the space shuttle during the early and mid-1980's will help to identify those MPS applications with commercial potential. At present, however, it is still too early to predict exactly which space processes will be truly commercially viable. It is generally agreed, however, that pharmaceuticals and electronics materials are the likeliest candidates for commercial-scale space manufacturing. Industry estimates of the potential value of space-processed materials range as high as $5–$10 billion annually by the 1990's. This commercial space processing activity will require space production facilities exceeding the capabilities of the space shuttle, hence the expected need for free flyers or a space station with MPS applications by the end of this decade.

Unfortunately, since the exact products and products required for commercial MPS have not yet been identified, it is difficult to predict the precise nature and benefits of MPS activities on a free flyer or space station. A space station would appear to have many significant advantages for commercial-scale space processing, particularly with regard to power availability, production space, mission duration, and manned interaction, but our knowledge in this area is very slim. Commercial space processing could become a major space station activity by the early 1990's, but much work needs to be done to verify this assertion. Much more MPS research needs to be done in order to focus in on the processes which do in fact have strong commercial potential, and study of MPS opportunities on free flyers and space stations is needed in order to identify the optimum design for commercial-scale processing facilities.

APPENDIX I

NASA Guidelines Regarding Early Usage of Space for Industrial Purposes

NASA, by virtue of the National Aeronautics and Space Act of 1958, is directed to conduct its activities so as to contribute to the preservation of the role of the United States as a leader in aeronautical and space science and technology and their applications.

Since substantial portions of the U.S. technological base and motivation reside in the U.S. private sector, NASA will enter into transactions and take necessary and proper actions to achieve the objective of national technological superiority through joint action with United States domestic concerns. These transactions and actions will be undertaken in the context of stated NASA program objectives and after a determination by the Administrator. They may include, but are not limited to (1) engaging in joint arrangements with U.S. domestic concerns in research programs directed to the development of enhancement of U.S. commercial leadership utilizing the space environment; (2) conducting research programs having as an end objective the enhancement of U.S. capability by developing space-related high-risk or long-lead-time technology; and (3) by entering into transactions with U.S. concerns designed to encourage the commercial availability of products of NASA space flight systems.

NASA incentives for these purposes may include, in addition to making available the results of NASA research, (1) providing flight time on the space transportation system on appropriate terms and conditions as determined by the Administrator; (2) providing technical advice, consultation, data, equipment and facilities to participating organizations; and (3) entering into joint research and demonstration programs where each party funds its own participation.

In making the necessary determination to proceed under this policy, the Administrator will consider the need for NASA funded support to commercial endeavors and the relative benefits to be obtained from such endeavors.

As a major areas for NASA enhancement of total U.S. capability, including the private sector, may become apparent from time to time, the factors to be considered by NASA prior to providing incentives may include, but not be limited to, some or all of the following considerations: (1) the public or social need for the expected technology development; (2) the contribution to be made to the maintenance of U.S. technological superiority; (3) possible benefits accruing to the public or the U.S. Government from sharing in results; (4) the enhanced economic exploitation of NASA capabilities such as the space transportation system; (5) the desirability of private sector involvement in NASA programs; (6) the merit of the research, development or application proposed; (7) the degree of risk and financial participation by the commercial concern; (8) the amount of proprietary data or background information to be furnished by the concern; (9) the rights in data to be granted the concern in consideration of its contribution; (10) the ability of the concern to project a potential market; (11) the willingness and ability of the concern to market and sell any resulting new or enhanced products on a reasonable basis; (12) the impact of NASA sponsorship on a given industry; (13) provision for a form of exclusivity in special cases when needed to promote innovation; (14) recoupment of the NASA contribution under appropriate circumstances; and (15) support of socioeconomic objectives of the Government.

ROBERT A. FROSCH
Administrator
June 25, 1979

APPENDIX II

Guidelines Regarding Joint Endeavors with U. S. Domestic Concerns in Materials Processing in Space

Background

NASA, by virtue of the National Aeronautics and Space Act of 1958, is directed to conduct its activities so as to contribute to the preservation of the role of the United States as a leader in aeronautical and space

science and technology, and their applications. In furtherance of these objectives, the Administrator of NASA on June 25, 1979, promulgated a statement of *NASA Guidelines Regarding Early Usage of Space for Industrial Purposes.* These guidelines recognize that "since substantial portions of the U. S. technological base and motivation reside in the U. S. private sector, NASA will enter into transactions and take necessary and proper actions to achieve the objective of national technological superiority through joint action with United States domestic concerns."

Materials Processing in Space (MPS) is an emerging technology which can potentially provide public benefits through applications in the private sector. However, in the foreseeable future, normal market incentives appear to be inadequate to bring about technological innovation in the private sector based on this technology. Therefore, in accordance with the above referenced Guidelines, NASA contemplates entering into joint endeavors with U. S. industrial concerns. Through these joint endeavors, NASA seeks, within the context of the MPS program objectives, to broaden the base of understanding of MPS technology, particularly with regard to its usefulness in the private sector where economic benefits may result. Present MPS program objectives are a) to understand the pervasive role of gravity in materials processing; b) to develop and demonstrate enhanced control of materials processes in weightless environment; c) to explore the unique nature of space vacuum for materials processing; and d) to foster commercial applications of MPS technology.

Nature of the Joint Endeavor

Joint endeavors in MPS will generally be for the purpose of 1) engaging in research programs directed to the development and/or enhancement of U. S. commercial leadership in the field of materials processing in space, and 2) encouraging commercial applications of MPS technology. Joint endeavors may cover ground-based research to create a sound scientific basis for investigations in space; the investigation of materials properties or phenomena and process technology in the unique environment of space; the making in space of exemplary materials to serve as a point of reference for ground-based materials and processes; and the application investigations and feasibility demonstrations of space-made or space-derived materials and processes.

In joint endeavors, NASA and the industrial concern share in the cost and risks of the endeavor. Terms and conditions, including the business arrangements, are negotiable within the limits of prevailing statutes and regulations and will be commensurate with the risks, involvement and investment of all the parties. NASA's intent is to offer as much latitude as practical in joint endeavor arrangements. Due to the experimental nature of the program, both technically and institutionally, each endeavor will be negotiated on a case-by-case basis. Endeavors are expected to vary in size, complexity, and arrangements to achieve diversity in the program. The number and/or size of the joint endeavors undertaken will depend upon the nature of the proposals received and resource availability. All joint endeavors will be subject to availability of appropriated funds, as well as NASA procedures regarding flight safety and verification.

NASA Provided Incentives

NASA incentives for these purposes may include, in addition to making available the results of NASA research, (1) providing flight time on the space transportation system on appropriate terms and conditions as determined by the Administrator; (2) providing technical advice, consultation, data, equipment and facilities to participating organizations; and (3) entering into joint research and demonstration programs where each party funds its own participation.

Factors to be Considered in Establishing Endeavors

To qualify for joint sponsorship, the offeror must be engaged in business in the U. S. in such a manner that any promising results from the endeavor will contribute principally to the U.S. technological position; the proposed joint endeavor must comport with one or more of the MPS program objectives as stated above; and the technical uncertainties and risk involved must be significant enough to warrant the government's participation.

The factors to be considered by NASA prior to providing incentives may include, but not be limited to, some or all of the following considerations: (1) the public or social need for the expected technology development; (2) the contribution to be made to the maintenance of U.S. technological superiority; (3) possible benefits accruing to the public or the U. S. Government from sharing in results; (4) the enhanced economic exploitation of NASA capabilities such as the space transportation system; (5) the desirability of private sector involvement in NASA programs; (6) the merit of the research, development or application proposed; (7) the degree of risk and financial participation by the commercial concern; (8) the amount of proprietary data or background information to be furnished by the concern; (9) the rights in data to be granted the concern in consideration of its contribution; (10) the ability of the concern to project a potential market; (11) the willingness and ability of the concern to market and sell any resulting new or enhanced products on a reasonable basis; (12) the impact of NASA sponsorship on a given industry; (13) provision for a form of process exclusivity in special cases when needed to promote innovation; (14) recoupment of the NASA contribution under appropriate circumstances; and, (15) support of socioeconomic objectives of the Government.

Administration

The Associate Administrator, Space and Terrestrial Applications, is delegated the authority to enter into negotiations and to approve MPS joint endeavors on behalf of the Agency. Before proceeding into comprehensive evaluation of a joint endeavor, a preliminary assessment will be made of the merits of the offer. (Joint endeavor offers which are too sketchy or ill-defined to establish that the basic idea contained in the offer has merit, is in accord with MPS program objectives, or that the organization is willing to make significant contribution to the endeavor, will not be evaluated in depth and will be handled as correspondence or advertising.) This preliminary assessment will be reviewed by the Associate Administrator, Space and Terrestrial Applications, or his designee, to determine if the proposed endeavor warrants further consideration from NASA's standpoint. If this determination is positive, further evaluation will be made. After such evaluation and discussions with the offeror, if the parties mutually agree to proceed with a joint endeavor, designated representatives of NASA will enter into detailed discussions and negotiations with the offeror regarding the technical and business aspects of the offer in an effort to consummate a mutually satisfactory joint endeavor agreement. Management of the MPS joint endeavor program will be carried out by the Division of Materials Processing in Space of the Office of Space and Terrestrial Applications.

Due to resource limitations and necessity for diversity in the program, normally only one offer will be accepted to apply a particular materials process in a given technical area. If substantially similar offers are received within any 45-day period, they will be evaluated and negotiated together. The one which provides the best total consideration for the Government will be accepted. Special consideration shall be given to small and minority businesses as appropriate.

ROBERT A. FROSCH
Administrator
August 3, 1979

PART IV: COMMUNICATIONS AND NAVIGATION

16

COMMUNICATIONS AND NAVIGATION: AN OVERVIEW

Robert R. Lovell

There can no longer be any doubt that the applications satellites, particularly communications and navigational satellites, have revolutionized the world in which we live. Today, two out of three overseas telephone calls travel via satellite, as do virtually all live overseas television transmissions. Each year tens of millions of airline reservations and trillions of dollars in electronic funds transfer are transmitted over the world's communications satellite systems. Navigational satellites keep ever-growing fleets of ships and jetliners operating at their maximum efficiency, with safe and direct routing patterns. Search and rescue operations are now directed from outer space and soon personal paging messages will become the newest satellite-delivered service.

It is increasingly true that the burgeoning number of applications satellites in geosynchronous orbit, as well as lower orbits, are becoming the world's electronic nervous system, which tie together a global economy and a global culture. Morris Desmond, the "pop" anthropologist, has said that the communications satellite is the greatest force for "supertribalization" that man has yet invented. John Naisbitt, in his book *Megatrends,* observes that Marshall McLuhan thought television would lead to the creation of the electronic global village, but Naisbitt argues that we now know the enabling technology was not television, but rather the communications satellite.

It is perhaps one of the more significant indexes of the success of applications satellites that to talk about them coherently we now are forced to be increasingly specific about their type, their operational frequencies, and the various services they perform.

Literally hundreds of communications and navigational satellites now provide more than a dozen different types of important services. There are experimental communications satellites; military communications satellites; search and rescue satellites; navigational satellites; and communications satellites that provide fixed-satellite services, mobile satellite services, and direct broadcast satellite services. There are also international and regional communications satellites; domestic communications satellites; data relay satellites; remote monitoring and relay satellites, and we will soon have operational space-to-space intersatellite services. In terms of use of different electromagnetic frequencies, we will soon be using the entire range, from high frequency (HF) to laser communications, in order to achieve earth-to-space, space-to-earth, and space-to-space communications.

Certainly, the many services that communications and navigational satellites provide have diversified. There has been a parallel diversification of the transmission and modulation techniques utilized and a growing number of types and sizes of satellites. These range from very small, low-mass satellites (like the OSCAR Amateur Radio Satellite or the new Hughes 399 communications satellite, which use only about 1/17th of a shuttle bay), up to the multiton Intelsat VI communications satellites, which require the full shuttle bay and a powerful perigee engine. We see satellites which are stabilized by body spinning, by three-axis body stabilization (using momentum and inertia wheels), as well as gravity-gradient beam-stabilized spacecraft. In the future we may likely see such new phenomena as clusters of satellites operating around a single nodal point in a so-called "halo" orbit, in geosynchronous orbit, as well as very large "condominium" applications satellite structures in the form of unmanned, serviceable, and replenishable space platforms with surface areas the size of football field; this topic will be discussed in more detail later.

The demands for multiple reuse of frequencies, the need to utilize ever-higher frequencies, and the burgeoning demand for ever-increasing wide-band services are pushing the communications satellite toward tomorrow with ever-increasing velocity. The new communications service requirements are not only exciting, but are also quite demanding on the transmission technologies. High-definition and multiple-rastered television, videophone, videoconferencing, and extremely high volumes of electronic mail, will all serve to stimulate very high-capacity, high-efficiency multiple reuse spacecraft in the coming decades.

The extent to which communications satellites have evolved in the last 20 years is nothing short of phenomenal. Prior to the advent of international communications satellites, there was a worldwide total of perhaps 300 transoceanic circuits, most of which were provided by low-capacity submarine cables (36 to 72 voice circuits in size), and the remainder of which were provided by unreliable troposcatter and HF radiotelephone circuits. Today, internationally available telecommunications circuits have increased by a factor of more that 200 times, to the equivalent of over 60 000 international telephone circuits (including international and regional TV relays, global and regional satellite systems, and submarine cable systems).

If the same trend line were to continue over the next 20 years, we would see in the international field alone a demand for the equivalent of over 12 million telephone circuits. If one further extrapolates the current demand for not only international services, but also for the even more rapid growth of communications and direct broadcasting satellite services at the domestic level, one would be forced to project future satellite requirements in excess of hundreds of millions of equivalent telephone circuits by the early twenty-first century!

In the chapters to follow, we will explore some of the many services and service organizations that have developed with this new technology. We will also look at the communications and navigation spacecraft development, which will give some insight into the ambitious goals, entrepreneurial spirit, and dedication of the engineers behind this new technology. One chapter is dedicated to a discussion of the enabling component technologies common to both navigation and communication satellites. Finally, we will look into the future, where emerging technologies present additional exciting opportunities in areas such as space power generation and relay, material production, and manned research facilities, in addition to continued growth of communication and navigation services.

The appropriate application of space technology continues to raise uncertainties. Realizing the potential for the public good is not automatically assured by the mere presence of technology. Space technology will inevitably arrive, but its ultimate utility, adoption, and effectiveness in a global society will be determined by economic, political, and institutional factors.

On a positive note, in the rapid evolution of satellite technology the lines between future concepts and operational programs appear to converge more rapidly each year. As far as satellite applications are concerned, today the sky is no longer the limit; tomorrow's prospects and challenges, stretching the limits, invite even more exciting opportunities.

The future of satellite communications is closely allied with a worldwide demand for information and communications. Only two decades ago, international communications was difficult, relying on HF radio and low-capacity submarine cables. Within two decades, the advent of the communication satellite—largely through the energy and imagination of Intelsat—brought telephone and television interconnection to not only the 109 member nations, but actually interconnected more than 150 nations. This must be considered a historic feat in a world which had long been a scene of isolation. During this period, several synergistic technologies were born and flourished; these included the development of the computer, the development integrated circuit leading to very large scale integration (VLSI), the development of television on a global scale with the ability to access events instantly, and the development of lasers and fiber-optic communication systems. The net result of these technologies, along with space sciences, have changed how the world functions and relates internally and internationally. Thus, it can be expected that the world of the future will see a "marriage" of these space and terrestrial technologies with lasers in space, interconnected with fiber-optic lines on earth and using computers and electronic circuits, to bring communications, education, information, and entertainment to all parts of the earth in an affordable manner.

17

COMMUNICATION SATELLITE APPLICATIONS

*Joseph N. Pelton**

Since the early 1960's, satellite applications have experienced phenomenal, perhaps even revolutionary, growth and expansion. Let's explore that development in several key sectors: international and regional satellite communications, domestic satellite communications, international and national direct-broadcast television services, international and domestic mobile communications services, tracking and data relay satellite services, search and rescue satellite services, and navigational satellite services.

INTERNATIONAL AND REGIONAL SATELLITE SERVICES

Intelsat

The International Telecommunications Satellite Organization (Intelsat) currently owns and operates a 15-satellite system that provides approximately two-thirds of the world's overseas telecommunications services and most overseas television broadcasts. It has 110 member countries and provides services to some 170 different countries, territories, and areas of independent sovereignty through almost 1700 international pathways. Intelsat's services include television, broadcast-quality radio, telephony, high-speed facsimile, and data communications services. Intelsat envisions rapid future growth in the next ten years in the field of international television networks (including high-definition television), lower power direct broadcast services, electronic mail, videoconferencing, and videophone services.

Intelsat earth stations range in size from the 30-m Standard A terminal operating in the *C*-band (that costs several million dollars to construct), down to the newly approved Standard D-1 terminal that is utilized with the Vista low-density telephony service to provide communications in rural and remote parts of the world at an estimated cost of approximately $125 000 per terminal. Intelsat has recently introduced a new microterminal Intelnet service for data distribution and networking that will operate to terminals down to 60 cm (2 ft) in diameter. These terminals (which cost approximately $2500) are capable of receiving data rates at 9.6 kbits/s and operate through the use of the spread-spectrum transmission mode. Intelligent microprocessors capable of decoding digitally encoded signals make possible operation with these very small terminals in a very high-noise environment.

The Intelsat system, at least today, is heavily dependent upon analog transmission techniques for a large percentage of its traffic—over 90 percent. It is anticipated, however, with the introduction of time division multiple access (TDMA) techniques during 1985–87 and, subsequently, 32 and 16 kbit/s voice processing, that today's analog FM traffic will be largely converted to digital transmission techniques and that, by the early-1990's, some 90 percent of all traffic on the Intelsat system (whether it be telephone, data, facsimile, videophone, videoconferencing of broadcast quality television, or even high-definition television) will be digital.

In addition to its international fixed satellite services and mobile maritime services, Intelsat is also making available space segment capacity under long-term leases for domestic services. As of the year end of 1985 there were nearly 30 countries utilizing over 40 transponders for domestic services, and the number of countries leasing capacity from Intelsat for domestic services is expected to increase to approximately 50 by the end of the 1980's (see the Appendix at the end of this chapter). Intelsat also provides dedicated (24-hour-a-day) international TV leases, and the number of such TV networks is expected to increase from 10 to approximately 40 by the end of the 1980's.

At the end of 1985, there were approximately 900 earth stations operating in the Intelsat network, but with the introduction of the Intelsat Business Service, the Vista low-density telephony service, and the microterminal Intelnet services, the number of earth stations is expected to experience an impulse-level increase in the next few years. This should dramatically increase the number of earth station to earth station pathways possible on a global basis.

The Intelsat system of the 1980's will consist of a combination of Intelsat V, V-A, V-B, and VI satellites. The Intelsat V, V-A and V-B satellites are three-axis body-stabilized spacecraft that operate in the *C*- and *K*-bands, with some especially equipped Intelsat V satellites operating in *L*-band. The Intelsat V satellites have a capacity of 12 000 high-quality telephone circuits, plus two color TV channels. The Intelsat V-A and V-B have a capacity of 15 000 telephone circuits plus TV, while the Intelsat VI satellite, to be launched in 1986, will have a capacity of approximately 40 000 telephone circuits plus two TV channels. In terms of data transmission capability, this satellite will be able to send the equivalent of the

* *The views expressed in this article are those of the author and do not necessarily reflect the official views of the Intelsat Organization.*

Encyclopedia Britannica some 20 times a minute! This satellite, which is the largest commercial communications satellite yet to be launched, will be a despun antenna stabilization design. (See Fig. 18-11.)

Intersputnik

The Intersputnik organization was first proposed by the Soviet Union on the occasion of the first U.N. Conference on the Peaceful Uses of Outer Space (held in Vienna, Austria in August 1968). In November 1971, nine countries signed the agreement, establishing the "Intersputnik" International System and Organization of Space Communications.

The original nine countries (Bulgaria, Cuba, Czechoslovakia, German Democratic Republic, Hungary, Mongolia, Poland, Rumania, and the U.S.S.R.), have now been joined by five other countries (Afghanistan, Laos, Syria, Vietnam, and South Yemen). A number of other countries, however, are planning or have established earth stations either for telephone or data communications or, more typically, for television reception. These countries, beyond the 14 Intersputnik signatories, include Algeria, Angola, Burma, India, Iraq, Korea (Democratic Republic of), Libya, Madagascar, Mozambique, Nicaragua, and Sri Lanka.

Some 20 earth stations now access Intersputnik capacity on leased Gorizont satellites, owned and operated by the Soviet Union. Originally, the Intersputnik organization operated on the Molnya 2 series of satellites. In 1977, the utilization of the non-geosynchronous Molnya 2 satellites was considered too expensive and the series was discontinued. The Molnya 2 satellites were replaced in the late-1970's by the geostationary Gorizont satellites in the Atlantic (140° W) and the Indian Ocean (33° E) regions. Gorizont satellites (weighing 2120 kg) are launched by the "Proton" launcher, from the Tyuratam-Baykonur range.

Based on three-axis stabilized platforms, the Gorizont satellites contain one 40-W transponder and five 15-W transponders, and use the 6/4 GHz C-band frequencies to deliver a variety of services. The satellites are employed to distribute TV programs to the 2.5 m earth stations of the Moskva TVRO system. The 40-W transponder is used for this particular task. The 15-W transponders are used to transmit voice and data to the 25-M and 12-M antennas of the Orbita stations located throughout the Soviet Union and in participating Intersputnik countries.

On the Gorizont-Statsionar 5 satellite, two 15-W transponders are leased to Intersputnik—one for radio-TV transmissions and one for phone-telex links. On the Gorizont-Statsionar 4 satellite, four transponders are available, but only three of them are leased—two for ratio-TV exchanges and one for phone-telex communications.

About five hours a day of TV are distributed on behalf of the Organization Internationale de Radiodiffusion et Television (OIRT), the East European television network that operates Intervision. The Intersputnik TV transmissions represent about 40 percent of TV transmissions to OIRT member countries. Experimental tests by Intersputnik have been planned with regard to the new U.S.S.R. Loutch satellite that operates in the K_u band (14/12 GHz) as part of the Intercosmos program. Further service development plans for the Intersputnik system, as presented in a recent official report, include the following:

- "increasing the size of service areas;
- increasing channel capacities through greater earth station reliability and performance and the installation of new equipment;
- transmitting new and more diverse types of information;
- installing less costly earth stations with smaller antennas;
- increasing earth station reliability;
- beginning of 11/14 GHz operation based on the results of work planned under the 'INTERCOSMOS' program."

Eutelsat

Eutelsat, initially established as Interim Eutelsat in May 1977, is an international organization headquartered in Paris, that was established to provide telecommunications services for Europe, as well as certain TV broadcast services for Europe and the Mediterranean basin countries. The Eutelsat system will offer a full range of telecommunications services, including telephone, data, and TV distribution services, through a combination of European Communications Satellites (ECS), designed and built by the European Space Agency (ESA), as well as through leased capacity obtained from the French Telecom satellites. The operational program has grown out of an experimental program of tests and demonstrations carried out with the Orbital Test Satellite (OTS) in the late-1970's and early 1980's.

Table 17-1 shows the member countries of Interim Eutelsat and their initial financial shares in the ECS space segment.

The first launch of an ECS satellite was successfully completed on an Ariane launch vehicle in June 1983 with two further launches taking place in 1984 and 1985. Eutelsat utilizes orbital slots at 6.5° W, 10° W, and 13° W to carry traffic in the K_u frequency band (14/12 GHz) with dual polarization for frequency reuse. Nine 80 MHz transponders (of the 14 available on-board the 1100 kg satellite) are used on the operational satellites. It is anticipated that the telephone traffic on the Eutelsat system will grow from approximately 2000 telephone circuits today to some 9500 circuits in 1992.

Eutelsat has also leased capacity from the French Telecom 1 satellite to provide intra-European services. Eutelsat has found that the greatest demand for satellite services is for television distribution which represents

TABLE 17-1

Country	ECS Share (%)
Austria	1.97
Belgium	4.92
Cyprus	0.97
Denmark	3.28
Finland	2.73
France	16.40
Germany (Fed. Rep.)	10.82
Greece	3.19
Ireland	0.22
Italy	11.48
Luxembourg	0.22
Netherlands	5.47
Norway	2.51
Portugal	3.06
Spain	4.64
Sweden	5.47
Switzerland	4.36
Turkey	0.93
United Kingdom	16.40
Yugoslavia	0.96
	100.00

the majority of all traffic carried. Like Intelsat, Eutelsat has introduced a customer-premise type of business communications in the 14/12 GHz band on a modified Eutelsat satellite. This service, known as a multiservice satellite system (SMS), is expected to grow from 30 Mbits/s digital traffic to more than 100 Mbits/s in the late-1980's. SCPC data channels of 64 kbits/s to 1.92 Mbits/s will be offered. The Eutelsat network will be known as Euronet and will include 3.5-m terminals for analog-based FDMA service, as well as 5-m TDMA terminals for digital services. The TDMA system will be a 120 Mbit/s, four-phase phase shift keyed (PSK) network, operating through 80 MHz transponders.

Participating countries include Austria, Belgium, Cyprus, Denmark, the Federal Republic of Germany, Finland, France, Greece, Ireland, Italy, Luxembourg, The Netherlands, Norway, Portugal, Spain, Sweden, Switzerland, Turkey, the United Kingdom, and Yugoslavia. Lease of capacity for domestic and/or regional television represents an ever growing portion of the Eutelsat revenues. TV distribution throughout the European Broadcasting Zone is also anticipated, with TV reception planned by the following Mediterranean basin countries: Algeria, Iraq, Iceland, Israel, Jordan, Lebanon, Libya, Liechtenstein, Malta, Morocco, Monaco, San Marino, Syria, Tunisia, and the Vatican City.

Arabsat

All of the Arab states within the Arab league concluded a formal arrangement among themselves in April 1976 to form the Arab Satellite communications Organization, with the purpose of implementing an Arab Communications Satellite System (Arabsat). Membership in the Arabsat system includes Algeria, Egypt, Iraq, Jordan, Kuwait, Lebanon, Libya, Morocco, Mauritania, Oman, Qatar, Saudi Arabia, Sudan, Syria, Tunisia, the United Arab Emirates, and the United Arab Republic.

The Arab satellite system will use a combination of 6- and 11-m earth stations operating in the *C*-band, as well as smaller (3-m diameter) antennas to operate in the *S*-band frequencies (2.5 GHz) for community antenna TV educational services. Both the operational and spare satellite (at 19° E and 26° E) were developed in 1984 and 1985 to provide a range of telephone, data, and television services on a regional basis, as well as for domestic services to countries participating in the Arabsat system. The initial satellite launch was only partially successful due to station-keeping problems. Accordingly the second Arabsat satellite launched in June 1985 was used as the operational satellite.

The Asean Regional Satellite System (Palapa)

The Indonesian government established a domestic satellite system, known as Palapa, to provide domestic services to a network of some 40 earth stations for this large nation, consisting of thousands of different islands. Subsequent to the establishment of this domestic network, it was decided to offer some of the spare telecommunications capacity existing on the system to neighboring countries in Southeast Asia which are members of the Asean group of countries. This offer included not only domestic services, but also certain types of regional telecommunications services (namely, for regional television and radio broadcasts, as well as for telecommunications services that connected rural and isolated portions of the Asean countries which were at the time not served by earth stations connected in the Intelsat system). This network, as of 1983, has been expanded with the launching of Palapa-B satellites (about twice the capacity of the initial Palapa-A satellites) at orbital positions of 83° E and 77° E for the operating and spare satellites, respectively. All of the Asean countries (namely, Indonesia, Malaysia, the Philippines, Singapore, and Thailand) are using the Palapa satellites for TV distribution. The Philippines, Malaysia, and Thailand are using several transponders for certain domestic telecommunications purposes as well. In the case of Thailand, domestic services are being derived from both Intelsat and Palapa satellites. Unlike the Eutelsat and Arabsat Systems, however, the regional telecommunications services being carried on Palapa are fairly limited in scope and involve significantly less traffic than the other two regional systems. The Palapa system operates exclusively in the *C*-band.

Other Regional Satellite Systems

In addition to these five international and regional satellite systems that are now operational or soon will be (Intelsat, Intersputnik, Eutelsat, Arabsat, and Palapa), a number of other regional systems are at various stages of planning or discussion. These include: (1) the Luxsat (Coronet) satellite system, that is intended to provide TV and perhaps other telecommunications services within Europe; (2) the European Business Satellite System sponsored by Scandia and several other Scandinavian corporations to provide business services in Europe; (3) the Afsat satellite system (studies sponsored by France, the U.K., and Italy, with financing potentially to be provided by the European Development Fund); (4) the AMS system, which was filed on behalf of the Israeli government, but which is essentially a multinational commercial enterprise designed potentially to provide service coverage to both Africa and Europe; and (5) various proposals concerning transpacific or transatlantic satellite systems. These include the proposals by Orion, International Satellite Inc., Cygnus, RCA, Pananasat, and Financesat concerning private systems for the North Atlantic; and proposals by Poufer Satellite Corporation, and Financesat in the Pacific Ocean region.

In addition, there are two regional systems to be provided through the Intelsat system. First, there is the already operational United Nations' Peacekeeping and Emergency Communications Services System that is uplinked from New York and Geneva on the Intelsat satellite system to provide peacekeeping and emergency communications services. This system was inaugurated in late 1984. The second system is to support the ASETA (the Andean Educational Satellite Telecommunications Agency) study of the possible lease of capacity for regional educational and other telecommunications services. This system, known as Project Condor, will be initiated through the base of capacity from Intelsat and then perhaps transferred to a dedicated satellite in the 1990's.

Finally, as the institutional framework governing space communications evolves during the 1980's, it is possible that a number of other satellite systems may come into being.

DOMESTIC COMMUNICATIONS SATELLITE SERVICES

If international and regional satellite systems have grown explosively, then domestic satellite systems' truly explosive growth can only be considered as a true phenomenon, akin to the videogame and microcomputer explosion of the late-1970's and early-1980's. The world's first domestic communications satellite system was that of the Soviet Union. It was known as Molnya and used three satellites in nearly polar 12-h orbits to achieve continuous 24-h coverages by exploiting to advantage the very northern latitude of the Soviet Union.

The first synchronous domestic communications satellite system was that of Canada, which became operational in 1973—some eight years later. The first satellite of the Telesat Canadian domestic system, known as ANIK-1, was launched in November 1972 and became operational the next year. This satellite, with 12 transponders, was launched on a Delta launch vehicle, with each transponder providing one television channel and operating into a network of earth stations that ranged in size from 8 to 30 m. This *C*-band ANIK satellite in many ways served as a prototype for an avalanche of communications satellites that was to follow. As can be seen in the Appendix, the number of dedicated satellite systems planned or in operation now exceeds 50 and, of these, nearly half are domestic fixed-satellites and direct-broadcast satellites for the United States.

Over time the domestic satellites became more sophisticated, in terms of migrating to higher frequencies, of developing multiple frequency reuse in the *C*-band, and of moving from body-spinning satellites to three-axis body-stabilization. The explosive growth of domestic communications in the United States and demand for higher-powered satellites in the U.S. market has led to the evolution of the typical satellite today as having 24 transponders, as opposed to 12, and with much higher power. The launch options have also increased. Most 34-transponder domestic satellites are launched on the uprated 3900 series Delta launch vehicle (that is being phased out of regular NASA service), or by the two new entrants into this field, as represented by the Ariane launch vehicle or the shuttle, with perigee engine, to boost the satellite to geosynchronous orbit. In the future, private launch vehicles from the U.S. or Europe may be available, as well as government-backed services from Japan, the Soviet Union, or even China or India. U.S. manufacturers, such as Hughes Aircraft Company, Ford Aerospace and Communications Corporation, RCA, TRW, and General Electric in particular, have to date captured the predominant share of domestic satellite systems launched in the U.S. and abroad. A number of manufacturers in Canada, Europe, and Japan, however, have showed increasing design skills and manufacturing capabilities in this area. These include NEC, Mitsubishi, and Toshiba in Japan; SPAR in Canada; and MATRA, Thompson CSF, British Aerospatiale (SNIAS), Selenia, ECTA, SAAB, and MBB in Europe.

In Europe, teams of aerospace companies have combined on major satellite programs to form consortia. These consortia, like MESH, CIFAS, and Eurosatellite, have proven to be an effective way of achieving the necessary integration of facilities, expertise, and "intellectual property" needed to compete in the international aerospace marketplace.

During the 1980's and 1990's, the continued very rapid growth of demand for conventional telecommunications services (particularly direct broadcast satellites), will undoubtedly push the field toward the use of higher and higher frequencies. Already, AT&T has conducted experiments with the 20/30 GHz band on its Telstar satellite, while American Satellite Corporation has formally filed to launch a satellite to operate in this frequency band in the mid-1980's. The joint NASA/RCA Advanced Communications Technology satellite (ACTS) project will not only operate in the 20/30 GHz band, but will also use on-board signal processing and regeneration, "electronic hopping" beams, and ground-commanded increases in dwell time and power to cope with the problems of precipitation attenuation in these very high-frequency ranges. The European Space Agency satellite Olympus, which features high-power spot beams and 20/30 GHz frequencies, represents the European attempt to stay competitive in this field.

There are also serious studies under way not only at NASA, but also at the European Space Agency (ESA) and within the Japanese government (Ministry of Science and Technology; NASDA), to explore the feasibility of the development of multipurpose space platforms that could operate at truly tremendous communications transmission rates. Although the explosive growth of dedicated communications satellite systems has been essentially focused in large industrialized countries with significant financial resources, such as Japan, Canada, the United States, the Soviet Union, France, Germany, and the United Kingdom, there have also been parallel developments in certain "industrializing" countries. A few large and relatively sophisticated "industrializing" countries (such as Indonesia, India, Brazil, and Mexico) already have their own systems. In the next few years other developing countries, such as the Republic of Korea, Nigeria, Saudi Arabia, Iran, and Pakistan, may also eventually proceed to provide dedicated satellite systems for their internal telecommunications needs.

More significantly, perhaps, is the much broader-based phenomenon of developing countries that have decided, in light of their limited financial resources, to lease capacity for domestic services. These countries have felt the need for an evolutionary strategy to foster telecommunications development which could be more readily accomplished by leasing communications satellite services from existing satellite systems. In this respect, the typical approach has been to lease capacity from the Intelsat system. Indeed, 25 countries are now leasing such capacity from Intelsat and this number is currently anticipated to double by the end of the 1980's or the early 1990's. In addition, the Philippines and Thailand have leased capacity from the Indonesian Palapa system. Indeed, some developed countries have also found that leasing capacity from Intelsat also makes sense. This has included, for instance, the lease by France, Spain, and Portugal of capacity to reach their off-shore territories. The United Kingdom and the Federal Republic of Germany have also leased higher powered spot beam transponders from Intelsat for services within Europe and, in particular, to obtain low-powered TV broadcase services.

This phenomenal growth of both domestic and international telecommunications satellite services has, of course, given rise to certain technical problems related to the efficient use of the geostationary orbital arc. In the United States, the FCC has moved to allocate frequencies and grant permission to seek orbital filings with the ITU for orbital spacing at locations that are closer and closer together. In the 1970's an orbital spacing of 5° was considered normal. This eventually shrunk to 3° and is now in the process of shrinking to 2° spacing. In a physical sense, communications satellites are now being located in geosynchronous orbit, some 770 mi (1130 km) apart.

DIRECT BROADCAST SATELLITE SYSTEMS

For many years there has been a clear and reasonably well understood distinction between telecommunications and broadcasting services. Telecommunications essentially have been two-way, interactive communications, while broadcasting has been a one-way, single-point, and multipoint service, without any interactive capability. Instead of involving two or a very limited group of people with active participation, broadcasting has traditionally involved a very large number of people acting as an audience. Just as the distinction between telephones, telecommunications, and computer services has been blurred in recent years, there has been a similar blurring between telecommunications and broadcasting services. New innovations such as videotext and teletext services and packet-switched data-broadcast networks have increasingly served to make the old definitions obsolete.

For many years there has been a clear distinction, at least in terms of ITU definitions, between direct broadcast satellite services (known as broadcast satellite services, or BSS), and fixed satellite services, or FSS. BSS frequencies were intended for point-to-multipoint TV and radio direct-to-the-home broadcasting and envisioned the use of very high-powered satellites, as well as frequencies in the higher ranges (i.e., K_u-band and above). In contrast, "conventional" telecommunications satellites or fixed-satellite services (FSS) have used lower power satellites and larger, more sophisticated antennas. Recently, however, these formal distinctions have not necessarily been observed in practice. Particularly in the United States, Canada, and the Soviet Union, head-start TV "distribution" services have begun using quite small antennas, and in many ways have started to resemble direct broadcast services, only they have been using FSS frequencies. This phenomenon has led to concern about premature saturation of the FSS frequencies in the *C*- and K_u-bands. Potential investors in true high-powered direct-broadcast services in the BSS frequencies allocated for these applications have been worried about these lower-cost planned investments—a concern that has grown as the planned investments in these systems have grown to billions of dollars.

The development of low-cost, high-performance communications satellite terminals that can receive from FSS satellites a "good quality" TV signal has created considerable turmoil in the DBS industry. It has, in particular, led to the demise of the Unisat DBS project in the United Kingdom and several such systems in the U.S., perhaps most notably the Comsat Satellite Television Corporation DBS system. It was at one time thought that to build a TV receive-only antenna for direct-to-the-home reception and to keep the antenna under $500 in cost would require, in the K_u-band, a down-link signal of at least 60 dB · W. Continuing breakthroughs in the antenna field, however, seem to indicate that receivers in this price range, through the use of solid-state technology, can receive acceptable TV signals from satellites with EIRP's for lower levels in the range of 42 to 47 dB · W. A recent article suggests that a figure of merit for the annual capital cost of an FSS transponder is about $350000, while the annual cost of a BSS transponder is about $3 million.

Direct broadcast satellite services, particularly in Europe, the United States, and Canada, are largely in competition with cable TV systems. From the economic perspective, if one can build a low-power and low-cost broadcast TV satellite system which provides an acceptable level of service from a space segment investment that is less by a factor of 5 to 10, then such an option cannot be easily overlooked.

Precise frequency plans, with strict orbital locations, were developed in 1977 for Region I (Europe, Africa, and the Middle East) and Region III (Asia and Australasia) of the world, in terms of strict allocation of frequencies for direct broadcast services, with STV channels allocated to each country. A much more flexible plan was subsequently developed in August 1983 for Region II (the Americas), with the U.S., Canada, and Mexico each being allocated 400 or more TV channels and even the smallest countries having a large number of TV channels reserved for them. The precedent of the Region II flexible planning approach to BSS satellites suggests that sufficient frequencies for the future will be available.

Aside from these technical, economic, and service definition issues, the direct broadcast satellite services also have an important political dimension. The United Nations' Committee on the Peaceful Uses of Outer Space (COPUOS) has been debating guidelines concerning the spillover of direct broadcast satellite systems from one country into another for many years. The key issue has been whether a country should be able to regulate direct broadcast systems into its own territory. The United States has pursued its traditional concerns with the First Amendment and protection of freedom of speech, and thus has strongly defended the idea of the right of uncensored direct broadcast satellite systems to broadcast to domestic and international audiences without constraint. Most other countries of the world have resisted such complete freedom and have strongly backed the adoption of guidelines and regulations that would restrict the unlimited right to direct broadcast satellites.

Despite the cost of capitalizing a direct broadcast satellite system (which can run anywhere from $2 to $10 million per transponder per TV channel per year), and despite the fact that the programming costs associated with continuous 24-hour-a-day TV transmissions run even higher (namely, in excess of $100 000 per high quality production per hour), the enthusiasm for such systems continues unabated. Indeed, in addition to plans for a number of direct broadcast satellite systems in the United States, the following countries have implemented or are planning to implement direct broadcast satellite systems: Canada (ANIK system), France (TDF satellite system), Federal Republic of Germany (TV Sat), Italy (Sarit satellite system), Japan (BS-1 and BS-2), Luxembourg (Luxsat satellite system), Sweden (Tele-X system), Switzerland (through its Helvesat satellite system), the United Kingdom (Unisat satellite system), and the U.S.S.R. (Statsonar-T)—which have either implemented or are all moving ahead to implement these technologically sophisticated systems.

In the developing world, a number of countries with large and geographically dispersed populations are moving to DBS satellite technology for educational and international purposes. These include India (through its Insat satellite system), Mexico (through its Morales system) and Saudi Arabia (through the Saudi Arabian satellite system).

As DBS technology matures, it may well become far more than just a one-way entertainment distribution system. Perhaps with the development of phased-array antennas a new capability will develop—namely, a return channel back to the satellite. This could allow DBS to become an interactive service capability. In the future, an interactive DBS with say a 9.6 kbit (second return) data channel could promote the development of videotext on demand, educational programming, etc. This could be central to DBS developing into a network for educational, library, and information services, stock market reports, electronic newspaper distribution, and other primary customer services delivered by second-generation DBS systems of the late 1980's or early 1990's.

MOBILE SATELLITE SERVICES

Mobile satellite communications include three service sectors: maritime, aeronautical, and land-mobile. The advent of highly efficient fiber-optic communications systems in the 1980's and 1990's will have a significant impact on long-distance fixed satellite communications services. Thus, mobile satellite communications services in the longer term will become increasingly important. In terms of long-term growth over the next 20 years, it is likely that mobile satellite services of all types will experience perhaps the greatest expansion of any of the communications services. Certainly, the development of higher powered satellite technologies coupled with new, more sophisticated mobile antenna systems (especially highly efficient and low-cost conformal phased-array systems) should allow the very rapid expansion of mobile services to a "mass market" service from the rather limited applications in today's world.

Maritime Satellite Services

The first mobile satellite service to be operationally introduced was that of maritime communications services. In 1976 the U.S. sponsored Marisat system was launched into orbit, initially in the Atlantic Ocean region but, subsequently, in the Pacific and Indian Ocean regions as well. This system, which operates in the UHF *L*-band frequencies, was designed as a hybrid project, with approximately half of the satellite capacity being devoted to U.S. military naval operations and the remaining half of the capacity being available for commercial services to shipping transport, offshore drilling rigs, and other maritime mobile communications service requirements.

This satellite system, which is substantially owned and operated by Comsat General (a wholly owned subsidiary of the Communications Satellite Corporation) continued its operations from the late 1970's through 1983.

Serious international negotiations began in the late 1970's to create a new international organization to provide international maritime satellite services. Inmarsat (International Maritime Satellite Organization) was brought into being in July 1979 with a membership of 28, which has now grown to 40. A listing of the current member countries in Inmarsat and their ownership shares is provided in Table 17-2.

The first generation of the Inmarsat space segment is in fact a combination of three different types of Marisat satellites, which are leased from the Comsat Corporation. These satellites provided the initial transitional space segment. The Marisat satellites, however, have been gradually replaced by a combination of Marecs satellites designed, built, and launched by the European Space Agency and maritime communications subsystems (MCS) that are available on specially equipped Intelsat V satellites. The three Marecs satellites and four MCS subsystems on Intelsat V satellites will constitute the first-generation Inmarsat system through 1988. At that time, Inmarsat plans to deploy a second-generation maritime mobile system that will provide a much higher-capacity system that will meet all foreseen maritime mobile communications services and may also potentially provide aeronautical communications services as well.

In addition to the Inmarsat system, which services most commercial maritime communications needs on a global basis, as well as certain services to offshore drilling units, there are several other maritime mobile communications systems now in operation. These include the Fltsatcom system to support U.S. Navy requirements and the Volna satellite system that supports the U.S.S.R.'s maritime mobile requirements. These systems all operate in the VHF *L*-band.

TABLE 17-2

Country	Investment Share (%)
United States	23.32851
U.S.S.R.[a]	14.07093
United Kingdom	9.87643
Norway	7.86633
Japan	6.98861
Italy	3.35017
France	2.88118
Germany, Federal Republic of	2.88118
Greece	2.88118
Netherlands	2.88118
Canada	2.61322
Kuwait	2.01012
Spain	2.01012
Sweden	1.87616
Australia	1.67518
Brazil	1.67518
Denmark	1.67518
India	1.67518
Poland	1.67518
Singapore	1.67518
China, People's Republic of	1.23480
Argentina	0.60305
Belgium	0.60305
Finland	0.60305
New Zealand	0.36223
Bulgaria	0.27162
Portugal	0.20580
Algeria	0.05000
Chile	0.05000
Egypt	0.05000
Iraq	0.05000
Liberia	0.05000
Oman	0.05000
Philippines	0.05000
Sri Lanka	0.05000
United Arab Emirates	0.05000
Tunisia	0.05000
Saudi Arabia	0.05000

[a]Includes the initial investment share of Byelorussian and Ukranian SSR's.

Aeronautical Communications Satellite Services

The history of aeronautical satellites is characterized by a series of starts and stops. Early in the history of Intelsat a series of studies was undertaken regarding the provision of such services through the Intelsat organization. At that time, however, there was disagreement between the European and U.S. user community. The U.S. airline carriers favored a VHF system because of the relatively low cost of retrofitting jet airliners for VHF antennas, while the European governments generally supported the use of the less congested UHF band, even if this might have involved larger and more sophisticated antenna systems.

When agreement could not be reached within the Intelsat organization, discussions outside of Intelsat were undertaken, and a joint venture agreement was signed in the early 1970's between the European Space Agency, Comsat General (the wholly owned subsidiary of Comsat), and the Canadian Government's Department of Communications, establishing a Memorandum of Understanding. This project, which was to have established a hybrid UHF/VHF pre-operational aeronautical communications satellite project, was ultimately cancelled as a result of the U.S. Congress not providing funds for the Federal Aviation Administration to lease capacity equivalent to the Comsat General share of the system. There was, at that time, considerable debate about the cost efficiency of such a service, particularly since operating in the UHF band would require rather expensive antenna installations on the jet airliners.

Most recently, the Intergovernmental Civil Aeronautics Organization (ICAO) is, through its ASTRA Panel, examining the feasibility of establishing an operational aeronautical communications satellite service (either through Inmarsat, Intelsat, or some other option). ICAO is also exploring the feasibility of delivering to the worldwide network of airports certain emergency communications and meteorological information. These studies continue in process, with the likelihood that by early 1986 a specific implementation plan will be devised to finally bring an aeronautical communications satellite system into operation after more than a decade of study and discussion about such a system. This much is clear, however; any future system will, of necessity, be a UHF or higher-frequency system, since the VHF band is fully congested.

As technology in this field matures and cost efficiency and the reliability of this service (both in terms of a direct data communications/facsimile service as well as voice service) evolves, it is indeed likely that by the late 1980's an operational service will have been established.

Land-Mobile Communications Systems

Land-mobile communications systems started much later than maritime and aeronautical communications services, with serious planning in this area only recently begun. It has certainly evolved very rapidly in the last few years. The Swedish TELE-X system, the European Space Agency's UHF mobile service, the U.S. based Geostar Satellite System for personalized "beeper" communications, the Mobilesat system for aeronautical and land mobile communications, the NASA and Canadian Department of Communications mobile program which represents commercial interests in the U.S. and Canada—all of these sources were formed in the last few years and are proposed to become operational within the next five years.

The M-SAT-X Swedish TELE-X projects have placed considerable emphasis on the development of conformal phased array antenna systems that could be easily mounted on cars and trucks for communications with vehicles.

The mobile communications systems come close to approximating the concept of the Dick Tracy mobile communicator radio wristwatch, which has been a vivid image for decades, but is now apparently becoming a reality. Certainly many experimental studies have been done that have demonstrated the feasibility of having, in the future, very powerful space-based satellite systems (e.g., space platforms) to provide space communications direct to personal communications devices.

The most intriguing aspect of land-mobile developments is that massive scales of utilization are examined, rather than design concepts that are aimed at communications with hundreds or even thousands of ground antenna systems. Some of the more advanced concepts would be aimed at the potential of serving millions of personal communications systems. It is possible that in the future one may see a merging of mobile communications satellite systems and certain broadcast applications, such as various proposals that have been made and indeed seriously studied by the Voice of America and the BBC, for a medium frequency or a VHF/UHF-based space broadcast radio system. Such a future satellite system could be designed to provide direct broadcast audio services to millions of radio receivers.

OTHER TYPES OF COMMUNICATIONS SATELLITE SYSTEMS

Most of the communications satellite systems that we have discussed up to this point have involved conventional fixed, mobile telecommunications, or broadcast satellite applications which are typically available to the general public and involve satellites that normally utilize the geosynchronous orbit. However, other types of communications satellite systems are emerging, which perform different functions and sometimes utilize different orbits. These include search and rescue satellites for detection of aircraft or ships in distress, tracking and data relay satellites, and remote data collection satellites. All of these satellites are support systems. They make possible a wide range of ground-based and satellite-based activities in such areas as meteorology, satellite tracking and command, earth resource sensing, and rescue work.

Search and Rescue Satellites

The rapid location of ships and aircraft in distress is certainly not a new requirement but, as the number of ships and aircraft in operation around the world has grown, the importance of a quick and effective distress and position location system has become increasingly important. Grim statistics have demonstrated that the speed with which a downed aircraft or a sunk ship with survivors can be located is crucial to the success of saving survivors. In the early 1970's the U.S. Government moved to require aircraft to carry emergency locator transmitters (ELT's) that emit a distress signal under emergency or accident conditions. On a parallel basis, ocean-going vessels were required to carry emergency-position-indicating radio beacons (EPIRB's) that could be activated either manually or by immersion in water. These emergency transmission devices, with continuous transmission capabilities of up to 48 h, operate in the 121.5 MHz and the 243 MHz bands, which are reserved for worldwide maritime and aeronautical distress calls.

Since the early 1970's the equipping of ships and aircraft in this manner has spread to a number of other countries and it is estimated that a quarter of a million ships and aircraft in several countries are now so equipped. Emergency transmissions from ELT and EPIRB's, however, were not especially effective unless specially equipped aircraft essentially knew where to look in the first place. In the mid-1970's, however, tests and demonstrations with satellite-based detection systems came under serious discussion and the U.S. National Aeronautics and Space Administration and the Canadian Department of Communications collaborated in a satellite-aided search and rescue system demonstration using the Nimbus-6 satellite. This was carried out in the fall of 1976. The project was subsequently expanded into a trilateral experimental program in December of 1977, when the French National Center for Space Studies (CNES) joined the effort. These tests and demonstrations used the Doppler (frequency shift) effect and triangulation calculations to pinpoint the location of emergency transmitters with a high degree of accuracy. As these experiments continued, increased emphasis was placed on the development of higher powered and more effective transmitters operating in the higher frequency band also reserved for distress signals in the 406 MHz band. This frequency was allocated exclusively for this purpose at the World Administrative Radio Conference in 1979, and thus will be far less affected by spurious "false alarms."

This new higher powered and higher accuracy system will increase the precision of location of accident sites from the current 10–20 km to a new, much more precise, range of 2–5 km. Also, the new system will transmit much more information, including whether the signal is from aircraft or ocean-going vessel, its country of origin, the nature of the distress, the elapsed time since the accident occurred, the identification of the vessel or aircraft and, if known, the exact location of the emergency.

Parallel to the U.S., French, and Canadian efforts, the Soviet Union initiated its own search-and-rescue satellite project, called Cospas, in the late 1970's. In the late 1970's, an agreement was reached between the U.S., Canada, France, the U.S.S.R. and Norway to undertake a pre-operational Cospas/Sarsat global project.

The first spacecraft launched within the framework of the joint Cospas/Sarsat project agreement was the Cospas-1 satellite, launched by the Soviet Union in June 1982. This satellite, which was launched into circular, near-polar orbit, with an inclination of 82° and a mean altitude of 1000 km, has been in operation since September 1982. This was followed by the launch in 1983 of the Cospas-2 by the Soviet Union, as well as the first Sarsat-equipped NOAH satellite (NOAA-8), known as NPAA-8. Since this preliminary Cospas/Sarsat system became operational, nearly 200 lives have been saved in terms of rescued passengers and crew members on aircraft and ships at sea. Five additional NOAA satellites (known as the ATN series) which are now being deployed are equipped with Sarsat instrumentation plus several more Soviet Union Cospas satellites. These satellite launches will provide full coverage and continuity of satellite services through the late 1980's and will allow time for final decisions to be made concerning the technical characteristics of the fully operational system, whether this system will fly on dedicated or multipurpose spacecraft, who will operate this system, how much it will cost, and how it will be funded. This much is clear: given the low cost of ELT and EPIRB units which now cost in the range of $150 to $300 per package, space-based search and rescue is a very cost-effective approach to coping with the serious problem of human safety. The cost effectiveness of the Cospas/Sarsat program is impressive, particularly when it is recognized that the cost of mounting one extensive search operation over a wide area can run into millions of dollars.

Tracking, Relay, and Data Collection Satellites

Another success story, in terms of the use of satellite technologies to provide a technologically and financially superior service over that of conventional methods, is that represented by tracking, relay, and data collection satellites. The United States, the Soviet Union, and the European Space Agency all have established worldwide networks of tracking and command stations to monitor and command satellites during launch and to send commands to maintain the satellites in orbit and to collect data for scientific application and manned spacecraft.

These ground-based systems have been in operation for a number of years. Other countries (such as Japan, China, India, and others) planning to develop their own space programs will likewise consider establishing such systems. Because these systems must be highly reliable and provide direct access to satellites, a ground-based tracking and data relay system requires a large number of earth-based systems.

To operate such extensive ground-based systems, a great amount of money is needed to establish and maintain a large crew of personnel available at all these sites. A network of three geosynchronous satellites with tracking and data capability, however, is capable of maintaining a global coverage system and, because of the lower operating cost associated with these space-based facilities, in the longer term they can mean net lower costs. Also, 100 percent coverage can be maintained and much greater flexibility in network control can be achieved. When such a system is established, of course other applications, such as the collection of data from remote sensing and monitoring stations for environmental and other purposes, become quite feasible.

With the cost of its ground network to support deep space and applications satellite programs approaching $350 million per year in the early 1980's, NASA has developed the TDRS satellite system, as perhaps the most technologically advanced satellite of this type. This satellite operates in both the *K*-band and the *S*-band frequencies, and offers 10 percent coverage for satellites in the altitude range of 1200 to 12 000 km and 85 percent coverage for satellites for locations below 1200 km. The high-gain phased-array system on the satellite allows very high-capacity data rate transfer and simultaneously supports 20 multiple access system users, each with data rates up to 50 kbits/s in the *S*-band. In the K_u-band in single access mode, the TDRS satellite can handle up to 300 Mbits/s. The TDRS system will thus be able to transmit and receive earth-imaging data from Landsat-D satellites at greater than 100 Mbits/s and can also support Spacelab missions for the transmission of scientific data at rates up to 50 Mbits/s. The TDRS system became operational in 1983 and supports the deep space network, the spaceflight tracking and data network, and special requirements such as the Spacelab and Landsat missions. The three TDRS satellites are positioned at longitudes 41° W, 99° W, and 171° W.

The TDRS satellite, which operates in a digital mode, is the first operational spacecraft to use satellite-switched time-division multiple-access techniques and thus, in addition to developing a new system capability, also represents an important research and development project for other satellite applications. It is thus likely that other spacecraft of the 1980's modeled after the TDRS satellite will be utilized for such applications as electronic mail, as well as for environmental monitoring from remote data collection systems (such as has been endorsed for worldwide implementation by the U.N. Economic Commission on Europe).

FURTHER INNOVATIVE NEW APPLICATIONS SATELLITE SYSTEMS

Beyond the search and rescue satellites and the tracking and data relay satellite programs, which represent new departures in the communications satellite applications field, it is of course quite likely that other new service capabilities will develop during the 1980's. Systems of this nature are likely to include the evolution from the first-generation mobile personal paging communications system, which can relay simple messages from personal communicators via satellite, to much more sophisticated systems that will allow personal two-way radio communications as represented by the Dick Tracy two-way wristwatch radio communicators. Such systems perhaps will be first developed for emergency police, fire, and military applications, but will then subsequently be developed for broader based public use. In this respect, the evolution of land-mobile communications systems beyond the existing programs, as represented by Insat and TELE-X (with its "trunk sat") will also move in the direction toward very mobile personal communications systems.

APPENDIX
GLOBAL SATELLITE SYSTEM GUIDE

International Satellite Systems	Members	Current Status
Intelsat Global Satellite System	110 members (170 users)	Operational (currently 15 satellites)
Inmarsat Maritime Mobile Satellite System	37 members	Operational (5 satellites)
Intersputnik Satellite System	15 members (19 users)	Operational (leased capacity, on U.S.S.R. Gorizont satellites)
U.N. Peacekeeping & Emergency Comm. System	Service to locations as needed (Intelsat Lease)	Operational

Regional Satellite Systems	Members/Intended Service Areas	Current Status
Andean Satellite System (Condor)	Andean countries	Under planning (Intelsat lease to be followed by dedicated Condor satellite.)
Arabsat Satellite System	18 members	Arabsat 1A and 1B launched
African Satellite System	Most African nations	Under consideration only
Eutelsat Satellite System	European countries	Operational
Palapa A&B Satellite System	ASEAN countries	Indonesian Domestic System/Regional services to ASEAN members/operational
AMS Satellite Systems	Service area (Europe, Africa, Middle East)	Under planning
Luxsat Regional Satellite System	Europe	Under planning

Domestic Satellite Systems—Dedicated Systems	Services	Current Status
Australian National Satellite Systems (Aussat)	FSS, CATV, DBS	1985 launch
Brazilian Domestic Satellite System (Brazilsat) (SBTS)	FSS	1985 launch
Canadian Domestic Satellite System (ANIK A-D)	FSS, DBS, domestic service to U.S.	Operational
Chinese Domestic Satellite System	FSS, BSS	Project under study
Colombian Domestic Satellite System (Satcol)	FSS	Project under study
French Domestic Digital Satellite System (Telecom)	FSS	Under construction for 1985 launch
French DBS Satellite (T.D.F.)	DBS	Under construction for 1986 launch
Germany (Fed. Rep. of) Domestic Digital and TV Distribution Satellite (Postsat)/DFS	FSS	Planned for 1987 and 1988
Germany (Fed. Rep. of) DBS Satellite (TV-Sat)	DBS	Planned for 1986
Indian Satellite System (Insat)	FSS, DBS, meterological package	Operational (temporarily out of service)
Indonesian Satellite System (Palapa A & B)	FSS	Operational
Iranian Satellite System (Zahreh)	—	Project cancelled
Italian Satellite System (Italsat)	FSS	Planned for 1986

APPENDIX - *continued*

GLOBAL SATELLITE SYSTEM GUIDE

DOMESTIC SATELLITE SYSTEMS—DEDICATED SYSTEMS

	Services	Current Status
Italian DBS Satellite System (Sarit)	DBS	Planned for 1985
Japanese Domestic Satellite System (CS-2a, 2b)	FSS	Planned for 1983
Japanese Domestic Broadcast System (BS-2a, 2b)	DBS	Planned for 1985
Korean Satellite (K_u band)	FSS	Under study
Mexican Domestic Satellite System (Morales)	FSS	
Saudi Arabian Broadcast Satellite System (SABS)	DBS	Planned for 1986
Swedish Domestic Satellite System (Tele-X)	FSS, DBS, Mobile	Planned for 1986
Swiss Domestic Broadcast Satellite (Helvesat)	DBS	Planned for 1986
United Kingdom Domestic (UNISAT) Satellite System	DBS, FSS	Program likely to be cancelled
United States Domestic Satellite Systems:		
Advanced Business Communications, Inc. (ABCI) (K_u-band)	FSS, lower power DBS	Planned for 1987
Alascom Inc.	FSS	Operational
Allstate Communications Company	FSS	Planned for late 1980's
American Satellite Company System (Continental & Fairchild) (C, K_u, K_a bands)	FSS	Operational (planned K_u and K_a Systems for 1986)
ARCO		Specialized carrier/operational/leased capacity
ASN, Inc.	FSS	Planned by late 1980's
Atlantic Transport Company	—	Specialized carrier/operational/leased capacity
Bonneville Satellite Corporation	—	Specialized carrier/operational/leased capacity
Cablesat General Corporation	—	Specialized carrier/operational/leased capacity
Columbia Communications Corp. (C, K_u bands)	FSS	Planned for late 1980's
Compact Video Services		Specialized carrier/operational/leased capacity
Comsat General Corporation (Comstar) (C, K_u bands)	FSS	Operational in C band (K_u band planned in 1988)
CBS Satellite Systems	DBS	Delayed
Digital Telesat, Inc. (K_u band) (Merrill Lynch)	FSS	Planned for 1988
Direct Broadcast Satellite Corporation	BSS	Planned for late 1980's
Eastern Microwave, Inc.		Specialized carrier/operational/leased capacity
Equatorial Communications Services (Equatorial, Aetna, Bank of America)	FSS	Planned for late 1980's
Federal Express (Electronic Mail Systems) (K_u band)	FSS	Planned for 1988
Ford Aerospace Satellite Services (Fordsat) (C, K_u Bands)	FSS	Planned for 1988
General Communications Inc.	—	Specialized carrier/operational/leased capacity
Geostar (Personal Communication)	MSS	Planned for late 1980's
Graphic Scanning	DBS	Under study
GTE Satellite Corp (GSTAR) (K_u band)	FSS, low-power DBS	Operational in 1984
GTE Spacenet Corp (Formerly SPCC) (K_u band)	FSS, low-power DBS	Operational in 1984
Hi-Net Communications Inc.	FSS	Planned for late 1980's
Hughes Communications, Inc. Galaxy (K_u band)	FSS	Operational
Hughes (K_a band)	FSS	Planned for late 1980's
ISACOM Inc.	—	Specialized carrier/operational/leased capacity
International Satellite Inc. (K_u band)	FSS	Under policy review by U.S. government
IT&E Overseas	—	Specialized carrier/operational/leased capacity
Martin Marietta Communications Systems (K_u band)	FSS	Planned for late 1980's
Meredith Corporation	FSS	Specialized carrier/operational/leased capacity
Metropolitan Communications Network Co.	—	Specialized carrier/operational/leased capacity
Midwestern Relay Company	—	Specialized carrier/operational/leased capacity
Mobile Satellite Corporation	MSS	Planned for late 1980's
National Exchange Inc.	—	Specialized carrier/operational/leased capacity
NETCOM International	—	Specialized carrier/operational/leased capacity
OAK Direct Broadcast Satellite System	DBS	Postponed
ORION Satellite Corporation (K_u band)	FSS	Under policy review by U.S. government
Phipstar Inc.	—	Specialized carrier/operational/leased capacity

APPENDIX - *continued*

GLOBAL SATELLITE SYSTEM GUIDE

DOMESTIC SATELLITE SYSTEMS—DEDICATED SYSTEMS

	Services	Current Status
Private Satellite Network, Inc. (PSN)	—	Specialized carrier/operational/leased capacity
Rainbow Satellite Inc. (K_u band)	FSS	Planned for 1986
RCA Americom, Inc.	FSS	Operational/K_u band system, planned for late 1980's
RCA DBS Satellite System	BSS	Planned for late 1980's
Satellite Business Systems (Comsat/IBM/Aetna) (K_u band)	FSS, low-power DBS	Operational
Satellite Communications Network, Inc.	—	Specialized carrier/operational/leased capacity
Satellite Data Broadcast Network, Inc.	—	Specialized carrier/operational/leased capacity
Satellite Television Corporation (STV)	BSS	Project cancelled
Southern Satellite System, Inc.	FSS	Planned for late 1980's
Starnet Corp.	—	Specialized carrier/operational/leased capacity
Systematics General Corporation	FSS	Project cancelled
SSS Satellite Syndicated System	DBS	Planned for late 1980's
Telstar Satellite System (AT&T)	FSS	Planned for 1984
Tracking, Data, Relay Satellite System (NASA)	Data relay, FSS, MSS	Operational
United Satellite Services Broadcasting (Hubbard)	DBS	Planned for mid-1987
U.S. Satellite Communications (lease on GSTAR)	Low-power DBS	Operational
United Satellite System, Inc.	BSS	Planned for 1988/89
United States Satellite Systems, Inc.	FSS	Planned for 1985
USAA Satellite Communication Co.	FSS	Planned for late 1980's
VideoStar Connections, Inc.	FSS	Planned for late 1980's
Video Satellite System	DBS	Planned for late 1980's
Western Union Satellite System (C, K_u band)	FSS	Operational (K_u band system planned for 1988/89
Western Union DBS Satellite System	DBS	Planned for mid-to-late 1980's
U.S.S.R. Domestic Satellite Systems:		
Ekran Broadcast Satellite System	DBS	Operational
Gorizont Satellite System	FSS	Operational
Loutch Satellite System	FSS	Operational
Molniya Satellite System	FSS	Operational
Raduga Satellite System	FSS, Military	Operational
Volna Satellite System	Maritime Mobile Services	Operational

DOMESTIC SATELLITE SYSTEMS — LEASED/SHARED USE SYSTEMS EXCEPT U.S. LEASED SYSTEMS

	Space Segment Facilities	Current Status
Algerian Satellite System	Intelsat lease	Operational
Angolean Satellite System	Intelsat lease	Planned for 1983
Argentine Satellite System	Intelsat lease	Operational
Australian Satellite System	Intelsat lease	Operational
Bangladesh Satellite System	Intelsat lease	Planned for 1986
Bolivian Satellite System	Intelsat lease	Planned for 1987
Brazilean Satellite System	Intelsat lease	Operational
Cameroon Satellite System	Intelsat lease	Planned for 1985
Chilean Satellite System	Intelsat lease	Operational
Chinese Satellite System	Intelsat lease	Operational
Colombian Satellite System	Intelsat lease	Operational
Denmark (Greenland) Satellite System	Intelsat lease	Operational
Ecuadorian Satellite System	Intelsat lease	Operational
French Overseas Territory (Reunion, Martinique) Satellite System	Intelsat lease	Operational
Germany (Fed. Rep. of) Satellite System Spot Beam 14/11 GHz)	Intelsat lease	Planned for 1985
Indian Satellite System	Intelsat lease	Operational
Iranian Satellite System (14/11 GHz)	Intelsat lease	
Israeli Satellite System (14/11 GHz)	Intelsat lease	Planned for 1984
Italian Satellite System (14/11 GHz) Spot Beam	Intelsat lease	Planned for 1985
Korean (Rep. of) Satellite System	Intelsat lease	Planned for 1986
Libyan Satellite System	Intelsat lease	Operational
Malaysian Satellite System	Intelsat lease	Operational
Mali Satellite System	Intelsat lease	Planned for 1986
Mauritanian Satellite System	Intelsat lease	Planned for 1986
Mexican Satellite System	Intelsat lease	Operational
Moroccan Satellite System	Intelsat lease	Operational
Mozambique Satellite System		
Niger Satellite System	Intelsat lease	Operational
Nigerian Satellite System	Intelsat lease	Operational

APPENDIX - *continued*
GLOBAL SATELLITE SYSTEM GUIDE

DOMESTIC SATELLITE SYSTEMS — LEASED/ SHARED USE SYSTEMS EXCEPT U.S. LEASED SYSTEMS	Space Segment Facilities	Current Status
Norwegian Satellite System	Intelsat lease	Operational
Oman Satellite System	Intelsat lease	Operational
Pakistan Satellite System	Intelsat lease	Planned for 1985
Papua New Guinea Satellite System	Aussat or Intelsat lease	Planned for 1985
Peruvian Satellite System	Intelsat lease	Operational
Philippine Satellite System	Palapa lease	Operational
Portugal (Azores) Satellite System	Intelsat lease	Operational
Saudi Arabia Satellite System	Intelsat lease	Operational
Solomon Island	Intelsat lease	Planned for 1985
Somalia Satellite System		
South Africa Satellite System	Intelsat lease	Planned for 1986
Spain (Canary Islands) Satellite System	Intelsat lease	Operational
Sri Lanka Satellite System	Intelsat lease	Planned for 1988
Sudan Satellite System	Intelsat lease	Operational
Thailand Satellite System	Intelsat and Palapa	Operational
United Kingdom (domestic) (14/11 GHz spot beam) Satellite System	Intelsat lease	Operational
Venezuelan Satellite System	Intelsat lease	Operational
Zairian Satellite System	Intelsat lease	Operational

MILITARY SATELLITE SYSTEMS	System Names	Current Status
NATO Military Satellite Communications Project	Phase I, II, III	Operational
U.K. Skynet Defense Communications Network	SKYNET	Operational
U.S. Military Satellite Communications Network:		
U.S.A.F. Communications Satellite System	Afsatcom	Operational
U.S. Defense Satellite Communications System	DSCS Phase I, II, III	Operational
U.S. Milstar Satellite System	Milstar	Operational in 1984
U.S. Naval Satellite Communications System	Marisat/Leasat	Operational
U.S. Fleetsatcom Communications System	Fltsatcom	Operational
U.S.S.R. Military Communications Satellite System:		
Cosmos Military Satellite System	Cosmos	Operational
Gals Military Satellite System	Gals	Operational
Volna Naval Communications Satellite System	Volna	Planned 1984

EXPERIMENTAL SATELLITE SYSTEMS	Sponsors	Current Status
Applications Technological Satellites	U.S. NASA	Virtually all out of service
Communications Technology Satellite	U.S./Canada	Out of service
European Mobile Communications Satellite	E.S.A.	Planned for 1987
Japanese Experimental Broadcast Satellite	Nasada (GE)	Planned for 1987
Japanese Experimental Communications Satellite	Nasada (FACC)	Operational
Lincoln Experimental Satellite Project	U.S. DOD	Out of service
M-Satellite Experimental Program (U.S./Canada)	NASA, DOC	Planned for late 1980's
NASA Advanced Communications Technologies Satellite Experiment	20/30 GHz	Planned for late 1980's
NASA Experimental/Prototype Space Platform		Planned for late 1980's
Orbital Test Satellite (OTS)	ESA Experimental Satellite	Operational
Sirio Experimental Communications Satellite	Italian	Operational
Symphonie Experimental Communications Satellite	French/German	Operational
Tacsatcom Experimental Military Satellite	U.S. military	Out of service

18

COMMUNICATIONS SPACECRAFT

Samuel W. Fordyce

PIONEERING EXPERIMENTAL PROGRAMS

The space age was opened with the Soviet Union's launch of Sputnik 1 in 1957, and the United States of America's launch of Explorer 1 in 1958. Communications were carried on satellites as early as 1958, when Project Score orbited a programmed message from President Eisenhower. During the late 1950's, there were differences of opinion about the function of communications satellites. There were proponents of passive satellites which merely reflected incident signals, and there were proponents of geosynchronous orbits as well as lower altitude orbits.

The advantages given for passive satellites included the ability to operate over a wide range of frequencies, to reflect multiple signals impinging upon the satellite simultaneously, and to operate without reliance on failure-prone electronic systems. The disadvantages cited were the limited traffic handling capability and the difficulties of maintaining the proper reflector shape in space.

The primary advantage of a geostationary orbit location is that the satellite appears to hover, or stay at a fixed position, in respect to an observer on earth. The earth stations do not have to track or continuously adjust the "look angle" after acquiring the satellite. A lower-altitude orbit, or indeed, any non-geosynchronous orbit requires the antennas to search and locate the satellite on each orbit as it "rises" over the horizon, and to hand over the satellite to other earth stations at different locations along the sub-satellite track in order to maintain continuous communications. The benefits of the lower-altitude orbits are that the launch vehicles can launch larger satellites, and the quarter-second time-delay involved in the trip to geostationary orbit and back is shortened. The environment of the geostationary orbit was unknown, and there was concern that the satellite would not survive long enough to be useful. There was also concern that a satellite could not be kept precisely on-station, without excursions which would take it out of the beamwidth of the non-tracking earth station antennas. In order to resolve these questions a series of experimental communications programs were attempted, several of which are described in this chapter. These brief descriptions are intended to outline some of the technological advances which were developed or demonstrated in these experimental programs.

In 1960, a large balloon (100 ft diameter) constructed of 0.5 mil Mylar with a thin coating of aluminum (shown in Fig. 18-1) was placed into a circular orbit at an altitude of 900 nmi. This balloon, called Echo 1, operating with earth stations using high-power transmitters, directional antennas, and sensitive receivers demonstrated communications over great distances. Deformations in the balloon soon introduced unwanted amplitude modulation at the spin rate of the balloon. A larger (135 ft diameter) balloon constructed of a thicker coating was orbited in 1964 (Echo 2). Although Echo 2 provided a larger reflecting surface, the traffic handling capacity (limited to a few voice circuits) of these passive reflectors was too small to compete successfully with commercial terrestrial communications systems. The development of the earth stations for these NASA programs, however, proved to be useful on subsequent active communications satellites.

In 1962, two active communications satellites (Telstars 1 and 2), sponsored principally by the Bell Telephone Laboratories, were placed into elliptical orbits. These satellites received signals in the ground-to-space link at 6.39 GHz, and after amplifying and down-converting them, re-transmitted them on a carrier frequency of 4.17 GHz. The spacecraft power was supplied by solar cells, and the communications payload was able to relay 600 voice channels between the participating earth stations.

Telstar 1, shown in Fig. 18-2, was a 34.5 in diameter sphere with solar cells covering most of the outer surface. The solar cells convert the photon energy radiated by the sun directly into electrical energy. In order to provide power when the satellite was shaded by the earth, a chargeable chemical battery was used as the secondary power source. Nickel-cadmium batteries were used because of their high charging efficiency and their

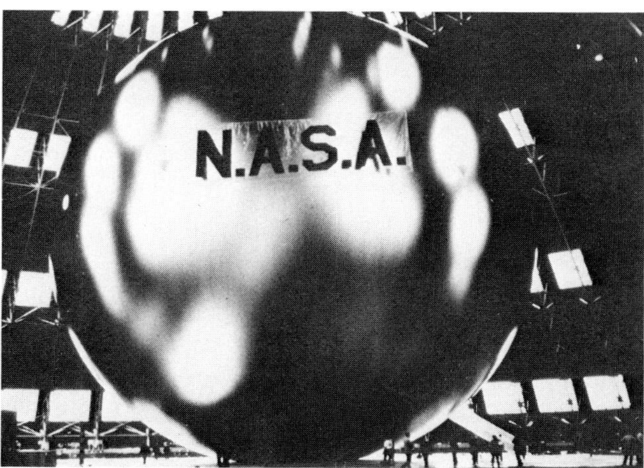

Fig. 18-1: The ECHO passive reflecting satellite.

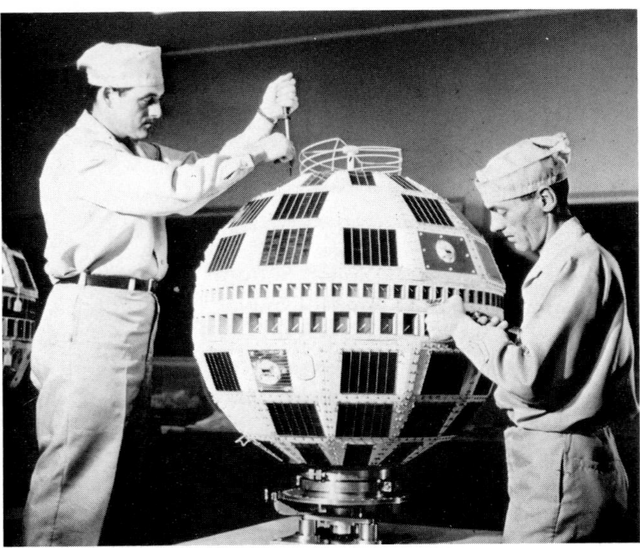

Fig. 18-2: The Bell System Telstar low-orbit satellite.

resistance to overcharge and deep discharge. The power capability of these solar cells and batteries was only 15 W.

A Thor/Delta launch vehicle placed this small satellite (weighing 170 lbs) into an elliptical orbit, with a perigee and apogee altitude of 514 × 3051 nmi. The satellite was spin-stabilized, rotating at 200 rpm about a north–south axis. This is a stable arrangement if the moment of inertia around the spin axis is larger than that around the axis perpendicular to the spin axis. However, various disturbing torques will disturb this stability and result in precession of the spin axis as well as nutation (vibrations of the spin axis) unless there are corrections. These require sensors to detect the unwanted movements, and reaction jets to correct them. Sensors can be used to detect the sun, the earth's horizons, and a bright star to provide reference directions.

The internal temperature of a communications satellite should be maintained within a range of 50–75° F. The efficiencies of the solar cells and batteries depend on the temperature, and temperature extremes can cause frozen valves or short lifetimes of the electronic components. In space the satellite receives heat from the sun, and generates heat from the inefficiencies of its active components, principally transmitters. This wasted heat must be radiated to space through radiators which can be varied to balance the heat lost with the heat gained to result in the proper equilibrium temperature. To distribute heat within the spacecraft, heat generating components can be mounted on "cold plates," and "heat pipes" can be used to distribute heat from the heat generators to the radiators. Local heaters are used to maintain the temperature of critical devices, and "superinsulation" provides thermal shields. Improperly designed thermal control systems have caused shortened lifetimes on many satellites when local cold or hot spots on the satellites have resulted in component or subsystem failures.

In order to understand the performance of satellites in space telemetry sensors detect the temperatures, power levels, accelerations, and many other critical characteristics and relay them to the tracking, telemetry, and command (TTC) earth station. The performance, position, and attitude of the satellite are monitored, and commands are sent up when it is necessary to reconfigure the on-board systems or adjust the orbit of the satellite.

Telstar's communications subsystem (the "transponder") used a solid-state receiver with a noise figure of 12.5 dB. The noise figure is a measure of the noise performance of an amplifier, and is defined as the signal-to-noise ratio at the input divided by that at the output. The received signal carrier frequency of 6390 MHz was down-converted to 4170 MHz, and amplified by a single traveling wave tube amplifier (TWTA) with a maximum power output of 4.5 W. The transponder had a bandwidth of 50 MHz, and was able to relay 600 voice circuits or one television signal between large earth stations. Earth stations in the U.S., the U.K., France, Italy, and Germany participated in experiments with Telstar. The communications transponder operated at the *C*-band frequencies which have become standard today. The antenna pattern was essentially omni-directional, which resulted in an effective isotropic radiated power (EIRP) of only 3 dB·W.

Telstar demonstrated many of the characteristics of the satellites which followed, but with one significant difference. Its low-altitude elliptical orbit was not geosynchronous, and would have required a network of many interconnected, tracking earth stations to provide continuous communications.

NASA's Syncom Program was intended to demonstrate geosynchronous communications. Syncom 2, shown in Fig. 18-3, was launched in July 1963. Its mass was only 86 lbs in orbit, which was the maximum capacity of the Thor/Delta launch vehicle of this time period. The satellite received in the 7350–7370 MHz band, and retransmitted in the 1800–1820 MHz band. The 2 W TWTA was able to relay several voice circuits or one television signal between large earth stations. Syncom 2 was geosynchronous, but had an inclination of 32°. Syncom 3, launched in August 1964, reduced the inclination to zero, so that its orbit was in the earth's equatorial plane. This eliminated the daily north–south excursions of Syncom 2, so that the satellite appeared to hover at a fixed point in the sky and became truly geostationary. This small satellite, with only 28 W of primary power, marked the beginning of the age of geostationary communications satellites.

Project Syncom demonstrated the feasibility of placing satellites into geostationary orbits and maintaining precise station-keeping and attitude control. Syncom 2 was spin-stabilized, and used a dual hydrogen peroxide system to provide thrust to correct the attitude and provide incremental velocity ("ΔV") to keep the satellite at its assigned station in space.

Geostationary satellites are usually first inserted into a low-altitude earth orbit by the launch vehicle in prepara-

Fig. 18-3: The NASA Syncom II satellite—the first geostationary satellite which was built by Hughes Aircraft.

tion for two major boosts in velocity. The perigee thrust adds a ΔV of approximately 9000 fps, which raises the apogee to the geostationary altitude of 22 340 mi. The apogee thrust adds a second velocity increment of approximately 5200 fps to circularize the orbit, and make the orbital plane parallel to the equatorial plane. While the perigee thrust engine is usually the responsibility of the launch vehicle, the apogee thrust engine is integrated with the spacecraft. On Syncom, and most spacecraft which followed, the apogee thrust was provided by a solid-propellant engine which was spinning to lessen velocity errors resulting from thrust malalignment. The spacecraft thermal control system must block the heat generated by this engine from damaging the spacecraft. Its attitude control system must change the spin rate after firing to that desired for the satellite in orbit, and align the spin axis in the north–south direction. The spacecraft antenna formed a "pancake" beam which concentrated the signal in the plane perpendicular to the spin axis. The EIRP of 6 dB·W permitted the transmission of a television signal between widely separated earth stations.

Demonstrations of both Telstar and Syncom were performed at the International Telecommunications Union's Extraordinary Administration Conference which was held in Geneva in the fall of 1963. Following studies conducted by the ITU's technical arm, the C.C.I.R., allocations were made of frequency bands assigned to the "Fixed Satellite Service" for point-to-point communications via satellite.

COMMERCIAL COMMUNICATIONS SATELLITES (1965–1971)

In August 1964 the International Telecommunications Satellite Organization (Intelsat), a global satellite cooperation, was created. The purpose of Intelsat was the production, ownership, management, and use of a global communications satellite system. The U.S. Signatory to Intelsat is the Communications Satellite Corporation (Comsat), which was incorporated in early 1963 under provisions of the Communications Satellite Act of 1962. Intelsat was formed under two international treaties in 1964 and reconstituted under revised agreements in 1973.

Encouraged by the success of Syncom 2, Intelsat contracted with Hughes Aircraft Corporation to build Intelsat I ("Early Bird"), shown in Fig. 18-4. Intelsat I weighed only 85 pounds in orbit. The spacecraft bus was closely modeled on Syncom 3, but the communications subsystem used the frequency band pioneered by Telstar. The solar array provided 45 W of power, which powered dual *C*-band transponders with bandwidths of 25 MHz. The dual TWTA's provided an output power of 6 W, which provided an effective isotropic radiated power ("EIRP") of 10 dB·W when focused by the "pancake" beam antenna. This small, low-powered satellite, parked in an orbit 22340 miles above the earth, was able to transmit television signals between widely separated earth stations (see Table 18-1).

The large gains of the earth station antennas (on the order of 60 dB, or one million) enable the small satellite with little antenna gain and a transmitter power of only 6 W to complete the link with a 5 or 6 dB margin over the carrier-to-noise threshold level of 10 dB. A threshold level of 10 dB provides a carrier signal which is ten times more powerful than the noise in the system, and is adequate to provide good quality voice and video signals (see Table 18-1).

Attitude control of the spacecraft was provided by spin-stabilization, with corrections from small hydrazine thrusters. The use of hydrazine in place of the cold gas or

Fig. 18-4: Intelsat I—Early Bird.

TABLE 18-1
CIRCUIT LINK CALCULATIONS FOR INTELSAT 1

EARTH-TO-SPACE ("UP") LINK		SPACE-TO-EARTH ("DOWN") LINK	
OUTPUT POWER (3.5kW)	35.4 dBW	EIRP	10.0 dBW
TRANS. FEED LOSS	− 1.0 dB	SPACE LOSS	−196.8 dB
ANTENNA GAIN	61.2 dB	ANTENNA GAIN	57.4 dB
SPACE LOSS	−200.6 dB	RECEIVING FEED LOSS	− 0.4 dB
SATELLITE ANTENNA GAIN	4.0 dB	RCVR. INPUT POWER	−129.7 dBW
FEED LOSSES	−1.5 dB	SYSTEM NOISE TEMP.	48.0 K
SATELLITE RCVR. POWER	−102.4 dBW	RCVR. NOISE DENSITY	−211.8 dBW
SATELLITE NOISE FIGURE	10.0 dB	CARRIER/NOISE DENSITY	82.1 dB
NOISE DENSITY (PER Hz)	−194.0 dBW	NOISE BANDWIDTH	4.9 MHz
CARRIER/NOISE (PER Hz)	91.6 dB	CARRIER/NOISE RATIO (4.9 MHz)	15.2 dB
SATELLITE NOISE BANDWIDTH	33.0 MHz		
CARRIER/NOISE (33 MHz)	16.3 dB	THRESHOLD MARGIN (OVER 10 dB)	5.2 dB
THRESHOLD MARGIN (OVER 10 dB)	6.3 dB		

hydrogen peroxide used in earlier spacecraft thrusters provided increased performance capabilities at lower weight than was possible previously. The communications capability was either 240 one-way voice circuits, or a single one-way television signal. The Intelsat 1, launched by a Thor/Delta launch vehicle in April 1965, had a design life of 1.5 years but remained in commercial service until 1969.

Intelsat 2 took advantage of the increased capabilities of the Thor/Delta launch vehicle to orbit a 192 lb satellite. The transponders include four 6-W TWTA's, which were able to relay 240 two-way voice circuits. Four Intelsat 2's were launched in 1966 and 1967. Both Intelsat 1 and 2 were built by the Hughes Aircraft Corporation.

Intelsat 3, built by TRW, Inc., incorporated a despun antenna which allowed the antenna beam to focus on the earth. This was a major improvement, for the antenna gain increased to 18 dB, and the 10 W TWTA's permitted an EIRP of almost 28 dB. The increased capability of the Thor/Delta launch vehicle permitted a spacecraft on-orbit weight of 330 lbs. Eight Intelsat 3's were launched between 1968 and 1970. Although this series was plagued with difficulties, including three launch vehicle failures, it enabled Intelsat to provide global service.

Intelsat IV was significantly larger than its predecessors, and used Atlas/Centaur launch vehicles to orbit the 1600 lb satellites. The spacecraft design was based on the Tactical Communications Satellite (Tacsat), a military communications satellite which was also built by Hughes Aircraft. The antenna and communications electronics are all mounted on a platform that is despun relative to the main body of the satellite in order to remain pointed at the earth. All of the other equipment is mounted within a large cylindrical satellite body, which spins to provide stabilization.

The communications sub-system used 12 transponders, with center frequencies spaced 40 MHz apart, to occupy the 500 MHz band allocated for satellite communications. These transponders, with effective bandwidths of 36 MHz (and 4 MHz of guard bands for isolation from adjacent transponders), and 6 W of output power, became standard for the industry.

The transmitters can be classified into two categories: a saturated amplifier which keeps the output level constant by varying the gain, and a linear amplifier whose output level is proportional to the input level. Linear amplifiers are better suited to multiple-access systems which receive simultaneous signals from many earth stations. The early satellites used saturated amplifiers, but later ones have used linear amplifiers.

In order to permit multiple signals to use the same channel without harmful interference, two basic modulation systems are used: FDMA (frequency division multiple access), in which each station has its assigned carrier frequency, and TDMA (time division multiple access), in which all stations use the same carrier frequency, but take turns transmitting.

The communications capability of these transponders was as high as 9000 one-way voice circuits, or 12 one-way television signals, depending on how the satellite was configured. The antennas, shown in Fig. 18-5, provided both global coverage beams (with 17° beamwidths) and zone beams (with 4.5° beamwidths).

Eight Intelsat 4 satellites were launched between 1971 and 1975, and many provided service into the 1980's.

COMMERCIAL COMMUNICATIONS SATELLITES (1972–1982)

The first commercial satellite used for domestic communications was Canada's ANIK A-1, launched in 1972. This satellite provided the transponder capability of Intelsat 4 on a smaller spacecraft "bus," with a mass of only 655 lbs in orbit, which was launched by a Thor/Delta launch vehicle. The satellite is spin-stabilized with a despun platform supporting a single antenna,

Fig. 18-5: Intelsat 4—The first high-performance satellite which also used spot beams.

Fig. 18-6: ANIK A—the first free world domestic satellite for Canada.

whose 3 × 8° beamwidth illuminates Canada. The concentration of the radiated energy into a single "footprint" on Canada permits an EIRP of 33 dB·W. This relatively high radiated power permits the reception of television signals by earth stations with receiving antennas as small as 8 to 15 ft in diameter. ANIK A satellites each carried 12 transponders, with bandwidths of 36 MHz. Each transponder could be used to transmit either 1000 one-way voice circuits or one television signal.

A picture of the ANIK A satellite is shown in Fig. 18-6. This spacecraft bus, built by Hughes Aircraft, was also used for Western Union's Westar, and Indonesia's Palapa satellites.

The first three-axis stabilized commercial communications satellite was RCA-s Satcom series, first launched in 1975. Satcom, shown in Fig. 18-7, has a boxlike body about 4 × 4 × 5 ft. Internally mounted momentum wheels provide the inertial stabilization and are used to correct the departures from the proper spacecraft attitude which are detected by the sensors. Hydrazine thrusters are used to "dump" excess momentum in the wheels. Solar arrays are deployed from the north and south face of the stabilized spacecraft. The ability to deploy these solar arrays permits 745 W of primary power from a 1010 lb satellite. All of the solar cells face the sun, and contribute to the array's power. The spin-stabilized satellites, which have body-mounted solar arrays, have the disadvantage that only about one-third of the solar cells are facing the sun. However, the

Fig. 18-7: Satcom—RCA's three-axis body stabilized 24-channel domestic satellite.

three-axis stabilized spacecraft have disadvantages in that they must stop the spinning required by the solid-propellant apogee engine, and provide thermal control systems to condition a spacecraft rotating once per day rather than 100 rpm.

In addition to the three-axis stabilization, and the deployable solar arrays, Satcom carried dual-polarized spacecraft antennas. Twelve transponders used horizontal polarization, and twelve other transponders used vertical polarization. The earth stations also had cross-polarized antennas to transmit and receive these signals. The use of cross-polarization provided enough discrimination so that the allocated frequency band could be used twice.

The cross-polarization permitted doubling of the use of the 500 MHz frequency band allocated in the *C*-band (6/4 GHz band), but the demand for satellite communications was growing rapidly, and an additional band at 14/12 GHz was allocated by the ITU. While the 6/4 GHz band was required to share with other radio communications services, including terrestrial microwave radio relay links, the 14/12 GHz band was an exclusive allocation. The 500 MHz allocated at the 14/12 GHz band looked attractive, but there were concerns about the extent that communications in this band would be interrupted by rainstorms.

The Communications Technology Satellite (CTS) was a joint effort by the Canadian Department of Communications and NASA. The primary purposes of this program were to determine the practicality of the 14/12 GHz band for satellite communications, and to experiment with broadcasting television signals from a high-powered satellite to small, low-cost terminals. Television broadcasts to receiver terminals using "dish" antennas as small as 40 cm in diameter were demonstrated. CTS operated successfully for several years after its launch in January 1976. CTS is shown in Fig. 18-8 while undergoing tests at NASA.

Fig. 18-8: Hermes—the joint U.S.-Canadian satellite which pioneered the use of the 14/11 GHz band and used 200 W TWTA for high-power DBS.

The success of CTS led to the first commercial use of the 14/12 GHz (K_u) band on Canada's ANIK B, which used RCA's Satcom spacecraft bus to carry a payload consisting of 12 *C*-band transponders with bandwidths of 36 MHz, plus 6 K_u-band transponders with bandwidths of 72 MHz each. The EIRP's were 36 dB·W at *C*-band, and 46.5 dB·W at K_u-band. ANIK B was launched in December 1978.

The first commercial satellite to operate solely in the K_u-band was SBS-1, shown in Fig. 18-9. SBS used Hughes Aircraft's HS-376 spacecraft bus, which was spin-stabilized, and a cylindrical telescoping skirt to provide additional solar cells for a compact launch configuration. The total primary power supply at the "beginning-of-life" (BOL) was 900 W. This supplied power for ten 43 MHz bandwidth TWTA's, which provided an EIRP of 40–44 dB W over the conterminious U.S. SBS-1 was launched successfully in 1980. The SBS earth stations were designed for time division multiple access (TDMA) operation. With TDMA the satellite transponder is used solely by a single specified carrier during a designated time period or slot. All of the network earth stations using a transponder are assigned time slots, so that they take turns transmitting to the satellite. These time slots are so short, and they are switched so rapidly, that any individual user is unaware that the circuit is being shared.

A third major band allocated by the ITU is the 30/20 GHz band, which has an allocated bandwidth of 2.5 GHz, five times the bandwidth originally allocated in either the 6/4 or the 14/12 GHz bands. The first use of

Fig. 18-9: SBS—the pioneering business systems satellite.

this band has been the Japanese CS (Sakura) series, built by Ford Aerospace under a subcontract from Melco. This small (750 lb) satellite, which uses both the 30/20 and 6/4 GHz bands, has approximately 50 percent more bandwidth than those satellites operating in either of the lower bands.

Another commercial satellite using dual bands is the Intelsat 5 series built by the Ford Aerospace and Communications Corp., and is shown in Fig. 18-10. The solar arrays on the spacecraft extend 50 ft from wing-tip to wing-tip when deployed. These arrays produce 1800 W of primary power at the beginning of life. Both the *C*- and K_u-bands are used with seven separate antennas operating with 27 TWTA's. The transponders vary in bandwidth from 34 to 241 MHz, and provide a total usable bandwidth of 2137 MHz through four-fold use at *C*-band and two-fold use at K_u-band. Both cross-polarization and multiple spot beams contribute to this frequency re-use. This large satellite (2200 lbs in orbit) can relay 12 000 simultaneous voice circuits plus two color television channels.

Large as the Intelsat 5 is, it is small compared to the Intelsat 6, shown in Fig. 18-11, which will weigh almost 4000 lbs in orbit. This large spin-stabilized satellite will have enough body-mounted solar arrays to produce 2.2 kW of electrical power. The communications capability is 33000 voice circuits. Fifty transponders operating in the *C*- and K_u-bands will be interconnectable. The antenna system includes 146 separated feeds, which can be reconfigured to form the desired beam "footprints." A microprocessor is incorporated into the spacecraft to control the interconnections, reconfigurations, and propulsion systems. The spacecraft carries an integrated bipropellant propulsion system for apogee boost, station-keeping, and attitude control. The bipropellant system provides fuel for 10 years of operations.

Table 18-2 shows the characteristics of all of the Intelsat satellites to date. The dramatic decrease in the costs of a circuit-year which was evident on each of the early generations has become more gradual on the later generations. In 1982 dollars, the annual circuit costs have decreased from $12480 for Intelsat 1, to $440 for Intelsat 6. The high costs of Intelsat 6 have spurred interest in future satellites which will be smaller and more specialized.

MOBILE COMMUNICATIONS SATELLITES (1966–1983)

NASA's Applications Technology Satellites (ATS) 1 and 3 successfully conducted experiments with aircraft in flight using the 149/136 MHz band; ATS-5 and 6 conducted experiments at *L*-band (1650/1550 MHz) with ships.

The maritime satellite service went into commercial operation in 1976 with three Marisat satellites providing global service to commercial ships and off-shore platforms. The ship terminals used a 40 W transmitter and a 4 ft diameter antenna on a stabilized platform to provide an EIRP of 36–38 dB·W in frequency band between 1537 and 1541 MHz (ratio of antenna gain to system noise temperature). The Marisat receiver has a G/T of −17 dB/K at the horizon. The link to the shore station uses 6/4 GHz. The down link to the ship uses a TWT with a maximum output power of 60 W transmitting between 1638.5 and 1642.5 MHz. The spacecraft antenna has a 4 cone helix providing a 20° beam with a gain of 14.4 dB. These small (720 lb) satellites, built by Hughes Aircraft, also incorporated lower frequency (306/254 MHz) transponders leased by the U.S. Navy.

The European Space Agency (ESA) sponsored development of a maritime communications satellite called

Fig. 18-10: Intelsat 5—some flight models included transponders at three frequency bands, built by Ford Aerospace.

Fig. 18-11: Intelsat 6—a giant high-capacity satellite being built by Hughes Aircraft for transoceanic service in the 1990's.

Marecs. This satellite, built by British Aerospace, used the same bands as Marisat. Its larger weight (940 lbs), and more powerful solar arrays (900 W) permitted 35 voice circuits to ships. Marecs A was launched successfully in December 1981.

A Maritime Communications Sub-system (MCS) package has been incorporated on three of the Intelsat 5 satellites currently in orbit. This *L*-band transponder provides 30 voice circuits for ship-to-ship or ship-to-shore communications.

The International Maritime Satellite Organization (Inmarsat) has ordered a series of satellites devoted exclusively to mobile communications in this band.

DIRECT BROADCASTING SATELLITES (1974–1984)

The first direct broadcast satellite (DBS) was NASA's ATS 6, built by the Fairchild Aircraft Company. This large (2790 lb) satellite was launched by a Titan 3C launch vehicle in May 1974. The spacecraft, shown in Fig. 18-12, had a 30 ft diameter antenna which was deployed in space. An 80 W transmitter operating at 860 MHz provided enough power to deliver television signals to small earth terminals using 10 ft diameter antennas. DBS experiments were conducted to small terminals in several thousand villages in India during 1976.

ATS 6 also had two 20 W transmitters operating in the 2.5–2.69 GHz (*S*-band) allocated for television distribution. Experiments in distributing educational television to small terminals in the U.S. were conducted in 1974 and 1975.

The first DBS to operate in the 12 GHz band was the CTS mentioned previously, and shown in Fig. 18-8. The 1 kW deployed solar arrays permitted a 200 W transmitter operating with a 28 in diameter antenna to produce an EIRP of 60 dB·W. A comparison of the signal

TABLE 18-2
GENERATIONS OF COMMUNICATIONS SATELLITES

	INTELSAT I	INTELSAT II	INTELSAT III	INTELSAT IV	INTELSAT IV-A	INTELSAT V	INTELSAT VI
YEAR OF FIRST LAUNCH	1965	1966	1968	1971	1975	1980	1986
HEIGHT (CM)	60	67	104	528	590	1570	1180
WEIGHT IN ORBIT (KG)	38	86	152	700	790	967	1670
ELECTRICAL POWER (KW)	0.04	0.075	0.120	0.400	0.500	1.200	2.264
CAPACITY (TELEPHONE CIRCUITS)	240	240	1,200	4,000	6,000	12,000	33,000
DESIGN LIFETIME (YEARS)	1.5	3	5	7	7	7	7
INVESTMENT COST PER CIRCUIT YEAR	$32,500	$11,400	$2,000	$1,200	$1,100	$800	$662
COST PER S/C ON ORBIT (MILLIONS OF $)	11.7	8.2	12.2	33.6	46.2	67.2	153
COST PER TRANSPONDER-YEAR (THOUSANDS OF '82 $)	12,480	4,180	1,710	1,020	620	530	440

Fig. 18-12: ATS 6—NASA's giant satellite of the 1970's which pioneered many new uses of communication satellites including education and medical communications.

margins on the space-to-earth link of the Intelsat I and the CTS is shown in Table 18-3.

The powerful transmitter and directional antenna of the CTS permitted delivery of a television signal to a 2 ft diameter antenna at the receiving earth station, while the weak transmitter and low directivity of the Intelsat I antenna required an 85 ft diameter receiving antenna.

Following the CTS, the Japanese ordered a broadcasting satellite called the BSE ("Yuri") from General Electric under a subcontract from Toshiba. This satellite was similar to the CTS, but used two 100 W TWTA transmitters. The BSE was launched in April 1978 by a Thor/Delta launch vehicle.

Television broadcasting on satellites operating in the fixed satellite service (11.7–12.2 GHz) band rather than in the DBS band (12.2–12.7 GHz) has begun recently in the U.S. The use of satellite master antenna television (SMATV) rather than DBS uses a Canadian satellite (ANIK-C), which has an EIRP of 49 dB·W rather than the 60 dB·W planned by DBS. This is enough power to reach receiving antennas with diameters of 4–6 ft, rather than 2 ft. These SMATV satellites could make inroads into the market envisaged by DBS proponents.

Another frequency band which has been allocated for television distribution is the 2.5–2.69 GHz band. Educational television experiments were conducted on the ATS 6 in the 2.5 GHz band or *S*-band as well as *C*-band. Both the Insat and the Arabsat (both built by Ford Aerospace) have television transmitters operating in the *S*-band.

TABLE 18-3
A COMPARISON OF THE SIGNAL MARGINS OF INTELSAT 1 AND CTS

	CTS	INTELSAT I
RADIATED POWER (EIRP) IN dBW	60	10
SPACE LOSS (dB)	−206	−197
EARTH STATION ANTENNA GAIN (dB)	35 (2')	57 (85')
RECEIVER INPUT POWER (dBW)	−111	−130
RECEIVER NOISE DENSITY (dBW/Hz)	−217	−212
CARRIER-TO-NOISE DENSITY (dB/Hz)	106	82
NOISE BANDWIDTH (MHz)	85	4.9
CARRIER-TO-NOISE RATIO (dB)	13.1	15.1

TRENDS IN FREQUENCY ALLOCATIONS AND SATELLITE EIRP (1984–1994)

The commercial communications satellites in orbit over North America in 1984 are illustrated in Fig. 18-13. Originally, these satellites were positioned 4–5° apart in longitude in the geostationary orbit. In order to accommodate the large number of applications, the Federal Communications Commission is making future assignments of U.S. satellites only 2° apart. While there is little danger of physical collision, there is concern about radio-frequency interference. The Earth station antennas must select the signals from the desired satellites, which are transmitting on the same frequencies. Small antennas may lack this discrimination, and pick up unwanted signal ("noise") from adjacent satellites.

In order to accommodate the increasing demand for use of the geostationary orbit, additional allocations were made in 1979 by the International Telecommunications Union (ITU). An additional 300 MHz were added to the *C*-band; 500 MHz were allocated to the K_u-band fixed satellite service (FSS); the DBS was moved out of the FSS and assigned its own band (18/12 GHz); and 2500 MHz were allocated in the 30/20 GHz (*K*) band. The use of cross-polarization provides the potential of doubling the use of these allocations. Spot beam antennas and spatial frequency reuse on the satellites provide the means to reuse the allocations. Another approach is the use of reverse links, i.e., reversing the up- and down-link allocations to provide another doubling of the communications capability. Finally, improvements in the utilization of the bandwidth can permit more voice circuits and television signals to share transponders.

The satellite trends of the 1970's have been towards larger spacecraft such as those for Intelsat. However, recent trends on domestic satellites have been toward smaller satellites in the delta-weight class due to the increasing launch and insurance costs, and large numbers of satellites procured from Hughes and RCA have resulted in production runs of standard spacecraft buses. Some of the U.S. domestic satellites have included both *C*- and K_u-band transponders. Weight limitations have constrained these "hybrid" satellites to full use of the *C*-band and half use of the K_u-band

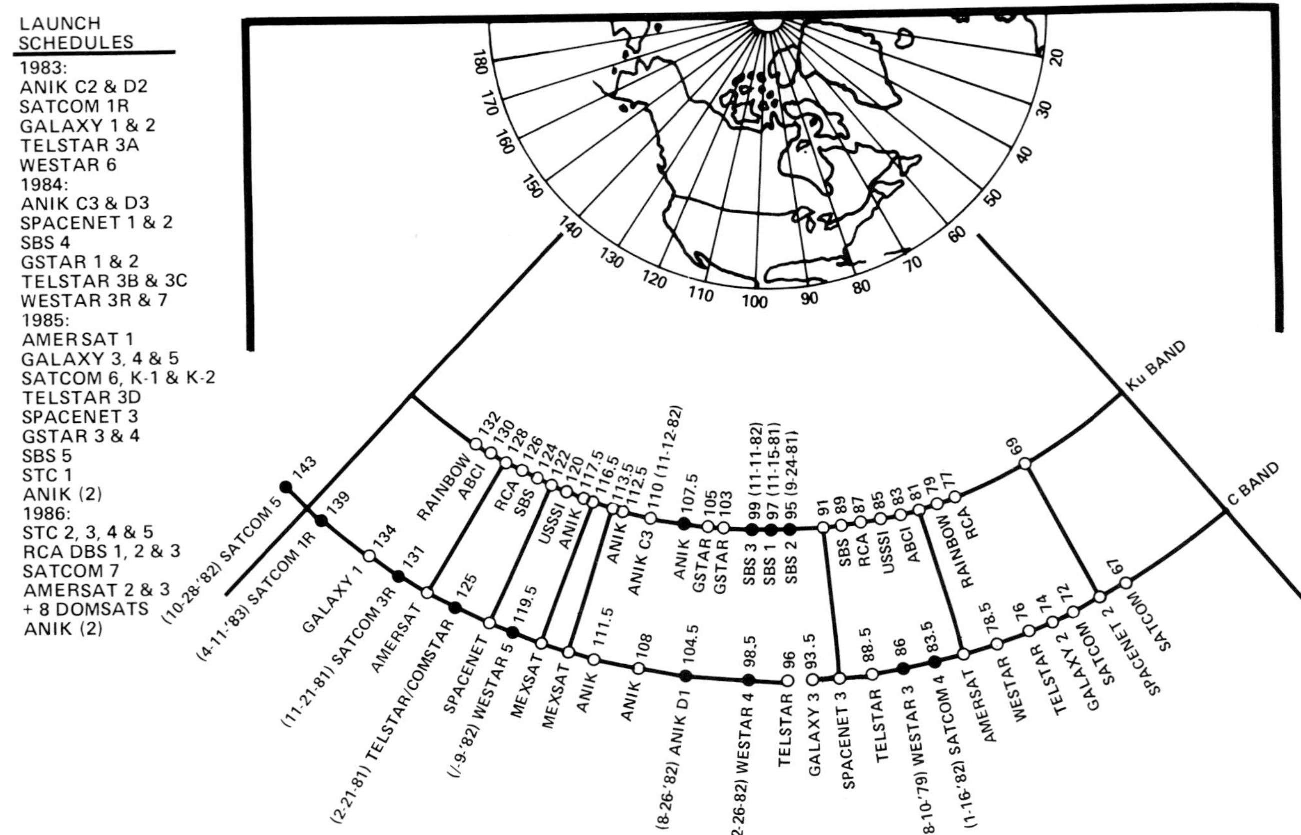

Fig. 18-13: Domestic communication satellites in orbit viewing the U.S. in 1984.

allocation. Future designs may be required to make full use of both bands to be granted a coveted orbital position and may require additional use of the 30/20 GHz band.

All modern domestic communication satellites are designed in accordance with the highly competitive factor of EIRP—or radiated power—which in turn determines both the total and per-channel bandwidth in the spacecraft, spacecraft and launch cost, the antenna size, G/T, and cost of receiving earth stations. EIRP's of 22–34 dB·W are typical of Intelsat EIRP's in pathways which use 10 and 30 antennas. In domestic satellite links, EIRP's in the 33–44 dB·W range serve pathways with antennas in the 3–10 m diameter range. In direct broadcast satellite systems, EIRP's in the 50–64 dB·W range are planned, making possible reception into very low-cost and rooftop earth stations.

The range of users between Intelsat Standard A earth stations (30 m) and DBS home earth stations (0.9 m) represents a range of costs between almost $10M for a standard A station and $250 for a DBS station. This is the economic motivation for increasing EIRP in the spacecraft to increase the earth segment size and lower costs per station, provided that the number of earth stations in the earth segment warrants the cost of increasing the communication satellite or DBS EIRP, which impacts satellite weight, cost, launch cost, and insurance cost.

DIRECT BROADCAST SATELLITES (1984–1994)

The modern 12 GHz direct broadcast satellite (DBS) has been designed according to standards set by WARC-77 to provide service into small (<1m diameter antennas) earth stations that would have costs which are attractive and competitive in the existing consumer entertainment marketplace.

Each television channel in a DBS requires a high-power transmission in the 60 dB · W range to transmit to a small earth station for successful competition with terrestrial television. The first DBS spacecraft on order in the United States (by Comsat from RCA Astro) was for a three-channel spacecraft. The primary power was 2 kW, which permits operation of three 200-W TWTA's. This is a relatively small spacecraft (1433 lbs). Larger high power DBS spacecraft are capable of transmitting six or more channels of television.

By the summer of 1982, six applications were made for high-power DBS in the U.S. However, after review of the high cost of programming, financing, and the severe competition with terrestrial cable networks, all applications were withdrawn or postponed by 1985.

Europe—which is a decade behind the U.S. in the use of CATV and cable—had already started major national DBS programs by 1980. France and West Germany set up a common office, Eurosatellite, to build the TDF-1 for France and TV-Sat for Germany. Also, Sweden contracted for TELE-X from Eurosatellite for DBS into Scandanavia. The United Kingdom started a British DBS, Unisat, but withdrew when the per-channel costs could not be underwritten by the BBC and the ITV. Italy is under contract to the European Space Agency to use one channel on the ESA satellite, Olympus, and Switzerland, Spain, Ireland, and Luxembourg have made their interest in DBS known.

All DBS satellites under construction in 1985 use high-power TWTA made by Thomson CSF in France and AEG-Telefunken in West Germany, giving Europe a technological edge in satellite technology competence. While high-power (>60 dB·W) EIRP direct broadcast satellites have been the subject of much international activity, Canada, followed by Australia, has proceeded in the direction of medium-power DBS (EIRP in the 48–56 dB·W range). Such DBS can use much lower power TWTA (in the 40 W range) and therefore make possible up to 16 DBS channels in a single 3000 lb satellite. However, medium-power satellites require earth station antenna diameters in the 1–1.3 m range to meet signal-to-noise ratio standards.

THE SWITCHBOARD IN THE SKY—ACTS (1989–1992)*

NASA's Advanced Communications Technology Satellite (ACTS) will be launched in 1989, beginning a two-year experiment (Fig. 18-14). ACTS will demonstrate reliable, efficient, point-to-point communications with new 30 and 20 GHz transmitter, receiver, and antenna technologies, and will also demonstrate advanced satellite-switched time division multiple access (SSTDMA) and baseband processor (BBP) on-board signal routing techniques usable in any satellite communications band. ACTS provides a suitable test bed to test the operation of a spacecraft designed to operate as a switching center in space.

The 30 GHz and 20 GHz satellite communications bands are not currently being used in the U.S. although they have been used by Japan since 1977. There is little spaceflight qualified hardware available at these frequencies, and conventional communications techniques used at *C*- and K_u-band (6/4 and 14/2 GHz, respectively) are unable to cope with the very high propagation losses that occur during intense rain at these (K_a-band) frequencies. Commercial communications companies prefer to compete for orbit/spectrum slots in the lower-communications bands rather than make the technology investment needed to open the new band. This is especially true when any techniques they develop are likely to be available to their competitors at greatly reduced development costs.

Fig. 18-14: The Advanced Communication Technology Satellite (ACTS) being built by NASA to test on-board switching and processing in space.

ACTS plans to demonstrate economic solutions to the 30/20 GHz variable propagation loss problem, provide hands-on experience to experimenters in a realistic U.S. domestic satellite operational environment, and develop space and ground hardware that can meet operational system requirements directly or by a defined process of scaling-up.

Opening the 30/20 GHz communications band greatly expands the U.S.'s orbit/spectrum resource. There is a 2.5 GHz bandwidth allocation available. Narrow-beam antennas can provide frequency reuse factors as great as 20, resulting in as much as 50 GHz communications bandwidth from a single satellite. Ground station antennas are inherently high gain and narrow-beam at these high frequencies, providing at least twice the number of orbit slots available at lower frequencies. These factors all combine to provide a 30/20 GHz communications capacity many times greater than the lower frequency bands combined.

The switchboard in-the-sky features of ACTS will be tested by two types of experimenter-provided terminals which will be used in the system. High-burst rate terminals will operate at 220 Mbit/s burst rates through a satellite-switched time division multiple access (SSTDMA) system. Low-burst rate terminals use either 11 or 27.5 Mbit/s burst rates through a baseband processing system.

A multiple beam antenna scans transmit and receive beams over areas covering about 10 percent of the continental U.S. (CONUS) as seen from geostationary orbit. The ACTS will combine the "hopping" of antenna beams from place to place in synchronism with the on-board switching center, routing information to these places and making possible tests of the most efficient approach to the use of each orbital slot.

MOBILE SATELLITES (MSAT-X) FOR THE 1990'S

A mobile satellite system is a satellite-based communications network that provides voice and data communications to mobile users throughout a vast geographical area. Such a system will be capable of providing mobile communications throughout the U.S., and will be especially valuable for remote or thinly populated areas that are not served by terrestrial radio systems. However, there are several major technology challenges inherent in land-mobile communications via satellite. Orbital slots for spacecraft, frequency spectrum for mobile satellite services, and spacecraft power are all limited station resources and will become more scarce as the market for mobile satellite service grows. Therefore, the challenges for MSAT-X are to develop ground-segment technologies and techniques which conserve power, spectrum, and orbit, at the lowest cost possible to the user to support future growth of the mobile satellite market.

To these ends, NASA and the U.S. industry have structured a three-phase development program. MSAT-A is aimed at development and test of advanced ground segment technology typical of what would be required for use with second-generation systems. This would prove out the basic system operation for a mobile satellite system including the ability to receive and transmit between moving vehicles and a satellite at frequencies in the 800 MHz band and/or 1.5 GHz band. MSAT-AA is a satellite of largely conventional design due for launch and test at the end of the 1980's. MSAT-B, with 20-m antennas, shown in Fig. 18-15, will have six spot beams and is due for operation in the mid-1990's. MSAT-C will use 55 m antennas which will provide up to 95 spot beams with the EIRP suitable to communicate with small terminals on moving vehicles anywhere in the U.S. MSAT-C will be designed to be assembled in space at the space station for operation after the year 2000.

THE SPACE STATION (1991–ON)

In 1992, the U.S. will assemble in space a manned space station, in response to a Presidental Directive made in January 1984. Fig. 18-16 illustrates one concept of the space station for use in low-earth orbit (LEO).

This space station is actually part of a family complex which includes the LEO manned space station, co-orbiting platforms, tethered platforms or spacecraft, free flyers, an unmanned polar orbiting platform, a staging base in the vicinity of the manned module, and service systems including the shuttle. By the year 2000, one or more geostationary platforms will be added to this family. Also included are the TDRS satellites and a global communication network capable of communicating with the tracking and data relay satellites (TDRS), the shuttle, or directly to the geostationary platform and to the manned modules when required. The manned space station in LEO will be in a 28.5° inclination accompanied by a family of co-orbiting platforms. The polar platform, also in LEO, will be able to see the portions of the earth not visible to the manned space station family and will include scientific packages for earth observations.

One of the principal features of the space station which impacts communication satellites and the use of space for communications, is the ability to bring parts of satellites from earth to the space station using the shuttle. These parts can then be assembled and tested at the space station before being transferred to final orbit.

Fig. 18-15: MSAT-B with its 20 antennas will test mobile communications to moving vehicles with rooftop antennas in the 1990's.

THE GEOSTATIONARY PLATFORM (1995–ON)

The space station as a staging and assembly base in low-earth orbit will bring about the era of the giant antennas in geostationary platforms which will open the way to very high levels of effective radiated power. In turn, this will produce corresponding technology developments in very small earth stations—down to the personal size. With the continued developments in very large-scale integration (VLSI) and computer technology, this era is expected to bring the spinoffs of the space

Fig. 18-16: The NASA space station being designed for assembly in low-earth orbit in the mid 1990's. The space station offers a unique staging area in space to assemble, test, and repair satellites.

station system to both commercial and scientific users in the forms of drastically reduced earth station costs.

If the demand for orbital positions continues, each position assignment may require a platform of payloads and antennas designed to make full use of allocated frequency bands. These payloads could belong to a number of diverse organizations who would rent space and "utilities" such as primary power, environmental control, station-keeping, and attitude control from the spacecraft "owner." The proponents of this plan envision an "economy of scale" in such a large spacecraft with common "utilities." Another school of thought agrees that full use should be made of orbital positions, but suggests this could be done better by a "cluster" of individual satellites which are co-located in the geostationary orbit. If interconnection between the diverse payloads is required, this could be done by "intersatellite links," rather than by hardwire connections on a single platform. Extensive allocations for intersatellite links have been made by the ITU, but operating systems have been confined to experimental or government-sponsored satellites.

Fig. 18-17 shows a concept of a geostationary platform based on a NASA-sponsored study done by General Dynamics and Comsat Laboratories during the 1978–1981 time frame. This platform can be built with structures carried in a single shuttle bay to the space station and has the capacity and interconnectivity of ten conventional satellites.

In the pre-space station era, satellite antennas have been relatively small in diameter (Intelsat 5's 4 GHz antenna is 92 in, and the largest of the tracking and data relay satellite antennas is 5 m) due to limitations imposed by the fairing of the launch vehicle or the width of the shuttle bay. By being able to bring satellite elements into low-earth orbit, then to assemble, deploy and test these elements at the space station, and then to transfer the platform supporting these elements to geostationary orbit, a new dimension is afforded to satellite design, capacity, and interconnectivity.

Fig. 18-17: A concept of a geostationary platform for the late 1990's which can provide high-power spot beams, on-board switching, and nodal operation.

Thus, the basic design concepts of communication satellites will drastically change from the "box" approach which was born in the 1960's based on available volume at the top of a three-stage rocket, to an extended array of structures in space designed for interconnection and conservation of the orbital arc resource.

19

NAVIGATION SATELLITES

George C. Weiffenbach

INTRODUCTION

Artificial satellites are the first practical means for attaining accurate and reliable all-weather navigation on a global basis. The art of navigation (defined as the method of determining position, course, and distance traveled) has been of intense practical interest since antiquity and has been characterized by steadily increasing demands for more accuracy and more availability, with a sharp acceleration in these demands during the last few decades. Historically, substantial resources have been invested to support and improve navigation, including the founding of major astronomical observatories, development of navigational instruments, establishment of major navigational systems, and emplacement of navigation aids. Despite these large investments, navigators' needs were never fully satisfied. Navigation by satellite is changing this situation.

The possibility of using artificial satellites for navigation was suggested before the first satellite launch. The advantages of placing a beacon in orbit are self-evident: one satellite in a high-inclination orbit at the right altitude can provide global coverage because it is visible at least twice a day from any point on earth, and this visibility makes it possible to use line-of-sight radio frequencies. Radio frequencies are all-weather, which accounts for the proliferation of ground-based radio frequency systems. However, ground-based systems are either line-of-sight and very short-range or, to achieve wider coverage, use frequencies that are refracted around the earth's curvature by the ionosphere. RF line-of-sight propagation velocities are typically stable and predictable, and thus provide stable and accurate navigation. Surface-to-surface radio propagation beyond a few hundred miles depends on the ionosphere and is subject to its unpredictable vagaries; this is an intrinsic impediment to high-accuracy navigation.

Although the superiority of satellite navigation was understood long before 1957, Sputnik demonstrated clearly and unequivocally that space had become a practical milieu; a satellite-based navigation system could be thought of as a real possibility.

1957–1958

When the U.S.S.R. announced the first Sputnik launch, there was tremendous interest in learning more about this first man-made satellite. One of the principal subjects of this curiosity was to find out, independently, just what the satellite's orbit was, and many estimates were quickly published. Most of the published orbits were based on either visual and photographic sightings or radio directional measurements. However, several laboratories used the pronounced Doppler shift of Sputnik's 20 MHz radio transmissions. The most thorough Doppler analysis was done at the Johns Hopkins University's Applied Physics Laboratory (APL), where it was found that enough information could be extracted from the Doppler shift during a single pass of the satellite over one receiver to fully determine the orbit to surprising accuracy. This result was published in *Nature* in May 1958.

The APL satellite-trackers were unaware at that time that the U.S. Navy was anticipating a serious navigation problem with its Fleet Ballistic Missile Submarines. But others at APL were aware of both this problem and the Sputnik studies; they recognized that the problem's solution might be found in a Doppler-based satellite navigation system.

The Navy's problem was that the submarines' inertial systems, though the best available at that time, exhibited unpredictable drift (characteristic of inertial systems) at rates that produced unacceptable navigation errors after a few days at sea. Since these submarines were to be out on patrol for months at a time, some means had to be found to periodically update the inertial systems. Furthermore, there were two strict conditions: the updates had to be accomplished without diverting the submarines from their patrol areas, and without disclosing the submarine's presence and location. The first condition was no problem in light of the world-wide coverage afforded by satellites. The second could not be met completely, but the Sputnik measurements had demonstrated that a submarine could obtain its updates passively through a few minutes exposure of a small non-directional antenna; the Navy considered this acceptable. They asked APL to design a satellite system to meet one explicit requirement: provide daily at-sea navigation fixes to the Ballistic Missile Submarines to an accuracy of 0.1 nm.

The Advanced Research Projects Agency, which had been given the initial responsibility for all DoD satellite programs by President Eisenhower, provided the first funds for this program in May 1958. A year later sponsorship was transferred to the Navy's Special Projects Office which was responsible for the Ballistic Missile Submarine fleet.

TRANSIT

The best description of the original concept of satellite Doppler navigation is an internal APL memo to the

Director by Dr. Frank McClure, chairman of the APL Research Center, who first suggested this navigation concept. This remarkably prescient memo reads as follows:

Yesterday (March 17, 1958) I spent an hour with Dr. Guier and Mr. Weiffenbach discussing the work they and their colleagues have been doing on Doppler tracking of satellites. The principal problem facing them was the determination of the direction which this work should take in the future. During this discussion it occurred to me that their work provided a basis for a relatively simple and perhaps quite accurate navigation system.

As you know, Guier, Weiffenbach *et al.* have been analyzing the shape of the Doppler signals from the Sputniks and Explorer I as recorded by Zink *et al.* in Mr. Riblet's group. They found that they were able to establish the parameters of a satellite's orbit with a surprising accuracy in view of the numerous difficulties that existed. In this work there were essentially eight free parameters which had to be determined. It is rather remarkable that the Doppler information received from a single pass seems to be sufficient to determine all of these parameters.

Bearing this success in mind, it occurred to me that the inverse problem, namely that of locating the observing station by analysis of the Doppler signal of a well-established satellite, would be much simpler and precision would be more easily obtained. In such a system one would use quite sophisticated equipment on land to determine the satellite orbit parameters with high precision. A receiving station would have knowledge of these orbit parameters and then Doppler observation of a pass would leave the receiving station with the problem of determining only two free parameters, namely its coordinates on the surface of the earth. Roughly it appears that an accuracy of one mile should not be too difficult to attain; this point, however, is being subjected to careful analysis.

A relatively simple antenna would suffice for getting the signal, no dishes or the like being involved. If the frequency used was up in the region of hundreds of megacycles, a satisfactory antenna should represent a surface object of extremely tiny radar cross-section which, in addition, would be unobservable from a reconnaissance satellite. Therefore the security of the position of the receiving station should be quite good.

While the computing required for determining all the parameters of an orbit from the observation of a single pass is an imposing one, it seems reasonable to believe that the two-parameter problem mentioned above should be considerably simpler, and one might hope that a relatively small computer could be designed which would do the job automatically. Thus the receiving station would be given its coordinates regularly on every pass.

While the possible importance of this system to the Polaris weapons system is clear and also its use for other Navy ships is evident, an extension of thinking in regard to peacetime use is to me quite exciting. One might envisage the following system: A satellite contains three transmitters; one of these is a highly stabilized oscillator for the purpose of providing Doppler information, another is a time signal which is corrected regularly as the satellite passes over the continent, and the third transmits the appropriate parameters of the satellite's orbit which are tape- or drum-recorded within the satellite and erased and corrected from time to time by the control stations on the continent. Thus the satellite is completely self-contained with respect to the necessary information for navigation. If a console can be designed which can use the three signals simultaneously, then the position of the antenna can be determined without reference to any other communication links. One could envision such a console on every ship at sea, giving the precise location of the ship on a one- or two-hour basis from a system of satellites, perhaps only ten in number. It occurs to me that the establishment of such a system might represent a wonderful cold war opportunity for the United States in providing world navigation to anyone who would make such a console a part of the equipment of his ship. This would put the United States in the position of being the first nation to offer worldwide service through its venture into outer space. This would be similar to the service that has been given by WWV from the Bureau of Standards.

It is also clear that it would be possible to add to the satellite navigation system communication links to our ships since a fourth and coded channel might be provided so that communication would be available to all ships through the same equipment. Again, erasable tape to store encoded messages would be used. As long as one-way transmission was all that was desired, the whole system would remain passive with respect to the ships at sea.

An immediate investigation is being made of the accuracy problem, and if this continues to look promising I feel that immediate attention should be given to the design of specialized receiving and computing equipment to do the job, to the nature of the equipment which should be aboard the satellites, to the number and selection of orbits which would be most useful, and to the nature and design of the land-based monitoring equipment.

I believe this could turn out to be one of the most important jobs APL could undertake. Since use of such a system on a worldwide basis would represent a considerable commercial venture, perhaps we ought to give attention to protecting the government from possible patent difficulties by appropriate disclosure. I therefore believe that the Patents Group should give some thought to how such protection can be obtained.

Transit was not the only navigation system considered by the Navy; in fact the most widely supported suggestion was to place a very simple CW radio beacon in an orbit at several thousand miles altitude and to navigate by observing this beacon with directional antennas in a manner analogous to the traditional sextant. (The Navy had already supported development of the millimeter-wave AN/SRN-4 radio sextant for all-weather navigation using natural radio noise from the sun and moon.) There were many arguments for this approach; the two more compelling reasons proferred were that the navigator's computations were simple and familiar, and that beacon positions could be predicted more accurately for higher rather than lower orbit altitudes (a consequence of the geodesy problem discussed below). All-weather operability was compromised to some degree by the attenuation of millimeter waves by heavy precipitation, a defect that could not be ameliorated by going to longer wavelengths because a larger receiving antenna would be needed to retain directional precision. The dish for the AN/SRN-4 sextant was already too large (~1 m diameter) for a submarine because of its large radar cross-section; this, added to the cost and complexity of a stable platform for the dish, was decisive in rejecting this approach. Submarine navigation, without satellites, by following ocean bottom topography was also considered. However, an active sonar system was required; this was an unacceptable compromise of submarine security. Many other possibilities were examined and rejected. The Navy selected Transit as the best prospect for submarine navigation.

There were questions to be answered and problems to be solved which fell into four categories: system design, hardware design and development, development of computational tools, and geophysics research.

One of the first questions that had to be addressed in the area of system design was the level of Doppler measurement accuracy needed for orbit determination and for navigation. A related question was the selection of a suitable geographic configuration for the ground tracking network. A complicating factor was the dependence of the accuracy requirements on the particular choice of orbit. These questions were resolved through simulation on a Univac 1103A computer (which used vacuum tubes!).

There was the universal problem in all space programs of developing reliable satellite hardware that could survive the rigors of launch; for Transit there was a specific requirement for a space-qualified frequency source with stability comparable to the best laboratory standards that were available at that time.

The geodetic question had two distinct but related components: position and gravimetric. The relative positions of points within continental areas had been determined by ground surveys to reasonable accuracies; geodetic datums had been established to high accuracy (a few meters) within several regions in Europe and North America, and to somewhat lesser accuracies elsewhere. However, the locations of the major datums with respect to one another were quite uncertain, as were the positions of islands. There was an urgent need for a unified world geodetic datum of known and high accuracy in which both satellite tracking stations and navigators could be positioned.

The gravimetric question was both more subtle and more uncertain. Gravimetry enters directly into geodetic positioning, because the geoid—defined as a gravitational equipotential—is the reference surface for geodetic location. The earth's gravity field is also the controlling factor in shaping satellite orbits. The mass density within the earth was known to be quite heterogeneous, and further, it had been determined that the overall figure of the earth exhibits a marked polar flattening; consequently the earth's external gravity field has a pronounced geographic structure. The orbital perturbations caused by this structure are substantial and can exceed several kilometers over a 24 h time span for a satellite at 1000 km altitude. These perturbations decrease with increasing satellite altitude (the earth looks more and more like an unstructured mass point) which was the argument put forward for high-altitude navigation satellite orbits.

Although precise orbit prediction has been a well-established tradition in navigation (the Nautical Almanac, now published in identical American and British editions, has a history of more than 200 years of uninterrupted annual publication), the accurate prediction of the future positions of a near-earth satellite is substantially more difficult than for natural celestial bodies. In addition to the gravity field problem there are surface forces that must be taken into account in calculating artificial satellite orbits, namely, residual atmospheric drag and the pressure of sunlight. (For comparison, the lunar mass/area ratio is some 7 orders of magnitude greater than that of a typical satellite; consequently, the moon's acceleration in response to solar radiation pressure is smaller by a factor of ten million.)

At that time no other satellite program had a requirement for accurate orbit prediction, nor was there a program to produce accurate geodesy at the right time in the Transit schedule. In the absence of other options the decision was made to solve the geodesy problem within the Transit program. This created formidable difficulties in developing computer programs: the program that was put together to calculate gravity field parameters was one of the largest ever written at that time. (To fit these programs into the largest computers available in about 1960 it was necessary to write all code in assembly language!)

A development plan was organized to attack the full range of problems that had been identified. The plan was based on a series of Transit experimental satellites; eight were launched, and six achieved orbit.

1959–1961

Transit 1-A (Fig. 19-1) was launched from Cape Canaveral on September 17, 1959. The third stage of the Thor Able-Star launch vehicle failed to ignite and 1-A

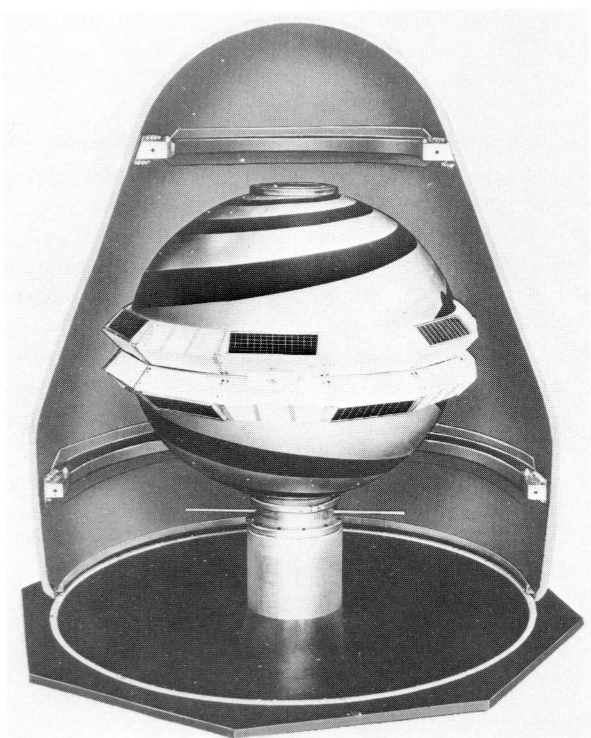

Fig. 19-1: Transit 1-A navigation satellite in flight shroud.

fell into the ocean just short of Ireland at the end of a 24-min flight. However, the satellite was in a ballistic trajectory for some 20 min, everything in the payload worked as expected, and Doppler data were acquired at two locations—enough to demonstrate in a comparison with radar tracking from the launch site that the Doppler-based trajectory was an order of magnitude more accurate than had been available with Sputnik (largely the result of more stable transmitted frequencies from 1-A).

A good deal of work had been accomplished in the 18 months between program start and the 1-A launch. The satellite itself had been designed, built, and tested. It was a 270-lb, 36-in diameter sphere with circularly polarized broad-band spiral antennas painted on its outer surface. The instrumentation was as simple as possible consistent with the mission objectives: 54/108 MHz and 162/216 MHz coherent transmitter pairs each controlled by a stable oscillator with passive thermal control, an Ag–Zn primary cell battery and a redundant and independent solar-cell/rechargeable Ni–Cd battery, and command and telemetry systems.

All of the basic orbit theory had been worked out and computer programs had been written for both orbit determination and navigation using Doppler data.

Six doppler tracking stations had been designed, built, and installed at APL in Maryland; Austin, Texas; Las Cruces, New Mexico; Newfoundland; and Farnborough, England. All the needed communications links had been arranged and the entire network carefully checked out.

Transit 1-B was launched April 13, 1960. This launch was successful, and 1-B was placed in a 51° inclination orbit with a 380 km perigee and 760 km apogee. (The intended orbit was 800 km altitude circular.) It operated for three months until a charge-limiting thermostatic switch failed and the Ni–Cd battery ran down. All of the launch objectives were met successfully.

Transit 1-B was essentially the same as 1-A, except that the 54/108 MHz Doppler pair was replaced by one at 54/324 MHz.

Most of the questions relating to system configuration and performance were settled by the information obtained from Transit 1-B. One example had to do with ionospheric propagation effects. It was known at the outset that ionospheric errors in the Doppler measurements would preclude accurate navigation. This error could be reduced by using higher satellite beacon frequencies, but this was undesirable because it made the design of the satellite more difficult. (The effective cross-section of a non-directional shipboard antenna decreases as the inverse square of the frequency. To maintain adequate signal for the navigator the power transmitted from the satellite would need to be increased correspondingly.) Furthermore, it was considered important to use solid-state devices whenever possible for all satellite electronics—and there were no suitable transistors available at that time that could operate effectively above 500 MHz.

The alternate scheme was to exploit ionospheric dispersion by using a coherent pair of frequencies; this method provides a first-order correction. The question in 1960 was the adequacy of a first-order correction in the frequency range of interest—100 to 500 MHz. The 1-B Doppler measurements allowed quantitative comparison between the 54/324 MHz and 162/216 MHz ionospheric correction terms over a wide range of ionospheric conditions. The results showed that first-order ionospheric corrections using the planned 150/400 MHz operational frequencies would be compatible with 0.1 nmi navigation.

There is one other practical consideration relating to the ionosphere; it is doubly refractive. The anisotropy caused by the presence of the geomagnetic field results in different refractive indexes for the two senses of circular polarization. A continuously changing satellite-ground station separation rotates the plane of polarization of a linearly polarized radio wave; this is termed Faraday rotation. If both transmitting and receiving antennas have linear polarization, the received signal will periodically undergo deep fading whenever the received signal and receiving antenna are cross-polarized. The simplest way to avoid Faraday fading is to transmit only one circularly polarized mode; this has been done in all Transit satellites.

Another important result of the Transit 1-B experiment was the confirmation that the earth's gravity field has a decided north–south asymmetry. The selection of different orbit inclinations for the experimental series had been made in the initial plan expressly for geodetic purposes; 1-B confirmed the wisdom of this plan when it showed the magnitude of gravity field orbit perturbations.

Transit 2-A was successfully launched on June 22, 1960 into a 66.7° inclination orbit with perigee of 630 km and apogee of 1080 km. (A Naval Research Laboratory satellite was launched piggy-back on 2-A, and the latter also contained two experiments–one from the Naval Ordnance Test Station in China Lake, California, and a cosmic-background receiver for the Canadian Telecommunications Research Establishment.)

Transit 2-A was very much like 1-A and 1-B except for orbit inclination, the removal of the redundant primary battery, and the addition of a digital clock for time dissemination to the Doppler tracking stations. It had a useful operating life of 2½ years.

Transit 3-A (Fig. 19-2) was launched on November 30, 1960. The launch vehicle malfunctioned and fell on Cuba. (Surprisingly, the Thor structure survived impact, killing a cow in the process. American newspapers carried front-page photographs of the intact first stage being paraded through Havana.)

Satellite 3-B successfully achieved orbit on February 21, 1961 at an inclination of 28.4°. Unfortunately, the Thor-Able-Star launch vehicle malfunctioned and the low perigee altitude of 180 km resulted in reentry on March 30, 1961.

The series of experimental Transit satellites ended with the successful launches of 4-A (Fig. 19-3), 4-B, and TRAAC on June 29, and November 15, in 1961. (TRAAC was launched piggy-back with 4-B.) The Starfish high-altitude fusion test of July 9, 1962 left behind an intense belt of energetic, trapped radiation that quickly degraded the 4-A, 4-B, and TRAAC solar cells and ended their operating lives. Nonetheless, all planned mission objectives were met successfully.

By the end of 1961 almost all of the elements of an operational system were available and had been tested in their operating environments: the tracking network was routinely collecting Doppler data, reducing it in real time and promptly sending the data to the APL computing center. Orbits were determined and predicted ephemerides were computed and sent to an injection station for transmission to the Transit experimental satellites where they were stored in magnetic-core memory for subsequent readout to the navigator. A prototype submarine navigation set (AN/BRN-3) had been built and was in test at APL. (Tests on submarines were to be started in mid-1962.)

1962–July 1964

Transit progressed from research and development to the operational stage during this period: the Naval Astronautics Group was established at Point Mugu, California, to run the system, navigation sets were installed on both submarines and surface ships, and operational satellites were in orbit.

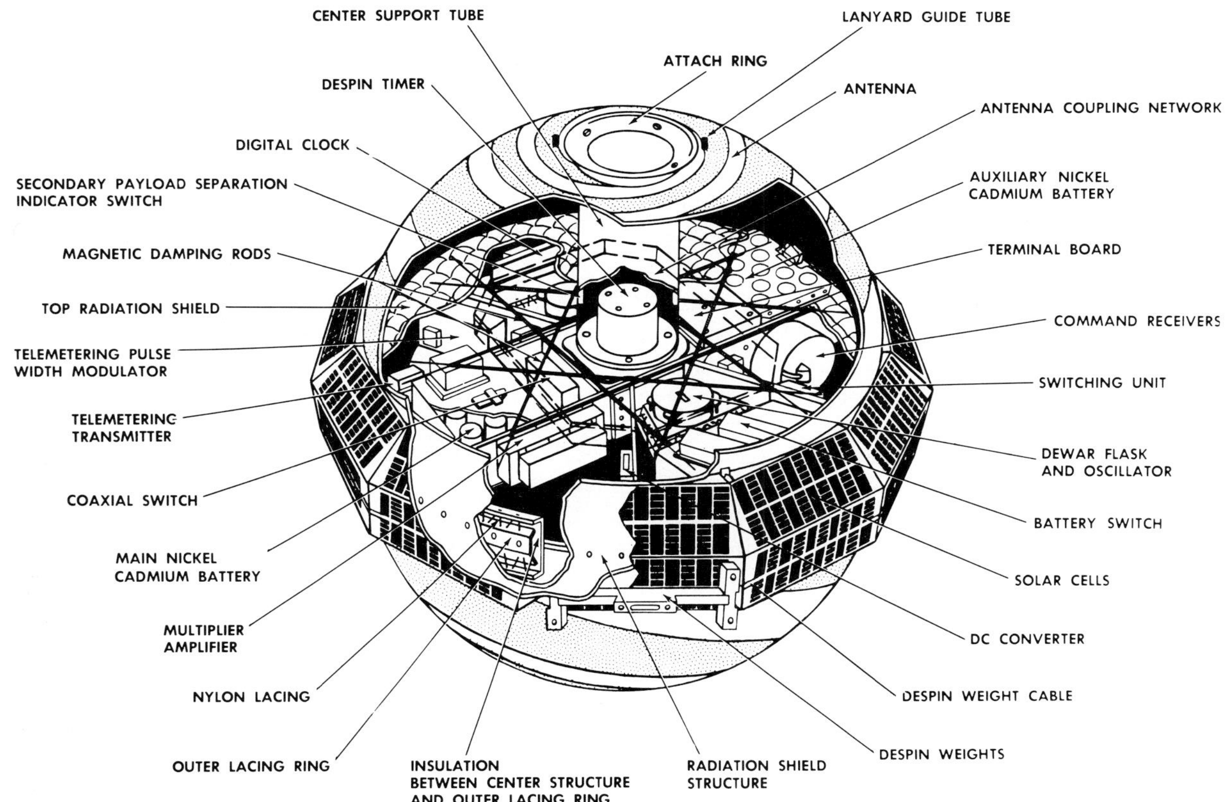

Fig. 19-2: Transit 3-A in cutaway view.

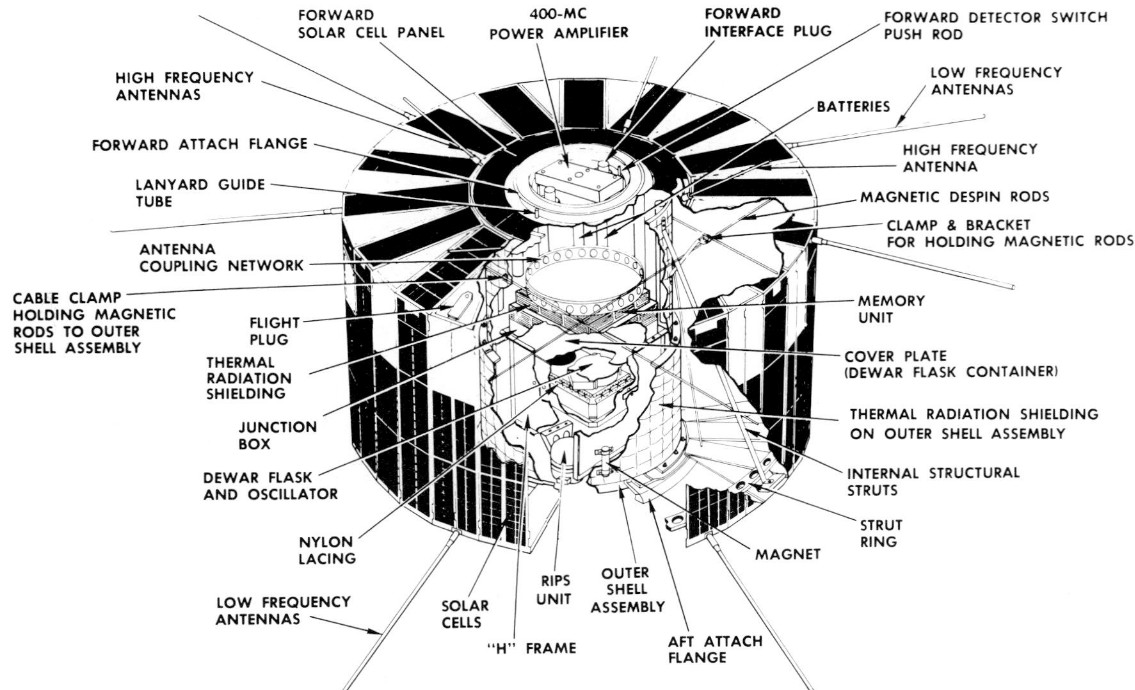

Fig. 19-3: Transit 4-A satellite in a cutaway view.

The Transit 5 satellites were developmental prototypes for the operational system. The Navy's decision to launch the operational satellites on the four-stage solid fuel Scout rocket required a total design change; weight had to come down to 120 lbs, and size had to be reduced drastically. At the same time more power was needed to support the operational electronics, and the spacecraft had to be more rugged to withstand the higher Scout thrust and vibration levels.

To avoid Faraday fading and also make the most efficient use of transmitter power, a navigation satellite should always radiate downward, i.e., the satellite attitude should be vertically stabilized. The gradient in the earth's field affords a simple means for doing this passively, as demonstrated by the moon. A body with unequal moments of inertia tends to align itself so that the axis of least moment of inertia sees the largest possible gradient. By mounting a long boom on a satellite with a mass at the opposite end, the gravity-gradient forces will align the boom vertically. (This arrangement is bi-stable, so the satellite can be either right-side down or upside down.) The difficult part of gravity-gradient stabilization is damping a satellite's oscillatory motion about the vertical. In Transit satellites this damping is done through long, thin, ferromagnetic, "hysteresis" rods, which absorb energy as they rotate in the ambient geomagnetic field. This was first demonstrated in orbit on Transit 5C-1.

By 1962 the detailed configuration of the operational system had been decided upon. Functionally, the system was to be identical to Frank McClure's March 17, 1958 concept. The constellation was to have four satellites in 1000 km altitude polar orbits, with equator crossings uniformly spaced in longitude. There were to be four operational tracking stations: East Coast, Midwest, West Coast, and Hawaii. Those in the continental U.S. were to have transmitters for injecting updated ephemerides into the satellite memories twice daily, and one was to have a colocated command, control, and computing center.

Seven Transit 5 prototype satellites were produced: three solar powered 5A's, three radioisotype powered 5BN's and 5Cl which, together with 5BN-2, became the first operational navigation satellites.

Despite two launch failures (5A-2 and 5BN-3) and numerous satellite problems (including the inversion of 5BN-1 after three months of maintaining a proper attitude—thereafter 5BN-1 transmitted to the entire universe, excluding the small volume enclosing the earth) routine operational use of Transit by U.S. Navy submarines started in 1964. Since that time there has always been at least one satellite available for operational use.

In 1958 the system error budget had set a target of 100 m accuracy in satellite position at the end of a 24 h prediction interval, and it was anticipated that this accuracy would be governed by geodetic errors. Experience with the first satellites showed that this was indeed the case; errors in fitting positional data to a calculated orbit were often as great as 1 km, and 24 h orbit predictions were frequently in error by 2 km. Positions obtained with different frequency pairs agreed to about 10 m, so it was clear that the orbit errors were predominantly geodetic in origin and did not originate from Doppler measurement errors.

By 1964, Doppler data from five satellites at four

widely separated inclinations were available for determining the gravity field: Transits 4-A (67°) and 4-B (32°), ANNA 21-B (50°), and Transits 5-E-1 and 5BN-2 (90°). These data were used to compute a gravity model and corrected tracking station positions; the resulting geodetic errors were estimated to contribute 75 m to orbit accuracy, and the accuracy of 24 h orbit predictions was better than 100 m.

A prototype AN/BRN-3 submarine navigation set was first installed on the USS Compass Island in February 1962 for at-sea tests. By the end of 1964 sets were in operational use on fleet ballistic missile submarines on patrol. Transit had progressed from a concept in 1958 to an operational system in six years; it fully met the submarine navigation requirements set by the Navy, exceeding the accuracy goal of 0.1 nmi positioning at sea.

A much simpler navigation set, the AN/SRN-9, had been developed for use on surface ships; the first prototype was installed on-board ship in 1963.

Fig. 19-4: Oscar 13 navigation satellite, not to be confused with Oscar amateur satellites.

1964–1970

For this time period, Transit underwent a series of improvements. Refinements in the gravity field model and tracking station coordinates reduced geodetic error contributions to orbit computations to less than 20 m. The gravity field model introduced into the system in 1969 was expressed in a full set of spherical harmonic coefficients for terms up to degree and order 15; this model was computed from Doppler tracking data from nine satellites.

"Oscar" (unfortunately also used for the amateur radio satellites) denotes the series of production satellites which closely followed the design of 5C-1. The original plan was to have the Oscars built at the Naval Avionics Facility, Indianapolis. After the failures of Oscars 1 and 2 to operate for more than a few days in orbit, it was decided that NAFI would complete Oscars 3, 5, and 7 while APL should build Oscars 4, 6, and 8–16. (NAFI satellites 5 and 7 were also short-lived; three failed to orbit.) RCA was selected to build serial numbers 17 on up.

Oscars 4, 6, 8, 9, and 10 were launched between June 1965 and August 1966, but all had power supply failures after about ten months in orbit. The failure was identified as open solar cell interconnections caused by thermal working. (The solar panels on these satellites are subjected to temperature variations between −80° and +60° C at orbital rate.) All of the Oscars launched after this problem was corrected have had many years of operational service; 13 (Fig. 19-4) and 14 are the longest lived with over 16 years of continuous operation in orbit, and are both alive and well at the time of this writing.*

The Naval Astronautics Group achieved an impressive level of performance in running operations, to the point that system reliability was remarkably high. This was a major stimulus in the appearance of commercial navigation sets in the marketplace during the early 1970's. The widespread interest in Transit for general navigation led APL to propose to the Navy development of a low-cost set with emphasis on cost reduction and simplicity of design rather than precision. This effort began in April 1969 and ultimately led to the development of low-cost sets by private industry (some are now available for less than $3000); this led in turn to a dramatic increase in the number of users.

By 1970 position accuracy at a fixed site using data from 40–50 Transit satellite passes had been reduced to less than 5 m. Because these position coordinates were obtained in a single global reference system—identical with the navigation reference—the Navy initiated the development of a portable geodetic receiver in 1968. This led to widespread application of TRANSIT to geodetic surveying in the 1970's.

The improvement in geodetic parameters left atmospheric drag as the main impediment to improving orbit predictions. The density of the atmosphere at Transit satellite altitudes varies in time by almost two orders of magnitude. These variations are induced by solar activity; they are not predictable. These drag effects were large enough that new satellite orbit parameters had to be injected into the operational satellite memories twice each day. For obvious reasons, the Navy wanted orbit predictions that would meet their accuracy requirements for at least one week in advance to reduce the dependence of the operational satellites on the ground system, and thus have satellite autonomy for seven days into the future without daily orbit injections.

Interestingly, the pressure of sunlight on a spacecraft exceeds atmospheric drag above some 700 km; unlike drag, sunlight does not continuously act in opposition to satellite velocity, can be reliably computed, and thus does not seriously degrade orbit prediction.

* *Note: Oscar 14 foiled in November '84; Oscar 13 is still operational as of 15 June 1985.*

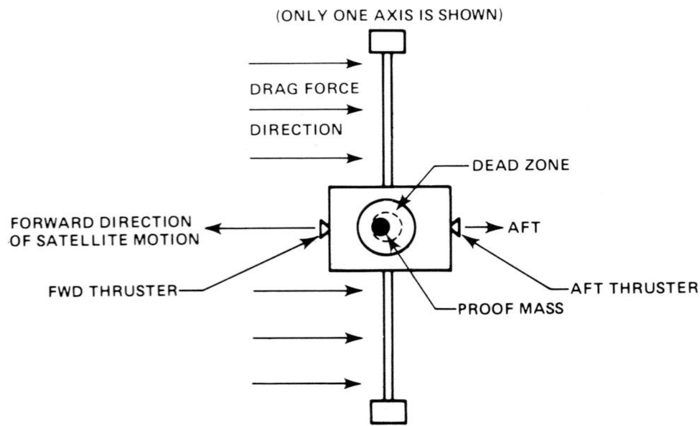

Fig. 19-5: Triad satellite which included the Lange "disturbance compensation system," Discos.

Fortunately, Lange[1] had conceived an ingenious method for eliminating drag accelerations. Lange proposed that a small satellite should be enclosed within a larger satellite. The small satellite—or proof mass—is thus shielded from all surface forces, namely, drag and sunlight. If the position and motion of the proof mass relative to the enclosing satellite is sensed, this information can be used to activate thrusters to keep the outer satellite centered on the proof mass. The outer satellite will then be constrained to the orbit of the proof mass which is essentially purely gravity-controlled.

The design of a satellite to test this concept was started in 1969 by APL and Stanford under Navy sponsorship.

1971–1984

The Lange disturbance compensation system (Discos), shown in Fig. 19-5, was implemented on the Triad satellite which was launched on a Scout vehicle from the Western Test Range on September 2, 1972. Triad was a 94 kg satellite with a 30 W radioisotope thermoelectric generator and gravity-gradient attitude control system.

The Triad proof mass was a 22 mm diameter, 111 g sphere made of a 70-percent gold, 30 percent platinum alloy to minimize magnetic susceptance and thus interaction with the geomagnetic field. Proof mass position was sensed by the capacitance between proof mass and three orthogonal pairs of capacitor plates that were part of the proof mass housing. Proof mass position controlled the firing of six orthogonal microthrusters using compressed Freon 14 as a cold-gas propulsion system.

[1] *B. Lange, "The drag free satellite," AIAA J., vol. 2, pp. 1590–1606, Sept. 1964.*

Discos operated successfully in orbit until the Freon fuel was exhausted after some 14 months; it produced a dramatic improvement in the accuracy of predicted orbits. Fig. 19-6 shows the in-orbit motion of the proof mass with its ±1 mm dead band. A secondary, and successful, objective was fine tuning of the Triad orbit by intentionally applying a bias to the capacitive pickoff electronics; this provides orbital station keeping at the 10 m level.

The development of a second-generation operational Transit followed the success of Triad. The major objective of this effort was to improve survivability. In the first instance this was through reduced vulnerability to high-altitude nuclear explosions. This was accomplished by selection of radiation hardened components, external shielding, circuitry designed for lower vulnerability, and event detectors to remove power from vulnerable circuits plus automatic restoration of navigation functions post EMP. Secondly, Discos orbit adjust and station-keeping was to prevent clustering of the satellites (all in polar orbits) at the poles to avoid simultaneous exposure and loss of the entire operational constellation in a single event. By extending the accuracy of satellite-stored ephemerides to seven days, Discos also reduced system dependence on vulnerable ground-based support elements.

Two second-generation prototype satellites were placed in orbit: TIP-II on October 12, 1975 and TIP-III on September 1, 1976. In both cases, the solar blades failed to deploy fully and power availability was severely curtailed. Although full spacecraft operation was not achieved, enough information was obtained to proceed with the operational satellites.

Three second-generation operational Transit satellites, called NOVA, have been built by RCA; NOVA 1 was launched on a Scout from WTR May 15, 1984, and is fully operational. NOVA 3 was successfully launched on October 12, 1984, which brings to seven the number of operating Transit satellites now in orbit.

Although it was designed to meet one specific requirement, the Transit system has been successful in a wide range of applications. It exceeded the U.S. Navy's

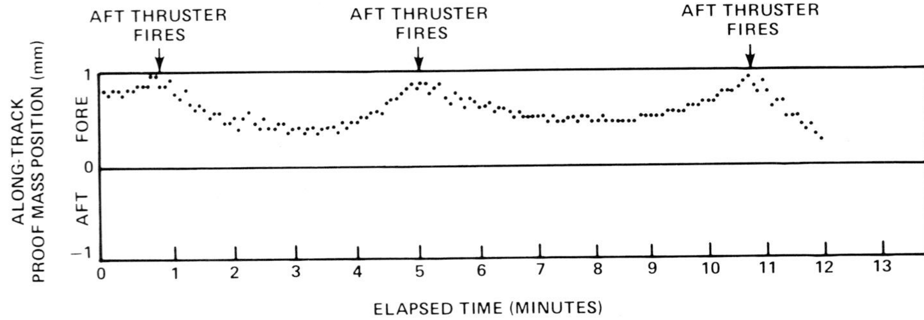

Fig. 19-6: Discos proof mass position versus time.

submarine navigation requirements—the sole objective of the system—and as incidental side benefits, it has been adopted as a primary geodetic standard by the U.S. and other countries, it is used in oil exploration, oceanographic research, and for shipboard navigation. More than 80 000 Transit receivers are now in use.

GLOBAL POSITIONING SYSTEM (GPS)

System Definition: The Global Positioning System (GPS) is a spaceboard radio navigation and nuclear burst detection system. The navigation system shall provide a signal environment which shall enable GPS navigation users to precisely determine three-dimensional position, velocity, and system time. The NUDET (Nuclear Detonation) Detection System (NDS) shall provide NDS receiver equipped users with the capability of near real time, global reporting and precision location of nuclear events. The two missions of the system, when discussed jointly, shall be known as GPS.

The space segment shall be composed of at least 18 operating Block 2 space vehicles (SV's) in 12 h circular orbits and shall include support equipment and a flight support computer program. The GPS shall be considered to be available to navigation users as long as users have at least four operating satellites in view. The GPS shall be capable of time data. Worldwide three-dimensional capability shall be achieved when the 18 SV's are operational. When this 18 SV constellation is in place, the GPS shall be capable of providing a three-dimensional position spherical error probable (SEP) of less than 15 m and a vertical position linear error probable (LEP) of less than 10 m for a set of navigation users uniformly distributed over the earth and uniformly distributed in time.

Two seminal programs emerged in the sixties that became the basis for the Global Positioning System (GPS): Air Force 621-B and Navy Timation. These programs were intended to address a basic shortcoming in Transit.

A Doppler shift is produced by a changing distance between transmitter and receiver; the navigator's movements make their contribution to this range rate. For a submarine with an excellent inertial system, the ship's movements are precisely known so their contribution to the observed Doppler can easily be accounted for. Typically, platform motions are less well known. As a rule of thumb for Transit, position error in nautical miles is one tenth of the platform velocity error in knots.

Another shortcoming is that Doppler measurements must necessarily be made over a significant time interval; for Transit this is several minutes. This is not a problem for ships, but can be a drawback for aircraft.

The objective of both Timation and 621-B was to develop a navigation satellite system based on simultaneous passive ranging to three or four satellites. In principle, the navigator can then make *instantaneous* fixes whose accuracy is independent of platform motion.

1969–1978

The Navstar Global Positioning System formally came into being with the merger, directed by the Deputy Secretary of Defense in 1973, of the Naval Research Laboratory's Timation program and the Air Force's 621B project. This developmental GPS program culminated in the launches of four developmental Block I satellites in 1978.

A GPS Joint Program Office was formed with personnel from the Air Force, Army, Navy, Marine Corps, and the Defense Mapping Agency; the Air Force was further designated as the lead agency to develop, test, acquire, and deploy the system. The intent was to provide continuous three-dimensional global positioning at the 10 m level. An important constraint was that all system ground components should be located within the coterminous U.S. this precluded geosynchronous orbits. The concept that was adopted (Fig. 19-7) called for eight satellites in each of three orbit planes for a total of 24. The orbits were to be circular with a 12 h period (more precisely, 4.79 s less than one half of a sidereal day—some 43 077.26 s) at an altitude of 20 190 km, so that a minimum of four satellites would always be in view at any point on the earth.

The key to all accurate range measurements (based on the speed of light) is extraordinary accuracy in measuring propagation time. To meet the 10 m accuracy goal in GPS while preserving passive ranging, it is necessary to maintain synchronization among the satellite clocks and user platform to within a few nanoseconds; unless the spacecraft are monitored continuously, this requires satellite clock stability of the order of one part in 10^{-14} per day. This could be attained only with atomic standards. Further, to avoid the need for an atomic standard in each and every navigation set, the system must continuously provide four in view; in this way the navigator can solve unambiguously for four parameters: latitude, longitude, altitude, and epoch.

The development of space-qualified precision clocks was a centerpiece of the Timation program; T-1 and T-2 spacecraft had quartz crystal clocks with stabilities of about 3×10^{-11}/day. T3/NTS-1 contained rubidium clocks and exhibited a stability of $5\text{–}10 \times 10^{-13}$/day.

Timations 1–3 were experimental satellites developed under the aegis of the Navy's Timation program. T-1 was an 85 lb satellite launched into a 500 nm altitude circular orbit with an inclination of 70°. It was one of several payloads launched from the Western Test Range on a Thor-Agena rocket.

Timation 1 was launched May 31, 1967 into a 70° inclination circular orbit with an altitude of 900 km. It was one of a cluster of satellites carried on a Thor-Agena D rocket. This 39 kg satellite was gravity-gradient stabilized, powered by solar cells, and transmitted a single navigation carrier at 400 MHz. T-1 demonstrated the feasibility of both aircraft and ship positioning through passive range measurements.

Timation 2 was also launched on a Thor-Agena D

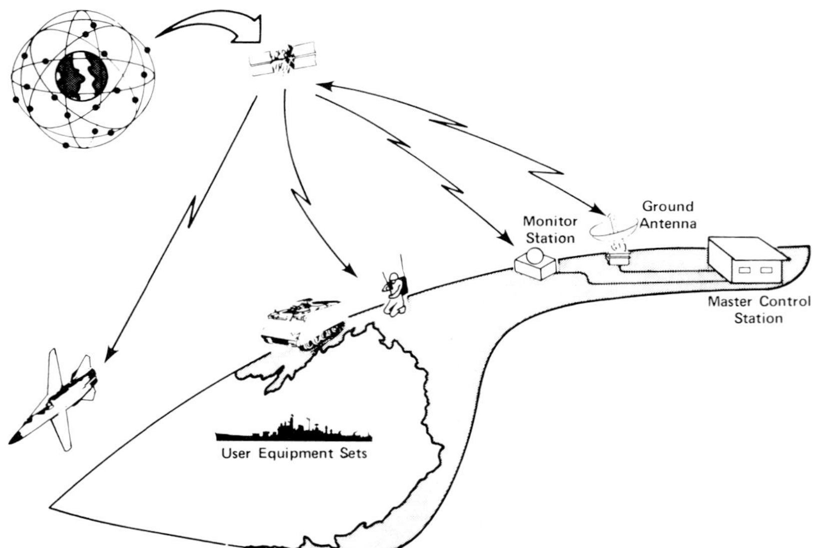

Fig. 19-7: The Global Positioning System (GPS)—Navstar.

into a similar orbit. The 57 kg satellite transmitted on two coherent frequencies at 150 and 400 MHz; this allowed correction for ionospheric propagation effects in a manner analogous to Transit.

The final satellite in the Timation series was launched July 14, 1974 on an ATLAS-F rocket after the formation of the GPS Joint Program Office; in consequence, the satellite was designated as Navigation Technology Satellite 1. This 293 kg spacecraft was placed in a nearly circular retrograde orbit at 125° inclination and 13 600 km altitude. NTS-1 was gravity-gradient stabilized, solar powered, transmitted both UHF and *L*-band navigation signals, and carried one quartz and two rubidium clocks. The *L*-band carrier was modulated with a pseudorandom noise ranging signal.

NTS-2 was the first satellite completely designed and built under the sponsorship of the Navstar GPS program. The 2000 km orbit had the 12 h period planned for the operational GPS spacecraft. The 430 kg, 400 W NTS-2 was placed into orbit by an Atlas-F rocket on June 23, 1977. It carried one quartz and two cesium clocks which demonstrated an in-orbit stability of $1\text{–}2 \times 10^{-13}$ per day. The navigation signals were broadcast on the L_1 and L_2 operational GPS frequencies.

A third technology satellite, NTS-3, was intended to carry advanced technology hydrogen maser clocks to determine the feasibility of operating GPS satellites independent of ground support for extended periods of time. However, NTS-3 was never launched.

Four Block 1 operational prototype NAVSTARS (Fig. 19-8) were launched in 1978: NDS-1 on February 22, NDS-2 on May 13, GSP-3 on October 7, and GPS-4 on December 11. All were launched on Atlas-F rockets into 12 h 63° inclination orbits. The orbital phases of these four satellites were selected so that all four were periodically in view of ground sites at the same time to support operational system tests of various kinds, including master control station (MCS) operations, a variety of developmental user sets, and orbit computation and predictions. (One of the more important tests concerned the use of this preliminary constellation as range instrumentation for the Navy's Trident Improved Accuracy Program; initial tests with one and two spacecraft were conducted in June and August of 1978. These tests were quite successful, and led to the routine use of GPS for determining Trident trajectory information; at the time of this writing some 43 developmental, shakedown, and operational readiness tests have been conducted.)

The Block 1 operational prototypes are 526 kg, vertically stabilized spacecraft transmitting two *L*-band navigation frequencies L_1 and L_2 at 1575.42 and 1227.6 MHz. All carry four redundant rubidium atomic clocks; from GPS-4 on, an additional cesium clock is added. The Block 1 design life is five years.

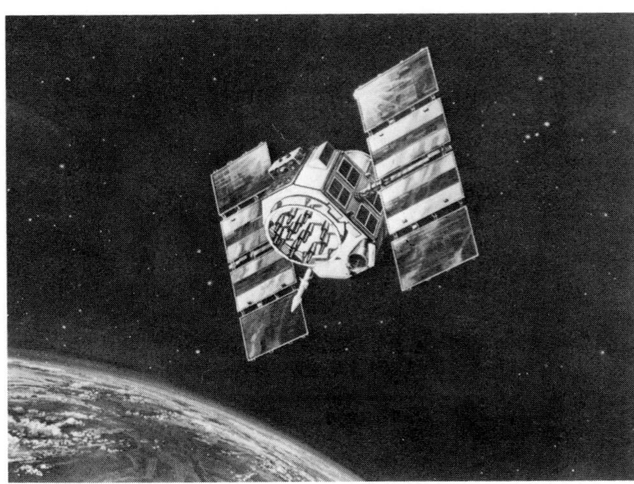

Fig. 19-8: A Navstar satellite.

1978–1988

By 1984, three additional Block I satellites were placed in orbit by Atlas-F rockets: GPS-5 on February 10, 1980, GPS-6 on April 26, 1980, and GPS-8 on July 14, 1983. A launch failure occurred with GPS-7 on December 18, 1981. Table 19-1 shows the status of the constellation as of February 1, 1984.

By 1986 up to six prototype satellites should remain operational for test purposes, providing two to three hours of system coverage per day for most parts of the world. Twelve satellites should be operational by the end of 1987, providing continuous, two-dimensional navigation worldwide, and 18 satellites should be operational by the end of 1988 (full operational capability) providing worldwide, continuous three-dimensional navigation (latitude, longitude, and altitude).

The 18 earth-orbiting satellites will be in six orbital planes spaced 60° apart in longitude and inclined to the equator at 55°. This is quite different from the polar orbit of Sarsat and the geostationary orbits of Geostar and the mobile satellites.

When fully operational, the system should be notable for its operational simplicity and low civil-user terminal cost ($500–$1500). A user fee of $370/year has been proposed by the Department of Defense. Civil accuracies will be limited to about 50 m; military accuracies will be better.

RADIOLOCATION FOR SEARCH AND RESCUE (SARSAT)

One area in which radiolocation satellites promise major achievement is in search and rescue, on land and at sea. The first such program, already working, is the COSPAS/Sarsat program. This international cooperative project uses satellite technology to detect and locate emergency or distress situations involving aircraft, vessels, or people. Still in its infancy, it has already helped to save close to 300 lives (as of August 1984). The U.S., Canada, France, and the Soviet Union developed the system. Norway, the U.K., Bulgaria, Finland, and others are participating.

The first satellite, COSPAS, was launched into polar orbit by the Soviet Union on June 30, 1982. Since then, two more COSPAS and the first Sarsat (search and rescue satellite-aided tracking) satellites were launched.

Within the U.S., the Sarsat program is a cooperative effort among NASA, the U.S. Coast Guard, the Federal Aviation Administration, the U.S. Air Force, and NOAA which uses the NOAA satellites.

The system operates as follows: The signals relayed by a distress transmitter on a vessel or aircraft are received by the polar orbiting spacecraft, and relayed to a local user terminal (LUT) where the signals are processed to determine the location of the distress transmitter. Although the electronic characteristics of these distress transmitters are identical, the aircraft units are referred to as ELT (emergency locating transmitter) and the marine units are called EPIRB (emergency position indicating radio beacon).

Two distinct experiments are being conducted: one on 121.5/243.0 MHz (the frequency of today's ELT's and EPIRB's) and the other on 406.0 MHz. The 406 MHz system is superior in performance and other qualities. The probability of detection of signals will exceed 95 percent and the location accuracy will be 2–5 km.

TABLE 19-1
GPS CLOCK STATUS: 2/1/84

SATELLITE	LAUNCHED	CLOCKS	STATUS
1	22 FEB 78	(R)▲ (R) (R)	"ON" VCXO MODE (MARGINAL USE)
2	13 MAY 78	(R) (R) (R)	OFF
3	6 OCT 78	(G)▲ (R) (G)	"ON"
4	12 DEC 78	(R) (R) (G)▲ [R]	"ON"
5	10 FEB 80	(G) (G) (G) [G]▲	OFF (REACTION WHEEL FAILURES)
6	26 APR 80	(G) (G) (G) [G]▲	"ON"*
8	14 JUL 83	(G) (G)▲ (G) [G]	"ON"

CODE:
○ —RUBIDIUM
□ —CESIUM
G —CLOCK IN SPEC.
R —CLOCK NOT IN SPEC.
▲ —CURRENT "ON" CLOCK
* —MOVEMENT OF 1-2 PARTS IN 10^{12} (1/29 & 1/30 1984)

21,384-84

THE GEOSTAR SYSTEM

Still another application of radionavigation and position location by satellite is the Geostar system. Geostar Corporation has applied for frequencies and a developmental license for a satellite system, in geostationary orbits that would provide navigation, position-location, and digital transfer services to subscribers throughout the continental U.S.

According to Geostar, each user would be equipped with Geostar terminals called automatic beacon transponders (ABT's). Each ABT will be capable of displaying one or more types of information: user position, speed, answer to a question, a warning message, or messages from other users. Geostar estimates that the pedestrian unit will cost about $450 and that monthly service charges will range from $30–40. Geostar claims its radio-positioning system will provide position information with accuracies of 1–7 m.

1990 AND BEYOND

By 1990 radionavigation and location users will have many options; terrestrial systems such as loran and Omega and satellite-based systems typified by GPS, Geostar, and Transit, and others being proposed. Of course, combinations of terrestrial and satellite systems will also be possible, such as mobile telephone, search and rescue, navigation, position-location, television broadcasting, and two-way data networking.

The capability of determining position will also extend to the space station and its family of moving platforms. Here, the GPS system will provide accurate position determination of each of the spacecraft, making possible a continual location determination of the full network or family of spacecraft relative to each other in addition to individual positions. This will well assure collision avoidance and aid in interspacecraft communications where timing and acquisition are important.

20

COMMUNICATIONS TECHNOLOGY

C. Louis Cuccia and Joseph Sivo

INTRODUCTION: ENABLING AND SUPPORTING TECHNOLOGY

In order to determine the optimum utilization and capacity of communications and DBS satellites, it is necessary to understand the role of the enabling and supporting technologies that establish these parameters. Two decades of technological developments have seen many advances in the design and utilization of enabling and support technologies.

The *enabling* technologies for communications are those technologies involved with antennas, power amplifiers, receivers, channelization (filter) devices, and on-board signal switching and processing. The basic *support* technologies are those which supply the power to the communications payload, and which point the antennas to the proper place on earth. It is these enabling and support communication technologies that will be discussed in this chapter.

Communications satellite technologies determine the number of channels and the per channel effective isotrophic radiated power (EIRP) required to reach an earth station on earth 22 000 miles away, into an antenna of required size and a receiver of specified sensitivity to provide a designated service. In a typical communication type of satellite, for example, 24 communication channels are provided, with each channel using a TWTA or solid-state amplifier having an output around 5 W, and producing EIRP's in the 30 dB·W range in a bent pipe mode in a common antenna beam. More advanced communications satellites will use fixed or electronically hopped antenna beams with on-board routing, based on technologies to be tested in the NASA ACTS satellite.

Historically, communications satellites of the early Intelsat variety had EIRP's in the 25 dB·W range requiring earth stations with 30 m antennas costing many millions of dollars. As the radiated power level of communications satellites increased in the 1970's into the 30–36 dB·W range, earth station antennas reduced in diameters to the 3–13 m range and costs reduced from the million dollar to the thousand dollar range, depending on amount and type of information transported. The availability of low-cost earth segments based on moving advanced technology and complexity into the space segment is the base line to the favorable economic environment in which communications satellites have developed.

The broadcast satellite, on the other hand, may have fewer channels. The high-power broadcast satellite, TV-SAT (Germany), for example, which was designed in accordance with standards defined in the 1977 WARC, will provide only 4–5 channels using a high-power TWTA in the 200 W power output range and a high gain antenna, to provide a narrow, contoured beam to illuminate only West Germany with an EIRP greater than 60 dB·W watts in each channel. However, economic pressures based on channel cost and the technological improvement in earth station sensitivity since 1977, are now leading to medium-power broadcast satellites with EIRP in the 50–55 dB·W range, which is driving satellite designs to 16 channels with 40 W power amplifiers and therefore reducing per channel costs. Broadcast satellites of both the high- and medium-power varieties can operate into 1 m antenna TVRO earth stations costing only a few hundred dollars.

Thus, satellite enabling and support technologies are developed and utilized in an economic environment which is impacted by system costs and user revenues. This chapter will be directed toward discussing the TIME-line development of enabling and support technologies, dealing with satellite radiated power, sensitivity, channelization, and on-board routing which reflect these economics and costs.

The Spaceborne Satellite Repeater

Since the earliest days of satellite communications, the spaceborne active-repeater has provided the basic functions for 1) receiving an up-link signal arriving at one antenna in one frequency band, 2) converting this signal to a second frequency, usually in the down-link frequency band, and 3) providing significant amplifier gain to produce from the down-link antenna sufficient effective radiated power to energize various earth terminals and to make possible the demodulation of one or more channels of information of specified quality from the receivers in these terminals. These functions are diagrammed in Fig. 20-1 which illustrates the route of information through an entire satellite link from the original data input, through the transmitting earth station, up and through a transponder channel on the satellite, and down to a receiving earth station which produces the transported information.

The antennas are the terminal points of the up link and the down link and as such, determine the interconnectivity of the space and earth segments. The power amplifiers with their transmitting antennas energize the up and down links, while the receivers with their receiving antennas establish the sensitivity to radiation from the transmitting antennas.

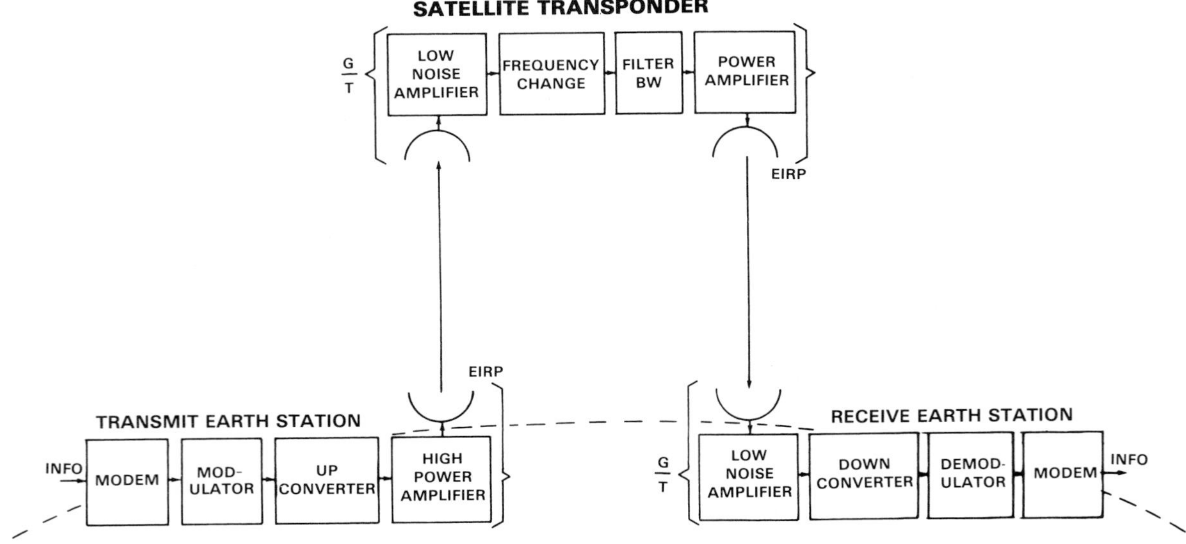

Fig. 20-1: Components of a satellite link including both earth and space segments.

The filters in the satellite determine the channelization. For modern communications satellites, 36 MHz bandwidths are commonly used for applications such as telephony and television distribution, and in, for example, the 500 MHz bandwidth in the 3.7–4.2 GHz down link, it is possible to situate twelve 36 MHz channels side by side with guard bands of 4 MHz between the channels. Modern communications satellites use filter multiplexers which provide the channelization required, and it is possible to double the number of channels to 24 simply by transmitting 12 channels in one polarization and 12 channels in another.

The communications satellite presently provides a "bent pipe" interconnection between transmitting and receiving earth stations, often within the same antenna beam. In order to conserve spectrum by reuse of the frequency bands used in satellite communications, the routing of information through multiple spot or contoured antenna beams from the satellite is being planned for future satellite systems based on technologies in development during the 1980's. These technologies which apply to both antennas and on-board switching systems provide for a nodal point in space where information can be transmitted from one up-link location to be switched, processed, and routed to other down-link locations (or the same location) as is illustrated in Fig. 20-2. This use of a satellite in principle provides for a switching center in space with the same basic functions as a conventional switching center such as an ESS-4 on earth. This function involves both multiple-beam antenna technologies and switching and signal processing technologies which have been developed by NASA for the ACTS satellite.

Communications Satellites Transponder Frequencies

Spacecraft transponder design and technologies are largely governed by the frequencies involved. There are six frequency bands which are now used for commercial satellite communications in the U.S. *L*-band (1.5 GHz) is used for maritime communications. *C*-band (6/4 GHz)

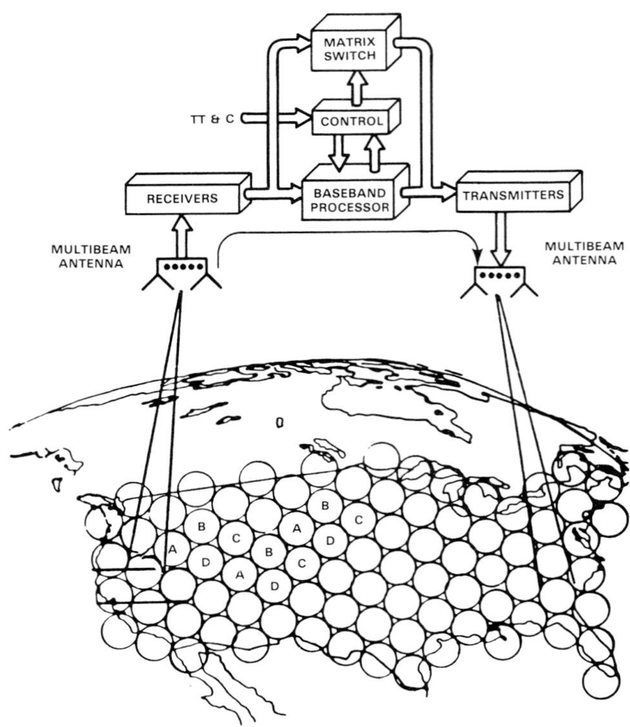

Fig. 20-2: Satellite transponder designed to provide a "switchboard in the sky" between earth areas.

is used for telephony, messages, TV distribution, audio distribution, and business communications, and is presently experiencing orbital congestion. X-band (7/8 GHz) is used primarily for military communications. K_u-band (11.7–12.2 GHz down link and 14–14.5 GHz up link) is used for data and telephony. The down-link band 12.2 to 12.7 GHz was recently approved by WARC-79 for direct-to-user television broadcast in Region 2. K-band (17.7–20.2 GHz down link and 27.5–30 GHz up link) is now perceived as an ideal frequency band for business and data communications. At optical frequencies, in the 0.8 μm range, intersatellite links can provide spacecraft-to-spacecraft communications.

In the 1960's and early 1970's, the use of the 4/6 and 7/8 GHz frequencies became standard, primarily using single frequency conversion transponders. As the move to frequencies above 10 GHz was made for satellite communications, the dual conversion transponder was readily adopted. Satellites such as Intelsat 5 and Japan CS double-converted from their K_u-band and K-band input frequencies to C-band as an intermediate frequency and back to output frequency. The European OTS satellite converted from 14 GHz to around 800 MHz and then back to 11 GHz. The future satellites designed for on-board signal processing and routing will provide direct reduction of data to baseband from their arriving modulated carriers, with the baseband information stored, processed, and routed before being applied for direct modulation at down-link frequencies, thus eliminating the need for frequency changing.

THE ENABLING COMMUNICATIONS TECHNOLOGIES OF COMMUNICATIONS SATELLITES

Fig. 20-3(a) provides a time line of the generic spacecraft enabling technologies. These technologies include the antenna, the power amplifiers, the low-noise amplifiers, LNA, switching systems, and on-board routing and intersatellite laser and millimeter-wave technologies. Fig. 20-3(b) lists the support technologies which complement the enabling technologies.

The space segment marketplace includes satellites optimized for bandwidth using the largest number of communications transponders at relatively low per-channel EIRP, or with satellites optimized for highest radiated power but with relatively small total operating bandwidth.

In the design of the satellite system, several parameters must be accounted for in determining satellite sensitivity (G/T) or satellite radiated power (EIRP) mentioned in Fig. 20-1. The parameters EIRP and G/T of a communications payload for either a communications satellite or a DBS cannot be attributed to a single technology, but rather to a combination of technologies. The satellite EIRP for example, is the combined contribution of both the transmit or down-link antenna and the output of a power amplifier, less any connection and filter losses; the antenna technology is only applicable if supported by a pointing or attitude control system which points the antenna to the area to be served.

The satellite sensitivity parameter G/T is also provided by a combination of technologies. The gain of the up-link antenna (or spot) and the noise temperature of the low-noise amplifier contribute to G/T which, in turn, establishes the signal-to-noise ratio at the final receiver demodulator.

Two key parameters which are filter dependent are the per channel bandwidth and the total satellite bandwidth—particularly for communication satellites. Table 20-1 lists the per channel bandwidth—and EIRP—of early and modern communications satellites. For comparison purposes, Fig. 20-4 shows how bandwidth has increased from 50 MHz on Intelsat 1 achieved in 1965 to 2300 MHz for Intelsat 5's, then increasing to 3800 MHz for Intelsat 6 in 1986, and to an excess of 6 GHz for advanced model spacecraft of the 1990's.

Fig. 20-3(a) and (b) has already shown the major enabling and support communications technologies in terms of major technological developments since 1963. The various advances in technologies listed in this figure are a result of the large number of generations and changes in technology which have been made in two decades, under the driving influence of six generations of Intelsat satellites, and the growth of domestic satellites in both the U.S. and Canada, leading to over 20 satellites in orbit by 1984, from the pioneering Syncom 2 sponsored by NASA in 1964.

These technological advances which have occurred in so short a time period include the development of fundamentally new technologies, such as the gallium arsenide field effect transistor, which did not exist except in laboratories at the microwave frequencies in the 1960's and early 1970's, and which became the main challenger for low-noise amplifier, power amplifier, and switch applications in the late 1970's.

Satellite Capacity

The capacity of a communications satellite is related to the number of channels provided and the EIRP and G/T associated for each of these channels. However, the capacity of a communications satellite is also dependent on the in-orbit dry mass weight, the percentage of that weight required to develop dc power (batteries and solar cells) and the percentage of that weight devoted to the communications payload which includes both the antennas and the transponders.

Fig. 20-5 and Table 20-2 relate satellite in-orbit dry mass weight to primary power and the total dry mass for satellites in common use. Note that Intelsat 5 has a dc power of around 1000 W. In most satellites, approximately 80 to 85 percent of the dc power is used to power the transponder. Also, due to the use of new lightweight materials such as graphite epoxy, the percent of the dry mass used for the communication function is now between 25 and 30 percent. As satellites become larger, it can be expected that the percentage of communications

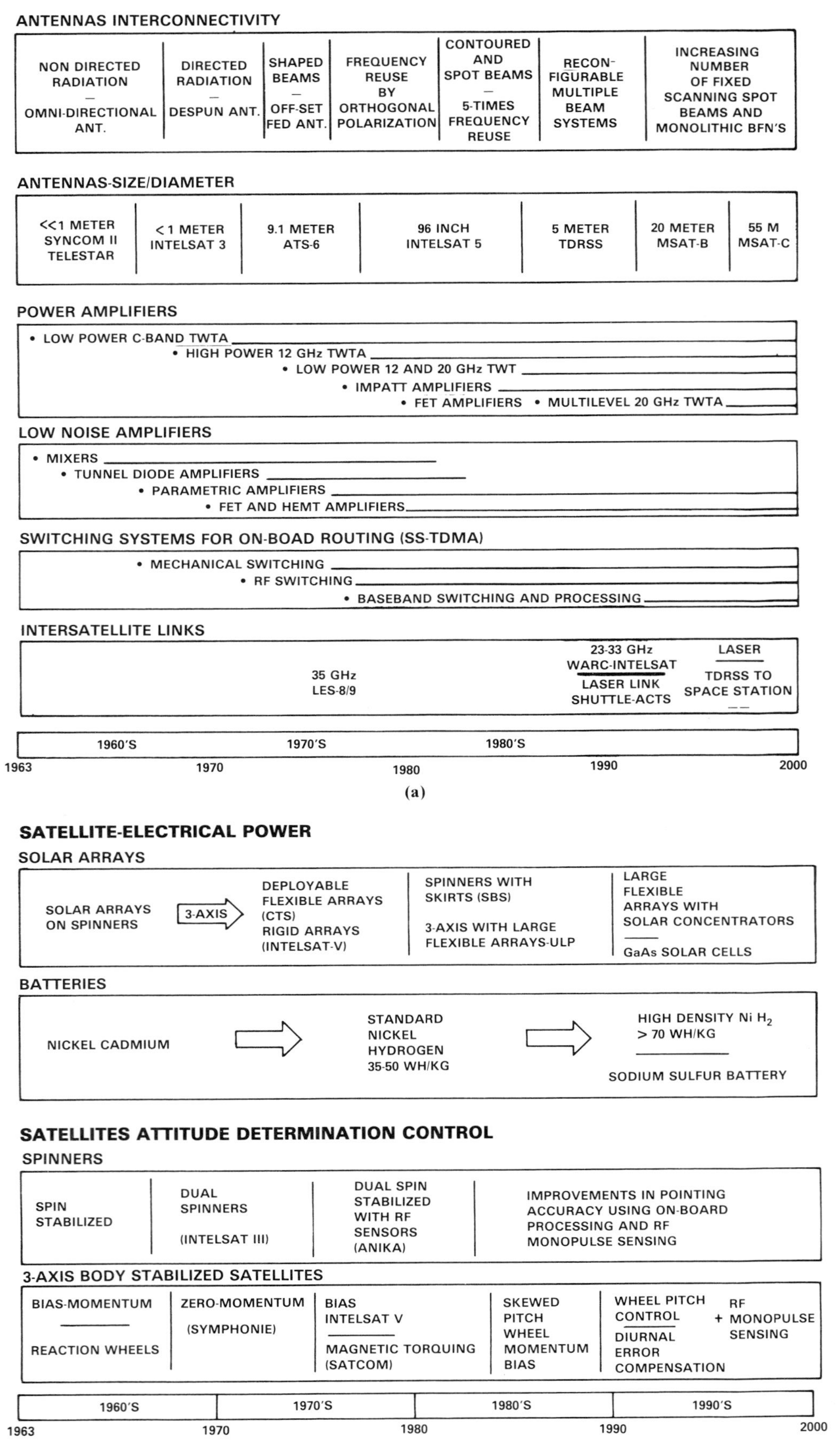

Fig. 20-3: (a) Evolution of critical enabling technologies of a communications satellite. (b) Evolution of critical supporting technologies of a communications satellite.

TABLE 20-1
CAPACITY (BW) AND EIRP OF EXISTING SATCOMS

PROGRAM	BANDWIDTH	EIRP
INTELSAT IV	432 MHz	22 dBW GLOBAL 33 dBW SPOT
INTELSAT IVA	720	22 dBW GLOBAL 26 dBW HEMI
INTELSAT V	2136	23.5 dBW GLOBAL 29 dB HEMI/ZONE 41 TO 44 dBW SPOT
SYMPHONIE	180	30 dBW
RCA SATCOM	816	33 dBW CONUS
COMSTAR	816	33 dBW CONUS
ANIK A/WESTAR A/PALAPA A	432	33 dBW
JAPAN CS	1600	29.5 dBW (C-BAND) TO 37 dBW (K-BAND)
SBS/ANIK C	430	40 dBW (AREA) 44 TO 48 dBW (SPOT)
ANIK D/PALAPA B/WESTAR B	816	33 dBW TO 35 dBW

payload will continue to increase. Table 20-3 lists some of the more important contributions to power and weight for three widely used satellites.

Satellite in-orbit dry mass weight is related to satellite cost and is the ultimate driver of affordability, cost of money, and the costs of launch and insurance. By the mid-1980's, the delta-class satellite (550–750 kg in-orbit dry mass) became the major source of competition in the marketplace because of its relatively low average communications satellite cost (around $30 million) and its low average launch cost (around $30 million). This led to an in-orbit per-satellite cost—including insurance and administrative costs—of around $70 million. An Atlas-Centaur class satellite (900–1200 kg in orbit dry mass) has a relatively high average communications satellite cost of about $50 to $100 million and an average launch cost of $50 to $75 million. This leads to an in-orbit per-satellite cost of over $120 million, including insurance and administrative costs. This cost difference due to spacecraft size and weight impacts on the selection of satellite capacity as the driving economic parameter in a highly competitive global marketplace.

ANTENNA TECHNOLOGY: KEY ENABLING TECHNOLOGY

The antenna for a communications satellite has gone through many technological developments, as illustrated in Fig. 20-6, and is today a very complex system. In the early 1960's communications satellites such as Telstar, Relay, and Intelsat 1 literally bathed space with radiation from simple antennas. Succeeding spinning satellites, starting with Intelsat 3, used despun antennas which concentrated the communication satellite beams primarily on the earth surface. Starting with satellites such as Intelsat 4, and CS, antennas produced shaped, contoured, or spot beams on the earth, which only illuminate the region or country of interest.

Now, shaped beam or spot beam technology for communications satellites is a requirement for most present and planned satellites. Intelsat 5, for example, provides five beams which provide global, hemisphere, zone, and spot coverage. Satcom 5 provides antenna beams contoured to include continental U.S., Alaska, and Hawaii. At the same time that shaped beam and spot beam antennas were under development, a major thrust on the use of new lightweight materials for communications satellite antennas was made. Materials such as graphite epoxy and Kevlar were employed to substantially reduce the weight of the growing size of antenna complexes now typical of new communication satellites, particularly those communication satellites such as Intelsat 5 and Japan's CS-2A which operate in more than one frequency band.

There are many types of antennas which can be used to produce shaped or multiple beams. The principle type of shaped beam or multiple beam antenna is the offset fed reflector antenna (Figs. 20-7 and 20-8) with multiple horns operated by a feed system which sums the radiation from the various horns to provide the radiation footprint or shape desired. This type of antenna system can be used with either single or dual offset feed systems and is widely used on most modern spacecraft.

Other antenna types include: 1) the offset reflector single-horn system in which the reflector is contoured to give the desired footprint; this type of antenna is used with the Japanese CS and on the K_u-band reflectors in Intelsat 5; 2) the multibeam constrained waveguide lens antenna developed by Riccardi for 7/8 GHz at Lincoln Labs, has been built for use at 7/8 GHz for DSCS-III and has the advantage of being able to provide beam nulls at

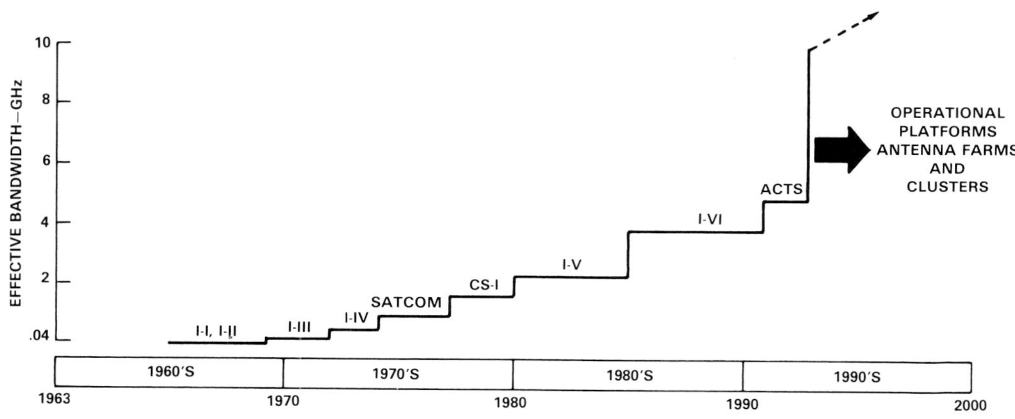

Fig. 20-4: The evolution of communications satellite bandwidth.

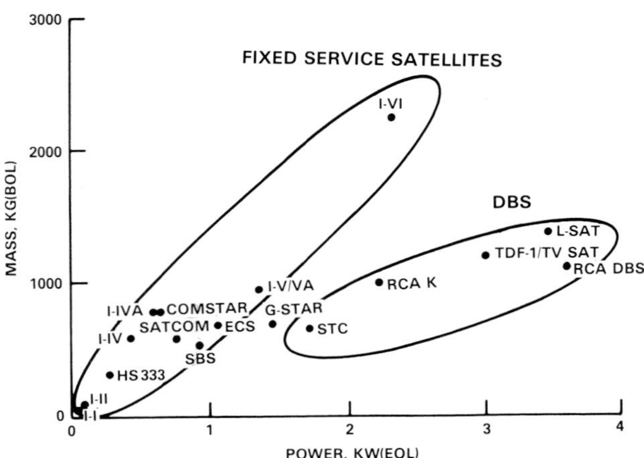

Fig. 20-5: Relationship between mass and power of fixed service and direct broadcast satellites.

TABLE 20-2
WEIGHT AND MASS OF REPRESENTATIVE SPACECRAFT

SATELLITE	WEIGHT IN ORBIT (Kg)	PRIMARY POWER (WATTS)	TYPE SPACECRAFT
INTELSAT I	38	40	SPINNER
INTELSAT II	152	120	SPINNER
INTELSAT IV	700	400	SPINNER
INTELSAT V	967	1,200	3-AXIS STABILIZED
ANIK A	240	320	SPINNER
PALAPA A	246	307	SPINNER
SYMPHONIE	230	780	3-AXIS
BSE (JAPAN)	317	1,010	3-AXIS
CS (JAPAN)	287	529	SPINNER
CTS	347	918	3-AXIS

selected points on command; 3) the phased array antenna now used in TDRSS; and 4) a proposed AT&T communications satellite by Reudink and Yeh was designed in the 1970's to employ a phased array feed system and reflector to produce several scanning spot beams to sweep the U.S. mainland. This satellite was the forerunner of the NASA ACTS satellite.

MULTIPLE BEAM ANTENNAS

Multiple beam offset fed reflector antennas have a unique ability to meet the ever-increasing demands on satellite antenna systems, by being able to accomplish such functions as: 1) improving EIRP over prescribed areas through pattern shaping; 2) allowing frequency reuse by both spatial and polarization diversity; and 3) reducing interference outside desired coverage areas, to meet new WARC requirements on both copolar and cross-polarized energy. Solutions to these problems generally result in larger, more complex antenna structures and systems, which have become an overriding factor in the design, weight, and cost of the entire satellite.

The multiple beam antenna (MBA) systems are capable of creating multiple simultaneous beams, each of which may be shaped from a number of smaller constituent beams by the principle of superposition. This principle is illustrated by current domestic satellite design in Fig. 20-7, showing how adjacent constituent beams added together in space can produce a single broader beam with a relatively flat top and steep "skirts." This allows more uniform coverage of the desired area, and more rapid decay of energy outside this area, to reduce interference while also improving efficiency.

Fig. 20-9 shows how a group of multiple beam antennas can be designed on a single spacecraft (e.g., Intelsat 5) to provide a set of beams which can either illuminate fixed areas, or designed to be electronically reconfigured or hopped as shown in Fig. 20-10 (e.g., ACTS).

SHAPED-REFLECTOR SHAPED-BEAM ANTENNAS

A new technology for producing shaped beams from a spacecraft is one which involves shaping or contouring

TABLE 20-3
MASS AND POWER DISTRIBUTION OF TYPICAL SATELLITES

	SATELLITE					
	COMSTAR		SATCOM		INTELSAT-V	
CHARACTERISTICS	24 CHANNEL SPINNER		24 CHANNEL 3-AXIS STABILIZED		26 CHANNEL 3-AXIS	
TOTAL DRY MASS/POWER	670/610		855/463		1003/1000	
SUBSYSTEM	MASS (kg)	% DRY S/C	MASS (kg)	% DRY S/C	KG	% OF DRY S/C
ANTENNA	61	9.2	51.6	6.8	66.7	8.3%
TRANSPONDER	139.4	20.9	176	20.7	167.8	20.8%
POWER	125.6	18.8	181.6	21.2	143.9	17.8%
ATTITUDE CONTROL	45.2	6.8	55	6.5	70.2	8.7%
SOLAR ARRAY	70.1		—			
THERMAL	28	4.2	21.7	2.5	27.8	3.4%
SUBSYSTEM	POWER (WATTS)	% TOTAL	POWER (WATTS)	% TOTAL	POWER	% TOTAL
COMMUNICATION	495	86	429	92	780	78%
TT&C	17.5	3	10	2.3	43	4.2%
ATTITUDE	24	4	14	3.7	49	4.8%

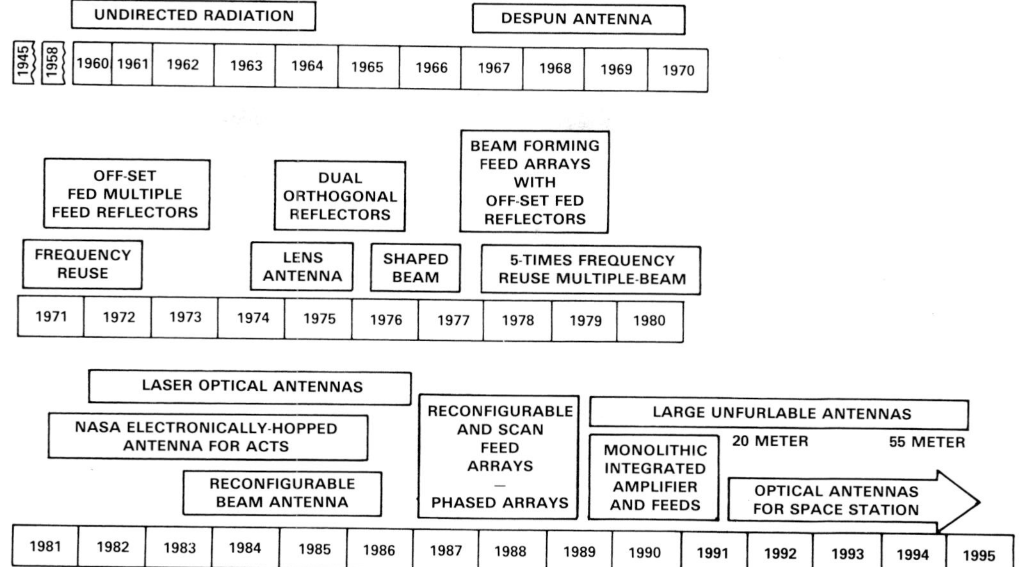

Fig. 20-6: Evolution of antenna technology in communications satellites.

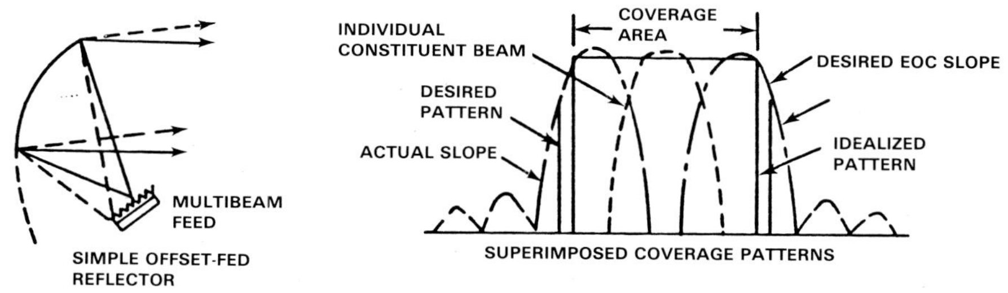

Fig. 20-7: Development of coverage areas by beam pattern superposition in multiple beam antennas.

an antenna reflector surface which is illuminated by a single (or multiple) offset feed horn, using the antenna reflector contoured to change the phases of the various rays produced by the antenna to produce a focused beam into the particular earth pattern or footprint. An important realization of this type of antenna is used in the Japan Communication Satellite (CS). This satellite produces antenna beams at 20 and 30 GHz contoured to fit the Japanese islands.

DRIVING THE OFFSET REFLECTOR MULTIPLE BEAM ANTENNA: THE BEAM FORMING NETWORK (BFN)

The multiple beam antenna using an offset reflector is the principal candidate for producing complex-contour footprints. Consider the multiple beam with many offset feeds which require the network by which the feeds produce the contoured footprint.

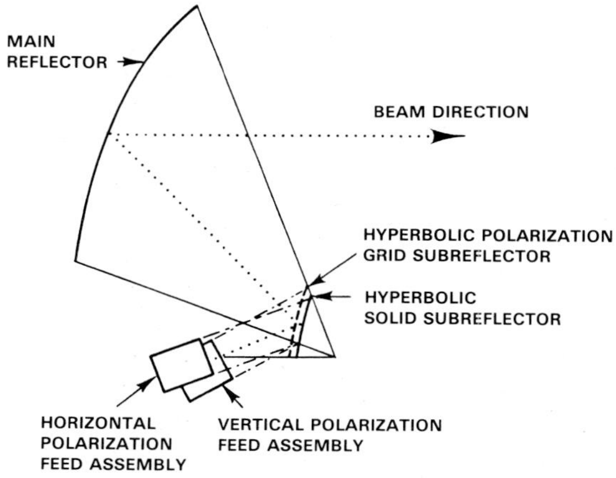

MAIN REFLECTOR

BEAM DIRECTION

HYPERBOLIC POLARIZATION GRID SUBREFLECTOR

HYPERBOLIC SOLID SUBREFLECTOR

HORIZONTAL POLARIZATION FEED ASSEMBLY

VERTICAL POLARIZATION FEED ASSEMBLY

Fig. 20-8: Possible antenna design for a 30/20 GHz antenna system.

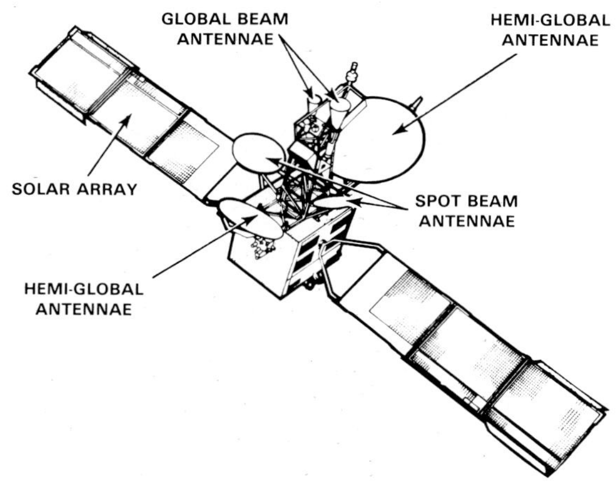

Fig. 20-9: The antennas of Intelsat 5.

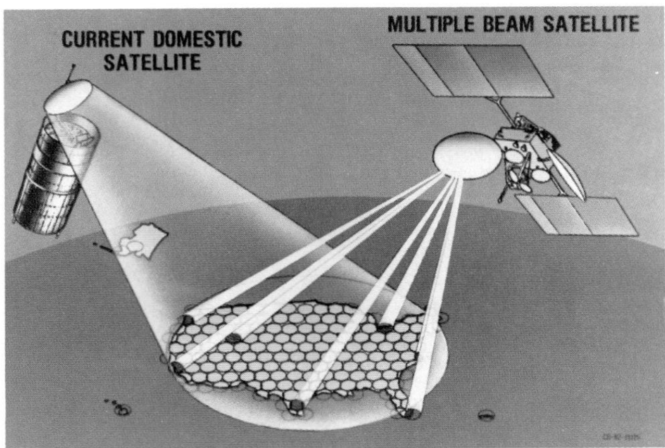

Fig. 20-10: Increasing capacity with frequency reuse by multibeam antennas.

This network is called the beam forming network (BFN) and is illustrated by the scan-beam BFN shown in Fig. 20-11. The BFN includes signal splitters and combiners, hybrid couplers, and devices known as variable power dividers or VPD which are used to provide a controlled power split of a signal between two output ports, and phase shifters which control the phases of the various signals—both making possible both beam contouring and beam hopping.

Future satellites will require even more complex antennas and feed networks to provide such features as reconfigurability—the ability to adjust beam shapes on command—and electronic hopping of spot beams, to meet changing user requirements or to avoid interference. For example, consider an antenna system with up to 256 beams for eliminating coverage in unused areas such as over oceans. Reconfigurability could be implemented by a BFN composed of a matrix of cascaded variable power dividers (VPD's). Full flexibility for each of the beams would require six BFN's with 255 VPD's in each, cascaded in eight levels, plus 256 six-way switches (one at each feed element to select the beam to which it is assigned), plus 256 phasors to control excitation phases. This would entail a total of 1530 VPD's and 1536 switches and phasors; if each weighed only an ounce the total BFN could weigh over 200 lbs, including interconnections. Its losses are projected in Table 20-4. These losses, as well as the size and weight of the BFN, can be reduced by simplifying the design, at the expense of some system flexibility.

An alternate form for the BFN, similar in principle to a phased array, uses separate amplifiers at each antenna element. This would avoid loss of sensitivity or EIRP due to the BFN losses and would enhance reliability by introducing "soft" failure rather than catastrophic single point failure modes.

MILLIMETER-WAVE AND LASER ANTENNA SYSTEMS

As transmit and receive frequencies move to higher frequencies, the antenna sizes required to provide a given gain become drastically reduced while demands on surface accuracy increase. One of the advantages of using the 30/20 GHz frequencies is the opportunity to use small 1 m 20 GHz spacecraft antennas to provide the same gain as a 5 m antenna at 4 GHz, resulting in a significant reduction in antenna weight and size which

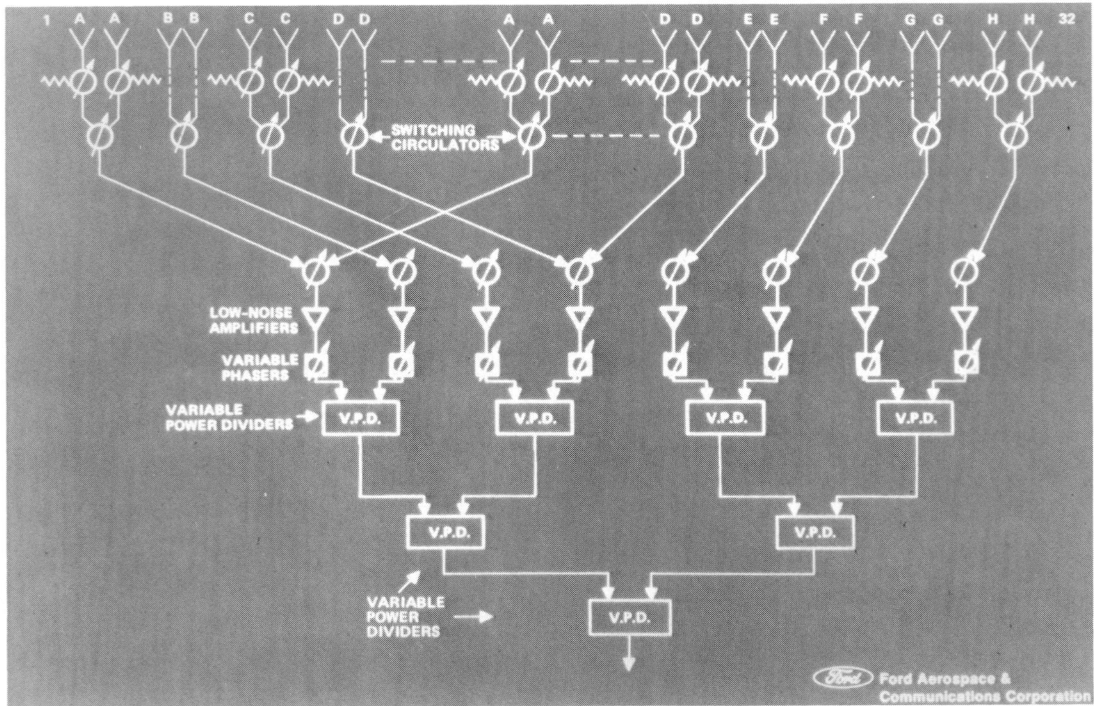

Fig. 20-11: Beam forming network (BFN) for a scan beam antenna.

TABLE 20-4
PROJECTED 256-BEAM BFN LOSSES

BAND, GHz	4/6	11/14	20/30
VPD LOSS(8), dB	1.6	2.4	3.6
SWITCH LOSSES, dB	0.5	0.8	1.2
PHASER LOSS, dB	0.4	0.5	0.8
CONNECTION LOSS, B	0.5	0.8	1.2
TOTAL LOSS, dB	3.0	4.5	6.8

can be used to increase transponder channelization or power.

However, as the transmit and receive frequencies move to optical frequencies, significant decrease in transceiver size is achieved. At 0.83 μm, the laser system can use small optical cassegrain telescopes (antennas), as shown in Figs. 20-12 and 20-13, and take advantage of giant strides in gallium arsenide laser diode technology for optical communications through fiber systems and a long history of optical telescope mirror development. Fig. 20-12 shows how a single optical telescope can be used to serve as a transmitter, receiver, and an aquisition and tracking system using frequency selective mirrors.

Table 20-5 illustrates a major advantage in size of a laser antenna system over a K_u-band antenna system for a 300 MBP's link showing the use of 7-10 in optical antenna diameters (with gains in excess of 100 dB) as compared to 12–16 ft K_u band antennas (with gains in the 50–55 dB range). The narrow beam width of an optical antenna of this type is a guard against interference; a beam originating in geostationary orbit from a laser transmitter with a 10 in reflector will have a beamwidth slightly larger than the width of 3 football stadiums.

ATTITUDE CONTROL TECHNOLOGY FOR SPACE ANTENNAS

In addition to designing a satellite to produce a downlink beam with a particular EIRP into an area of illumination, it is necessary to stabilize the position of

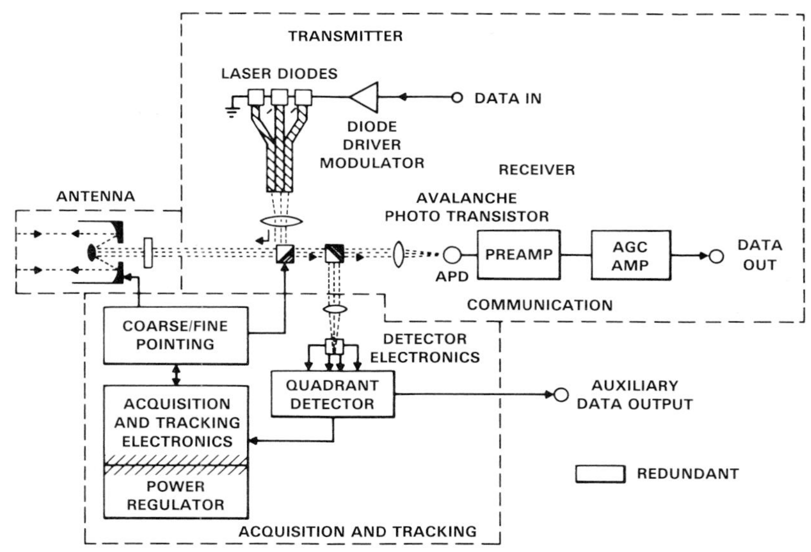

Fig. 20-12: Laser transmit/receive module using a common optical antenna system (telescope).

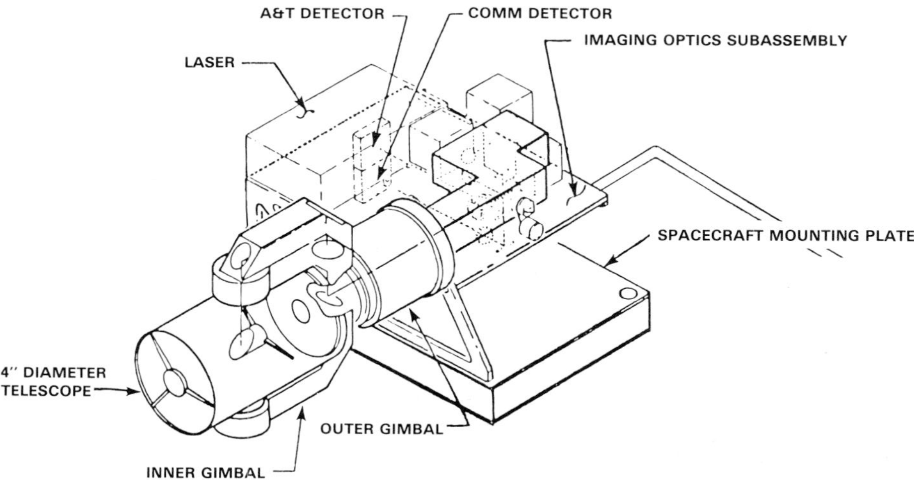

Fig. 20-13: Laser communication transceiver using a gimballed 4 in telescope.

TABLE 20-5
LOW-EARTH ORBITER TO TDRS COMMUNICATION LINK COMPARISON: LASER VERSUS K_u BAND

LINK CHARACTERISTICS

- 300 MBPS DATA RATE
- 10^{-5} BIT ERROR PROBABILITY (UNCODED)

	LASER	Ku BAND
LEO TRANSMITTER POWER	0.25 WATTS	10 WATTS
LEO ANTENNA DIAMETER	7 INCHES	12 FEET
TDRS ANTENNA DIAMETER	10 INCHES	16 FEET
LINK MARGIN	6 DB	6 DB

the satellite so that the antenna beam does not move away from its destination. Accordingly, communication satellites use attitude control systems which are designed according to the specifications shown in Table 20-6, with Fig. 20-14 showing the progression in time from the 0.3/0.2 roll/pitch specifications of 1970 to the 0.01/0.005 specs achievable in 1985: the parameters are in degrees.

Table 20-7 shows the width in miles of a satellite footprint (circular) at the equator from a satellite directly overhead. Note that a spot beam with a 0.5° 3 dB beamwidth will have an illumination width of 190 mi.

Table 20-8 shows that a shift of 0.1° of the boresight axis will result in a shift of the footprint by 38 mi. This may be trivial for a very wide footprint, but for a narrow-beam footprint designed for single city operation and required to provide isolation from a neighboring contoured beam, this movement may be enough to produce major interference into the neighboring area.

Having shown the effect of beam motion as a result of a shift in boresight axis, it is important to realize that the primary parameter for maintaining satellite position to minimize such a shift is the satellite pointing accuracy which is based on attitude control technology. This required support technology is as important to the communications payload as the primary enabling technology of the multiple beam antenna. This support technology includes earth, sun, and star sensors which determine the satellite's position, momentum wheels to gyroscopically control position, and hydrazine jets which are energized by the attitude control system to provide thrusts to offset changes in the satellite's position as reported by the sensors. The cells of hydrazine fuel for the thrusters are a major contribution to satellite weight, and the rate of expenditure of hydrazine fuel determines spacecraft lifetime.

PAYLOAD WEIGHT

Satellite payload weight represents the mass and weight of both the antenna and the transponders. In more advanced satellites such as Intelsat 5, the ratio is around 28 percent. With new materials, new technologies, and methods of construction and propulsion, this ratio could exceed 50 percent on a geostationary platform using shuttle enabling techniques of assembly in low-earth orbit to permit higher spacecraft mass ratios with the objective of committing most of this weight to the giant antenna. This then contributes antenna gain to a very large number of spot beams (or contoured beams) with very high EIRP, with the spot beams interconnected by an on-board switching center. Since spot beam antenna gains can be very large, the RF power required to drive a spot beam could be very low, by replacing RF gain with structure gain. For example, consider a 12 GHz broadcast satellite with a 2.5 ft diameter (36 dB) antenna and a 200 W TWTA; EIRP will be 59 dB·W. However, a DBS with a 30 ft antenna and a 1 W amplifier will give the same EIRP and require far less power from the solar array, assuming comparable coverage.

TESTING ANTENNAS IN SPACE

The capability of assembly, test, and deployment of satellites in low-earth orbit, in addition to the continued use of expendable launch vehicles with enhanced payload capability over the payloads of the 1970's, will lead to a space network of satellites and platforms occupying space from the space station orbit height of around 250 mi to and beyond the geostationary arc.

The space station will be placed in low-earth orbit in

TABLE 20-6
SUMMARY OF ATTITUDE CONTROL SPECIFICATIONS FOR GEOSTATIONARY SATELLITE APPLICATIONS

APPROACH	ACCURACY	COMMENTS
EARTH AND SUN SENSORS	.05° - .1° PITCH, ROLL .25° - 1° YAW	• MOST COMMON APPROACH AT PRESENT
MONOPULSE RECEIVER	.05° - .1° PITCH, ROLL	• REQUIRES DEDICATED EARTH STATION DECOUPLES ANTENNA FROM SPACECRAFT ERRORS
STAR SENSOR (POLARIS, CANOPUS)	.02° - YAW, ROLL	
STAR MAPPER WITH GYROS AND ON-BOARD COMPUTATION USING EPHEMERIS DATA RECEIVED FROM THE GROUND OR FROM THE GLOBAL POSITIONING SATELLITE	.003° - ALL AXES	• ON-BOARD COMPUTATION COMPLEXITY

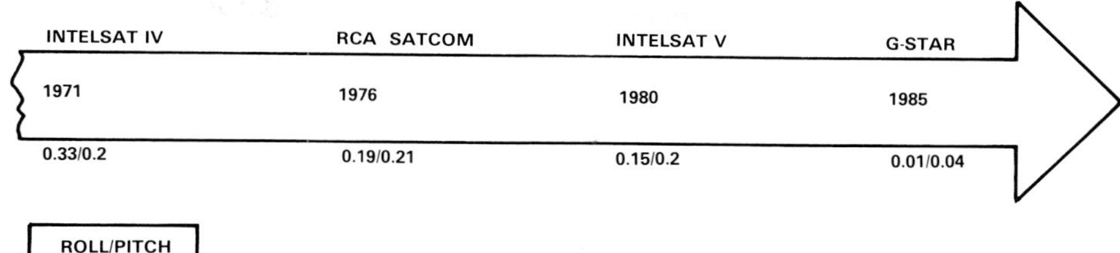

Fig. 20-14: Evolution of satellite pointing accuracy capability.

the mid-1990's. It will include a staging area where a large antenna can be assembled or unfurled from units brought into low-earth orbit by the shuttle. Today, in the pre-space station era, satellite antennas are relatively small in diameter (Intelsat 5's 4 GHz antenna is 92 in, and the largest of the TDRS antennas is 5 m) due to limitations imposed by the fairing of the launch vehicle or the width of the shuttle bay. By being able to bring large antenna elements into low-earth orbit, assemble, deploy, and test them at the Space Station, and then to transfer a giant antenna to geostationary orbit, a new dimension is afforded to satellite design, capacity, and interconnectivity in which the antenna is the key. Fig. 20-15 illustrates the test of a giant antenna at the space station before transfer of the antenna system or platform to geostationary orbit. The development of antenna testing technology for the space environment will be a by-product of the space station.

SATELLITE AMPLIFIERS

The amplifiers on a spacecraft contribute to both satellite sensitivity and satellite radiated power. These amplifiers utilize the majority of the satellite dc power developed by the solar arrays, and, with the antennas, have been the objective of intensive technology developments in low-noise devices and achieving power during the last two decades, where a major thrust for spacecraft has been TWTA efficiency and long life to minimize the dc power utilization.

SATELLITE RECEIVER TECHNOLOGY

Satellite receiver technology has experienced dramatic growth in concert with other space-related technologies. Seven types of low-noise receiver frontends have been or are in use in communication and DBS satellites:

1) Tunnel diode amplifiers
2) Mixer/low-noise post-amplifier complexes
3) Parametric amplifiers
4) FET amplifiers
5) HEMT amplifiers
6) Monolithic GaAs low-noise microwave integrated circuit
7) Photo transistors for laser receivers.

Fig. 20-16 shows the evolution of low-noise amplifiers from the maser which was used for very early ground-based receivers for ECHO, to the modern field effect transistor FET amplifier. Fig. 20-17 provides the general noise temperature ranges as a function of frequency for FET amplifiers (room temperature), the parametric amplifier, and the mixer in 1985.

The tunnel diode amplifier with its 4–6 dB noise figure in the 6 and 15 GHz frequency ranges was widely used in early Intelsat satellites and in Symphonie, and was proven to be a stable low-noise amplifier. This amplifier is still used at 15 GHz in flight models of Intelsat 5.

Mixer technology is now providing mixers with conversion losses in the 2.5–6 dB range at frequencies from 2 to 60 GHz. Such a mixer operating into a post low-noise amplifier having a noise figure of, say, 2 dB will provide an overall noise figure of 5–7 dB. The mixer/post-amplifier combination was used at 30 GHz for the Japan Communication Satellite CS-2A to provide a 10.5 dB noise figure based on a 6 dB mixer operating into a wideband 3-5 GHz FET amplifier with a 4.5 dB noise figure. By 1985, a FET receiver amplifier

TABLE 20-7
FOOTPRINT WIDTH FOR VARIOUS ANTENNA BEAMWIDTHS

CIRCULAR SPOT BEAM 3-dB BEAMWIDTH	WIDTH IN MILES CIRCULAR FOOTPRINT
4°	1536 MILES
3°	1152 MILES
2°	768 MILES
1°	384 MILES
0.5°	190 MILES
0.4°	153.4 MILES
0.3°	115.2 MILES
0.2°	76.8 MILES
0.1°	40 MILES

TABLE 20-8
BEAM DISPLACEMENT FOR VARIOUS BEAM POINTING ERRORS

BEAM POINTING ERROR	MOTION IN MILES OF BEAM BORESIGHT
2°	768 MILES
1.5°	576 MILES
1°	384 MILES
0.5°	192 MILES
0.25°	95 MILES
0.2°	76.7 MILES
0.15°	57.6 MILES
0.1°	38.4 MILES
0.05°	19.2 MILES

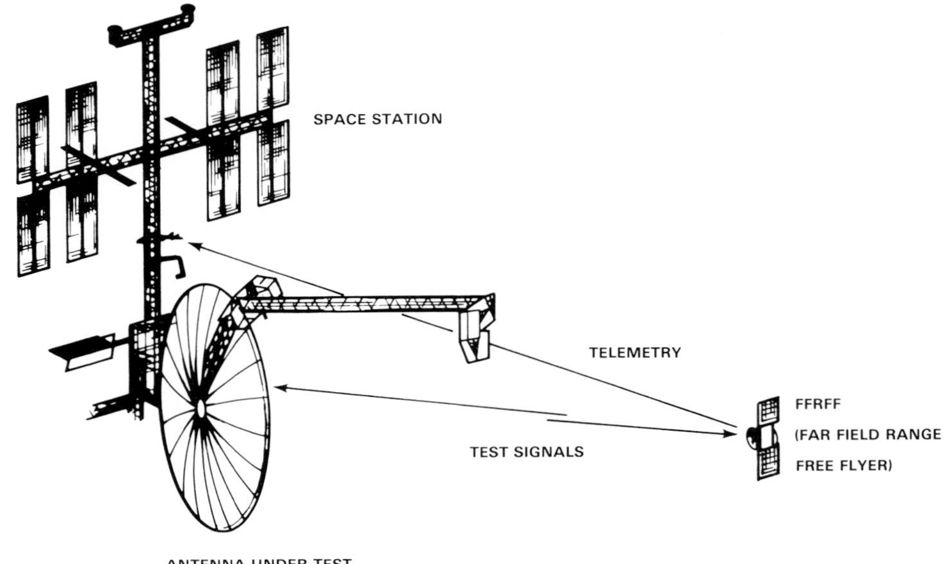

Fig. 20-15: Testing an antenna in space at the space station.

with 4 dB noise figure or less up to 30 GHz, can replace the mixer/post amplifier with an improvement of more than 6 dB in noise figure.

During the 1960's and early 1970's, the parametric amplifier was looked upon as "Peck's bad boy" whose need to be constantly tweaked mitigated against its consideration as a space device. However, the advent of the stable long-life Gunn oscillator pump and computer-aided design gave rise to a paramp which could be reliably operated in spacecraft. Several space satellite communication paramps have been developed in Japan, Europe, and the U.S. A 14 GHz paramp by GTE Telecommunications of Italy in CTS amassed more than 20 000 before CTS was retired. A 30 GHz paramp was under development by GTE-Telecommunications for L-SAT under contract to the European Space Agency, and a 30 GHz paramp was under development by Fujitsu in Japan for CS-3.

The most exciting new device development for communication satellite low noise frontends, however, is the FET amplifier which does not require the pump power of a paramp, and which gives a noise figure almost as low as that of an uncooled paramp, particularly if it can be cooled to some temperature around −40° C.

Fig. 20-18 shows the evolution of communications satellite low-noise amplifiers from 1963 to the year 2000. As shown, the FET now dominates the 6 GHz communication satellite receiver technology—having replaced the tunnel diode amplifier. At 14 GHz, the FET is now in competition with the parametric amplifier and will undoubtedly become the standard K_u-band communication satellite receiver technology. At 30 GHz (and

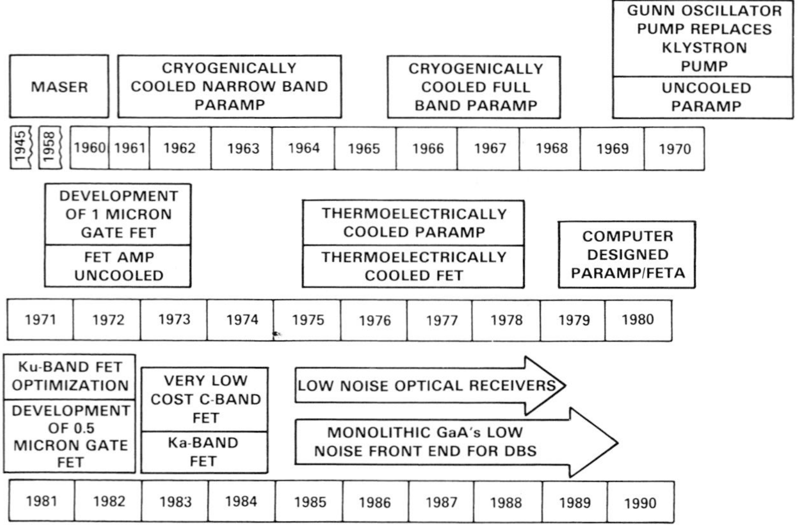

Fig. 20-16: Evolution of low-noise devices for use in space.

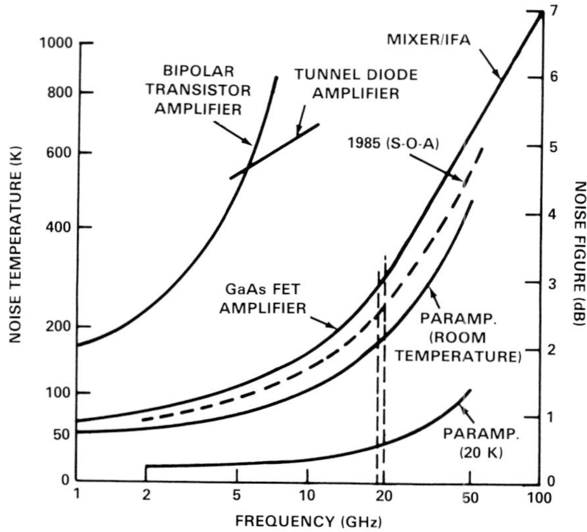

Fig. 20-17: Noise figures of candidate space low-noise amplifiers.

44 GHz for DOD's Milstar), the FET in 1985 is in competition with both the paramp and the mixer. However, a new competing technology has entered the scene—the HEMT (high-electron-mobility transistor), using gallium and arsenide but including other materials such as aluminum. The HEMT is capable of noise figures of less than 3 dB at 30 GHz and around 5 dB at 60 GHz, and is a viable candidate for spacecraft receivers above 18 GHz.

At optical frequencies two types of receiver amplifiers are available: the photo multiplier and the avalanche photo transistor, which are particularly sensitive in the 800–860 nm band. Also, two types of receivers are involved: direct detection and heterdyne detection.

The silicon avalanche photo diode has undergone intensive development of materials and processing to achieve a higher ratio of electron to hole multiplication and has achieved a very low-excess noise factor of 3.48 and an 80 percent quantum efficiency.

SPACE POWER AMPLIFIERS

Traveling wave tube amplifiers (TWTA's) and solid-state amplifiers are the two major candidates for power amplication sockets in communications satellite transponders. Solid-state power amplifiers (SSPA) have the advantages of a simpler power supply and better linearity with a lower power backoff than required by TWTA's. The reliability of SSPA's is also higher. TWTA's, on the other hand, have greater power capability, especially at higher frequencies, and can be built with greater efficiency—particularly at the high power levels used by DBS (100–200 W).

Fig. 20-19 shows the evolution of satellite power amplifiers from the first satellites to modern communication satellites and DBS. For *C*-band and K_u-band communication links, the solid-state power amplifier is the heir apparent for those sockets replacing TWTA's; while for higher power applications, e.g., DBS at 12 GHz, and millimeter-wave power (10–30 W) at 20 GHz, the TWTA will no doubt continue to be the standard amplifier for the next decade.

The TWTA has been the baseline component for the communication satellite since Bell Telephone Laboratories first used a 10 W *C*-band TWT in Telstar two decades ago. The excellent life history of the TWT in space built by Hughes Electron Dynamics Division, who supplied TWT's for most of the succeeding commercial communication satellites, provided a reliable source of power amplification in the transponder which did more to confirm and establish the practicality of satellite communications than any other single device aboard the communication satellite.

The early U.S. domination of the 4 GHz space TWT technology led the European Space Agency's predecessor, ESRO, to fund 11 GHz TWT programs in both Thomson CSF and AEG-Telefunken at the 20 W level, a decision which has had far reaching consequences in

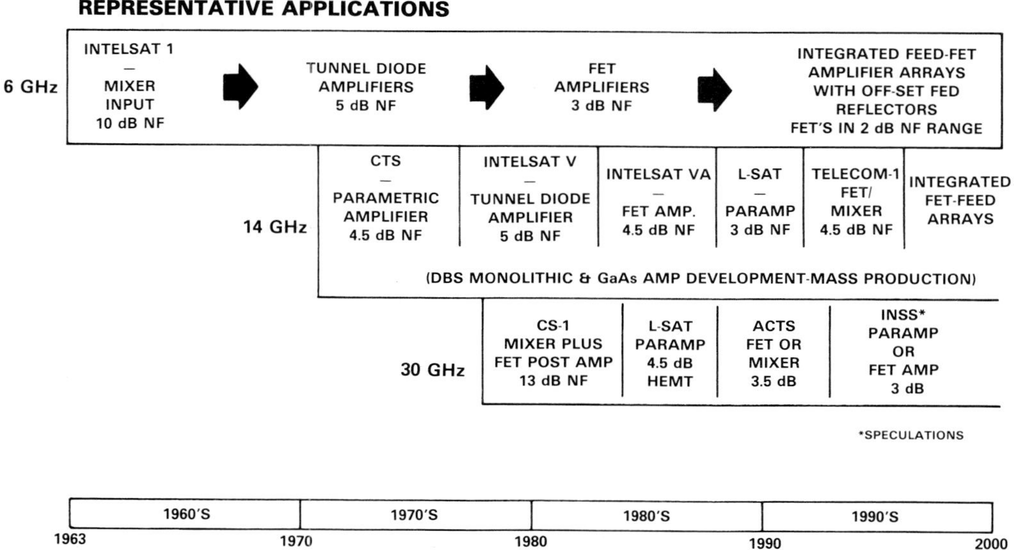

Fig. 20-18: Evolution of low-noise amplifiers at three frequencies.

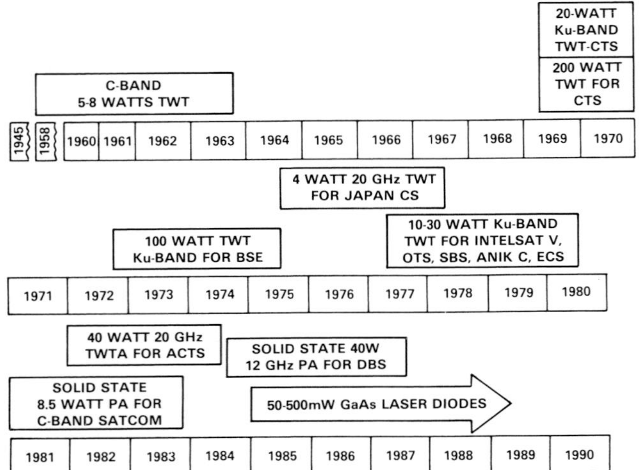

Fig. 20-19: Evolution of space power amplifier/device development.

establishing Europe as a major TWT supplier for space communication satellites, and in particular, those addressing the 11/14 GHz frequencies. Also, in the early 1970's, the Japanese national space agency, NASDA, funded NEC to develop space TWT for 4 and 19 GHz for ultimate use in the Japan CS, thereby creating another important space TWT technology base in the world. The U.S.S.R. entered the space TWT development arena in the 1960's, producing 50 W 4 GHz TWT for the Molnya satellites, and in the early 1970's, a 300 W space klystron at 716 MHz was used for Statsionar-T.

Fig. 20-20 shows a time chronology of space TWT and solid-state power amplifiers, illustrating the variety of devices and power levels which have developed over two decades at 4 GHz, 11/12 GHz, and 20 GHz.

TWT's are electron-tube devices that propagate the input signal down a helix surrounding a focused electron beam. The electron beam interacts with energy traveling along the helix, resulting in amplification of the signal.

TWT's offer several important advantages. They are very-wide-bandwidth devices capable of operation over the entire down-link band. A single TWT can supply sufficient output power for a down-link beam and is capable of multiple power at levels permitting an adaptive link power control. TWT's offer relatively high efficiency (35–50%) and can be used to amplify more than one channel at a time. Considerable development is still required to produce an operational TWT capable of operating at new frequency and power levels while still meeting the design lifetime criterion of 10 years.

The elements of TWT technology still considered to be critical include low-loss helix structures, the electron-beam-forming components, and the ability to extract heat from the helix at high power. Fig. 20-21 shows the 20 GHz TWT now under development by Hughes for the NASA ACTS Satellite Program.

Another type of TWTA uses coupled RF cavities instead of a helix as the slow-wave structure. Such TWTA's are capable of much higher power over restricted bandwidths, although much larger and heavier than comparable helix TWTA's, and are used for, say, very high-power 12 GHz amplifiers in broadcast satellite applications at the 150–750 watt level; the helix type of power TWTA at 12 GHz is built for power levels up to 125 W.

TWTA's for use at millimeter waves are the principal candidates for satellite systems, due to the relative lack of significant power of competing technologies such as FET's and impatt amplifiers above 18 GHz. TWTA's with power levels in the 4 W level were developed at 20 GHz for the Japan CS-2 series. However, 20 GHz TWTA's at the 30–50 W level will be used in the NASA ACTS, ESA's Olympus, and the DOD Milstar satellites.

At frequencies up to 44 and 60 GHz, the impatt amplifier with multiple diodes has provided up to 10 W of saturated output power; however, low efficiency and

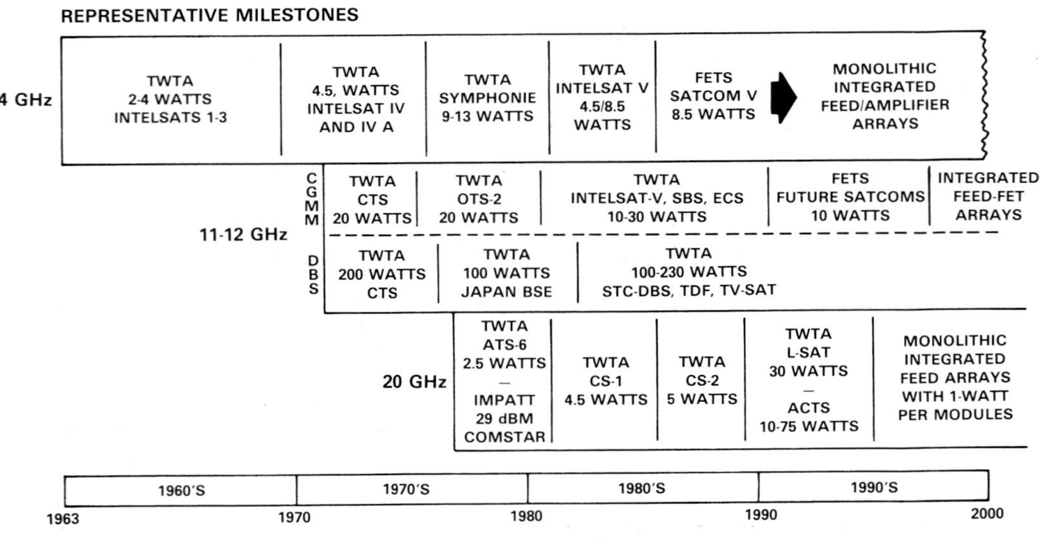

Fig. 20-20: Evolution of space amplifier development at three frequencies.

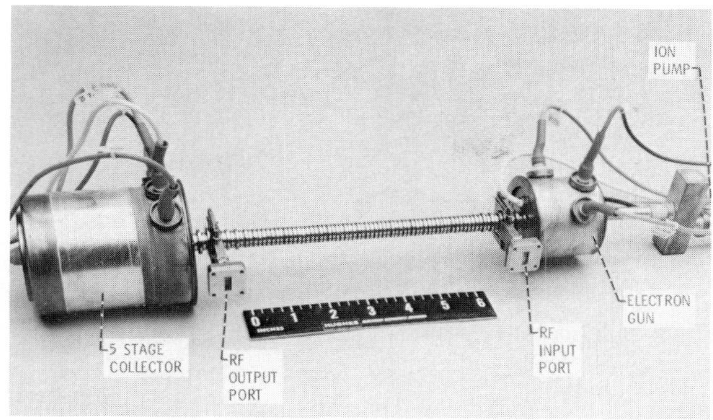

Fig. 20-21: 20 GHz space-type traveling wave tube for ACTS as developed by Hughes Electron Dynamics Division for NASA.

reduced bandwidth have reduced its competitive position with the TWTA.

SOLID-STATE POWER AMPLIFIERS

The linearity characteristics of the FET amplifier are superior to those of the TWTA, although efficiency must be traded for linearity, a choice not available with the TWTA. The FET amplifier also contributes an additional equivalent 2.0 dB link margin for the communication channel with multiple carriers. For linear power amplification, the TWTA requires 6–8 dB backoff to achieve a carrier to third-order intermodulation distortion ratio of 25 dB; for the FET amplifier, 2–3 dB power backoff achieves the same kind of linearity. This linear characteristic makes FET amplifier technology more suitable for multiple carrier and some digital communications system. As for AM-to-PM conversion, the typical value for the TWTA is 6–7°/dB, but the FET amplifier runs about 3–4°/dB, which gives the FET amplifier great advantage in digital modulation links.

In addition to providing superior RF performance, the FET solid-state amplifier also offers additional weight and size savings in the spacecraft design. For a 5-W *C*-band transmitter, the FET amplifiers, for example, weigh about 1.4 pounds with volume equal to 50 in^3. For a TWTA, the weight goes up to 3.0 lbs with volume equal to 150 cubic inches.

Solid-state power amplifiers (SSPA's) based on gallium arsenide field-effect transistors have been developed and flown since 1979. Satcom-V by RCA employed 8.5 W SSPA's. The Telstar 301 series employs 5.5 W SSPA's, and Intelsat 6 will use 1.8 and 3.2 W SSPA's for its low-power transponders. The expectation is that SSPA's will provide longer life and greater reliability than traveling wave tube amplifiers. The significantly greater efficiency of TWTA's is offset by the superior linearity of SSPA's and the ability to operate closer to saturation in multiple carrier operation.

In the future, the performance margin between SSPA's and TWTA's will narrow. The ability to produce flight quality SSPA's will be extended to K_u band and eventually K_a band. It is possible to build 16 W amplifiers at 15 GHz in 1985 with 15 percent efficiency. Likewise, it is possible to build 3 W amplifiers at 20 GHz with 12 percent efficiency. This technology will move ahead rapidly; it should provide operationally useful SSPA's at 2 GHz by 1990.

The lower efficiency of the SSPA is not necessarily a problem in communication satellite design. For example, RCA demonstrated that it was possible to include a small increase in solar array size to make up for SSPA low efficiency within the mass margins of RCA's Satcom satellite.

OPTICAL TRANSMITTERS

At optical frequencies and in particular in the 800–860 nm range, from 5 mm to 500 mW are required from laser transmitters to close intersatellite links (using phototransistor receivers) when using 10 in optical antennas and modulation rates up to 500 Mbits/s. At these frequencies, gallium arsenide laser diodes, operated singly or in arrays, have provided 30–100 mW as manufactured by Hitachi, RCA, General Optronics, and Spectra Diode Labs. This development has benefited from synergistic developments with the laser recording industry and the rapidly growing terrestrial fiber-optic networks. The existence of such available diodes with proven reliability makes intersatellite link technologically possible—opening the door to system considerations of acquisition, pointing accuracy, and modulation technology.

PRIMARY POWER TECHNOLOGY

A communication satellite requires primary or direct current (dc) power to operate all its electronic equipment, including the transponder, and the attitude control system. The power sources can include solar arrays made up of tens of thousands of solar cells, batteries, and power control electronics systems.

Direct current power requirements will vary with the size and weight of the communication satellite. Some typical primary powers starting with early and generic satellites as derived from solar cells are displayed with the time chronology of spacecraft dc power in Fig. 20-22.

A satellite has to develop a prescribed amount of dc power which is a function of the area and cell efficiency of the solar array. Most of this power is available for conversion to RF power which must then be divided up among all the down-link channels. The conversion efficiency of dc power to RF power is in the range of 35–50 percent depending on the type of power amplifier used.

Broadcasting satellites which normally use high-power amplifiers achieve a conversion efficiency toward the high end of this range, while communication-type power amplifiers are usually at the low end. Taking into account the other uses of electrical power in a satellite, the ratio of RF power to total prime dc power falls into the range of 20–30 percent.

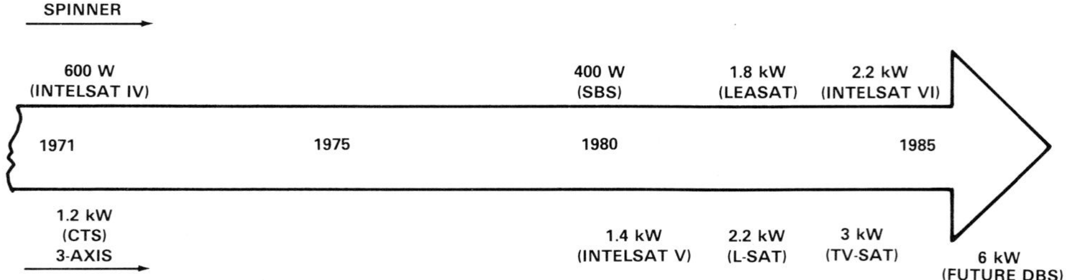

Fig. 20-22: Evolution of satellite dc power capability from solar arrays.

Table 20-3 includes the important relationships between spacecraft in-orbit dry weight (mass in kilograms) and spacecraft dc power. Note that the mass/power ratios for DBS and communications satellites are relatively linear.

In general, between 75 and 80 percent of satellite dc power is used to power the communications payload. Table 20-3 illustrates this by showing the mass and average dc power breakdown for Intelsat 5 showing 780 W of the 1004 W total allocated to the transponders.

Satellite prime power capacity has grown and will continue to grow as a result of improvements in launch vehicle capability, lightweight spacecraft subsystems, and improvements in solar cell technology which will see gallium arsenide solar cells used by the end of the decade.

FILTERS AND MULTIPLEXERS FOR COMMUNICATIONS SATELLITES

The TWT has been a pacing design item in size, weight, and efficiency and materially determines the satellite size weight and dc power, which in turn determines or specifies the booster requirements and cost to launch the communications satellite into geosynchronous orbit.

However, in the first decade of satellite design, up through Intelsat 4 (early 1970's) in the Intelsat system, the filters of 12-channel spacecraft proved to be a major size and weight constraint on the spacecraft, and for all 4/6 GHz satellites with 12 or more channels built in that time period, the input and output filter multiplexers became the pacing item.

Several developments for 4 and 11 GHz input/output multiplexers as time-lined in Fig. 20-23 took place in the 1973–1980 period which greatly impacted on spacecraft filter design and manufacture, as follows.

- The dual-mode elliptic filter by Atia and Williams at Comsat Laboratories, which provided channel characteristics in small lightweight filters using only a few cavities to replace the large and heavy Chebyshev waveguide filters.
- Filters in graphite epoxy by RCA Limited in Canada (now COM-DEV), for the early 24-channel RCA Satcom communication satellites. This development greatly reduced filter weight for the Chebyshev filters and led the way to reducing the weight limitations produced by all *C*-band filters.
- Dual-mode elliptic filters developed by Hughes Aircraft for Intelsat 4-A, which were also manufactured by AEG Telefunken, Germany, on a "build-to-print" basis.
- Linear phase filters developed by GEC-Marconi for the European Space Agency.
- Four GHz dual-mode contiguous band filters in graphite epoxy developed by Ford Aerospace for use in Intelsat 5.
- Dual-mode elliptic filters for 11 GHz developed by COM-DEV for Anik C and SBS.

Fig. 20-24 present typical characteristics of Chebychev and dual-mode elliptic filters showing the ability to get better isolation from guard bands between closely spaced channels for the latter filters.

Table 20-9 lists the channel bandwidths and the minimum guard bands of typical communication satellites. It has been the U.S. and Canada that have pursued maximum channelization and minimum guard bandwidths, which is baseline to conservation of bandwidth.

The technologies of communication satellite filters and multiplexers have gone through many changes and upgrades at *C*-band and K_u-band and were achieving a state of maturation by the mid 1980's even though the technology could admit to new advances such as the dielectrically loaded filter by Ford Aerospace which once again greatly reduced filter size and weight. However, filter technology above 15 GHz can now be

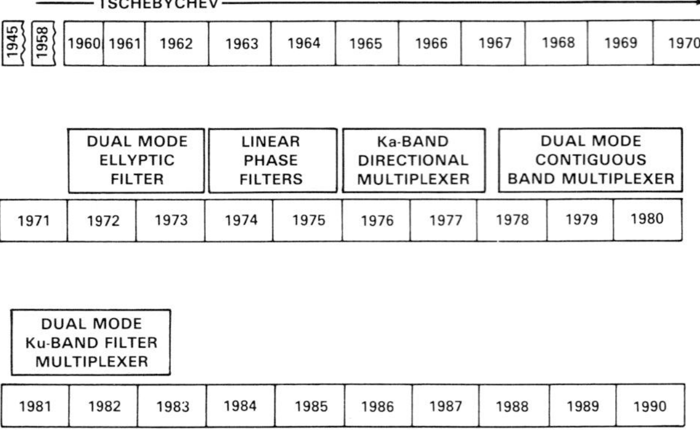

Fig. 20-23: Evolution of satellite filter technology.

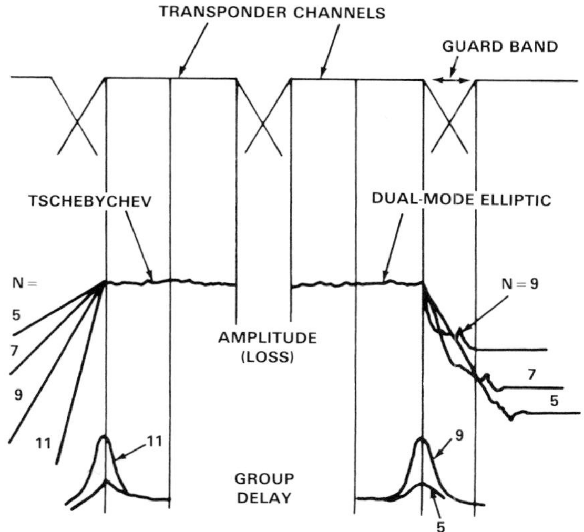

Fig. 20-24: Basic rolloff characteristics of Chebyshev and elliptic filters.

considered to be at the limiting edge of mechanical tolerances, materials technology and manufacturability.

THE SIGNIFICANCE OF CHANNEL BANDWIDTH

The filter bandwidth is the key to the effective utilization of a channel—it must provide a pathway for one or more information signals which are developed by a user. The information signal must be transformed into an electrical equivalent and then used to modulate a carrier, which is then applied from the modulator modem to the IF input to the transmitting path of the earth station. After passing through the entire satellite system, the modulated carrier is applied from the receiving path IF to a receive demodulator modem which recovers the information from the carrier.

The process of transmitting information through a satellite channel determined by a filter makes use of the radio spectrum, a valuable earth resource which is as subject to conservation efforts as are energy and food sources. An active FCC in the U.S. and an International Telecommunication Union on a worldwide basis are among those agencies charged with the responsibility of preventing misuse of the spectrum by both interference and by inefficient use of bandwidth.

Conservation of bandwidth involves maximizing the amount of data or information which can be transmitted through a given unit of bandwidth. Conservation of bandwidth includes:

- reduction in the number of bits used to represent data;
- use of bandwidth-efficient modulation technique (increase number of bits transmitted per Hertz of bandwidth);
- maximizing the number of bits included in a given bandwidth from the standpoint of reducing degradation due to non-linear devices and achieving an optimization of transmitted power, receiver sensitivity, and bandwidth efficiency;
- the use of channelization with minimum guardband bandwidth;
- the linearization of channels.

The figure of merit of the "efficiency" of bandwidth utilization refers to the *number of bits per unit of bandwidth* for a particular modulation technique. Bits/hertz is now becoming a familiar part of the communication system lexicon and has, obviously, considerable economic connotations to a user who wishes to purchase or lease a portion or all of a channel bandwidth.

Conservation of bandwidth in terms of carrier-bandwidth reduction is virtually meaningless unless the data bandwidth has also been minimized. Eventually, the total channel bandwidth optimization is a combination of both. Consider the following equation for, say, a voice signal having 3.1 kHz analog bandwidth.

TABLE 20-9
TYPICAL EXAMPLE OF CHANNEL BANDWIDTHS OF SELECTED GEOSTATIONARY SATELLITES

SYSTEM	FREQUENCY BAND (GHz)	USEABLE CHANNEL BANDWIDTHS (MHz)	MINIMUM GUARD BAND IN TERMS OF 0/0 CHANNEL BW (%)
DSCS-2	3/8	7, 50, 100	20-30
NATO-3	7/8	17/50/85	20-30
INTELSAT IV	4/6	36	10
INTELSAT IVA	4/6	36	10
INTELSAT V	4/6, 11/14	34,36,41,72,77,241	8.3 TO 11
AN1K	4/6	34	11
WESTAR	4/6	36	10
RCA SATCOM*	4/6	36	11
JAPAN CS	20/30	200	50-60
ATT COMSTAR*	4/6	40	17.6
OTS (ESA)	11/14	5,40,120	100
SYMPHONIE	4/6	80	32

*DUAL LINEAR POLARIZATION

$$\begin{bmatrix}\text{Total BW}\\ \text{for a}\\ \text{3.1 kHz BW}\\ \text{voice}\end{bmatrix} = \begin{bmatrix}\text{Baseband bits}\\ \text{per second}\\ \text{of voice in}\\ \text{a 3.1 kHZ band}\end{bmatrix} \times \begin{bmatrix}\text{Bits per hertz}\\ \text{of the}\\ \text{modulation}\\ \text{technique}\end{bmatrix}$$

Thus, for example, a voice channel requiring 1000 bits/s from a vocoder as compared to the Spade voice standard of 64 kbit/s, each in a modulation system using two bits per hertz, will require only 2 kHz of radio frequency bandwidth for the vocoder transmission, as compared to 128 kHz of bandwidth for the spade system. This comparison must, of course, be measured against useful voice quality for each system.

THE BASEBAND INFORMATION

Fifty years ago, the only information signals to be transmitted over any medium—wire or wireless—were voice and telegraph. At that time the definition of the voice spectrum occurring in the 3.1 kHz band was established. By the 1950's the transmission of NTSC video and color TV, all in a baseband of 0–4.5 MHz, also became standard for the U.S., Canada, and Japan. (PAL and SECAM standards were adopted in other parts of the world.) During this period, the multiplexing of the human voice into groups of from 12 voice or message channels to master super groups of up to 960 voice or message channels both matured and became a global capability in a marketplace which involved the participation of many companies. By the 1960's, the computer and the digital representation of voice were producing requirements to transmit bit streams from kilobits to megabits, and the demand for transport paths of the numerous data streams, many of which are listed in Table 20-10, grew rapidly. In the 1980's the 56 kbit/s data stream and the $T1 = 1.544$ Mbit/s data streams have been the most widely used.

Fortunately, worldwide CCITT and CCIR activity has forced a data rate commonality where possible, to prevent a tower of Babel of information data rates from occurring; however, Europe and the U.S./Canada have significantly different generic hierarchies which fortunately can be commonly clocked.

Computer data streams in the 1200–9600 bits/s range are standard the world over. In the areas of digital representation of voice and video, as shown in Figs. 20-25 and 20-26, it is clear that with time the standard 64 kbit/s digital voice representation is being reduced to less than 10 kbits/s and is headed for 2.4 kbits/s, which

TABLE 20-10
TYPICAL INFORMATION SIGNALS IN USE

HUMAN VOICE	3.1 Kh_3		
525 LINE VIDEO	4.5 MHz		
TELETYPE (TTY)	50-75 bps		
MODEMS	1200 bps 2400 bps 4800 bps 9600 bps		
DIGITIZED VOICE – PCM (8000 SAMPLES/SEC; 7 BITS/SAMPLE)	56 Kpbs		
DIGITIZED VOICE – PCM WITH ADDED HOUSEKEEPING DATA (SPADE)	64 Kbps		
DIGITIZED VOICE – DELTA MODULATION	32 Kbps (U.S.) 28 Kbps (JAPAN)		
– DPCM	32 Kbps		
– MILITARY	16 Kbps		
– PREDICTIVE CODING	9.6 Kbps		
SYNTHESIZED VOICE – VOCODER	4.8 Kbps 2.4 Kbps 1.2 Kbps		
GLOBAL DIGITAL HIERARCHIES	USA-CANADA	JAPAN	EUROPE
T1	1.544	1.544	2.048
T1C	6.312	6.312	8.448
T2	44.736	32.064	34.368*
T3	274.176	97.728	139.264
T4	––	400.352	FROM 560 TO 840
600-CHANNEL MASTER GROUP DIGITIZED (USA)	T3		
400-CHANNEL SPECIAL MATER GROUP (TELSAT)	25.74 Mbps		
NTSC DIGITAL COLOR TV			
PCM	89.472 Mbps (TWO T3)		
DPCM – U.S.	T3		
DPCM – JAPAN	22 Mbps		
TELECONFERENCE VIDEO			
AT&T	T1		
AMERICAN SATELLITE	772 Kbps		
WIDCOM/NEC	56 Kbps		
SLOW SCAN VIDEO	8 Kh_3		
ISDN – U.S. BOC			
INTERFACE	80/144 Kbps		
BUSINESS	56 Kbps		
RESIDENTIAL	64 Kbps		

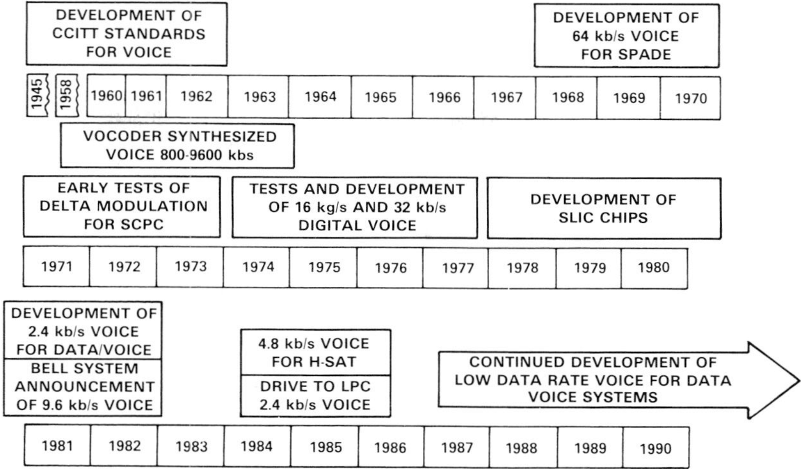

Fig. 20-25: Evolution of digital representation of the human voice.

will be a major contribution to bandwidth conservation. Also in the video area, the standard NTSC video digital bit stream of 89.5 Mbit/s (two *T*3) has given way to much lower rates by redundancy elimination and by bandwidth reduction through encoding techniques, leading to videoconference quality color TV at less than 1 Mbit and color TV at reduced quality at 56 kbits/s where considerations of integrated data systems and low cost circuits are required.

MODULATION TECHNOLOGY

The principal format of information passing through a satellite is analog, representing both voice and/or television. Analog voice and television (Fig. 20-26) represent the bulk of modern satellite transmission, although the transmission of both in digital format is increasing with each year.

Analog modulation techniques were already mature before the first geostationary satellite Syncom 2 was launched. They included FM and single sideband modulation, and frequency division multiplexed (FDM) voice, all veteran techniques of high-capacity terrestrial radio links. Fig. 20-27 shows the time line of the introduction of these essentially analog terrestrial techniques into satellite links, starting with FDM/FM and with SCPC-FM introduced in 1975 into the Alaskan bush system. Single sideband was introduced by both AT&T and RCA in the 1982–1983 time frame.

The introduction of digital modulation techniques into satellite communication links did not seriously start until the mid-1970's. Since then, digital modulation

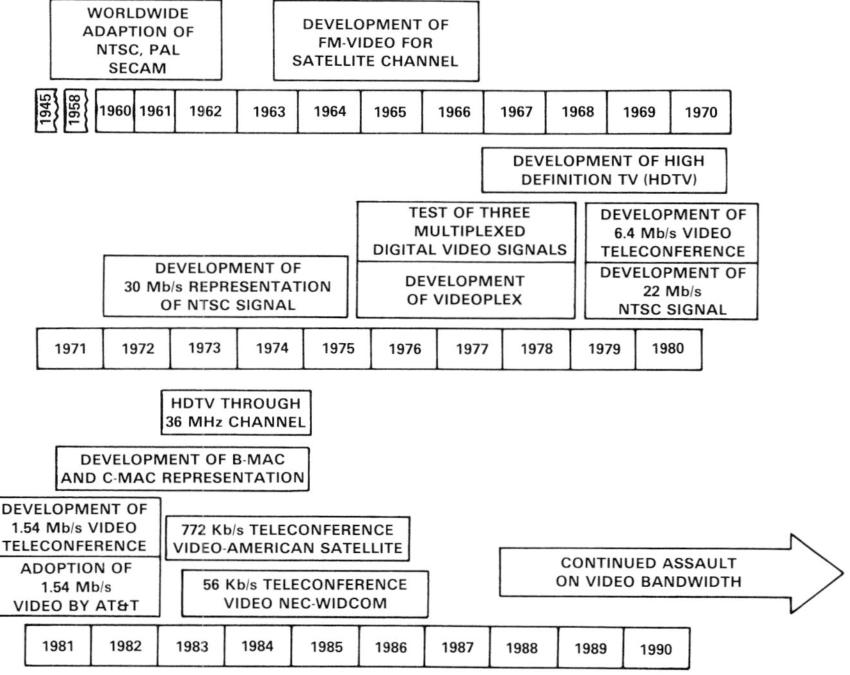

Fig. 20-26: Evolution of digital representation of a broadcast quality and teleconference quality video signal.

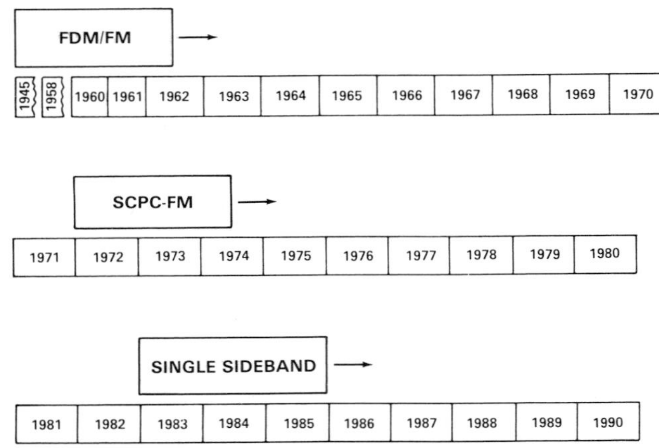

Fig. 20-27: Evolution of analog modulation techniques which carry messages or data.

using biphase phase-shift-keyed (PSK) modulation and quadriphase modulation, and its derivative, staggered QPSK (SQPSK) are now standard in the world today.

Other derivations include 8θ PSK, 16θ PSK, MSK, and multilevel modulation. Two bits/Hz is used for QPSK in Intelsat-5 and 3 bits/hertz is now standard for 8θ PSK for many terrestrial radio systems. Workers in this art are experimenting with more advanced combined multi-dimensionally coded amplitude and phase shift modulation which has already achieved a bandwidth efficiency of 4 bits/Hz and can be extended to as much as 10 bits/hertz and beyond. Table 20-11 lists some of the achievements in use as reported in the literature of satellite communications. 4D-QAM was developed at Comsat Labs and has been given extensive satellite tests.

The higher order QAM modulation techniques up to 256 QAM are more suited to terrestrial links since such short links (typically 20–30 mi) have a much lower space loss ($\approx$145 dB at 4 GHz) than the space loss in the 22 000 mi down link ($\approx$200 dB at 4 GHz) of communication satellites and can therefore provide a much higher energy per bit that is required. However, in the future era of giant antennas with high EIRP spot beams, this limitation may not exist.

In the case of multiple access or multiple users in a transponder, two types of access are used: FDMA which arrays the users side-by-side in frequency, and TDMA which arrays the users in succession in time. FDMA is very dependent on channel linearity and bandwidth for capacity, while TDMA is dependent on network timing.

Consider the case where only a *single* user will use the channel. Thus, single-access users can use one of the following *maximum occupancy techniques:*

- one FM television signal modulated with frequency deviation to occupy a full 37 MHz channel;
- a 900 voice channel FDM/FM carrier consisting of three master groups combined (multiplexed) into a super master group;
- companded 900 voice channel SCPC-FM (this can be doubled by use of additional companding);
- 900 voice channel SCPC-PCM (64 kbit/s PCM/voice);
- 960 voice channel TDMA using 64 kbit/s PCM based on 15 accesses;
- 1920 voice channels using TDMA/PCM/DSI;
- 1920 voice channels using TDMA/CVSD;
- 3840 voice channels using TDMA/CVSD/DSI;
- 7800 voice channels using single sideband into 30 m earth stations.

TABLE 20-11
BANDWIDTH EFFICIENCY FOR DIFFERENT MODULATION TECHNIQUES

BANDWIDTH EFFICIENCY	MODULATION TECHNIQUE	
TYPE MODULATION	BITS/SEC/Hz	WHERE USED
2 θPSK	0.50	MARISAT TTY
QPSK QPSK QPSK	1.00 1.12 1.53	SBS SPADE (INTELSAT) TELSAT
SQPSK	1.30	BELL SYSTEM 3 TRANSMISSION VIA 11 GHz RADIO
QPSK (BANDWIDTH LIM)	2.00	INTELSAT V
SMSK/2	2.00	F. AMOROSO (HUGHES)
FFSK (MFSK)	2.20	CTS CANADA EXPERIMENT
QPSK	2.25	MICROWAVE ASSOCIATES 11 HGz TERRESTRIAL RADIO
8 θPSK	3.00	COLLINS TERRESTRIAL RADIO FOR 11 GHz
16-QAM 16-QAM	3.00 3.37	NEC 6 GHz TERRESTRIAL RADIO CINCINNATI ELECTRONICS
16 θPSK	4.00	COMSAT LABS/JAPAN NTT/ECL
16 APSK	4.00	JAPAN NTT/ECL (MIYAUCHI)
4 D-QAM	4.00	COMSAT LABS (WELTI)
RBQPSK	4.75	BRITISH POST OFFICE
16 QPR	4.7	COLLINS
32 QAM	5	COLLINS
64 QAM	6	COLLINS/NEC
256 QAM	10	NTT/ECL

Many of these types of *single access* occupancy are rarely used except for the case of TV-FM, TDMA, TDMA/DS1, and single sideband modulated carriers. However, the list illustrates the wide variety of carrier and baseband techniques which can be used.

In 1985, ten modulation and access techniques were in use.

- *FM message:* most economical for heavy point-to-point trunking.
- *FM Video:* for network and cable video.
- *SCPC(PSK or FM):* most cost-effective for thin route communications.
- *TDM/TDMA:* most cost-effective for large star networks.
- *TDM/FDMA:* most cost-effective for low duty cycle point-to-point systems.
- *TDMA:* most cost-effective for large mesh networks including those with high traffic fluctuations.
- *Spread Spectrum:* most cost-effective for data distribution into large-quantity small antennas with low interference.
- *Single Sideband:* most economical for high capacity trunking and maximum transponder voice message capacity.

Single access through a transponder channel was first realized in a practical sense by the distribution of TV in the U.S. in the 1970's for CATV. In the 1980's, the introduction of single sideband techniques made possible up to 7800 voice or message channels through a 36 MHz channel—an improvement by almost a factor of 10 over the occupancy provided by a single FDM/FM carrier. Also, the demand of long distance telephony has introduced a demand for single access–multiple user communications using TDMA-QPSK transmission; for example, GTE's Sprint at around 45-Mbit/s through a 36 MHz channel.

When more than one access is introduced into a transponder channel, it is not always possible to realize the maximum occupancy level promised by many modulation techniques—particularly when more than one carrier sits side-by-side in the channel requiring operation at backoff for the power amplifier to avoid interference.

Studies by Comsat Labs have shown that in the practical utilization of a 36 MHz channel by multiple FDM/FM users, the theoretical capability of 960 message single-access FDM/FM reduces to less than 500 messages/voices. Where time division multiple access TDMA is used, the single access theoretical capability of TDMA/QPSK of about 900 messages/voices reduces to about 850 due to the need for "overhead," preamble, and housekeeping bits.

THE TECHNOLOGIES OF SPACE INTERCONNECTIVITY

The communication and broadcast satellite of the 1980's is essentially a "bent pipe" path between a sender and a receiver. During the 1990's, the communication satellite will become a nodal point in space for interconnecting a large number of users in common areas or between remote areas. This space–earth interconnectivity will integrate with the growing cable and fiber-optic terrestrial links and will also utilize intersatellite links between nodal satellites.

The basic technologies of space interconnectivity include the following:

- on-board switching and signal processing centers at both RF and baseband;
- electronically hopped antenna beams which can move from point to point in synchronism with data moving through the on-board switch;
- laser links to interconnect nodal satellites;
- geostationary platforms using giant antennas, capable of assembly at the space station before transfer to geostationary orbit, which are capable of many spot beams which are interconnected by a switch and computer center on the platform.

These technologies are already under development in the 1980's by NASA and are programmed for use on such communication satellites of the 1990's as ACTS and MSAT-C which serve the growing demand for customer premise business communications, and the new business area of mobile and aircraft communications.

Ground-switched interconnectivity of telephony and data has been in use for many decades in a wide variety of currently operating commercial communications systems. Indeed, second- and third-generation systems employing vastly simplified protocols and hardware are coming into being. Incorporation of the switch in the satellite (SS-TDMA) is a logical evolution for space interconnectivity taking full advantage of the multiple beam capability of the satellite and the use of on-board switches for interconnecting multiple spot beams.

The key enabling technology for space interconnectivity is the matrix switch at RF and/or the baseband switch whose joint evolution is shown in Fig. 20-28. In 1988 SS-TDMA will be introduced in Intelsat 6 and will have wide applications in national and military communications systems during the 1990's. Intelsat 6 will have a 6×6 RF switch. Matrix switches—using funding by Intelsat, NASA, and ESA—have been developed at NEC in Japan, at Plessey in Europe, and Ford Aerospace, Hughes, G.E., Sanders, and Comsat Labs in the U.S. Baseband matrix switches have been developed by TRW and Motorola in the U.S., and Signaal in Holland. The associate new enabling technology is the electronically-hopped-beam antenna which is synchronized with the RF matrix or baseband switch.

In order to reduce the risk of using the principal technologies of space interconnectivity by commercial spacecraft manufacturers in actual flight programs, NASA has a program to place an experimental satellite, ACTS (Advanced Communications Technology Satellite), in orbit by 1990 to test out the basic interconnectivity technologies listed above. These tech-

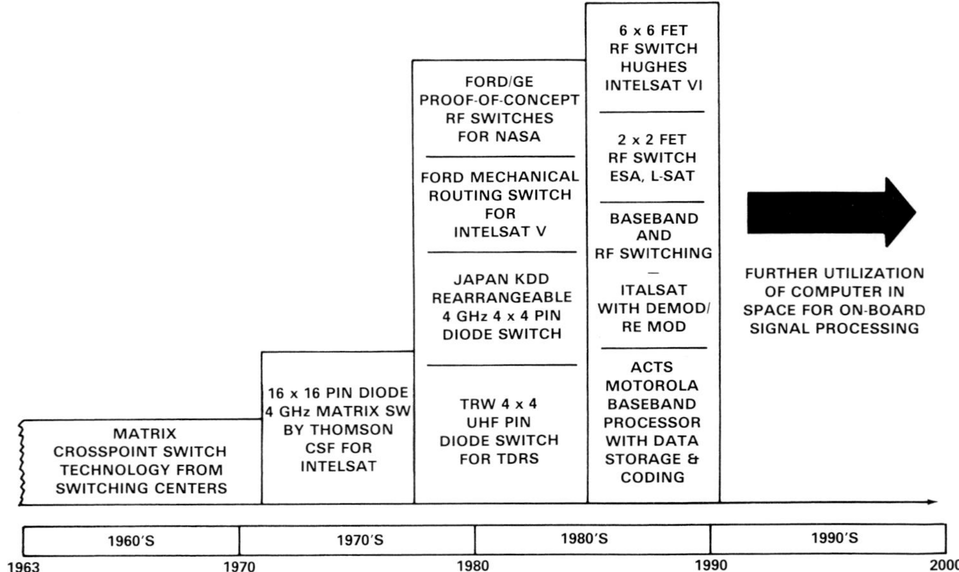

Fig. 20-28: Evolution of on-board switching technology.

nologies are illustrated in Fig. 20-29, which shows the various proof-of-concept technology models or components relating to on-board switching and antenna hopping. Fig. 20-30 shows a laser experiment from the shuttle to ACTS, in about 1990, which will also contribute the verification of new interconnection technologies for intersatellite laser links.

Both RF and baseband matrix switching will be tested by ACTS. RF matrix switching will contribute switch simplicity, while baseband processing in turn permits

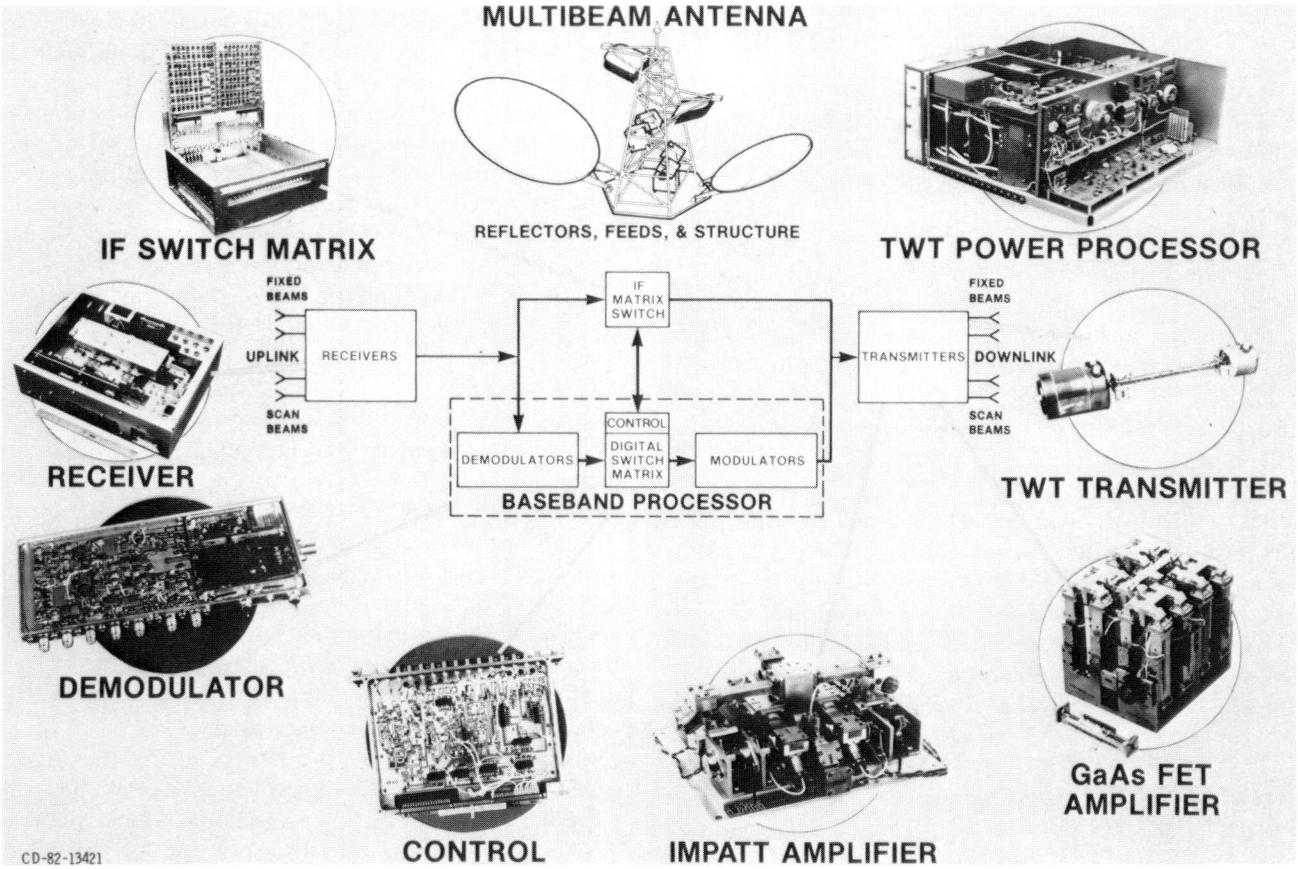

Fig. 20-29: Array of new technologies developed for NASA to make possible a flight experiment of "switchboard in the sky" operation.

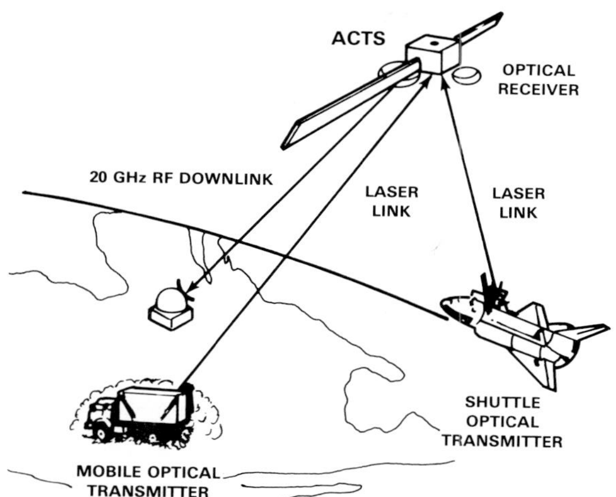

Fig. 20-30: Optical satellite link planned between the shuttle, the ACTS satellite, and ACTS stations on earth.

improved link performance through separate optimization of up- and down-link parameters, for adaptive compensation for rain fades with forward error correction (FEC) coding and for error detection and correction, data rate control, and up-link power control.

The ACTS spacecraft transmit antenna is 3 m in diameter with a corresponding 2 m receive antenna; both form 0.4° spot beams having low sidelobes to permit demonstration of frequency reuse. A typical hopped beam coverage pattern is shown in Fig. 30-31.

Three fixed spot beams for high data rate trunking are interconnected with an IF matrix switch. For the lower data rate systems, two scanning beams are used in conjunction with a baseband processor which employs demodulation, buffering, switching, and remodulation along with forward error correction to process messages to and from low-cost customer premise terminals.

GIANT SPACE ANTENNAS AND THE SPACE STATION

In the mid-1990's, the space station will serve as a staging base in low-earth orbit at which large satellites can be assembled, tested, and transferred into geostationary orbit. This staging and assembly base capability opens up the era of the giant space antenna with diameters from 20 to 100 m, which will be brought up into low-earth orbit by the shuttle and deployed at the space station. Fig. 20-32 shows a concept of MSAT-C which will use a 55 m antenna and which is being considered for mobile satellite service in the late 1990's. This satellite will provide up to 90 separate spot beams onto the continental U.S., with on-board switching between spots, and is indicative of the potential of this new technology for interconnectivity and for the optimum use of the radio spectrum through spatial reuse.

Giant space antenna technology development which will make a geostationary platform possible is already under way in the U.S., Europe, and the U.S.S.R. Significant developments in the 1970's include a 9.1 m antenna on NASA's ATS-6 satellite and a 10 m antenna which was deployed from the Salut 6 by the U.S.S.R. At present, in the U.S., Lockheed has a 55 m wrap rib space antenna under development for JPL, while the Harris Corporation has a 15 m hoop column space antenna under development for NASA-Langley Research Center. A 20 m space antenna also is reported under development in Europe at the European Space Agency.

The giant antenna which may be the entire platform or may be a component of a system or a cluster is a

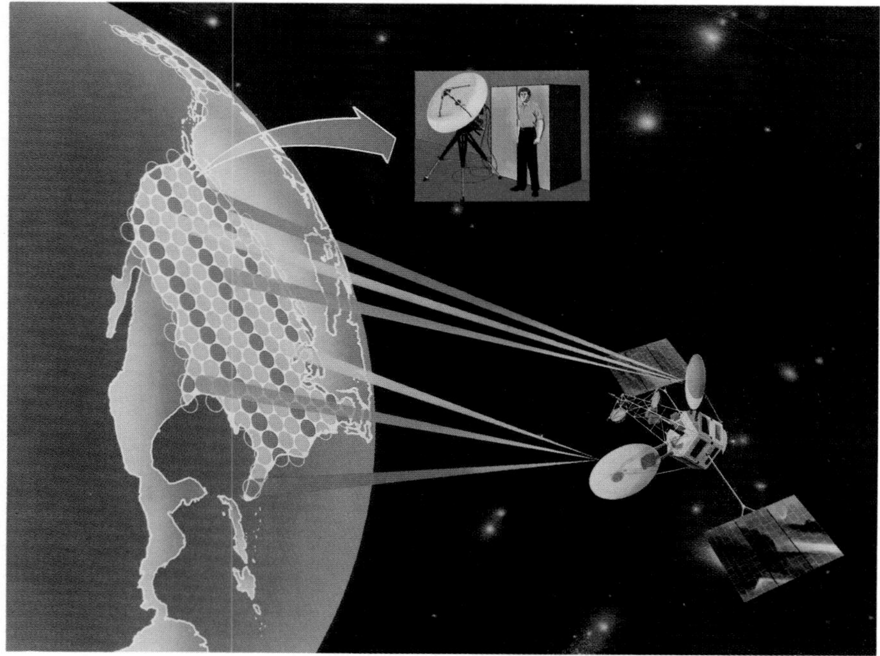

Fig. 20-31: Artist's concept of a satellite with on-board switching and electronically hopped beams for interconnecting small areas of the U.S..

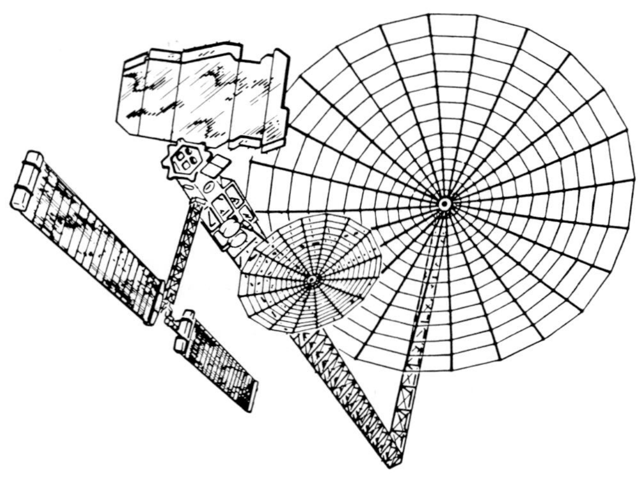

Fig. 20-32: MSAT-C with a 55 m antenna and feed system for interconnecting 95 spot beams, illustrating a satellite design which can be assembled at the space station in the 1995–2000 time frame and transferred to geostationary orbit.

structure unique to the space environment. Unfurlable space antennas larger than 10 m are difficult to deploy and test in the earth's gravity. Optimally, the zero gravity of space will be utilized to permit deployment without the weight stress that would deform the antennas. It also is believed that it is more cost-effective for such antennas to be deployed and tested at the space station (after being transferred from earth via the shuttle and before being committed to GEO) rather than in GEO. These large antennas could be transferred from LEO to GEO, using OTV's at low thrust acceleration (which could result in a transfer time of months).

Thus, while the new era of space communications predicted for the year 2000 looms in the seemingly distant future, the enabling technology of this era—on-board switching, agile multiple beam feeds, giant antenna technology including techniques for test and assembly in space at the space station, and space laser communications—are already in the early development and proof-of-concept stage.

21

FUTURE DEVELOPMENTS

Thomas F. Rogers

In suggesting important advances to be expected in the satellite communications area, it is necessary both to define the time frame and to outline the expected technological advances that could take place.

Here we are interested in changes envisioned beyond the end of this decade and throughout the following decade. To predict changes over the next few years requires relatively little insight, for the technology is already reasonably well in hand and the political-economic-market forces are seemingly well enough understood to strongly suggest their acceptance. And to attempt to predict changes for the interval beginning after the end of this century is to adopt a science fiction construction inasmuch as, over the time span of a generation, important technological and societal changes border on the unpredictable.

There are several technological changes that can be expected with some confidence. The country's shuttle fleet will long since have "shaken down," and flight schedules and prices will be stabilized. Bi-weekly flights will have become routine, as will the provision of assembly, checkout, and maintenance services by technicians both within the orbiter bay and external to the orbiter. Very large and sophisticated physical-electronic-chemical assemblages will be able to be dealt with, and sent on their way to appropriate orbits from complexes of appropriate infrastructure ("space station elements") located in low-earth-orbit.

Defense research, development, testing, and evaluation efforts will have solved the basic problems attendant upon the efficient generation and accurate and reliable transmission through the earth's atmosphere and near-space (of at least megawatts average) of microwave and optical electricity in the form of collimated beams; also, they will have developed satisfactory in-space nuclear power sources of 100 kW–1 MW average.

NASA, DOD, and the private sector will have learned how to deploy and maintain very high-speed, high-capacity, flexible "computer-switches" in space, where they can be expected to operate reliably over lifetimes that match those of their related RF and other satellite elements—decades, if need be.

Stationary "clusters" or "focal points" of infrastructure will be located both in the earth's atmosphere at altitudes of some 20 mi and in space in geostationary orbit. These locations will be provided with large amounts of electrical power, chemical fuel, heat dissipation, stability and telemetry, and will be served by maintenance crews. They will be used to accommodate equipments designed to serve a wide variety of interests in communications, navigation, position-fixing, and remote sensing.

And, as a consequence of having such very stable, sophisticated, and high-power "focal points" available, very short radio wavelengths, and very narrow-beam and dynamically pointable radiation and reception apertures will be used at these "points" for communications. Therefore, surface and near-surface equipments, both fixed and mobile, will be of very small size and low unit capital cost; and greater allowable operating tolerance of departures from optimum circuit performance will permit low equipment installation, operations, and maintenance costs.

Technological changes will not be confined to the U.S. European countries, Canada, India, Japan, and perhaps other countries, and even the U.S.S.R., will conduct vigorous commercial-industrial space-related research and development testing and evaluation programs, and will be in a position to offer equipment and services to the communications market. Indeed, this competition may well force greater and more rapid technological change than is suggested here.

NEW SERVICES

In this context then, at least from the technological capability point of view, some fundamentally new space-related communication services could come into being. The ones chosen to be described here are not those that come to mind when thinking of "linear extrapolations" from present technology or present services, but rather, are qualitatively different and more speculative.

Direct Audio Broadcasting

Direct international audio broadcasting from space to fixed and mobile surface and near-surface receivers could be available to every country in the world in the form of a relatively low-cost, equitable, reliable common-user (carrier) service. Small, electrically tuned, low-cost fixed and mobile receivers would replace today's relatively large, complex, mechanically tuned, short-wave receivers. Governments would avail themselves of this service to replace their present nationally owned surface-based short-wave broadcasting transmitters. And both these governments and commercial interests would be offered the prospects of enormous new markets, serving at least the 400 million short-wave listeners with reliable high-quality broadcasting services, and stimulating wholly new international markets with novel forms of international programming. In terms of international political, cultural, and economic impact, this communications advance could well be one of the most important advances that the world has yet seen.

Private Service Networks

For decades, the trend in long-haul voice and later data communications has been towards the concentration of myriads of local low-capacity circuits into bundles of circuits to be handled by local surface switches and high-capacity long-haul trunk circuits. The past few years have seen the first modest reversal of this trend: communications equipment, designed to utilize surface or space-related long-haul wide-band microwave circuits, is now being installed on commercial–industrial "customer premises." In so doing, the local concentrator-switch-distributor communications nodes are being bypassed. This qualitatively different trend will continue and, in due course, the "customer premise" will become the individual residence, the apartment, the automobile, etc. Given the power and versatility available in the elevated (upper atmosphere and in-space) transceiver-switch "focal points," all radio communications would be able to revert, essentially, to "point-to-point," with the switching taking place far above the earth in the long-haul circuits themselves. Long-haul trunk circuits and networks, as presently conceived with their local surface exchanges, will no longer be nearly as important. This fundamental change could have a first-order impact upon the present telephone network and plant—perhaps even a zero-order impact.

Low-Cost Navigation and Position Determination

For some 25 years, a few military ships have been provided, via signals from military satellites, with a reliable navigation and position-fixing service. For the past decade, a growing number of civilian users around the world also have been able to avail themselves of this service by using these signals. But the surface equipments have been relatively large and costly, the service slow and two-dimensional, the user is oftentimes required to intervene in the calibration and calculation processes, and the accuracy provided is of value to only a modest number of users. By the end of this decade advances in both space and surface segment design should allow surface and near-surface navigation and position-fixing equipment to become very small in size, light in weight, and of relatively low capital cost, and would automatically provide very great and immediate three-dimensional position and velocity accuracies and very precise time signals. Almost anywhere around the world, automobiles, trucks, ships, boats, aircraft, trains—indeed pockets and pocketbooks—could be potential locations for such receiver-calculator-display devices. And these devices would be able to be coupled to other local sources of information so as to be able to make prompt and accurate predictions, and to supply directions of importance to the traveler. A potentially large market could be opened with the offering of a new, useful, and relatively inexpensive travel service.

Space Power Generation and Relay

For nearly 40 years, communications has looked to free-space microwave propagation to allow an increasing amount of information and data to be carried reliably over increasing distances. The next decade could see the use of millimeter radio wavelength and optical wavelength free-space beams as well to transmit not only audio and video signals, but also large amounts of raw electricity. Collimated electromagnetic beams could carry megawatts of electricity from the earth's surface to the upper atmosphere and to geostationary orbit "focal points" at a cost competitive to that of solar and nuclear in-space sources. By the end of the century, this method could also be used to provide great amounts of energy to reusable orbital transfer vehicles (low-earth orbit, geostationary orbit, lunar orbit) and, through the use of in-space reflectors, to distribute electrical energy about the world. New "space electricity" utilities could come into being, and countries with enormous potential renewable energy resources (hydro and solar) could begin to treat electricity as an international and inter-continental exportable commodity—just as nonrenewable coal and oil are today.

General Space Travel

The next decade should see low-earth orbit visited by at least hundreds, and probably thousands of "ordinary" people each year. They would spend short periods of time (a week or so) in space lodges, perhaps utilizing modified shuttle external tank pressure vessels as the basic in-orbit structure, and would pursue individual and group activities as they rotate about the earth in the absence of the influence of gravity. By the end of this century, or at least in the first decade of the next, the general public could also have its first opportunity to stay in geostationary orbit and even visit the surface of the moon. They would want audio and video communications between themselves and their lodge or settlement, and with the earth. For the first time, space, which has served the general public here at the earth's surface through the use of orbiting satellites, would become a new domain for the general public to begin to reside in, and to be served there by space communications. And a wholly new market for commercial space-related telecommunications services would begin to form—one that, in time, would encompass the solar system.

There are very important economic and institutional corollaries to all of the above. The president of the Aerospace Industries Association observed in December 1983, that ". . . there has been a considerable acceleration . . . in [the] building of commercial communications satellites . . . that's just the beginning of an indicated boom; worldwide projections show enormous increases in demand for satellite communications services between now and the end of the century." And a large strategic defense initiative research and development program could offer the opportunity for the creation of new private sector goods, services, and processes. Particularly with the creation of new services and the stimulation of new markets such as suggested here, it seems reasonable to project that the space-related communications business (broadly defined) could well continue to grow at an annual rate of 15–20 percent per year, compounded. If so, by the end of this century the annual U.S. sales volume would grow from 1983's near $2 billion to $20–40 billion per year (1984 dollars).

By then, such a sales volume could yield total federal, state, and local U.S. tax revenues of $5–10 billion per year, thereby, in effect, reducing the net public cost of our federal expenditures on civilian space activities (if they were to continue at the present dollar level) perhaps to zero. And private sector space research, development, testing and evaluation expenditures, as well, could reach $2 billion per year.

Since the mid-1960's, which saw the creation of the defense global satellite communication system-network and the beginning of commercial international satellite communications services, the great value of space to the communications world has become increasingly appreciated. Beginning late in this decade a new and converse appreciation will begin to form—that of the great value of the space-related communications business to space. For growth in communications, navigation, position-fixing, and energy distribution commercial services could begin to generate sufficient sales, during the next decade, to allow the private sector, through tax revenues, to offset the entire cost of a public space science, exploration, and technology development program of at least today's dimensions. This would mark an extraordinary and long awaited reversal of institutional roles.

GLOBAL INTERCONNECTIVITY BY THE YEAR 2000

The acquisition of space infrastructure elements and a new generation of communications, navigation and position-fixing satellites and geostationary "focal points," as well as scientific and commercial earth-observation satellites, will result in a large and variegated array of satellites from low-earth orbit to geostationary orbit as shown in Fig. 21-1. This artist's concept attempts to illustrate the proliferation of space users who will require a space communication network, not only to provide space broadcast and two-way space–earth communications, but also communications between spacecraft, and between scientific instruments in space and data processing systems on earth.

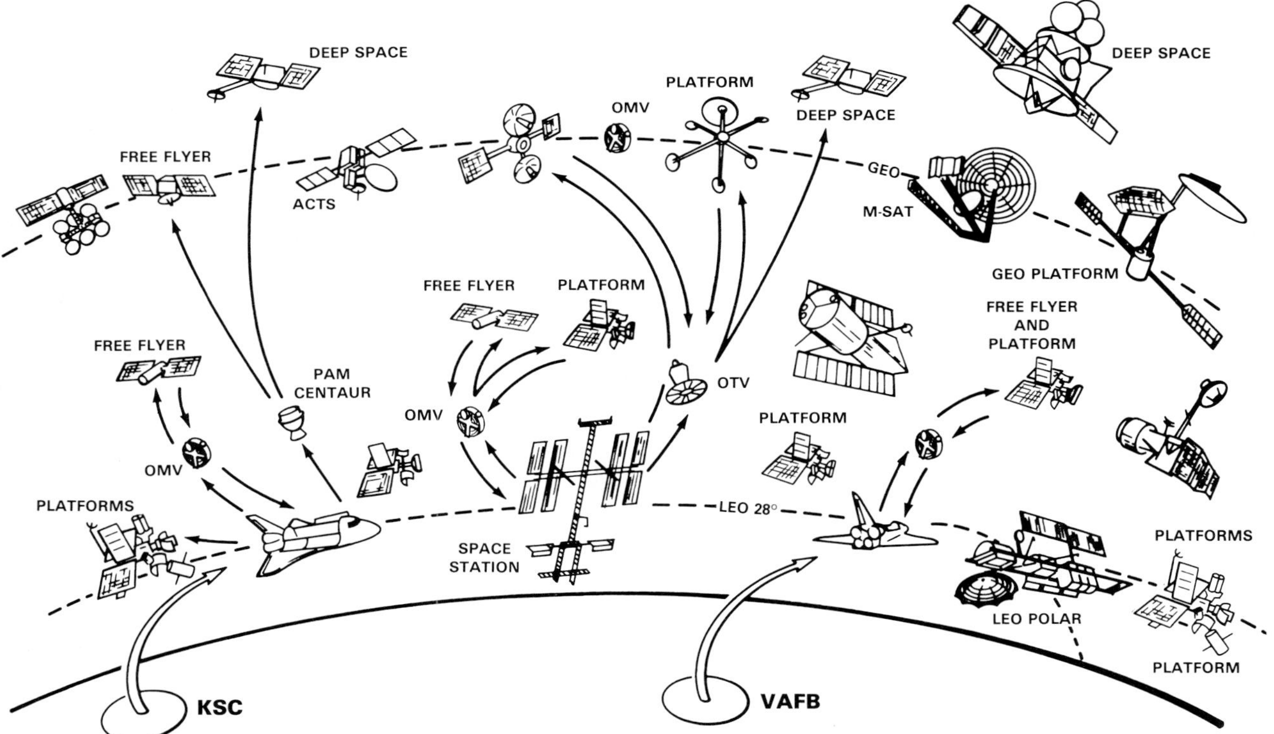

Fig. 21-1: Concept of the proliferation of space users by the year 2000 showing the geostationary satellites and platforms joined by low- and medium-orbit satellites for earth observation, navigation, position-fixing meteorology, and numerous scientific and commercial applications.

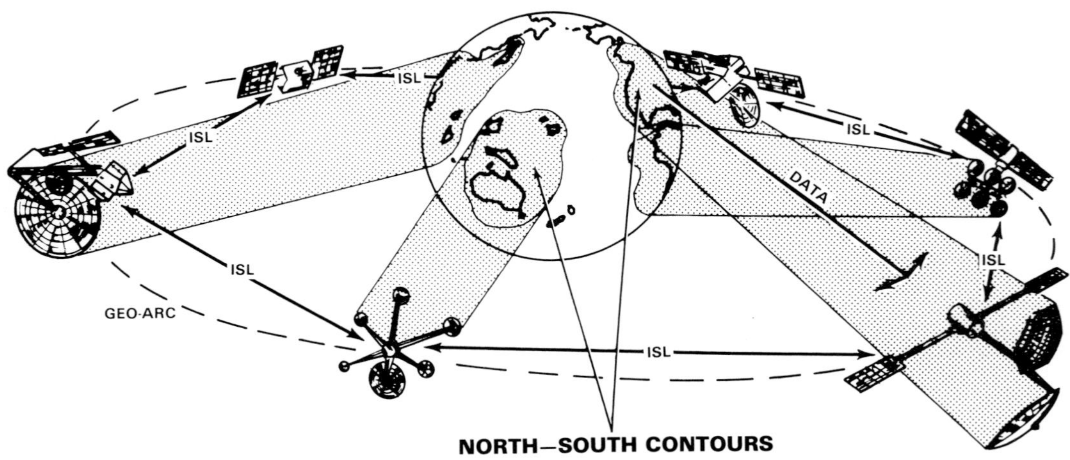

ISL = INTERSATELLITE LINK USING LASERS

Fig. 21-2: Concept of world communications interconnectivity in the year 2000 using intersatellite optical laser communication links around the geostationary arc, which are accessed by millimeter-wave circuits between the arc and earth.

One outgrowth of the need for improved communications links between spacecraft will be the development of intersatellite optical links which will not only take advantage of small equipment size and high data rate capability, but also will synergistically interface with microwave radio and fiber-optic cable terrestrial links being installed all over the earth. Fig. 21-2 shows one concept of space-related long-haul communications by the year 2000. It illustrates one possible scenario of global interconnection using optical links between satellites in the geostationary orbital arc—comprising a continuous communication path around the entire arc. This path can be accessed by up links and/or down links using appropriate spot beams between the arc and earth surface "footprints." The data transmitted up to a satellite in the arc, would travel around the arc by intersatellite links to a satellite over the destination, and then be beamed down to the earth.

INDEX

A

Abbott, C. G., 9
Active microwave measurements, 126
Advanced Communications Technology Satellite, 190–191, 211
Advanced Solar Observatory (ASO), 52, 54
Advanced very high radiometer, 125
Advanced X-ray Astrophysics Facility (AXAF) 5, 34, 48–50, 54
Alaskan pipeline, 46
Alborn, H. 169
Alfven, H. 7
Alfven velocity (alfven speed) 10, 12
Alfven waves, 10, 14, 16
Alpha Centauri, 43
American Satellite Corporation, 190
American Society of Photogrammetry, 120
American Society for Testing Materials (ASTM), 175
Ames Research Center, 61
Amplifiers, 242
Antennas, 232–236, 249
  Deep space net, 65
  Laser, 234–235
  Millimeter-wave, 234–235
  Multiple-beam, 232–234
  Shaped-beam, 232–233
  Space antennas, 235–236
  Technology, 231–232
  Testing, 236–237
Apollo program, 4, 23, 55, 106, 153, 160–161, 173, 176
Apollo-Soyuz mission, 55, 63, 153
Applications Technology Satellite (ATS), 13
Arabsat, 189
Arecibo radar telescope, 25
A ring, 27
Armstrong, Neil, 3
Asean Regional Satellite System, 189
Astro-C X-ray Satellite, 53
Astronomy, 31, 41–42, 48
Astrophysics, 31–54
  Distance scale, 43
  Observatories, 46–50
  See also Advanced X-ray Astrophysics Facility
Atlas/Centaur launch vehicles, 204
Atlas rocket, 20
Atmosphere remote sensing
  See Meteorological remote sensing
Atmospheric dust, 146
Atmospheric dynamics, 143
Atmospheric Explorer series, 16
Auroral current system, 17
Auroral heating, 17

B

Baseband information, 244–245
Beam forming network, 233–234
Bell Telephone Laboratories, 201
Bethe, Hans, 36
Beyond the Atmosphere, 3
Blossom missions, 56
Bragg crystal spectrometer, 50
Burnight, T. R., 31
Burst and Transient Source Experiment (BATSE), 48

C

California Department of Forestry, 105
Cameras, 46
Carrington, Richard, 7
Carruthers, John, 105
CCD imaging spectrometer, 50
Cepheid variables, 43, 46
Channel bandwidth, 243–244
Chapman, S., 7
Clark, Walter, 87
Climate monitoring and prediction, 145–146
Coastal zone color scanner, 128
Color film, development of, 87–88
Comets, 29
Communication, satellite application, 187–199
  Arabsat, 189
  Asian Regional Satellite system, 189
  Direct broadcast satellites, 191–192
  Domestic communications satellite services 190–191
  Eutelsat, 188–189
  Intersputnik, 188
  Mobile satellite services, 192–194
  Satellite services, 187–188
Communications satellites, 17, 204–207
  Transponder frequencies, 228
Communications spacecraft, 201–213
  Commercial communications satellites, 203–208
  Direct broadcasting satellites, 208–209
  Frequency allocations, 209, 211
  Mobile satellites, 211–213
  Pioneering experimental programs, 201–203
Compton telescopes, 39
Comsat Satellite Television Corporation, 191
Controlled Ecological Life Support Systems (CELSS), 63
Coordinate Resource Planning Program of U.S. Forest Service, 105
Copernicus, 33
Corona
  See Solar corona.
Cosmic Background Explorer (COBE), 34
Cosmic Discovery, 42
Cosmic plasma processes, 16
Cosmic rays, 39–40
COSMOS flight, 61, 62, 65
Cygnus X-1, 31, 36

D

Deep Space Network (DSN), 20
  Antennas, 65
Deep Space Probes, 3
Defense Meteorological Satellite Program (DMSP), 13, 140
Department of Defense, 55
Desmond, Morris, 185
Direct audio broadcasting, 251
Direct broadcast satellite, 191–192, 208–211
du Hauron, Louis Ducos, 87
Dynamic Explorer mission, 140

E

Earth's gravity field, 138
Earth's magnetic field, 12, 136
  See also Magnetosphere.
Earth Satellite Corporation, 94
Eddington limit, 44
Eisenhower, Dwight D., 201
Ektachrome, 87–88
Electrically Scanning Microwave Radiometer (ESMR), 12
Electromagnetic radiation, 37–39
Electrooptical devices, 92

Energetic Gamma-Ray Experiment Telescope (EGRET), 48
Environmental Research Satellite, 34
Eutelsat, 188
European Center for Medium Range Forecasts, 144
European Space Agency, 4, 13, 53, 130, 176, 191
Evapotranspiration, 148
Exobiology, 63–65
Explorer I, 3, 14
Extravehicular Activity (EVA), 63

F

Faint object camera, 46
Faint object spectrograph, 46
Filters, 242–243
Find guidance sensor, 46
Fisheries, 127
Flight experiment program, 65–68
Frequency allocation, 209–210
F ring, 27

G

Galileo spacecraft, 25, 27
Galston, Arthur, 61
Gamma Ray Observatory (GRO), 5, 34, 48, 54
Gelles, S. H., 169, 175
Gemini program, 55, 106
Geoid, 124, 138
Geophysical remote sensing, 133–142
  Geophysical structure, 133–135
  Crustal composition, 135–136
  Imaging of earth's gravity field, 138–140
  Imaging of polar aurora, 140–141
  Satellite mapping, 136–138
Geophysical Research mission, 124
Geopic image enhancement, 94
GEOSAT mission, 127, 130
Geostar system, 225
Geostationary platform, 212–213
Giacconi, Ricardo, 31
Giacobini–Zinner comet, 29
Global Atmospheric Research Program, 145
Global biology, 65
Global interconnectivity, 253
Global Positioning System (GPS), 223–225
Goddard, George, 87
Goddard Space Flight Center, 46, 47
Godowsky, 87
Gravitational physiology, 62
Gravity
  Earth's field, 138
  Role in astrophysics, 35–37
  Waves, 40

H

Hackucho X-ray satellites, 53
Hale telescope, 47
Halley's comet, 29
Hannig device, 168
Harwit, Martin, 42, 49, 54
Hawkeye series, 13
Hazard, Cyril, 43
Heflex experiment, 60
High-Energy Astronomy Observatory (HEAO) series, 4, 34, 43, 48, 49
High resolution spectrograph, 46
Hintori X-ray satellite, 53
Hubble, Edwin P., 43, 46, 47
Hubble Space Telescope, 4, 33, 34, 46, 53, 54
Hughes Aircraft Corporation, 203, 204
Hughes 399 Communications satellite, 185

I

Ice, 148
  See also Sea ice
Infrared Astronomy satellite, 5, 34, 38, 50, 51
Infrared Space Observatory, 53
Institute for Space and Aeronautical Science (Japan), 18
Intelsat, 187, 203, 204
International Cometary Explorer (ICE), 29
International Council of Scientific Unions (ICSU), 18
International Geophysical Year, 140
International Geosphere and Biosphere Program (IGBP), 18
International Halley Watch, 29
International Solar-Terrestrial Physics (ISTP) Program, 18
International Telecommunication Union's Extraordinary Administration conference, 203
Interplanetary Monitoring Platform, 11
Intersputnik, 188
Ionization densities, 16
IRAS telescope, 38
ISEE-3 International Sun–Earth Explorer (ISEE-3) spacecraft, 11

J

Japan Petroleum Exploration Co., Ltd., 94
Jet Propulsion Laboratory, 20, 173
Johnson, Lyndon, 78
Jovian Ring, 28
Jupiter, 25, 28

K

Kelvin, Lord, 7, 10
Kerwin, Joseph, 55
Kuiper, Gerard, 19
Kyokko satellite, 140

L

Land remote sensing, 77–122
  Application areas, 81–85
  Image enhancement, 87–122
  Imaging from space, 78
  Space photography, 79–81
  Usefulness of, 85–87
Land resources inventories, 79
Landsat, 73, 93, 94, 99, 105, 129, 135
  Multispectral scanner, 78, 93, 97–100, 108
Large deployable reflector telescope, 35, 51
Lewis Research Center, 160
Life support systems, 62–63
Lincoln, Abraham, 78, 120
Low-gravity experiments, 155–158
Lowell, Percival, 19
Luxsat satellite system, 189

M

Magnetic field of earth, 12, 16, 136
Magnetosphere, 11, 12–17
Magsat mission, 137
The Manual of Remote Sensing, 118, 120
Marine geophysics, 129
Marine resources, 127
Mariner program, 2, 3, 20–23
Mars, 3–5, 20, 21, 22, 23
Marshall Space Flight Center, 48, 153, 160
Martin Marietta Corporation, 22
Materials processing, 153
  Apollo–Soyuz experiments, 166–168
  Commercialization, 177–182
  Preskylab experiments, 160–162
  Skylab experiments, 162–165

Space Processing Applications Rocket (SPAR) experiments, 168–170
Fluid behavior, 156–157
Liquid shape behavior, 157–158
Maunder, E., 7
Maunder minimum, 8
Maxwell, James Clerk, 87
Maxwell Montes mountains, 25
McDonnell Douglas Aerospace Corp., 175
McLuhan, Marshall, 185
Megatrends, 185
Mercury, 4–5
Mercury Program, 3, 55, 106
Meteorological remote sensing, 143–150
Atmospheric dust and volcanic aerosols, 146
Climate monitoring and prediction, 145–146
Evapotranspiration and soil moisture, 148
Man's impact on climate, 149–150
Sea surface temperature, 146
Snow and ice, 148
Microgravity environment, 159
Phenomena, 173–176
Shuttle flight program, 174–176
Microgravity Science and Applications Division (NASA), 173, 174
Microwave measurements, 126
Microwave sensing techniques, 126
Mobile communications satellites, 192–194, 207–208
MSAT–X, 211–212
Modulation technology, 245–247
Morey–Holten, Emily, 61
Multiplexers, 242–243
Thomas Mutch Memorial Station, 22

N

Nadar, 71
Naisbitt, John, 185
NASA, 3–5, 13, 18, 22, 25, 29, 46, 158, 173–175, 177–178, 202
Guidelines, 180–182
Life Science Program, 55–68
Policy statements, 179–180
NASA–USDA Forestry Remote Sensing Laboratory, 91
National Academy of Sciences, 3
National Bureau of Standards, Micrometrology Group, 175
National Research Council Astronomy Survey Committee, 5
Naval Research Laboratory, 48
Navigation satellites, 216–226, 252
Geostar system, 225
Global Positioning System, 223–225
Search and rescue, 225
Neptune, 4–5
Neutrinos, 8, 40–41
Newell, Homer, 3
News services, 251
Nimbus 7 spacecraft, 97, 128
NOAA weather satellite, 97, 125

O

Ocean remote sensing, 123–131
Marine resources, 127–128
Microwave measurements, 126
Ocean color, 128
Radar altimetry, 127
Sea ice, 125–126
Ocean topography experiment, 124
Optical combiner, 90
Optical transmitters, 241
Orbiting Geophysical Observatory (OGO), 13
Orbiting Solar Observatories (OSO–I), 3, 10, 33–34, 45
Orgel, Leslie, 64
Oriented Scintillation Spectrometer Experiment (OSSE), 48
OSCAR Amateur Radio Satellite, 185

P

Pace, 61
Passive microwave measurements, 126
Physiology, 56–60
Pioneer project, 4, 11, 14, 19, 25, 28
Venus mission, 14, 23, 24
Planck's constant, 37
Planck spectrum, 37–38
Planet formation, 45
Plasma instability, 16
Plasma processes, 14, 16
Plasma sheet, 17
POGO spacecraft, 137
Polar aurora, satellite imaging of, 140
Pollutants, remote sensing of, 128
Position determination, 252
Primary power technology, 241–242
Private service networks, 252
Proxima Centauri, 43
Pulsars, 43–45

Q

Quasars, 31, 43–45

R

Radar, 126, 127
Radar altimetry, 124, 125, 127
Radiolocation, 225
Radiometers, 125, 126
Radio telescope, Arecebo, 25
Rayleigh's Law, 106
Remote sensing, 3, 26, 71–76, 123, 127, 133, 135, 143
Geophysical, 133–142
Land applications, 77–122
Meteorological, 143–150
Ocean applications, 123–131
Ritchey–Cretien design, 46
Roentgen Satellite (Rosat), 53

S

Salk Institute, 64
Salyut Earth–Orbiting mission, 57, 61, 65, 176
Satellite amplifiers, 237
Satellite capacity, 229–231
Satellite payload weight, 236
Satellite receiver technology, 237–239
Satellite remote sensing
See Remote sensing
Satellite systems, new applications of, 196–199
Saturn, 25
Scanning multifrequency microwave radiometer (SMMR) 126
Scanning sensors, 71
Schmidt, Maartin, 43
Schwabe, S. H., 8
Scorpius X–1 X–ray source, 39
Sea ice, 125
Sea surface temperature, 146
Search and Rescue Satellite (SARSAT) 194–195, 225
Seasat mission, 123, 130, 135, 139
Sensors, 46
SETI (Search for Extraterrestrial Intelligence) program, 64
Shaped–reflector shaped–beam antennas, 232–233
Shuttle/Centaur launch, 29
Shuttle imaging radar, 133, 134
Shuttle Orbiter Medical System, 66

Skylab, 55, 59, 153, 173, 176
Small Astronomy Satellite, 4, 31, 34
Smithsonian Institution, 9
Snow, 148
Soil moisture, 148
Solar
  See also Sun
Solar cells, 17
Solar corona
  heating, 45
  "holes", 12
Solar flares, 7, 10, 11
Solar machine, 8
Solar magnetic field, 11
Solar Maximum Mission (SMM), 9, 33, 42
Solar Maximum Year, 10
Solar minimum conditions, 12
Solar optical telescope, 33, 41, 52
Solar photosphere, 11
Solar plasma, 11
Solar processes, 8–10
Solar seismology, 9
Solar storm, 9
Solar system, 5, 19–29
Solar wind, 12, 13, 16
Solid-state power amplifiers, 24
Solrad mission, 3
Solrad satellite, 33
Soyuz, 153
Space adaption, 57, 58, 59
Space antennas, 249
  Altitude control for, 235–236
Space astronomy, 31, 41–42
Space biology, 60–62
Spaceborne Satellite Repeater, 227
Spacecraft, 7, 13
Spacecraft thermal control blankets, 17
Space Infrared Telescope Facility (SIRTF) 34, 51, 54
Space interconnectivity, technologies of, 247–249
Spacelab, 56, 65, 66, 68
Space medicine, 56–60
Space photography, 79, 102–106
Space power amplifier, 239–241
Space power generation and relay, 252
Space Processing Applications Rockets (SPAR), 174, 178
Space Science, 3
Space shuttle, 3, 18, 33
Space station, 3, 4, 66, 212
Space Telescope Science Institute, John Hopkins University, 46
Space travel, 252
Spectrographs, 46
Spectrometers, 50
SPOT Image Inc., 116
SPOT satellite, 72, 108, 113, 114, 115, 116
Sputnik I, 201
Star formation, 45
Stokes' parameters of polarization, 41
Suborbital systems, 3
Sun, 7-18.
  See also Solar
Sun-earth connection, 17–18
Sun-earth plasma couplings, 10–12
Sunspots, 8
Suspended sediments, remote sensing of, 128
Synthetic aperture radar (SAR) 126, 127
Synthetic stereo, 93

T

Telesat Canadian domestic system (ANIK-1), 190
Telescopes, 4, 33, 34, 35, 38, 39, 46, 47, 48, 51, 53, 54
Telstar, 190, 203
Temperature sounding, 144
Tenma X-ray satellite, 53
Thematic Mapper (TM), 71, 72, 78
Theta Aurora, 140
Thor/Delta launch vehicle, 202, 204
Tiros satellites, 13, 143
TOPEX (Ocean Topography Experiment), 124
Tousey, R., 31
Tracking, relay, and data collection satellites, 195–196
TRW, Inc., 204

U

Ulysses mission spacecraft, 12
U.N. Conference on the Peaceful Uses of Outer Space, 188
Uniformitarian principle, 64
Upper Atmosphere Research Satellite, 18
Uranus, 4–5, 28
U.S. Department of Defense, 126
U.S. Forest Service, 105
U.S. Navy/NASA/NOAA joint mission, 127

V

Valles Marineris canyon, 21
Van Allen, 14
Van Allen belts, 3, 4, 137
Vanderhoff, John, 174
Vega, 50
Vela program, 13
Venera landings, 25
Venus, 3–5, 20, 23, 24, 25
Venus Radar mission, 5
Viking missions, 4, 21, 22, 29, 56, 63
Volcanic aerosols, 146
Voyager mission, 4, 11, 19, 26, 28, 63
V-Z rocket, 31

W

Washburn, Bradford, 87
Water quality attributes, color coding of, 102
Weather forecasts, 143–145
Weather predictability, 144
Weather remote sensing
  See Meteorological remote sensing
Weather systems, 143
Wide field camera, 46
Wind speed determination, 144
Wolter type I mirror pairs, 49

Y

Young's theory of vision, 87

Editors' Biographies

John H. McElroy
Senior Editor

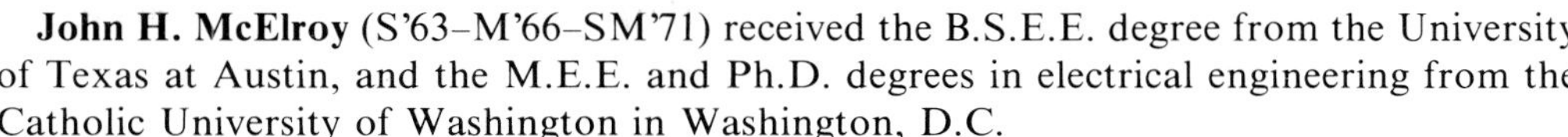

John H. McElroy (S'63–M'66–SM'71) received the B.S.E.E. degree from the University of Texas at Austin, and the M.E.E. and Ph.D. degrees in electrical engineering from the Catholic University of Washington in Washington, D.C.

He joined the staff of the Space and Communications Group of Hughes Aircraft Company in 1985. He served as the Assistant Administrator of the National Oceanic and Atmospheric Administration (NOAA) for Environmental Satellite, Data, and Information Services from 1982 to 1985, where he was responsible for the nation's civil operational earth observation satellites. Prior to joining NOAA, he worked as NASA Headquarters from 1966 to 1982. At NASA, he served as Director of Communications and Information Systems Programs and directed laser research at the Goddard Space Flight Center, and later became Deputy Director of the Space Flight Center.

Dr. McElroy is a Senior Member of the IEEE.

Franklin D. Martin
Part I Editor

Franklin D. Martin received the A.B. degree in physics and math in 1966 from Pfeiffer College in Misenheimer, NC, and the Ph.D. degree in 1971 from the University of Tennessee. A native of China Grove, NC, he now resides in Ft. Washington, MD.

He is presently Director of Space and Earth Sciences at the Goddard Space Flight Center at NASA Headquarters. Since joining NASA in 1974, he has held a variety of positions including Manager of Advanced Programs in Astrophysics, Solar Terrestrial, and Lunar Divisions. He was appointed to the position of Deputy Director of the Astrophysics Division in 1978, and served as Director of that division from 1979 to 1983. Prior to joining NASA, he served as Physicist with the Naval Oceanographic Office, and also as an Aerospace Engineer for Lockheed in Houston, TX, where he was involved in the science mission support for Apollo flights 15, 16, and 17.

Dr. Martin has served on a variety of committees such as the SES Performance Review Board and the Space Science Steering Committee, which reviewed the selection of investigations flown on NASA missions. He has also served as the Assistant Editor for the Geophysical Research Letter and has chaired a number of agency studies. He received the NASA Exceptional Service Medal in 1982.

Ralph Bernstein
Part II Editor

Ralph Bernstein (S'55–M'57–SM'68–F'83) was born in Germany in 1933. He received the B.S. degree in electrical engineering from the University of Connecticut in 1956 and the M.S. degree in electrical engineering from Syracuse University in 1960.

He is Senior Technical Staff Member at the IBM Palo Alto Scientific Center in Palo Alto, CA. He joined the center in 1979 and is currently involved in image processing science, applications, and systems development. He is a Principal Investigator on the NASA Landsat-4 and -5 programs. He joined the IBM Federal Systems Division (FSD) in 1956. During his FSD career, he conducted research and analyses and developed systems for aircraft and submarine navigation and control, satellite simulation, geophysical data processing and control, advanced waste treatment control, and image processing. He was responsible for developing the first computerized shipboard oceanographic data processing and control system for the Woods Hole Oceanographic Institution. He was also a Principal Investigator on the NASA Landsat-1 and -2 satellite programs and led the development of advanced algorithms and computer programs for digitally processing satellite image data. He edited the IEEE book *Digital Image Processing for Remote Sensing* and has contributed to several books including *Geoscience Instrumentation*, the *American Society of Photogrammetry Manual of Remote Sensing*, *Computer Methods for the 80's in the Mineral Industry*, and *Engineering Solutions to Pollution Problems*. He has numerous publications in geoscience, automatic control, system simulation, digital image processing, navigation, oceanography, and data management. He received the NASA Medal for Exceptional Scientific Achievement, the NASA OSO-III Project Team Group Achievement Award, an IBM Outstanding Contribution Award, the IBM General Manager's Award for Significant Achievement, an award for co-developing the IBM Personal Computer Image Processing System (PCIPS), and the Tau Beta Pi Eminent Engineer Award for 1985.

Ralph Bernstein is a Fellow of the IEEE, and a member of Tau Beta Pi and the American Society of Photogrammetry. He is past Chairman of the IEEE Washington Geoscience and Remote Sensing chapter, and is currently Chairman of the IEEE Geoscience and Remote Sensing Society of the San Francisco area. His professional responsibilities include past member of the Space Science Board of the National Academy of Science (NAS), member of the NAS Space Applications Board, past Chairman of the NAS SSB Committee on Data Management and Computation, and Chairman of the NAS SAB Committee on Practical Applications of Remote Sensing from Space. He is a consultant to NASA and a member of their Space Applications Advisory Committee.

Louis R. Testardi
Part III Editor

Louis R. Testardi received the A.B. degree in physics from the University of California, Berkeley, in 1955. He commenced graduate studies at the University of Rome, and received the M.S. degree in 1960 and the Ph.D. degree in 1963 in physics from the University of Pennsylvania.

He was a member of the research staff at the Electric Storage Battery Company in Philadelphia, PA, from 1957 to 1959; and a member of The Franklin Institute in Philadelphia from 1959 to 1962. He was a Member of the Technical Staff of Bell Laboratories in Murray Hill, NJ, from 1963 to 1980. He was also a Visiting Professor in the Physics Department at Princeton University from 1976 to 1977. In 1980, he joined the NASA Headquarters and later became the Director of the Materials Processing in Space Program. In 1982 he became Chief of the Metallurgy Division at The National Bureau of Standards. He is the author of over 100 technical articles in the fields of physical properties of solids, superconductivity, thin films, materials science, microgravity science, acoustics, and nondestructive evaluation.

Dr. Testardi is a Fellow of The American Physical Society.

Robert R. Lovell
Part IV Editor

Robert R. Lovell earned the B.S. and M.S. degrees in engineering at the University of Michigan, and studied management at Syracuse University.

He has worked for the National Aeronautics and Space Administration for 23 years. Since 1980, he has served as Director of the Communications Division, Office of Space Science and Applications, where he manages a program in advanced communications satellite technology research and development. His responsibilities include providing leadership within NASA to formulate, advocate, and execute a broad range of communication programs in support of NASA, other government agencies, and U.S. industry.

Mr. Lovell is a member of the American Institute for Aeronautics and Astronautics (AIAA) and Chairman of the Board of Advisors to EASCON, as well as a member of several U.S. Government advisory committees.